Systems Analysis
and Design

NINTH EDITION

GLOBAL EDITION

Kenneth E. Kendall

RUTGERS UNIVERSITY
School of Business–Camden
Camden, New Jersey

Julie E. Kendall

RUTGERS UNIVERSITY
School of Business–Camden
Camden, New Jersey

PEARSON

Boston   Columbus   Indianapolis   New York   San Francisco   Upper Saddle River
Amsterdam   Cape Town   Dubai   London   Madrid   Milan   Munich   Paris   Montréal   Toronto
Delhi   Mexico City   São Paulo   Sydney   Hong Kong   Seoul   Singapore   Taipei   Tokyo

Editor in Chief: Stephanie Wall
Executive Editor: Bob Horan
Senior Acquisitions Editor, Global Edition: Steven Jackson
Director of Editorial Services: Ashley Santora
Senior Editorial Project Manager: Kelly Loftus
Editorial Assistant: Ashlee Bradbury
Director of Marketing: Maggie Moylan
Executive Marketing Manager: Anne Fahlgren
Marketing Manager, International: Dean Erasmus
Marketing Assistant: Gianna Sandri

Senior Managing Editor: Judy Leale
Production Project Manager: Ilene Kahn
**Senior Manufacturing Controller,
    Production, Global Edition:** Trudy Kimber
Creative Director: Blair Brown
Art Director: Steve Frim
Interior Designer: Jill Lehan
Cover Designer: Jodi Notowitz
Cover Art: © evv - Shutterstock

Pearson Education Limited
Edinburgh Gate
Harlow
Essex CM20 2JE
England
and Associated Companies throughout the world

Visit us on the World Wide Web at:
www.pearson.com/uk

ISBN-13: 978-0-273-78710-5
ISBN-10: 0-273-78710-1

British Library Cataloguing-in-Publication Data
A catalogue record for this book is available from the British Library

ARP impression 98

Typeset in Times 10/12 by Element LLC
Printed and bound by Ashford Colour Press Ltd.

The publisher's policy is to use paper manufactured from sustainable forests.

To the memory of Julia A. Kendall and the memory of Edward J. Kendall, whose lifelong example of working together will inspire us forever.

CONTENTS

Design the Output for the App 350 / Design the Output a Second Time for a Different Orientation 350 / Design the App's Logic 351 / Create the User Interface Using Gestures 351 / Protect Your Intellectual Property 352 / Market Your App 352

Output Production and XML 352
Ajax 353

12 DESIGNING EFFECTIVE INPUT 362

Good Form Design 363
Making Forms Easy to Fill In 363 / Meeting the Intended Purpose 366 / Ensuring Accurate Completion 366 / Keeping Forms Attractive 366 / Controlling Business Forms 366

Good Display and Web Forms Design 367
Keeping the Display Simple 367 / Keeping the Display Consistent 368 / Facilitating Movement 368 / Designing an Attractive and Pleasing Display 368 / Using Icons in Display Design 368

Consulting Opportunity 12.1 *This Form May Be Hazardous to Your Health* 369

Consulting Opportunity 12.2 *Squeezin' Isn't Pleasin'* 370

Graphical User Interface Design 370 / Form Controls and Values 373 / Hidden Fields 373 / Event-Response Charts 374 / Dynamic Web Pages 376 / Three-Dimensional Web Pages 376 / Ajax (Asynchronous JavaScript and XML) 378 / Using Color in Display Design 380

Website Design 380

13 DESIGNING DATABASES 389

Consulting Opportunity 13.1 *Hitch Your Cleaning Cart to a Star* 390

Databases 390

Data Concepts 391
Reality, Data, and Metadata 391 / Files 396 / Relational Databases 397

Normalization 399
The Three Steps of Normalization 399 / A Normalization Example 400 / Using an Entity-Relationship Diagram to Determine Record Keys 408 / One-to-Many Relationships 409 / Many-to-Many Relationships 409

Guidelines for Master File/Database Relation Design 410
Integrity Constraints 410

PREFACE

NEW TO THIS EDITION

The ninth edition of *Systems Analysis and Design* includes extensive changes inspired by the swift transformations in the IS field over the past three years, and they are included as a response to the thoughtful input of our adopters, students, and reviewers. Many innovative and upgraded features are incorporated throughout this new edition. In particular:

- New coverage of how systems analysts and organizations can participate in open source communities (Chapter 1)
- Expanded coverage of the analyst role in ERP (enterprise systems) (Chapter 2)
- New in-depth coverage of project management techniques (Chapter 3)
- Expanded coverage of when to use cloud services versus purchasing hardware and software (Chapter 3)
- New coverage of time estimation techniques for project management (Chapter 3)
- New coverage of the work breakdown structure (WBS) for project management (Chapter 3)
- New material on designing corporate and ecommerce sites to include Web 2.0 technologies and social media (Chapter 11)
- Innovative treatment of designing apps for smartphones and tablets (Chapter 11)
- Expanded coverage of designing input for intranets, the Web, smartphones, and tablets (Chapter 12)
- New material on the relationship of business intelligence to data warehouses, big data, business analytics, and text analytics (Chapter 13)
- Innovative coverage on designing gesture-based interfaces for smartphones and tablets (Chapter 14)
- Additional material on designing alerts, queries, and notices for smartphones and tablets (Chapter 14)
- Innovative handling of designing two-dimensional (2D) codes such as Microsoft Tags and QR codes for input (Chapter 15)
- New material on how service-oriented architecture and cloud computing are changing the nature of information systems design (Chapter 16)
- Expanded coverage of ERP systems and their relationship to cloud computing (Chapter 16)

DESIGN FEATURES

Figures have a stylized look in order to help students more easily grasp the subject matter.

Conceptual diagrams are used to introduce the many tools that systems analysts have at their disposal. This example shows the differences between logical data flow diagrams and physical data flow diagrams. Conceptual diagrams are color coded so that students can distinguish easily among them, and their functions are clearly indicated. Many other important tools are illustrated, including use case diagrams, sequence diagrams, and class diagrams.

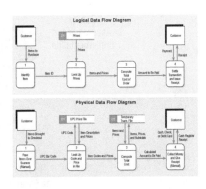

Computer displays demonstrate important software features that are useful to the analyst. This example shows how a website can be evaluated for broken links by using a package such as Microsoft Visio. Actual screen shots show important aspects of design. Analysts are continuously seeking to improve the appearance of the screens and Web pages they design. Colorful examples help to illustrate why some screen designs are particularly effective.

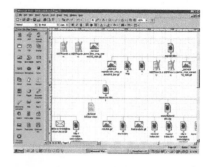

Paper forms are used throughout to show input and output design as well as the design of questionnaires.

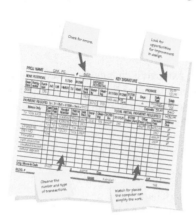

Blue ink is always used to show writing or data input, thereby making it easier to identify what was filled in by users. Although most organizations have computerization of manual processes as their goal, much data capture is still done using paper forms. Improved form design enables analysts to ensure accurate and complete input and output. Better forms can also help streamline new internal workflows that result from newly automated business-to-consumer (B2C) applications for ecommerce on the Web.

Tables are used when an important list needs special attention or when information needs to be organized or classified. In addition, tables are used to supplement the understanding of the reader in a way that departs from how material is organized in the narrative portion of the book. Most analysts find using tables a useful way to organize numbers and text into a meaningful "snapshot."

This example of a table from Chapter 3 shows how analysts can refine their activity plans for analysis by breaking them down into smaller tasks and then estimating how much time it will take to complete them. This book is built on the idea that systems analysis and design is a process that integrates the use of many tools with the unique talents of the systems analyst to systematically improve business through the implementation or modification of computerized information systems. Systems analysts can grow in their work: by taking on new IT challenges, whether they are posed by designing for multiple platforms, new types of users, or implementing cloud-based systems, and by keeping up to date in their profession through the application of new methods, software, and alternative tools.

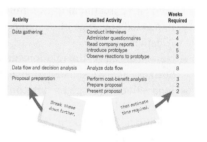

A BRIEF TOUR OF THE NINTH EDITION

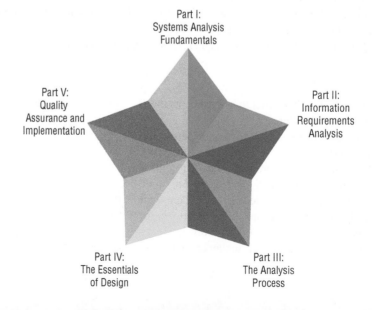

Systems analysis and design is typically taught in one or two semesters. This book may be used in either situation. The text is appropriate for undergraduate (junior or senior) curricula at a four-year university, graduate school, or community college. The level and length of the course can be varied and supplemented by using real-world projects, HyperCase, The CPU Case online, or other materials available on the Instructor Resource Center.

The text is divided into five major parts: Systems Analysis Fundamentals (Part I), Information Requirements Analysis (Part II), The Analysis Process (Part III), The Essentials of Design (Part IV), and Quality Assurance and Implementation (Part V).

Part I (Chapters 1–3) stresses the basics that students need to know about what an analyst does and introduces the three main methodologies of the systems development life cycle (SDLC), agile approaches, and object-oriented analysis with UML, along with reasons and situations for when to use them. Part I introduces the three roles of a systems analyst—consultant, supporting expert, and agent of change—along with ethical issues and professional guidelines for serving as a systems consultant. There is also material on virtual teams and virtual organizations, and the concept of human–computer interaction (HCI) is introduced. The use of open source software (OSS) and how analysts and organizations can participate in open source communities is also introduced. Chapter 2 includes how to initially approach an organization by drawing context-level data flow diagrams, using entity-relationship models, and developing use cases and use case scenarios. It views the

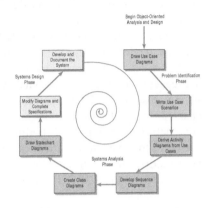

organization as a system through the description of enterprise systems (ERP). Chapter 3 focuses on project management. It introduces new material on when to use cloud services versus purchasing hardware and software. Expanded coverage of project management techniques is also included, including new time estimation techniques for project management. Chapter 3 also includes new material to help students approach projects using the work breakdown structure (WBS). Creating a problem definition, developing a project charter, and determining feasibility are also covered. Chapter 3 guides students in professionally writing and presenting an effective systems proposal, one that incorporates figures and graphs to communicate with users.

Part II (Chapters 4–6) emphasizes the use of systematic and structured methodologies for performing information requirements analysis. Attention to analysis helps analysts ensure that they are addressing the correct problem before designing a system. Chapter 4 introduces a group of interactive methods, including interviewing, Joint Application Design (JAD), listening to user stories, and constructing questionnaires. Chapter 5 introduces a group of unobtrusive methods for ascertaining information requirements of users. These methods include sampling, investigating hard and archival data, and observation of decision makers' behavior and their physical environment. Chapter 6 on agile modeling and prototyping is innovative in its treatment of prototyping as another data-gathering technique that enables the analyst to solve the right problem by getting users involved from the start. Since agile approaches have their roots in prototyping, this chapter begins with prototyping to provide a proper context for understanding, and then takes up the agile approach. The values and principles, activities, resources, practices, processes, and tools associated with agile methodologies are presented.

Part III (Chapters 7–10) details the analysis process. It builds on the previous two parts to move students into analysis of data flows as well as structured and semistructured decisions. It provides step-by-step details on how to use structured techniques to draw data flow diagrams (DFDs). Chapter 7 provides coverage of how to create child diagrams; how to develop both logical and physical data flow diagrams; and how to partition data flow diagrams. Chapter 8 features material on the data repository and vertical balancing of data flow diagrams. Chapter 8 also includes extensive coverage of Extensible Markup Language (XML) and demonstrates how to use data dictionaries to create XML. Chapter 9 includes material on developing process specifications. A discussion of both logical and physical process specifications shows how to use process specifications for horizontal balancing. Chapter 9 also covers how to diagram structured decisions with the use of structured English, decision tables, and decision trees. In addition, the chapter covers how to choose an appropriate decision analysis method for analyzing structured decisions and creating process specifications.

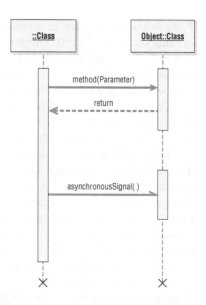

Part III concludes with Chapter 10 on object-oriented systems analysis and design. This chapter includes an in-depth section on using Unified Modeling Language (UML). There is detailed coverage of the use case model, creating the class model diagram with UML, sequence diagrams, creating gen/spec diagrams, use case scenarios, and activity diagrams. Through several examples and Consulting Opportunities, this chapter demonstrates how to use an object-oriented approach. Consulting Opportunities, diagrams, and problems enable students to learn and use UML to model systems from an object-oriented perspective. Students learn the appropriate situations for using an object-oriented approach. This chapter helps students to decide whether to use the SDLC, the agile approach, or object-oriented systems analysis and design to develop a system.

Part IV (Chapters 11–14) covers the essentials of design. It begins with designing output because many practitioners believe systems to be output driven. The design of Web-based forms is covered in detail. Particular attention is paid to relating output method to content, the effect of output on users, and designing good forms and screens. Chapter 11 considers output, including Web displays, audio, and electronic output such as Web pages, email, and RSS feeds. Designing

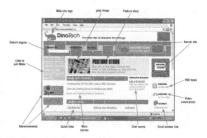

a website for ecommerce purposes is emphasized, and the importance of adding Web 2.0 technologies and social media to corporate and ecommerce websites is explored. Designing apps for smartphones and tablets is included, along with storyboarding, wireframing, and mockups. Output production and XML are covered.

Chapter 12 includes innovative material on designing for smartphones and tablets as well as designing Web-based input forms and other electronic forms design. Also included is computer-assisted forms design. Chapter 12 also features in-depth coverage of website design, including guidelines on when designers should add video, audio, and animation to website designs. There is detailed consideration of how to create effective graphics for corporate websites and ways to design effective onscreen navigation for website users.

Coverage of intranet and extranet page design is also included. Consideration of database integrity constraints has been included as well, in addition to how the user interacts with the computer and how to design an appropriate interface. The importance of user feedback is also found in Part IV. How to design accurate data entry procedures that take full advantage of computer and human capabilities to assure entry of quality data is emphasized here.

Chapter 13 demonstrates how to use an entity-relationship diagram to determine record keys, as well as providing guidelines for file/database relation design. Students are shown the relevance of database design for the overall usefulness of the system, and how users actually use databases. The concepts of business intelligence (BI) and its relationships to big data, business analytics, and text analytics are also introduced in the context of data warehouses.

Chapter 14 emphasizes human–computer interaction (HCI), especially as it relates to interface design. It discusses the importance of HCI in designing systems that suit individuals and assisting them in achieving personal and organizational goals through their use of information technology. The concept of usability is introduced, so that systems analysis students can knowledgeably incorporate HCI practices into their designs. Chapter 14 introduces material on how to design gesture-based (multitouch) interfaces for smartphones and tablets, as well as designing alerts, notices and queries. Material on designing easy onscreen navigation for website visitors is also included. The chapter presents innovative approaches to searching on the Web, highlights material on GUI design, and provides innovative approaches to designing dialogs. Chapter 14 articulates specialized design considerations for ecommerce websites. Mashups, new applications created by combining two or more Web-based application programming interfaces, are also introduced. Chapter 14 also includes extensive coverage on how to formulate queries, all within the framework of HCI.

Part V (**Chapters 15 and 16**) concludes the book. Chapter 15 focuses on designing accurate data entry procedures and includes material on managing the supply chain through the effective design of business-to-business (B2B) ecommerce. It includes suggestions for incorporating two-dimensional codes, such as QR codes and Microsoft Tags, into data entry designs. It also considers the usefulness of RFID for automatic data collection. Chapter 16 emphasizes taking a total quality approach to improving software design and maintenance. In addition, material on system security and firewalls is included. Testing, auditing, and maintenance of systems are discussed in the context of total quality management. This chapter helps students understand how service-oriented architecture (SOA) and cloud computing combined with ERP are significantly altering the landscape of information systems design. In addition,

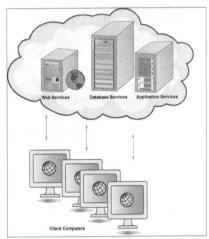

students learn how to design appropriate training programs for users of the new system, how to recognize the differences among physical conversion strategies, and how to be able to recommend an appropriate one to a client. Chapter 16 also presents techniques for modeling networks, which can be done with popular tools such as Microsoft Visio.

Material on security and privacy in relation to designing ecommerce applications is included. Coverage of security, specifically firewalls, gateways, Public Key Infrastructure (PKI), Secure Electronic Transaction (SET), Secure Sockets Layer (SSL), virus protection software, URL filtering products, email filtering products, and virtual private networks (VPN) is included. In addition, current topics of interest to designers of ecommerce applications, including the development and posting of corporate privacy policies, are covered.

Important coverage of how the analyst can promote and monitor a corporate website is included in this section, which features Web activity monitoring, website promotion, Web traffic analysis, and audience profiling to ensure the effectiveness of new ecommerce systems. Techniques for evaluating the completed information systems project are covered systematically as well.

This ninth edition contains an updated **Glossary** of terms and a separate list of updated **Acronyms** used in the book and in the systems analysis and design field.

PEDAGOGICAL FEATURES

Chapters in this ninth edition contain:

- **Learning Objectives** at the beginning of each chapter
- **Summaries** that tie together the salient points of each chapter while providing an excellent source of review for exams
- **Keywords and Phrases**
- **Review Questions**
- **Problems**
- **Group Projects** that help students work together in a systems team to solve important problems that are best solved through group interaction
- **Consulting Opportunities,** now with more than 50 minicases throughout the book
- **Mac Appeal** columns that inform students on design software available on the Mac and iPhone
- **HyperCase Experiences**

CONSULTING OPPORTUNITIES

This ninth edition presents more than 50 Consulting Opportunities, and many of them address significant and emerging topics arising in information systems, including designing systems from an HCI perspective, ecommerce applications for the Web, cloud computing decisions, and using UML to model information systems from an object-oriented perspective. Consulting Opportunities can be used for motivating thoughtful in-class discussions or assigned as homework or take-home exam questions.

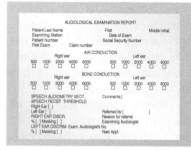

Because not all systems work demands extended two- or three-year projects, our book contains many Consulting Opportunities that can be solved in 20 to 30 minutes of group discussion or individual writing. These minicases, written in a humorous manner to enliven the material, require students to synthesize what they have learned up to that point in the course, ask students to mature in their professional and ethical judgment, and expect students to articulate the reasoning that led to their systems decisions.

HYPERCASE EXPERIENCES

HyperCase® Experiences that pose challenging student exercises are present in each chapter. HyperCase 2.9 has organizational problems featuring state-of-the-art technological systems. HyperCase represents an original virtual organization that allows students who access it to become immediately immersed in organizational life. Students will interview people, observe office environments, analyze their prototypes, and review the documentation of their existing systems.

HyperCase 2.9 is Web-based, interactive software that presents an organization called Maple Ridge Engineering (MRE) in a colorful, three-dimensional graphics environment. HyperCase permits professors to begin approaching a systems analysis and design class with exciting multimedia material. Carefully watching their use of time and managing multiple methods, students use the hypertext characteristics of HyperCase on the Web to create their own individual paths through the organization.

Maple Ridge Engineering is drawn from the actual consulting experiences of the authors of the original version (Raymond Barnes, Richard Baskerville, Julie E. Kendall, and Kenneth E. Kendall). Allen Schmidt joined the project for version 2.0 and has remained with it. Peter Schmidt was the HTML programmer, and Jason Reed created the images for the initial Web version.

Each chapter contains HyperCase Experiences that include assignments (and even some clues) to help students solve difficult organizational problems including developing new systems, merging departments, hiring of employees, security, ecommerce, and disaster recovery planning they encounter at MRE. HyperCase has been fully tested in classrooms and was an award winner in the Decision Sciences Institute Innovative Instruction competition.

EXPANDED WEB SUPPORT

Systems Analysis and Design, ninth edition, features Web-based support for solid but lively pedagogical techniques in the information systems field:

- The website, located at **www.pearsonglobaleditions .com/kendall**, contains a wealth of critical learning and support tools, which keep class discussions exciting.

- **HyperCase 2.9** is an award-winning, interactive organization game. Students are encouraged to interview people in the organization, analyze problems, drill down into and modify data flow diagrams and data dictionaries, react to prototypes, and design new input and output.

- **Entire Central Pacific University (CPU) case now online** In keeping with our belief that a variety of approaches are important, the entire Central Pacific University (CPU) case, accompanied by partially solved Student Exercises, is now available online. There is an episode to accompany each chapter of the ninth edition. The CPU case makes use of Microsoft Access, Microsoft Visio, and the popular CASE tool Visible Analyst by Visible Systems, Inc., for the sample screen shots and the student exercises. The CPU case takes students through all phases of the systems development life cycle. The CPU case has been fully tested in classrooms around the world with a variety of students over numerous terms. The case is detailed, rigorous, and rich enough to stand alone as a systems analysis and design project spanning one or two terms.

- **Student Exercises based on the online CPU case** This running case gives students an opportunity to solve problems on their own, using a variety of tools and data that users of the book can download from the Web containing Microsoft Visio, Microsoft Access, and Visible Analyst exercises specifically keyed to each chapter of the book. Partially solved problems and examples stored in Microsoft Access and Visible Analyst files allow students to develop a Web-based computer management system.

EXPANDED INSTRUCTOR SUPPLEMENTAL WEB SUPPORT

Extended support for instructors using this edition can be found at the official website located at **www.pearsonglobaleditions.com/kendall**. Resources include:

- **Instructor's Manual**—The Instructor's Manual contains answers to problems, solutions to cases, and suggestions for approaching the subject matter.

- **Solutions to Student Exercises**—These exercises are based on the ongoing CPU case, with solutions and examples stored in Visible Analyst files and Microsoft Access files.

- **PowerPoint Presentations**—The PowerPoints feature lecture notes that highlight key text terms and concepts. Professors can customize the presentation by adding their own slides or by editing the existing ones.

- **Test Item File**—The Test Item File is an extensive set of multiple-choice, true/false, and essay-type questions for each chapter of the text. Questions are ranked according to difficulty level and referenced with page numbers from the text. The Test Item File is available in Microsoft Word format and as the computerized Prentice Hall TestGen software, with course management system conversions.

- **TestGen**—Pearson Education's test-generating software is available from **www.pearsonglobaleditions.com/kendall**. The software is PC/Mac compatible and preloaded with all the Test Item File questions. You can manually or randomly view test questions and drag and drop to create a test. You can add or modify test-bank questions as needed.

- **Image Library**—This collection of the text art is organized by chapter. This collection includes all the figures, tables, and screenshots from the book. These images can be used to enhance class lectures and PowerPoint slides.

- **CourseSmart eTextbooks Online**—CourseSmart eTextbooks were developed for students looking to save on required or recommended textbooks. Students simply select their eText by title or author and purchase immediate access to the content for the duration of the course using any major credit card. With a CourseSmart eText, students can search for specific keywords or page numbers, take notes online, print out reading assignments that incorporate lecture notes, and bookmark important passages for later review. For more information or to purchase a CourseSmart eTextbook, visit **www.coursesmart.co.uk**

ACKNOWLEDGMENTS

The field of information systems was transforming astonishingly as we were writing the ninth edition of *Systems Analysis and Design*. We are thrilled that this edition is being published at the right time for us to capture many of these innovations in systems development.

One major change is the rapidly increasing use of the Web as a platform for information systems. Cloud computing will dramatically change the way that analysts approach designing systems solutions.

Another major change addressed in this edition is the emergence of smartphones and tablets as corporate platforms for IT. With the advent of BYOD (bring your own device) systems, analysts face new challenges in developing successful and secure systems that can easily traverse multiple platforms.

Throughout the book you will learn and apply numerous techniques, methods, tools, and approaches to help visually capture a system. But when the time comes to interpret what is happening in the organization and to develop meaningful information systems from the application of rules to your analysis, your training combines with creativity to produce a system that is in some ways a surprise: It is structured, yet intuitive, multilayered, and complex, in keeping with the character of the organization and uniquely reflective of you as a systems analyst and a human being.

The artist Richard Kalina, who created the colorful collage on the cover of the ninth edition, writes, "*P3 Vega* is inspired by the representation of scientific phenomena, ranging from astronomy, chemistry and physics to cybernetics and information theory. The painting is not a literal depiction, but rather an abstracted map or chart, a way of thinking and observing. *P3 Vega* is a set of interlocking connections, a network of circular nodes joined by colored lines. It feels stable but shifting—a static depiction of a changeable state. I am trying to find a visual corollary to the beauty that underlies logical systems, and to make something beautiful, hopefully, in the process." We hope that you as a student will also appreciate, through this book, the beauty that underlies logical systems.

It is, in fact, our own students who deserve recognition for this new edition because of their feedback and recommendations for improvements and requests for increased depth in certain topics. Students told us that they quickly put to use the new material on designing apps and interfaces for smartphones and tablets as well as the material on new project management techniques and cloud computing. We are indebted to their quest to continually improve their skills. We want to thank our coauthor, Allen Schmidt, who once again worked with us on the HyperCase 2.9 and *CPU Case Episodes* for all his dedication, insight, and humor during our collaboration. He is a superb human being. Our appreciation also goes to Peter Schmidt and Jason Reed for their improvements to the early HyperCase. In addition, we want to thank the other two original authors of HyperCase, Richard Baskerville and Raymond Barnes, who contributed so much.

We would like to thank our ninth edition production team, especially our executive editor, Bob Horan, whose intelligence and tranquil demeanor are always inspiring. We are also grateful to Kelly Loftus, who is our extremely capable senior project manager for MIS, for her composed competency and for her enthusiasm in keeping the project going. Ilene Kahn, our production project manager, also deserves thanks for helping us succeed in making this a robust, inclusive, and accurate revision. Their help and keen interest in our book facilitated the completion of this project in a smooth and timely manner.

We also appreciate the encouragement and support of the entire Rutgers community, including our chancellor, Wendell Prittchett, and our colleagues and staff in the School of Business–Camden and throughout all of Rutgers. They have been very enthusiastic about this edition as

Julie and Ken Kendall personally thank all of our friends in the theatre and the performing arts. Here are the Kendalls at the 2012 Tony Awards afterparty with Tony-Award winning Actor James Corden (*right*). Photo by Anita & Steve Shevett.

well as the many translations and versions of *Systems Analysis and Design* available in Spanish, Chinese, English for the Indian subcontinent, and Indonesian.

All the reviewers for the ninth edition deserve our thanks as well. Their thoughtful comments and suggestions helped to strengthen the book. They are:

Ron Davis, University of North Alabama
Chang-tseh Hsieh, University of Southern Mississippi
Sukgon Kim, Northern Illinois University
Angela Marsh, University of Arkansas–Monticello
Keng Siau, Missouri University of Science and Technology
Mead Bond Wetherbe, Jr., Texas Tech University

Many of our colleagues and friends have encouraged us through the process of writing this book. We thank them for their comments on our work. They include: Ayman Abu Hamdieh, Macedonio Alanis, Michel Avital, the Ciupeks, Roger T. Danforth, Gordon Davis, John Drozdal, EgoPo, Matt Germonprez, Nancy V. Gulick, Andy Hamingson, Blake Ives, Richard Kalina, Colleen Kelly-Lawler, Carol J. Latta, Ken and Jane Laudon, Josh Lawler, Lars Mathiassen, Joel and Bobbie Porter, Caryn Schmidt, Marc and Jill Schniederjans, Gabriel Shanks, Detmar W. Straub, Jr., the Vargos, Merrill Warkentin, Brian Warner, Jeff and Bonnie Weil, Arlene and Paul Wolfling, Brett Young, Ping Zhang and all of our friends and colleagues in The Drama League, The Actors Fund, the American Theatre Wing, The New York Marriott Marquis, the Association for Information Systems, the Decision Sciences Institute, IFIP Working Group 8.2, and all those involved in the PhD Project (founded by the KPMG Foundation), which serves minority doctoral students in information systems.

Our heartfelt thanks go to the memory of Julia A. Kendall and to the memory of Edward J. Kendall. Their belief that love, goals, and hard work are an unbeatable combination continues to infuse our every endeavor.

CHAPTER 1

Systems, Roles, and Development Methodologies

LEARNING OBJECTIVES

Once you have mastered the material in this chapter you will be able to:

1. Understand the need for systems analysis and design in organizations.

2. Realize what the many roles of a systems analyst are.

3. Comprehend the fundamentals of three development methodologies: SDLC, the agile approach, and object-oriented systems analysis and design.

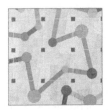

Organizations have long recognized the importance of managing key resources such as people and raw materials. Information has now moved to its rightful place as a key resource. Decision makers now understand that information is not just a by-product of conducting business; rather, it fuels business and can be the critical factor in determining the success or failure of a business.

To maximize the usefulness of information, a business must manage it correctly, just as it manages other resources. Managers need to understand that costs are associated with the production, distribution, security, storage, and retrieval of all information. Although information is all around us, it is not free, and its strategic use for positioning a business competitively should not be taken for granted.

The ready availability of networked computers, along with access to the Internet and the Web, has created an information explosion throughout society in general and business in particular. Managing computer-generated information differs in significant ways from handling manually produced data. Usually there is a greater quantity of computer information to administer. Costs of organizing and maintaining it can increase at alarming rates, and users often treat it less skeptically than information obtained in different ways. This chapter examines the fundamentals of different kinds of information systems, the varied roles of systems analysts, and the phases in the systems development life cycle (SDLC) as they relate to human–computer interaction (HCI) factors; it also introduces computer-aided software engineering (CASE) tools.

Need for Systems Analysis and Design

Systems analysis and design, as performed by systems analysts, seeks to understand what humans need to analyze data input or data flow systematically, process or transform data, store data, and output information in the context of a particular organization or enterprise. By doing thorough analysis, analysts seek to identify and solve the right problems. Furthermore, systems analysis and design is used to analyze, design, and implement improvements in the support of users and the functioning of businesses that can be accomplished through the use of computerized information systems.

Installing a system without proper planning leads to great user dissatisfaction and frequently causes the system to fall into disuse. Systems analysis and design lends structure to the analysis and design of information systems, a costly endeavor that might otherwise have been done in a haphazard way. It can be thought of as a series of processes systematically undertaken to improve a business through the use of computerized information systems. Systems analysis and design involves working with current and eventual users of information systems to support them in working with technologies in an organizational setting.

User involvement throughout a systems project is critical to the successful development of computerized information systems. Systems analysts, whose roles in the organization are discussed next, are the other essential component in developing useful information systems.

Users are moving to the forefront as software development teams become more international in their composition. This means that there is more emphasis on working with software users; on performing analysis of their business, problems, and objectives; and on communicating the analysis and design of the planned system to all involved.

New technologies also are driving the need for systems analysis. Ajax (Asynchronous JavaScript and XML) is not a new programming language but a technique that uses existing languages to make web pages function more like a traditional desktop application program. Systems analysts will increasingly need to build and redesign web pages that utilize Ajax technologies. New programming languages, such as the open source Web framework *Ruby on Rails* (*"Rails"* for short), which is a combination programming language and code generator for creating Web applications, will require more analysis.

Roles of a Systems Analyst

A systems analyst systematically assesses how users interact with technology and how businesses function by examining the inputting and processing of data and the outputting of information with the intent of improving organizational processes. Many improvements involve better support of users' work tasks and business functions through the use of computerized information systems. This definition emphasizes a systematic, methodical approach to analyzing—and potentially improving— what is occurring in the specific context experienced by users and created by a business.

Our definition of a systems analyst is necessarily broad. An analyst must be able to work with people of all descriptions and be experienced in working with computers. An analyst plays many roles, sometimes balancing several at the same time. The three primary roles of a systems analyst are consultant, supporting expert, and agent of change.

Systems Analyst as Consultant

A systems analyst frequently acts as a systems consultant to humans and their businesses and, thus, may be hired specifically to address information systems issues within a business. Such hiring can be an advantage because outside consultants can bring with them a fresh perspective that other people in an organization do not possess. It also means that outside analysts are at a disadvantage because an outsider can never know the true organizational culture. As an outside consultant, you will rely heavily on the systematic methods discussed throughout this text to analyze and design appropriate information systems for users working in a particular business. In addition, you will rely on information systems users to help you understand the organizational culture from others' viewpoints.

Systems Analyst as Supporting Expert

Another role that you may be required to play is that of supporting expert within a business for which you are regularly employed in some systems capacity. In this role, an analyst draws on professional expertise concerning computer hardware and software and their uses in the business.

CONSULTING OPPORTUNITY 1.1

Healthy Hiring: Ecommerce Help Wanted

"You'll be happy to know that we made a strong case to management that we should hire a new systems analyst to specialize in ecommerce development," says Al Falfa, a systems analyst for the multioutlet international chain Marathon Vitamin Shops. He is meeting with his large team of systems analysts to decide on the qualifications that their new team member should possess. Al continues, saying, "In fact, they were so excited by the possibility of our team helping to move Marathon into an ecommerce strategy that they've said we should start our search now and not wait until the fall."

Ginger Rute, another analyst, agrees, saying, "The demand for website developers is still outstripping the supply. We should move quickly. I think our new person should be knowledgeable in system modeling, JavaScript, C++, and Rational Rose and familiar with Ajax, just to name a few."

Al looks surprised at Ginger's long list of skills but then replies, "Well, that's certainly one way we could go. But I would also like to see a person with some business savvy. Most of the people coming out of school will have solid programming skills, but they should know about accounting, inventory, and distribution of goods and services, too."

The newest member of the systems analysis group, Vita Ming, finally breaks into the discussion. She says, "One of the reasons I chose to come to work with all of you was that I thought we all got along quite well together. Because I had some other opportunities, I looked very carefully at what the atmosphere was here. From what I've seen, we're a friendly group. Let's be sure to hire someone who has a good personality and who fits in well with us."

Al concurs, continuing, "Vita's right. The new person should be able to communicate well with us, and with business clients, too. We are always communicating in some way, through formal presentations, drawing diagrams, or interviewing users. If they understand decision making, it will make their job easier, too. Also, Marathon is interested in integrating ecommerce into the entire business. We need someone who at least grasps the strategic importance of the Web. Page design is such a small part of it."

Ginger interjects again with a healthy dose of practicality, saying, "Leave that to management. I still say the new person should be a good programmer." Then she ponders aloud, "I wonder how important UML will be?"

After listening patiently to everyone's wish list, one of the senior analysts, Cal Siem, speaks up, joking, "We'd better see if Superman is available!"

As the group shares a laugh, Al sees an opportunity to try for some consensus, saying, "We've had a chance to hear a number of different qualifications. Let's each take a moment and make a list of the qualifications we personally think are essential for the new ecommerce development person to possess. We'll share them and continue discussing until we can describe the person in enough detail to turn a description over to the human resources group for processing."

What qualifications should the systems analysis team be looking for when hiring their new ecommerce development team member? Is it more important to know specific languages or to have an aptitude for picking up languages and software packages quickly? How important is it that the person being hired has some basic business understanding? Should all team members possess identical competencies and skills? What personality or character traits are desirable in a systems analyst who will be working in ecommerce development?

This work is often not a full-blown systems project, but rather it entails a small modification or decision affecting a single department.

As the supporting expert, you are not managing the project; you are merely serving as a resource for those who are. If you are a systems analyst employed by a manufacturing or service organization, many of your daily activities may be encompassed by this role.

Systems Analyst as Agent of Change

The most comprehensive and responsible role that the systems analyst takes on is that of an agent of change, whether internal or external to the business. As an analyst, you are an agent of change whenever you perform any of the activities in the systems development life cycle (discussed in the next section) and are present and interacting with users and the business for an extended period (from two weeks to more than a year). An agent of change can be defined as a person who serves as a catalyst for change, develops a plan for change, and works with others in facilitating that change.

Your presence in the business changes it. As a systems analyst, you must recognize this fact and use it as a starting point for your analysis. Hence, you must interact with users and management (if they are not one and the same) from the very beginning of your project. Without their help, you cannot understand what they need to support their work in the organization, and real change cannot take place.

If change (that is, improvements to the business that can be realized through information systems) seems warranted after analysis, the next step is to develop a plan for change along with the people who must enact the change. Once a consensus is reached on the change that is to be made, you must constantly interact with those who are changing.

As a systems analyst acting as an agent of change, you advocate a particular avenue of change involving the use of information systems. You also teach users the process of change because changes in the information system do not occur independently; rather, they cause changes in the rest of the organization as well.

Qualities of a Systems Analyst

From the foregoing descriptions of the roles the systems analyst plays, it is easy to see that a successful systems analyst must possess a wide range of qualities. Many different kinds of people are systems analysts, so any description is destined to fall short in some way. There are some qualities, however, that most systems analysts seem to display.

Above all, an analyst is a problem solver. He or she is a person who views the analysis of problems as a challenge and who enjoys devising workable solutions. When necessary, an analyst must be able to systematically tackle the situation at hand through skillful application of tools, techniques, and experience. An analyst must also be a communicator capable of relating meaningfully to other people over extended periods of time. Systems analysts need to be able to understand humans' needs in interacting with technology, and they need enough computer experience to program, to understand the capabilities of computers, to glean information requirements from users, and to communicate what is needed to programmers. They also need to possess strong personal and professional ethics to help them shape their client relationships.

A systems analyst must be a self-disciplined, self-motivated individual who is able to manage and coordinate other people, as well as innumerable project resources. Systems analysis is a demanding career, but, in compensation, an ever-changing and always challenging one.

The Systems Development Life Cycle

Throughout this chapter we have referred to the systematic approach analysts take to the analysis and design of information systems. Much of this is embodied in what is called the systems development life cycle (SDLC). The SDLC is a phased approach to analysis and design which holds that systems are best developed through the use of a specific cycle of analyst and user activities.

Analysts disagree on exactly how many phases there are in the SDLC, but they generally laud its organized approach. Here we have divided the cycle into seven phases, as shown in Figure 1.1. Although each phase is presented discretely, it is never accomplished as a separate step. Instead, several activities can occur simultaneously, and activities may be repeated.

FIGURE 1.1

The seven phases of the systems development life cycle (SDLC).

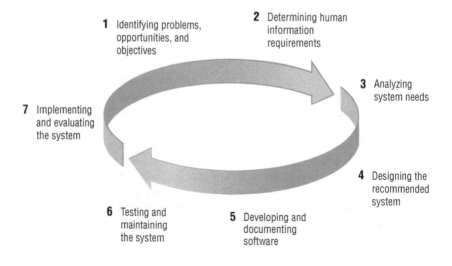

Incorporating Human–Computer Interaction Considerations

In recent years, the study of human–computer interaction (HCI) has become increasingly important for systems analysts. Although the definition is still evolving, researchers characterize HCI as the "aspect of a computer that enables communications and interactions between humans and the computer. It is the layer of the computer that is between humans and the computer" (Zhang, Carey, Te'eni, & Tremaine, 2005, p. 518). Analysts using an HCI approach are emphasizing people rather than the work to be done or the IT that is involved. Their approach to a problem is multifaceted, looking at the "human ergonomic, cognitive, affective, and behavioral factors involved in user tasks, problem solving processes and interaction context" (Zhang, Carey, Te'eni, & Tremaine, 2005, p. 518). HCI moves away from focusing first on organizational and system needs and instead concentrates on human needs. Analysts adopting HCI principles examine a variety of user needs in the context of humans interacting with information technology to complete tasks and solve problems. These include taking into account physical or ergonomic factors; usability factors that are often labeled cognitive matters; the pleasing, aesthetic, and enjoyable aspects of using the system; and behavioral aspects that center on the usefulness of the system.

Another way to think about HCI is to think of it as a human-centered approach that puts people ahead of organizational structure or culture when creating new systems. When analysts employ HCI as a lens to filter the world, their work will possess a different quality than the work of those who do not possess this perspective.

Your career can benefit from a strong grounding in HCI. The demand for analysts who are capable of incorporating HCI into the systems development process keeps rising, as companies increasingly realize that the quality of systems and the quality of work life can both be improved by taking a human-centered approach at the outset of a project.

The application of human–computer interaction principles tries to uncover and address the frustrations that users voice over their use of information technology. These concerns include a suspicion that systems analysts misunderstand the work being done, the tasks involved, and how they can best be supported; a feeling of helplessness or lack of control when working with the system; intentional breaches of privacy; trouble navigating through system screens and menus; and a general mismatch between the system designed and the way users themselves think of their work processes.

Misjudgments and errors in design that cause users to neglect new systems or that cause systems to fall into disuse soon after their implementation can be eradicated or minimized when systems analysts adopt an HCI approach.

Researchers in HCI see advantages to the inclusion of HCI in every phase of the SDLC. This is a worthwhile approach, and we will try to mirror it by bringing human concerns explicitly into each phase of the SDLC. As a person who is learning systems analysis, you can also bring a fresh eye to the SDLC to identify opportunities for designers to address HCI concerns and ways for users to become more central to each phase of the SDLC. Chapter 14 is devoted to examining the role of the systems analyst in designing human-centered systems and interfaces from an HCI perspective.

Identifying Problems, Opportunities, and Objectives

In this first phase of the systems development life cycle, an analyst is concerned with correctly identifying problems, opportunities, and objectives. This stage is critical to the success of the rest of the project because no one wants to waste subsequent time addressing the wrong problem.

The first phase requires that the analyst look honestly at what is occurring in a business. Then, together with other organizational members, the analyst pinpoints problems. Often others will bring up these problems, and they are the reason the analyst was initially called in. Opportunities are situations that the analyst believes can be improved through the use of computerized information systems. Seizing opportunities may allow the business to gain a competitive edge or set an industry standard.

Identifying objectives is also an important component of the first phase. The analyst must first discover what the business is trying to do. Then the analyst will be able to see whether some aspect of information systems applications can help the business reach its objectives by addressing specific problems or opportunities.

The people involved in the first phase are the users, analysts, and systems managers coordinating the project. Activities in this phase consist of interviewing user management, summarizing the knowledge obtained, estimating the scope of the project, and documenting the results. The output of this phase is a feasibility report that contains a problem definition and summarizes the objectives. Management must then make a decision on whether to proceed with the proposed project. If the user group does not have sufficient funds in its budget or if it wishes to tackle unrelated problems, or if the problems do not require a computer system, a different solution may be recommended, and the systems project does not proceed any further.

Determining Human Information Requirements

The next phase the analyst enters involves determining the human needs of the users involved, using a variety of tools to understand how users interact in the work context with their current information systems. The analyst will use interactive methods such as interviewing, sampling and investigating hard data, and using questionnaires, along with unobtrusive methods, such as observing decision makers' behavior and their office environments, and all-encompassing methods, such as prototyping.

The analyst will use these methods to pose and answer many questions concerning HCI, including questions such as, "What are the users' physical strengths and limitations?" In other words, "What needs to be done to make the system audible, legible, and safe?" "How can the new system be designed to be easy to use, learn, and remember?" "How can the system be made pleasing or even fun to use?" "How can the system support a user's individual work tasks and make them more productive in new ways?"

In the information requirements phase of the SDLC, the analyst is striving to understand what information users need to perform their jobs. At this point, the analyst is examining how to make the system useful to the people involved. How can the system better support individual tasks that need to be done? What new tasks are enabled by the new system that users were unable to do without it? How can the new system be created to extend a user's capabilities beyond what the old system provided? How can the analyst create a system that is rewarding for workers to use?

The people involved in this phase are the analysts and users, typically operations managers and operations workers. The systems analyst needs to know the details of current system functions: the who (the people who are involved), what (the business activity), where (the environment in which the work takes place), when (the timing), and how (how the current procedures are performed) of the business under study. The analyst must then ask why the business uses the current system. There may be good reasons for doing business using the current methods, and these should be considered when designing any new system.

Agile development is an object-oriented approach (OOA) to systems development that includes a method of development (including generating information requirements) as well as software tools. In this text, it is paired with prototyping in Chapter 6. (There is more about object-oriented approaches in Chapter 10.)

If the reason for current operations is that "it's always been done that way," however, the analyst may wish to improve on the procedures. At the completion of this phase, the analyst should understand how users accomplish their work when interacting with a computer and begin to know how to make the new system more useful and usable. The analyst should also know how the business functions and have complete information on the people, goals, data, and procedures involved.

Analyzing System Needs

The next phase that the systems analyst undertakes involves analyzing system needs. Again, special tools and techniques help the analyst make requirement determinations. Tools such as data flow diagrams (DFDs) to chart the input, processes, and output of the business's functions, or activity diagrams or sequence diagrams to show the sequence of events, illustrate systems in a structured, graphical form. From data flow, sequence, or other diagrams, a data dictionary is developed that lists all the data items used in the system, as well as their specifications.

During this phase the systems analyst also analyzes the structured decisions made. Structured decisions are those for which the conditions, condition alternatives, actions, and action rules can be determined. There are three major tools for analyzing structured decisions: structured English, decision tables, and decision trees.

MAC APPEAL

At home and in our visits to university campuses and businesses around the world, we've noticed that students and organizations are increasingly showing an interest in Macs. Therefore, we thought it would add a little bit of interest to show some of the Mac options available to a systems designer. At the time we're writing this book, about one out of seven personal computers purchased in the United States is a Mac. Macs are quality Intel-based machines that run under a competent operating system and can also run Windows, so in effect, everything that can be done on a PC can be done on a Mac. One way to run Windows is to boot directly into Windows (once it's installed); another is to use virtualization, using software such as VM Fusion, which is shown in Figure 1.MAC.

Adopters of Macs have cited many reasons for using Macs, including better security built into the Mac operating system, intelligent backup using the built-in Time Machine, the multitude of applications already included, the reliability of setup and networking, and the ability to sync Macs with other Macs and iPhones. The most compelling reason, we think, is the design itself.

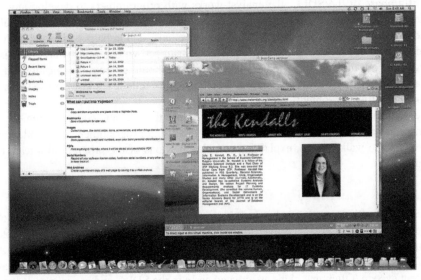

FIGURE 1.MAC

Running Windows on a Mac using virtualization called VM Fusion. (Screenshot of Apple Desktop. Reprinted with permission of Apple Inc.; Screenshot from YOJIMBO. Copyright © 2012 by Bare Bones Software, Inc. Reprinted by permission; Screenshot from www .thekendalls.org. Copyright © by Kenneth and Julie Kendall. Reprinted with permission.)

At this point in the SDLC, the systems analyst prepares a systems proposal that summarizes what has been found out about the users, usability, and usefulness of current systems; provides cost–benefit analyses of alternatives; and makes recommendations on what (if anything) should be done. If one of the recommendations is acceptable to management, the analyst proceeds along that course. Each systems problem is unique, and there is never just one correct solution. The manner in which a recommendation or solution is formulated depends on the individual qualities and professional training of each analyst and the analyst's interaction with users in the context of their work environment.

Designing the Recommended System

In the design phase of the SDLC, the systems analyst uses the information collected earlier to accomplish the logical design of the information system. The analyst designs procedures for users to help them accurately enter data so that data going into the information system are correct. In addition, the analyst provides for users to complete effective input to the information system by using techniques of good form and web page or screen design.

Part of the logical design of the information system is devising the HCI. The interface connects the user with the system and is thus extremely important. The user interface is designed with the help of users to make sure that the system is audible, legible, and safe, as well as attractive and enjoyable to use. Examples of physical user interfaces include a keyboard (to type in questions and answers), onscreen menus (to elicit user commands), and a variety of graphical user interfaces (GUIs) that use a mouse or touch screen.

The design phase also includes designing databases that will store much of the data needed by decision makers in the organization. Users benefit from a well-organized database that is logical to them and corresponds to the way they view their work. In this phase the analyst also works with users to design output (either onscreen or printed) that meets their information needs.

Finally, the analyst must design controls and backup procedures to protect the system and the data, and to produce program specification packets for programmers. Each packet should contain input and output layouts, file specifications, and processing details; it may also include decision trees or tables, Unified Modeling Language (UML) or data flow diagrams, and the names and functions of any prewritten code that is either written in-house or using code or other class libraries.

Developing and Documenting Software

In the fifth phase of the SDLC, the analyst works with programmers to develop any original software that is needed. During this phase the analyst works with users to develop effective documentation for software, including procedure manuals, online help, and websites featuring frequently asked questions (FAQs) or Read Me files shipped with new software. Because users are involved from the beginning, phase documentation should address the questions they have raised and solved jointly with the analyst. Documentation tells users how to use software and what to do if software problems occur.

Programmers have a key role in this phase because they design, code, and remove syntactical errors from computer programs. To ensure quality, a programmer may conduct either a design or a code walkthrough, explaining complex portions of the program to a team of other programmers.

Testing and Maintaining the System

Before an information system can be used, it must be tested. It is much less costly to catch problems before the system is signed over to users than after. Some of the testing is completed by programmers alone, some of it by systems analysts in conjunction with programmers. A series of tests to pinpoint problems is run first with sample data and eventually with actual data from the current system. Often test plans are created early in the SDLC and are refined as the project progresses.

Maintenance of the system and its documentation begins in this phase and is carried out routinely throughout the life of the information system. Much of the programmer's routine work consists of maintenance, and businesses spend a great deal of money on maintenance. Some maintenance, such as program updates, can be done automatically via a vendor site on the Web. Many of the systematic procedures the analyst employs throughout the SDLC can help ensure that maintenance is kept to a minimum.

Implementing and Evaluating the System

In this last phase of systems development, the analyst helps implement the information system. This phase involves training users to handle the system. Vendors do some training, but oversight of training is the responsibility of the systems analyst. In addition, the analyst needs to plan for a smooth conversion from the old system to the new one. This process includes converting files from old formats to new ones or building a database, installing equipment, and bringing the new system into production.

Evaluation is included as part of this final phase of the SDLC mostly for the sake of discussion. Actually, evaluation takes place during every phase. A key criterion that must be satisfied is whether the intended users are indeed using the system.

It should be noted that systems work is often cyclical. When an analyst finishes one phase of systems development and proceeds to the next, the discovery of a problem may force the analyst to return to the previous phase and modify the work done there.

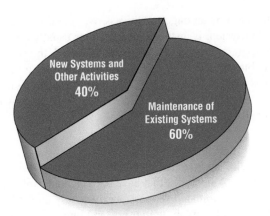

FIGURE 1.2

Some researchers estimate that the amount of time spent on system maintenance may be as much as 60 percent of the total time spent on systems projects.

The Impact of Maintenance

After the system is installed, it must be maintained, meaning that the computer programs must be modified and kept up to date. Figure 1.2 illustrates the average amount of time spent on maintenance at a typical MIS installation. Estimates of the time spent by departments on maintenance have ranged from 48 to 60 percent of the total time spent developing systems. Very little time remains for new systems development. As the number of programs written increases, so does the amount of maintenance they require.

Maintenance is performed for two reasons. The first of these is to correct software errors. No matter how thoroughly a system is tested, bugs or errors creep into computer programs. Bugs in commercial PC software are often documented as "known anomalies," and they are corrected when new versions of the software are released or in an interim release. In custom software (also called *bespoke software*), bugs must be corrected as they are detected.

The other reason for performing system maintenance is to enhance the software's capabilities in response to changing organizational needs, generally involving one of the following three situations:

1. Users often request additional features after they become familiar with the computer system and its capabilities.
2. The business changes over time.
3. Hardware and software are changing at an accelerated pace.

Figure 1.3 illustrates the amount of resources—usually time and money—spent on systems development and maintenance. The area under the curve represents the total dollar amount spent. You can see that over time, the total cost of maintenance is likely to exceed that of systems development. At a certain point it becomes more feasible to perform a new systems study because the cost of continued maintenance is clearly greater than the cost of creating an entirely new information system.

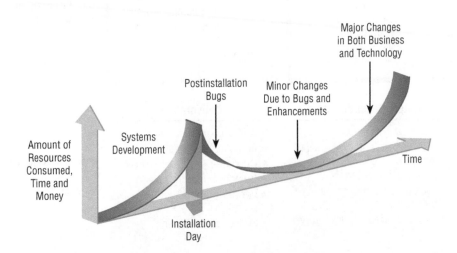

FIGURE 1.3

Resource consumption over the system life.

In summary, maintenance is an ongoing process over the life cycle of an information system. After the information system is installed, maintenance usually takes the form of correcting previously undetected program errors. Once these are corrected, the system approaches a steady state, providing dependable service to its users. Maintenance during this period may consist of removing a few previously undetected bugs and updating the system with a few minor enhancements. As time goes on and the business and technology change, however, the maintenance effort increases dramatically.

Using CASE Tools

Analysts who adopt the SDLC approach often benefit from productivity tools, called computer-aided software engineering (CASE) tools, that have been created explicitly to improve their routine work through the use of automated support. Analysts rely on CASE tools to increase productivity, communicate more effectively with users, and integrate the work that they do on the system from the beginning to the end of the life cycle.

All the information about the project is stored in an encyclopedia called the CASE repository, a large collection of records, elements, diagrams, screens, reports, and other information (see Figure 1.4). Analysis reports may be produced using the repository information to show where the design is incomplete or contains errors.

FIGURE 1.4

The repository concept.

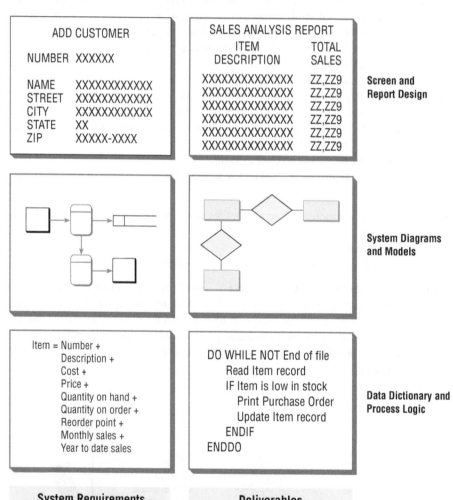

Visible Analyst (VA) is one example of a CASE tool that enables systems analysts to do graphical planning, analysis, and design in order to build complex client/server applications and databases. Visible Analyst and software products like Microsoft Visio or OmniGraffle allow users to draw and modify diagrams easily.

Analysts and users alike report that CASE tools afford them a means of communication about the system during its conceptualization. Through the use of automated support featuring onscreen output, clients can readily see how data flows and other system concepts are depicted, and they can then request corrections or changes that would have taken too much time with older tools. CASE tools can also help support the modeling of an organization's functional requirements, assist analysts and users in drawing the boundaries for a given project, and help them visualize how the project meshes with other parts of the organization.

The Agile Approach

Although this text tends to focus on SDLC, the most widely used approach in practice, at times an analyst will recognize that his or her organization could benefit from an alternative approach. Perhaps a systems project using a structured approach has recently failed, or perhaps the organizational subcultures, composed of several different user groups, seem more in step with an alternative method. We cannot do justice to these methods in a small space; each deserves and has inspired its own books and research. By mentioning these approaches here, however, we hope to help you become aware that under certain circumstances, your organization may want to consider an alternative or a supplement to structured analysis and design and to the SDLC.

The agile approach is a software development approach based on values, principles, and core practices. The four values are communication, simplicity, feedback, and courage. We recommend that systems analysts adopt these values in all projects they undertake, not just when adopting the agile approach.

In order to finish a project, adjustments often need to be made in project management. In Chapter 6 we will see that agile methods can ensure successful completion of a project by adjusting the important resources of time, cost, quality, and scope. When these four control variables are properly included in the planning, there is a state of balance between the resources and the activities needed to complete the project.

Taking development practices to the extreme is most noticeable when one pursues practices that are unique to agile development. In Chapter 6 we discuss four core agile practices: short releases, the 40-hour workweek, hosting an onsite customer, and using pair programming. At first glance these practices appear extreme, but as you will see, we can learn some important lessons from incorporating many of the values and practices of the agile approach into systems analysis and design projects.

Developmental Process for an Agile Project

Activities and behaviors shape the way development team members and customers act during the development of an agile project. Two words that characterize a project done with an agile approach are *interactive* and *incremental*. By examining Figure 1.5, we can see that there are five distinct stages: exploration, planning, iterations to the first release, productionizing, and maintenance. Notice that the three red arrows that loop back into the "Iterations" box symbolize incremental changes created through repeated testing and feedback that eventually lead to a stable but evolving system. Also note that there are multiple looping arrows that feed back into the productionizing phase. These symbolize that the pace of iterations is increased after a product is released. The red arrow is shown leaving the maintenance stage and returning to the planning stage, so that there is a continuous feedback loop involving customers and the development team as they agree to alter the evolving system.

EXPLORATION. During exploration, you will explore your environment, asserting your conviction that the problem can and should be approached with agile development, assemble the team, and assess team member skills. This stage will take anywhere from a few weeks (if you already know your team members and technology) to a few months (if everything is new). You also will be actively examining potential technologies needed to build the new system. During this stage you

FIGURE 1.5

The five stages of the agile modeling development process show that frequent iterations are essential to successful system development.

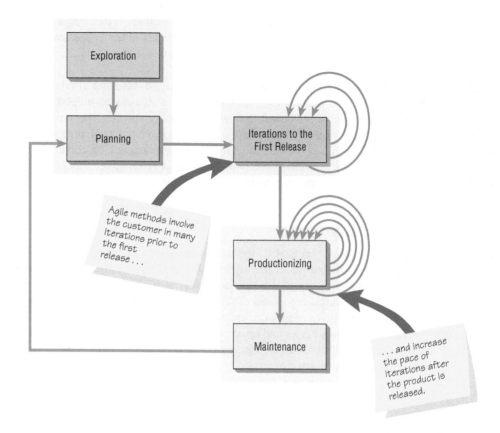

Agile methods involve the customer in many iterations prior to the first release . . .

. . . and increase the pace of iterations after the product is released.

should practice estimating the time needed for a variety of tasks. In exploration, customers also are experimenting with writing user stories. The point is to get the customer to refine a story enough so that you can competently estimate the amount of time it will take to build the solution into the system you are planning. This stage is all about adopting a playful and curious attitude toward the work environment, its problems, technologies, and people.

PLANNING. The next stage of the agile development process is called planning. In contrast to the first stage, planning may take only a few days to accomplish. In this stage you and your customers agree on a date anywhere from two months to half a year from the current date to deliver solutions to their most pressing business problems (you will be addressing the smallest, most valuable set of stories). If your exploration activities were sufficient, this stage should be very short.

The entire agile planning process has been characterized using the idea of a *planning game*, as devised by Kent Beck, the father of Extreme Programming. The planning game spells out rules that can help formulate the agile development team's relationship with their business customers. Although the rules form an idea of how you want each party to act during development, they are not meant as a replacement for a relationship. They are a basis for building and maintaining a relationship.

So, we use the metaphor of a game. To that end we talk in terms of the goal of the game, the strategy to pursue, the pieces to move, and the players involved. The goal of the game is to maximize the value of the system produced by the agile team. In order to figure the value, you have to deduct costs of development, and the time, expense, and uncertainty taken on so that the development project can go forward.

The strategy pursued by the agile development team is always one of limiting uncertainty (downplaying risk). To do that they design the simplest solution possible, put the system into production as soon as possible, get feedback from the business customer about what's working, and adapt their design from there. Story cards become the pieces in the planning game that briefly describe the task, provide notes, and provide an area for task tracking.

There are two main players in the planning game: the development team and the business customer. Deciding which business group in particular will be the business customer is not always easy because the agile process is an unusually demanding role for the customer to play. Customers decide what the development team should tackle first. Their decisions will set priorities and check functionalities throughout the process.

ITERATIONS TO THE FIRST RELEASE. The third stage in the agile development process is composed of iterations to the first release. Typically these are iterations (cycles of testing, feedback, and change) of about three weeks in duration. You will be pushing yourself to sketch out the entire architecture of the system, even though it is just in outline or skeletal form. One goal is to run customer-written functional tests at the end of each iteration. During the iterations stage you should also question whether the schedule needs to be altered or whether you are tackling too many stories. Make small rituals out of each successful iteration, involving customers as well as developers. Always celebrate your progress, even if it is small, because this is part of the culture of motivating everyone to work extremely hard on the project.

PRODUCTIONIZING. Several activities occur during the productionizing phase. In this phase the feedback cycle speeds up so that rather than receiving feedback for an iteration every three weeks, software revisions are being turned around in one week. You may institute daily briefings so everyone knows what everyone else is doing. The product is released in this phase, but it may be improved by adding other features. Getting a system into production is an exciting event. Make time to celebrate with your teammates and mark the occasion. One of the keys to the agile approach, which we heartily embrace, is that it is supposed to be fun to develop systems!

MAINTENANCE. Once a system has been released, it needs to be kept running smoothly. New features may be added, riskier customer suggestions may be considered, and team members may be rotated on or off the team. The attitude you take at this point in the developmental process is more conservative than at any other time. You are now in a "keeper of the flame" mode rather than the playful one you experienced during exploration.

Object-Oriented Systems Analysis and Design

Object-oriented (O-O) analysis and design is an approach that is intended to facilitate the development of systems that must change rapidly in response to dynamic business environments. Chapter 10 helps you understand what object-oriented systems analysis and design is, how it differs from the structured approach of the SDLC, and when it may be appropriate to use an object-oriented approach.

Object-oriented techniques often work well in situations in which complicated information systems are undergoing continuous maintenance, adaptation, and redesign. Object-oriented approaches use the industry standard for modeling object-oriented systems, called Unified Modeling Language (UML), to break down a system into a use case model.

Object-oriented programming differs from traditional procedural programming in that it examines objects that are part of a system. Each object is a computer representation of some actual thing or event. Objects may be customers, items, orders, and so on. Objects are represented by and grouped into classes that are optimal for reuse and maintainability. A class defines the set of shared attributes and behaviors found in each object in the class.

The phases in UML are similar to those in the SDLC. Since those two methods share rigid and exacting modeling, they happen in a slower, more deliberate pace than the phases of agile modeling. An analyst goes through problem and identification phases, an analysis phase, and a design phase, as shown in Figure 1.6.

Although much of the specifics are discussed in Chapters 2 and 10, the following steps give a brief description of the UML process:

1. Define the use case model.

   In this phase the analyst identifies the actors and the major events initiated by the actors. Often the analyst will start by drawing a diagram with stick figures representing the actors and arrows showing how the actors relate. This is called a use case diagram (Chapter 2), and it represents the standard flow of events in the system. Then an analyst typically writes up a use case scenario (Chapter 2), which describes in words the steps that are normally performed.

FIGURE 1.6

The steps in the UML
development process.

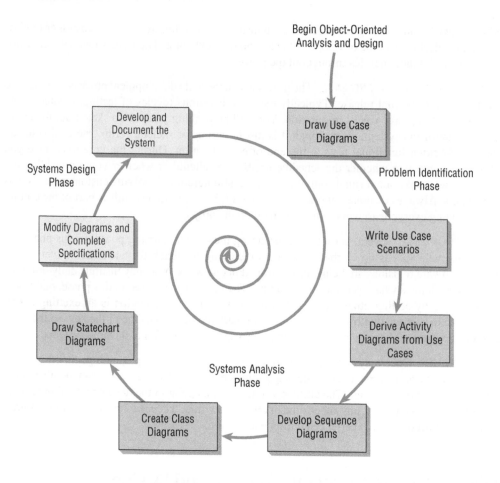

2. During the systems analysis phase, begin drawing UML diagrams.

   In the second phase (Chapter 10), the analyst draws activity diagrams, which illustrate all the major activities in the use case. In addition, the analyst creates one or more sequence diagrams for each use case, which show the sequence of activities and their timing. This is an opportunity to go back and review the use cases, rethink them, and modify them if necessary.

3. Continuing in the analysis phase, develop class diagrams.

   The nouns in the use cases are objects that can potentially be grouped into classes. For example, every automobile is an object that shares characteristics with other automobiles. Together they make up a class.

4. Still in the analysis phase, draw statechart diagrams.

   The class diagrams are used to draw statechart diagrams, which help in understanding complex processes that cannot be fully derived by the sequence diagrams. The statechart diagrams are extremely useful in modifying class diagrams, so the iterative process of UML modeling continues.

5. Begin systems design by modifying the UML diagrams. Then complete the specifications.

   Systems design means modifying the existing system, and that implies modifying the diagrams drawn in the previous phase. These diagrams can be used to derive classes, their attributes, and methods (methods are simply operations). An analyst will need to write class specifications for each class, including the attributes, methods, and their descriptions. The analyst will also develop method specifications that detail the input and output requirements for each method, along with a detailed description of the internal processing of the method.

6. Develop and document the system.

   UML is, of course, a modeling language. An analyst may create wonderful models, but if the system isn't developed, there is not much point in building models. Documentation is critical. The more complete the information you provide the development team through documentation and UML diagrams, the faster the development and the more solid the final production system.

Object-oriented methodologies often focus on small, quick iterations of development, sometimes called the *spiral model*. Analysis is performed on a small part of the system, usually starting with a high-priority item or perhaps the item that has the greatest risk. This is followed by design and implementation. The cycle is repeated with analysis of the next part, design, and some implementation, and this is repeated until the project is complete. Reworking diagrams and the components themselves is normal. UML is a powerful modeling tool that can greatly improve the quality of your systems analysis and design and the final product.

Choosing Which Systems Development Method to Use

The differences among the three approaches described earlier are not as big as they seem at the outset. In all three approaches, the analyst needs to understand the organization first (Chapter 2). Then the analyst or project team needs to budget time and resources and develop a project proposal (Chapter 3). Next, they need to interview organizational members and gather detailed data by using questionnaires (Chapter 4) and sample data from existing reports and observe how business is currently transacted (Chapter 5). The three approaches have all of these activities in common.

Even the methods themselves have similarities. The SDLC and object-oriented approaches both require extensive planning and diagramming. The agile approach and the object-oriented approach both allow subsystems to be built one at a time until the entire system is complete. The agile and SDLC approaches are both concerned about the way data logically moves through the system.

So given a choice to develop a system using an SDLC approach, an agile approach, or an object-oriented approach, which would you choose? Figure 1.7 provides a set of guidelines to help you choose which method to use when developing your next system.

Developing Open Source Software

An alternative to traditional software development in which proprietary code is hidden from the users is called open source software (OSS). With OSS, many users and programmers can study, share, and modify the code, or computer instructions. Rules of this community include the idea that any program modifications must be shared with all the people on the project.

Development of open source software has also been characterized as a philosophy rather than simply as the process of creating new software. Often those involved in open source

Choose	When
The Systems Development Life Cycle (SDLC) Approach	• systems have been developed and documented using SDLC • it is important to document each step of the way • upper-level management feels more comfortable or safe using SDLC • there are adequate resources and time to complete the full SDLC • communication of how new systems work is important
Agile Methodologies	• there is a project champion of agile methods in the organization • applications need to be developed quickly in response to a dynamic environment • a rescue takes place (the system failed and there is no time to figure out what went wrong) • the customer is satisfied with incremental improvements • executives and analysts agree with the principles of agile methodologies
Object-Oriented Methodologies	• the problems modeled lend themselves to classes • an organization supports the UML learning • systems can be added gradually, one subsystem at a time • reuse of previously written software is a possibility • it is acceptable to tackle the difficult problems first

FIGURE 1.7

How to decide which development method to use.

software communities view it as a way to help societies change. Widely known open source projects include Apache for developing a Web server, the browser called Mozilla Firefox, and Linux, which is a Unix-like open source operating system.

However, it would be an oversimplification to think of open source software as a monolithic movement, and it does little to reveal what type of users or user analysts are developing open source software projects and on what basis. To help us understand the open source movement, researchers have recently categorized open source communities into four community types—ad hoc, standardized, organized, and commercial—along six different dimensions—general structure, environment, goals, methods, user community, and licensing. Some researchers argue that open source software is at a crossroads and that the commercial and community open source groups need to understand where they converge and where the potential for conflict exists.

Why Organizations Participate in Open Source Communities

Organizations participate in open source communities for a variety of reasons. One is the rapidity with which new software can be developed and tested. It is faster to have a committed group of expert developers develop, test, and debug code than it is to have one isolated team working on software development. This cross-fertilization can also be a boon to creativity.

Another reason to participate is the benefit of having many good minds work with innovative applications. Yet another reason for participating in an open source community might be the potential for keeping down development costs.

Organizations might also seek to participate in an open source community due to a desire to bolster their own self-image and contribute something worthwhile to the larger software development community; they may want to contribute generously or altruistically to a higher good beyond developing profitable proprietary software, and doing so may make them appear as "good guys" to the external public.

While organizations may ask or even require that their software developers become involved in one or more open source projects or communities, individual developers must interact with the community in a meaningful and knowledgeable way in order to first prove themselves worthy members of the group, and then strike up and maintain relationships that are mutually beneficial.

Organizations and the design teams within them interact with open communities. Companies are taking advantage of an entire range of options that help them strike a harmonious equilibrium so that their contributions to the open community and their differentiation from the open community become clear strategically. Reasons for contribution to and differentiation from include cost, managing resources, and the time it takes to bring a new product to the market.

The Role of the Analyst in Open Source Software

As an analyst you may find yourself, at the request of your chief employer, participating in an open source community. One widely known open source community is that surrounding the Linux kernel. This is a large, mostly virtual community of developers who all have different levels or types of participation and who all have different reasons for being involved. Other well-known open source projects include Mozilla Firefox, Android, Apache projects, and many more. Even NASA, the U.S. National Aeronautics and Space Administration, has a lively open source community (see http://ti.arc.nasa.gov/opensource/).

One reason your company may ask you to participate in an open source community is curiosity about what the software benefits to the organization might be. This may be a result of a sort of bandwagon effect, for when it becomes known that competitors are already participating, your organization may want to get involved. With competitors actively participating in an open community, an organization may calculate that it is something that should at least be investigated seriously, not dismissed summarily. Another reason your company might ask you to participate as a developer in an open community is to achieve what researchers have labeled "shared design." *Shared design* means that while you are participating in the open source community, you are at the same time employed by an organization that wants to leverage your participation in the open source community to incorporate open source software designs into proprietary products, processes, knowledge, and IT artifacts that it is developing and that it hopes to eventually sell as a product that is differentiated from what the open source community has produced. Through a process of shared design the IT artifact is imbued with both community and organizational structures, knowledge, and practices.

HYPERCASE® EXPERIENCE 1

"Welcome to Maple Ridge Engineering, what we call MRE. We hope you'll enjoy serving as a systems consultant for us. Although I've worked here five years in different capacities, I've just been reassigned to serve as an administrative aide to Snowden Evans, the head of the new Training and Management Systems Department. We're certainly a diverse group. As you make your way through the company, be sure to use all your skills, both technical and people oriented, to understand who we are and to identify the problems and conflicts that you think should be solved regarding our information systems."

"To bring you up to date, let me say that Maple Ridge Engineering is a medium-sized medical engineering company. Last year, our revenues exceeded $287 million. We employ about 335 people. There are about 150 administrative employees as well as management and clerical staff like myself; approximately 75 professional employees, including engineers, physicians, and systems analysts; and about 110 trade employees, such as drafters and technicians."

"There are four offices. You will visit us through HyperCase in our home office in Maple Ridge, Tennessee. We have three other branches in the southern United States as well: Atlanta, Georgia;

Charlotte, North Carolina; and New Orleans, Louisiana. We'd love to have you visit when you're in the area."

"For now, you should explore HyperCase using either Firefox, Safari, or Microsoft Internet Explorer."

"To learn more about Maple Ridge Engineering as a company or to find out how to interview our employees, who will use the systems you design, and how to observe their offices in our company, you may want to start by going to the website www.pearsonglobaleditions.com/kendall. Then click on the link labeled **HyperCase**. At the HyperCase display screen, click on **Start**, and you will be in the reception room for Maple Ridge Engineering. From this point, you can start consulting right away."

This website contains useful information about the project as well as files that can be downloaded to your computer. There is a set of Visible Analyst data files, and there is another set of Visio data files that match HyperCase. They contain a partially constructed series of data flow diagrams, entity-relationship diagrams, UML diagrams, and repository information. The HyperCase website also contains additional exercises that may be assigned. HyperCase is designed to be explored, and you should not overlook any object or clue on a web page.

Summary

Systems analysis and design is a systematic approach to identifying problems, opportunities, and objectives; to analyzing human- and computer-generated information flows in organizations; and to designing computerized information systems to solve problems. Systems analysts are required to take on many roles in the course of their work. Some of these roles are (1) an outside consultant to business, (2) a supporting expert within a business, and (3) an agent of change in both internal and external situations.

Analysts possess a wide range of skills. First and foremost, an analyst is a problem solver, someone who enjoys the challenge of analyzing a problem and devising a workable solution. Systems analysts require communication skills that allow them to relate meaningfully to many different kinds of people on a daily basis, as well as computer skills. Understanding and relating well to users is critical to their success.

Analysts proceed systematically. The framework for their systematic approach is provided in what is called the systems development life cycle (SDLC). This life cycle can be divided into seven sequential phases, although in reality the phases are interrelated and are often accomplished simultaneously. The seven phases are identifying problems, opportunities, and objectives; determining human information requirements; analyzing system needs; designing the recommended system; developing and documenting software; testing and maintaining the system; and implementing and evaluating the system.

The agile approach is a software development approach based on values, principles, and core practices. Systems that are designed using agile methods can be developed rapidly. Stages in the agile development process are exploration, planning, iterations to the first release, productionizing, and maintenance.

A third approach to systems development is called object-oriented analysis design. These techniques are based on object-oriented programming concepts that have become codified in UML, a standardized modeling language in which objects that are created include not only code about data but also instructions about the operations to be performed on the data. Key diagrams help analyze, design, and communicate UML-developed systems. These systems are usually developed as components, and reworking the components many times is typical in object-oriented analysis and design.

Analysts will have increasing opportunities to participate in open source development communities, often through their primary organization. There are many reasons organizations are participating in development of open source. One of the well-known open source communities maintains the Linux kernel and is supported by the Linux Foundation.

Keywords and Phrases

agent of change
agile approach
agile methods
Ajax
bespoke software
computer-assisted software engineering (CASE)
CASE tools
exploration phase
human–computer interaction (HCI)
iterations to the first release phase
maintenance phase
object-oriented systems analysis and design

open source communities
open source software (OSS)
planning game
planning phase
productionizing phase
prototyping
systems analysis and design
systems analyst
systems consultant
systems development life cycle (SDLC)
Unified Modeling Language (UML)

Review Questions

1. List the advantages of using systems analysis and design techniques in approaching computerized information systems for business.
2. List three roles that a systems analyst is called upon to play. Provide a definition for each one.
3. What personal qualities are helpful to a systems analyst? List them.
4. List and briefly define the seven phases of the systems development life cycle (SDLC).
5. What are CASE tools used for?
6. Explain what is meant by *agile approach*.
7. What is the meaning of the phrase *the planning game*?
8. What are the stages in agile development?
9. Define the term *object-oriented analysis and design*.
10. What is UML?
11. What is open source software?
12. What is the role of a systems analyst in the development of open source software?
13. List two reasons an organization may want its analysts to participate in an open source community.

Selected Bibliography

Beck, K., and C. Andres. *Extreme Programming Explained: Embrace Change,* 2nd ed. Boston, MA: Addison-Wesley, 2004.

Coad, P., and E. Yourdon. *Object-Oriented Analysis,* 2nd ed. Englewood Cliffs, NJ: Prentice Hall, 1991.

Davis, G. B., and M. H. Olson. *Management Information Systems: Conceptual Foundation, Structure, and Development,* 2nd ed. New York: McGraw-Hill, 1985.

Germonprez, M. and B. Warner. "Commercial Participation in Open Innovation Communities." In *Managing Open Innovation Technologies.* Edited by E. Lundström, J.SZ, M. Wiberg, S. Hrastinski, M. Edenius, P. J. Ägerfalk. Berlin: Springer-Verlag, 2012.

Kendall, J. E., K. E. Kendall, and S. Kong. "Improving Quality Through the Use of Agile Methods in Systems Development: People and Values in the Quest for Quality." In *Measuring Information Systems Delivery Quality.* Edited by E. W. Duggan and H. Reichgelt, pp. 201–222. Hershey, PA: Idea Group Publishing, 2006.

Laudon, K. C., and J. P. Laudon. *Management Information Systems,* 12th ed. Upper Saddle River, NJ: Pearson Prentice Hall, 2012.

www.visible.com/Products/index.htm. Last accessed May 28, 2012.

Lee, G., and R. Cole. "From a Firm-Based to a Community-Based Model of Knowledge Creation: The Case of the Linux Kernel Development." *Organization Science,* Vol. 14, 2003, pp. 663–649.

Yourdon, E. *Modern Structured Analysis.* Englewood Cliffs, NJ: Prentice Hall, 1989.

Zhang, P., J. Carey, D. Te'eni, and M. Tremaine. "Integrating Human–Computer Interaction Development into the Systems Development Life Cycle: A Methodology." *Communications of the Association for Information Systems,* Vol. 15, 2005, pp. 512–543.

The CPU Case Episode and accompanying student files are available online at www.pearsonglobaleditions.com/kendall.

CHAPTER 2

Understanding and Modeling Organizational Systems

LEARNING OBJECTIVES

Once you have mastered the material in this chapter you will be able to:

1. Understand that organizations and their members are systems and that analysts need to take a systems perspective.

2. Depict systems graphically, using context-level data flow diagrams, entity-relationship models, use cases, and use case scenarios.

3. Recognize that different levels of management require different systems.

4. Comprehend that organizational culture impacts the design of information systems.

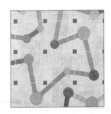

To analyze and design appropriate information systems, systems analysts need to comprehend the organizations they work in as systems shaped through the interactions of three main forces: the levels of management, design of organizations, and organizational cultures.

Organizations are large systems composed of interrelated subsystems. The subsystems are influenced by three broad levels of management decision makers—operations, middle management, and strategic management—that cut horizontally across the organizational system. Organizational cultures and subcultures influence the way people in subsystems interrelate. These topics and their implications for information systems development are considered in this chapter.

Organizations as Systems

Organizations and their members are usefully conceptualized as systems designed to accomplish predetermined goals and objectives through people and other resources that they employ. Organizations are composed of smaller, interrelated systems (departments, units, divisions, etc.) serving specialized functions. Typical functions include accounting, marketing, production, data processing, and management. Specialized functions (smaller systems) are eventually reintegrated through various ways to form an effective organizational whole.

The significance of conceptualizing organizations as complex systems is that systems principles allow insight into how organizations work. To ascertain information requirements properly and to design appropriate information systems, it is of primary importance to understand the organization as a whole. All systems are composed of subsystems (which include information systems); therefore, when studying an organization, we also examine how smaller systems are involved and how they function.

Interrelatedness and Interdependence of Systems

All systems and subsystems are interrelated and interdependent. This fact has important implications both for organizations and for those systems analysts who seek to help them better achieve their goals. When any element of a system is changed or eliminated, the rest of the system's elements and subsystems are also significantly affected.

For example, suppose that the managers of an organization decide not to hire administrative assistants any longer and to replace their functions with networked PCs. This decision has the potential to significantly affect not only the administrative assistants and the managers but also all the organizational members who built up communications networks with the now-departed assistants.

All systems process inputs from their environments. By definition, processes change or transform inputs into outputs. Whenever you examine a system, check to see what is being changed or processed. If nothing is changed, you may not be identifying a process. Typical processes in systems include verifying, updating, and printing.

Another aspect of organizations as systems is that all systems are contained by boundaries separating them from their environments. Organizational boundaries exist on a continuum ranging from extremely permeable to almost impermeable. To continue to adapt and survive, organizations must be able first to import people, raw materials, and information through their boundaries (inputs) and then to exchange their finished products, services, or information with the outside world (outputs).

Feedback is one form of system control. As systems, all organizations use planning and control to manage their resources effectively. Figure 2.1 shows how system outputs are used as feedback that compares performance with goals. This comparison in turn helps managers formulate more specific goals as inputs. An example is a U.S. manufacturing company that produces red-white-and-blue weight-training sets as well as gunmetal-gray sets. The company finds that one year after the Olympics, very few red-white-and-blue sets are purchased. Production managers use this information as feedback to make decisions about what quantities of each color to produce. Feedback in this instance is useful for planning and control.

The ideal system, however, is one that self-corrects or self-regulates in such a way that decisions on typical occurrences are not required. An example is a supply chain system for production planning that takes into account current and projected demand and formulates a proposed

FIGURE 2.1

System outputs serve as feedback that compares performance with goals.

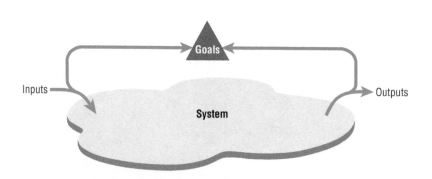

CONSULTING OPPORTUNITY 2.1

The E in Vitamin E Stands for Ecommerce

"Our retail shops and mail-order division are quite healthy," says Bill Berry, one of the owners of Marathon Vitamin Shops, "but to be competitive, we must establish an ecommerce website." His father, a co-owner, exclaims, "I agree, but where do we start?" The elder Berry knew, of course, that it wasn't a case of setting up a website and asking customers to order off the website. He identified eight different parts to ecommerce and realized that they were all part of a larger system. In other words, all the parts had to work together to create a strong package. His list of elements essential to ecommerce included the following:

1. Attracting customers to an ecommerce website
2. Informing customers about products and services offered
3. Allowing customers to customize products online
4. Completing transactions with customers
5. Accepting payment from customers in a variety of forms
6. Supporting customers after the sale via the website
7. Arranging for the delivery of goods and services
8. Personalizing the look and feel of the website for different customers

Bill Berry read the list and contemplated it for a while. "It is obvious that ecommerce is more complex than I thought," he says. You can help the owners of Marathon Vitamin Shops in the following ways:

1. Make a list of the elements that are interrelated or interdependent. Then write a paragraph stating why it is critical to monitor these elements closely.
2. Decide on the boundaries and ultimate scope of the system. Write a paragraph expressing an opinion on which elements are critical for Marathon Vitamin Shops and which elements can be explored at a later date.
3. Suggest which elements should be handled in-house and which should be outsourced to another company that may be better able to handle the job. Justify your suggestions in two paragraphs, one for the in-house jobs and one for the outsourced tasks.

---

solution as output. An Italian knitwear manufacturer that markets its clothing in the United States has just such a system. This company produces most of its sweaters in white, uses its computerized inventory information system to find out what colors are selling best, and then dyes sweaters in hot-selling colors immediately before shipping them.

Feedback is received from within the organization and from the outside environments around it. Anything external to an organization's boundaries is considered to be an environment. Numerous environments, with varying degrees of stability, constitute the milieu in which organizations exist.

Among these environments are (1) the environment of the community in which the organization is physically located, which is shaped by the size of its population and its demographic profile, including factors such as education and average income; (2) the economic environment, influenced by market factors, including competition; (3) the political environment, controlled through state and local governments; and (4) the legal environment, issuing federal, state, regional, and local laws and guidelines. Although changes in environmental status can be planned for, they often cannot be directly controlled by the organization.

Related and similar to the concept of external boundary permeability is the concept of internal openness or closedness of organizations. Openness and closedness also exist on a continuum because there is no such thing as an absolutely open or completely closed organization.

Openness refers to the free flow of information within the organization. Subsystems such as creative or art departments often are characterized as open, with a free flow of ideas among participants and very few restrictions on who gets what information at what time when a creative project is in its infancy.

At the opposite end of the continuum might be a defense department unit assigned to work on top-secret defense planning affecting national security. Each person needs to receive clearance, timely information is a necessity, and access to information is on a "need to know" basis. This sort of unit is constrained by numerous rules.

Using a systems overlay to understand organizations allows us to acknowledge the idea of systems composed of subsystems; their interrelatedness and their interdependence; the existence of boundaries that allow or prevent interaction between various departments and elements of other subsystems and environments; and the existence of internal environments characterized by degrees of openness and closedness, which might differ across departments, units, or even systems projects.

Virtual Organizations and Virtual Teams

Not all organizations or parts of organizations are visible in a physical location. Entire organizations or units of organizations can now possess virtual components that permit them to change configurations to adapt to changing project or marketplace demands. Virtual enterprises use networks of computers and communications technology to bring people with specific skills together electronically to work on projects that are not physically located in the same place. Information technology enables coordination of these remote team members. Often virtual teams spring up in already-established organizations; in some instances, however, organizations of remote workers have been able to succeed without the traditional investment in a physical facility.

There are several potential benefits to virtual organizations, such as the possibility of reducing costs of physical facilities, more rapid response to customer needs, and helping virtual employees to fulfill their familial obligations to growing children or aging parents. Just how important it will be to meet the social needs of virtual workers is still open to research and debate. One example of a need for tangible identification with a culture arose when students who were enrolled in an online virtual university, with no physical campus (or sports teams), kept requesting items such as sweatshirts, coffee mugs, and pennants with the virtual university's logo imprinted on them. These items are meaningful cultural artifacts that traditional brick-and-mortar schools have long provided.

Many systems analysis and design teams are now able to work virtually, and in fact, many of them marked the path for other types of employees to follow in accomplishing work virtually. Some applications permit analysts who are providing technical assistance over the Web to "see" the software and hardware configuration of the user requesting help, in this way creating an ad hoc virtual team composed of the analyst and user.

Taking a Systems Perspective

Taking a systems perspective allows systems analysts to start broadly clarifying and understanding the various businesses with which they will come into contact. It is important that members of subsystems realize that their work is interrelated. Notice in Figure 2.2 that the outputs from the production subsystems serve as inputs for marketing and that the outputs of marketing serve as new inputs for production. Neither subsystem can properly accomplish its goals without the other.

Problems occur when each manager possesses a different picture of the importance of his or her own functional subsystem. In Figure 2.3 you can see that the marketing manager's personal

FIGURE 2.2

Outputs from one department serve as inputs for another, such that subsystems are interrelated.

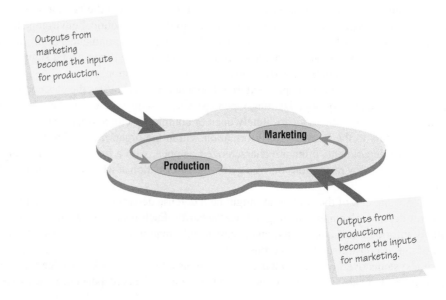

Outputs from marketing become the inputs for production.

Outputs from production become the inputs for marketing.

FIGURE 2.3

A depiction of the personal perspective of functional managers shows that they feature their own functional area as central to the organization.

How a Marketing Manager May View the Organization

How a Production Manager May See the Organization

perspective shows the business as driven by marketing, with all other functional areas interrelated but not of central importance. By the same token, the perspective of a production manager positions production at the center of the business, with all other functional areas driven by it.

The relative importance of functional areas as revealed in the personal perspectives of managers takes on added significance when managers rise to the top through the ranks, becoming strategic managers. They can create problems if they overemphasize their prior functional information requirements in relation to the broader needs of the organization.

For example, if a production manager is promoted but continues to stress production scheduling and performance of line workers, the broader aspects of forecasting and policy making may suffer. This tendency is a danger in all sorts of businesses: where engineers work their way up to become administrators of aerospace firms, college professors move from their departments to become deans, or programmers advance to become executives of software firms. Their tunnel vision often creates problems for the systems analyst trying to separate actual information requirements from desires for a particular kind of information.

Enterprise Systems: Viewing the Organization as a System

Enterprise system, or *enterprise resource planning (ERP) system*, is a term used to describe an integrated organizational (enterprise) information system. Specifically, ERP is software that helps the flow of information between the functional areas in the organization. It is a customized system that, rather than being developed in-house, is usually purchased from one of the software development companies well known for its ERP packages, such as SAP or Oracle. The product is then customized to fit the requirements of a particular company. Typically, the vendor requires an organizational commitment in terms of specialized user or analyst training.

ERP has taken root in many large companies and is spreading to small and medium-sized enterprises as well. Thankfully, most of the lessons we as systems analysts have learned are highly applicable to implementing enterprise systems. However, you as an analyst need to recognize and act on the similarities and differences between implementing networked information systems and implementing an ERP.

One of the major differences is that rather than redesigning business processes based on a logical analysis of those processes and how they support the business strategy, and then choosing the IT to support those processes, a large installation of ERP can reverse this by requiring the implementation of new business processes that are embedded in the technology provided. Oftentimes you will be part of an internal systems team whose chief job is to look after the interface between legacy systems and the ERP being installed.

Enterprise implementations are complex and intense endeavors that result in tremendous organizational change. They can affect every aspect of the organization, including design of employees' work, the skills required to become competent in one's job, and even the strategic positioning of the company. What is clear is that implementing an ERP has become increasingly complex. Many issues present important hurdles to clear if the ERP installation is to be declared a success; these include user acceptance, integration with legacy systems, and the supply chain; upgrading functionality (and complexity) of ERP modules; reorganizing work life of users and decision makers; expanded reach across several organizations (as part of the IT supply chain); and strategic repositioning of the company adopting an ERP.

Some areas to take particular note of include looking at the critical success factors (CSFs) and the interplay among them when different ERP implementations, upgrades, and conversions are approached. In addition, it's good to consider what heuristics or rules of thumb have been developed over time on the organizational, team, and even individual levels for how to do a successful ERP upgrade or initial implementation.

New research has shown that while ERP implementations can result in making employees more effective and efficient two or three years after the installation, it is in the planning and early stages of ERP that a deep analysis examines how the new enterprise system will affect the daily lives and jobs of employees involved in the new system (such as discussed in Chapter 14 on HCI and usability issues). ERP systems are best considered as real game changers when it comes to the job design of employees (including systems analysts) and as a moving force that can also alter the strategic approaches to change that organizations take.

Depicting Systems Graphically

A system or subsystem as it exists within a corporate organization may be graphically depicted in several ways. The various graphical models show the boundaries of the system and the information used in the system.

Systems and the Context-Level Data Flow Diagram

The first model is the context-level data flow diagram (also called an environmental model). Data flow diagrams focus on the data flowing into and out of the system and the processing of the data. These basic components of every computer program can be described in detail and used to analyze a system for accuracy and completeness.

As shown in Figure 2.4, the context-level data flow diagram employs only three symbols: (1) a rectangle with rounded corners, (2) a square with two shaded edges, and (3) an arrow. Processes transform incoming data into outgoing information, and the content level has only one process, representing the entire system. The external entity represents any entity that supplies or receives information from the system but is not a part of the system. This entity may be a person, a group of people, a corporate position or department, or other systems. The lines that connect the external entities to the process are called data flows, and they represent data.

An example of a context-level data flow diagram is found in Figure 2.5. This example represents the most basic elements of an airline reservation system. The passenger (an entity) initiates a travel request (data flow). The context-level diagram doesn't show enough detail to indicate exactly what happens (it isn't supposed to), but we can see that the passenger's preferences and the available flights are sent to the travel agent, who sends ticketing information back to the process. We can also see that the passenger reservation is sent to the airline. The context-level data flow diagram serves as a good starting point for drawing the use case diagram (discussed later in this chapter).

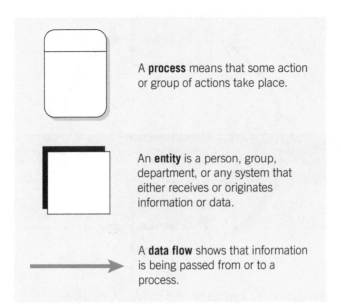

FIGURE 2.4

The basic symbols of a data flow diagram.

In Chapter 7 we see that a data flow contains much information. For example, the passenger reservation contains the passenger's name, airline, flight number(s), date(s) of travel, price, seating preference, and so on. For now, however, we are concerned mainly with how a context level data flow diagram defines the boundaries of the system. In the preceding example, only reservations are part of the process. Other decisions that the airline would make (for example, purchasing airplanes, changing schedules, pricing) are not part of this system.

The context-level data flow diagram is one way to show the scope of the system, or what is to be included in the system. The external entities are outside the scope and something over which the system has no control.

Systems and the Entity-Relationship Model

Another way a systems analyst can show the scope of a system and define proper system boundaries is to use an entity-relationship model. The elements that make up an organizational system can be referred to as *entities*. An entity may be a person, a place, or a thing, such as a passenger on an airline, a destination, or a plane. Alternatively, an entity may be an event, such as the end

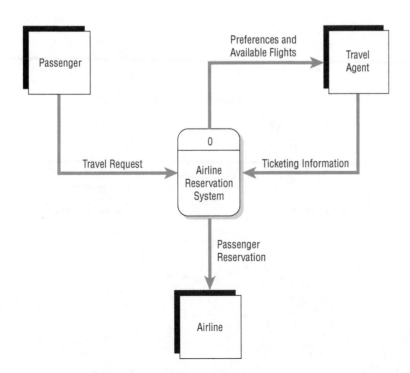

FIGURE 2.5

A context-level data flow diagram for an airline reservation system.

FIGURE 2.6

An entity-relationship diagram showing a one-to-one relationship.

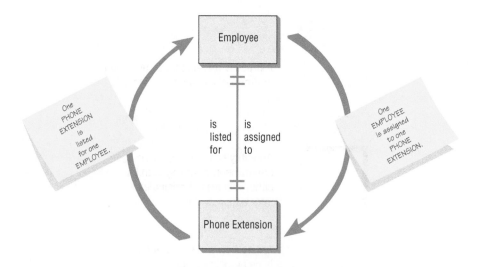

of the month, a sales period, or a machine breakdown. A relationship is the association that describes the interaction among the entities.

There are many different conventions for drawing entity-relationship (E-R) diagrams (with names like crow's foot, arrow, or Bachman notation). In this book, we use crow's foot notation. For now, we assume that an entity is a plain rectangular box.

Figure 2.6 shows a simple entity-relationship diagram. Two entities are linked together by a line. In this example, the end of the line is marked with two short parallel marks (| |), signifying that this relationship is one-to-one. Thus, exactly one employee is assigned to one phone extension. No one shares the same phone extension in this office.

The red arrows are not part of the entity-relationship diagram. They are present to demonstrate how to read the entity-relationship diagram. The phrase on the right side of the line is read from top to bottom as follows: "One EMPLOYEE is assigned to one PHONE EXTENSION." On the left side, as you read from bottom to top, the arrow says, "One PHONE EXTENSION is listed for one EMPLOYEE."

Similarly, Figure 2.7 shows another relationship. The crow's foot notation (>—+) is obvious on this diagram, and this particular example is a many-to-one example. As you read from left to right, the arrow signifies, "Many EMPLOYEES are members of a DEPARTMENT." As you read from right to left, it implies, "One DEPARTMENT contains many EMPLOYEES."

FIGURE 2.7

An entity-relationship diagram showing a many-to-one relationship.

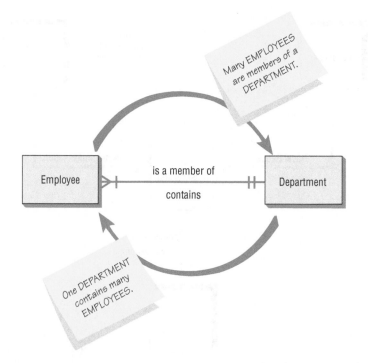

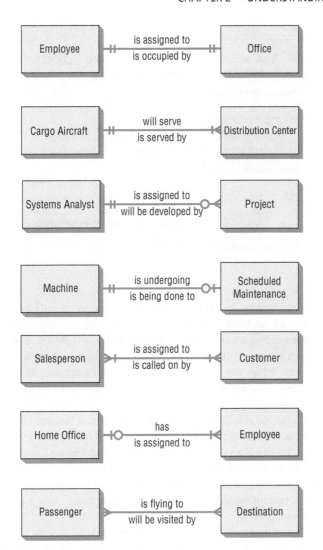

FIGURE 2.8

Examples of different types of relationships in E-R diagrams.

Notice that when a many-to-one relationship is present, the grammar changes from "is" to "are," even though the singular "is" is written on the line. The crow's foot and the single mark do not literally mean that this end of the relationship must be a mandatory "many." Instead, they imply that this end could be anything from one to many.

Figure 2.8 elaborates on this scheme. Here we have listed a number of typical entity relationships. The first, "An EMPLOYEE is assigned to an OFFICE," is a one-to-one relationship. The second one is a one-to-many relationship: "One CARGO AIRCRAFT will serve one or more DISTRIBUTION CENTERs." The third one is slightly different because it has a circle at one end. It can be read as "A SYSTEMS ANALYST may be assigned to MANY PROJECTs," meaning that the analyst can be assigned to no projects [that is what the circle (O), for zero, is for], one, or many projects. Likewise, the circle (O) indicates that none is possible in the next relationship. Recall that the short mark means one. Therefore, we can read it as follows: "A MACHINE may or may not be undergoing SCHEDULED MAINTENANCE." Notice that the line is written as "is undergoing," but the end marks on the line indicate that either no maintenance (O) or maintenance (I) is actually going on.

The next relationship states, "One or many SALESPEOPLE (plural of SALESPERSON) are assigned to one or more CUSTOMERs." It is the classic many-to-many relationship. The next relationship can be read as follows: "The HOME OFFICE can have one or many EMPLOYEEs" or "One or more EMPLOYEEs may or may not be assigned to the HOME OFFICE." Once again, the I and O together imply a Boolean situation—in other words, one or zero.

The final relationship shown here can be read as, "Many PASSENGERs are flying to many DESTINATIONs." Some people prefer this symbol [>—+]to indicate a mandatory "many"

FIGURE 2.9

Three different types of entities
used in E-R diagrams.

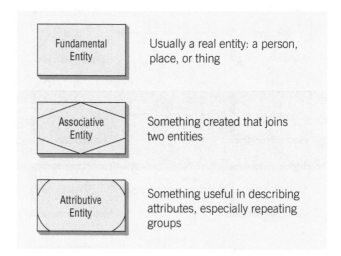

condition. (Would it ever be possible to have only one passenger or only one destination?) Even so, some CASE tools such as Visible Analyst do not offer this possibility because the optional one-or-many condition as shown in the SALESPERSON–CUSTOMER relationship will do.

Up to now we have modeled all our relationships using just one simple rectangle and a line. This method works well when we are examining the relationships of real things such as real people, places, and things. Sometimes, though, we create new items in the process of developing an information system. Some examples are invoices, receipts, files, and databases. When we want to describe how a person relates to a receipt, for example, it becomes convenient to indicate the receipt in a different way. There are three different types of entities: the fundamental entity, the associative entity, and the attributive entity are depicted in Figure 2.9.

An associative entity can exist only if it is connected to at least two other entities. For that reason, some call it a *gerund,* a *junction,* an *intersection,* or a *concatenated entity.* This wording makes sense because a receipt wouldn't be necessary unless there were a customer and a salesperson making the transaction.

Another type of entity is an attributive entity. When an analyst wants to show data that are completely dependent on the existence of a fundamental entity, an attributive entity should be used. For example, if a library has multiple copies of the same book, an attributive entity can be used to designate which copy of the book is being checked out. The attributive entity is useful for showing repeating groups of data. For example, suppose we are going to model the relationships that exist when a patron gets tickets to a concert or show. The entities seem obvious at first: "a PATRON and a CONCERT/SHOW," as shown in Figure 2.10. What sort of relationship exists?

FIGURE 2.10

The first attempt at drawing an
E-R diagram.

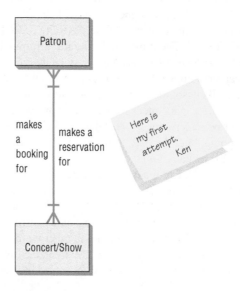

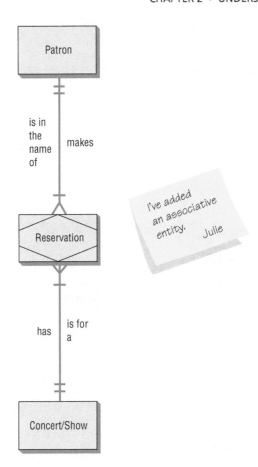

FIGURE 2.11

Improving the E-R diagram by adding an associative entry called RESERVATION.

At first glance, we see that the PATRON gets a reservation for a CONCERT/SHOW, and the CONCERT/SHOW can be said to have made a booking for a PATRON.

The process isn't that simple, of course, and the E-R diagram need not be that simple either. The PATRON actually makes a RESERVATION, as shown in Figure 2.11. The RESERVATION is for a CONCERT/SHOW. The CONCERT/SHOW holds the RESERVATION, and the RESERVATION is in the name of the PATRON. We added an associative entity here because a RESERVATION was created due to the information system required to relate the PATRON and the CONCERT/SHOW.

Again this process is quite simple, but because concerts and shows have many performances, the E-R diagram is drawn once more in Figure 2.12. Here we add an attributive entity to handle the many performances of the CONCERT/SHOW. In this case the RESERVATION is made for a particular PERFORMANCE, and the PERFORMANCE is one of many that belong to a specific CONCERT/SHOW. In turn the CONCERT/SHOW has many performances, and one PERFORMANCE has a RESERVATION that is in the name of a particular PATRON.

To the right of this E-R diagram is a set of data attributes that make up each of the entities. Some entities may have attributes in common. The attributes that are underlined can be searched for. The attributes are referred to as keys and are discussed in Chapter 13.

Systems designers often use E-R diagrams to help model a file or database. It is even more important, however, that a systems analyst understand early both the entities and relationships in the organizational system. In sketching out some basic E-R diagrams, the analyst needs to:

1. List the entities in the organization to gain a better understanding of the organization.
2. Choose key entities to narrow the scope of the problem to a manageable and meaningful dimension.
3. Identify what the primary entity should be.
4. Confirm the results of steps 1 through 3 through other data-gathering methods (investigation, interviewing, administering questionnaires, observation, and prototyping), as discussed in Chapters 4 through 6.

FIGURE 2.12

A more complete E-R diagram, showing data attributes of the entities.

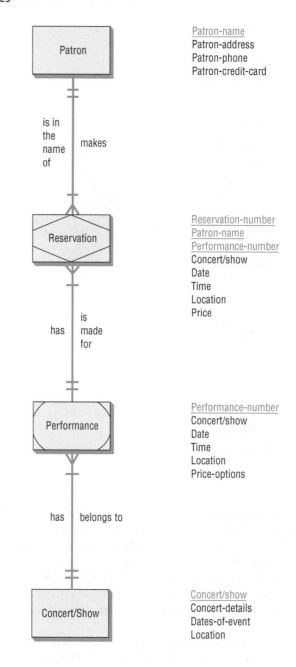

It is critical that a systems analyst begin to draw E-R diagrams upon entering an organization rather than waiting until the database needs to be designed because E-R diagrams help the analyst understand what business the organization is actually in, determine the size and scope of the problem, and discern whether the right problem is being addressed. The E-R diagrams need to be confirmed or revised as the data-gathering process takes place.

Use Case Modeling

Originally introduced as a diagram for use in object-oriented UML, use cases are now being used regardless of the approach to systems development. A use case diagram can be used as part of the SDLC or in agile modeling. The word *use* is pronounced as a noun ("yoos") rather than a verb ("yooz"). A use case model describes *what* a system does without describing *how* the system does it; that is, it is a logical model of the system. (Logical and conceptual models will be further discussed in Chapter 7.) A use case model reflects the view of a system from the perspective of a user outside the system (that is, the system requirements).

MAC APPEAL

Microsoft Visio makes it easy for a systems analyst to draw E-R diagrams as well as most of the other diagrams found in this book, but it is available only for PCs. Mac users have an alternative, OmniGraffle Professional. OmniGraffle is easier to use than Microsoft Visio because its drag-and-drop interface is smoother and more intuitive.

It also features a "smart guide" that uses pop-up distance markers to help position the symbols in the correct places. Many symbols, like those used in E-R diagrams, are built in, but OmniGraffle also lets the user search a third-party library called Graffletopia to find UML and other specialized symbols.

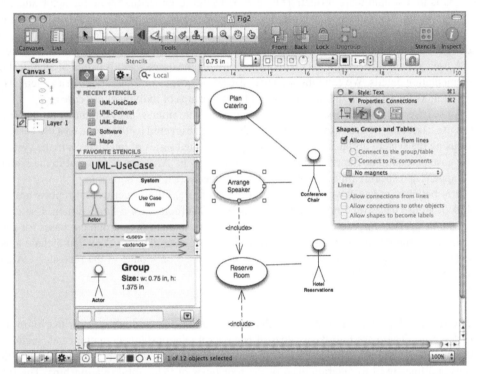

FIGURE 2.MAC

OmniGraffle from The Omni Group is an easy-to-use and powerful drawing package. (Screenshot from OmniGraffle, a registered trademark of the Omni Group. Graphic reprinted with permission.)

An analyst develops use cases in a cooperative effort with the business experts who help define the requirements of the system. A use case model provides an effective means of communication between the business team and the development team. A use case model partitions the way the system works into behaviors, services, and responses (the use cases) that are significant to the users of the system.

From the perspective of an actor (or user), a use case should produce something that is of value. Therefore, the analyst must determine what is important to the user and remember to include it in the use case diagram. For example, is entering a password something of value to the user? It may be included if the user has a concern about security or if it is critical to the success of the project.

Use Case Symbols

A use case diagram contains the actor and use case symbols, along with connecting lines. Actors are similar to external entities; they exist outside the system. The term *actor* refers to a particular role of a user of the system. For example, an actor may be an employee but also may be a customer at the company store. Even though it is the same person in the real world, it is represented as two different symbols on a use case diagram because the person interacts with the system in different roles. The actor exists outside the system and interacts with the system in a specific way. An actor can be a human, another system, or a device such as a keyboard or Web connection. Actors can initiate an instance of a use case. An actor may interact with one or more use cases, and a use case may involve one or more actors.

Actors may be divided into two groups. Primary actors supply data or receive information from the system. Some users directly interact with the system (system actors), but primary actors may also be businesspeople who do not directly interact with the system but have a stake in it. Primary actors are important because they are the people who use the system and can provide details on what the use case should do. They can also provide a list of goals and priorities. Supporting actors (also called secondary actors) help to keep the system running or provide other services. These are the analysts, programmers, people who run the help desk, and so on.

Sometimes it is useful to create an actor profile that lists the actors, their background, and their skills in a simple table format. This may be useful to understand how the actor interacts with the system. An example is an Order Processing Specialist. The profile would be, "A routine user of the software, familiar with minor features, order exceptions, and order customization." It is also useful to list the actors and their goals and priorities. Each goal may become a use case.

A use case provides developers with a view of what the users want. It is free of technical or implementation details. We can think of a use case as a sequence of transactions in a system. The use case model is based on the interactions and relationships of individual use cases.

A use case always describes three things: an actor that initiates an event, the event that triggers a use case, and the use case that performs the actions triggered by the event. In a use case, an actor using the system initiates an event that begins a related series of interactions in the system. A use case is used to document a single transaction or event. An event is an input to the system that happens at a specific time and place and causes the system to do something.

It is better to create fewer use cases rather than more. Often queries and reports are not included; 20 use cases (and no more than 40 or 50) are sufficient for a large system. Use cases may also be nested, if needed. Some use cases use the verb *manage* to group use cases for adding, deleting, and changing into another, lower-level, use case diagram. You can include a use case on several diagrams, but the actual use case is defined only once in the repository. A use case is named with a verb and a noun.

Use Case Relationships

Active relationships are referred to as behavioral relationships and are used primarily in use case diagrams. There are four basic types of behavioral relationships: communicates, includes, extends, and generalizes. Notice that all these terms are action verbs. Figure 2.13 shows the arrows and lines used to diagram each of the four types of behavioral relationships. The four relationships are described next.

Relationship	Symbol	Meaning
Communicates	———————	An actor is connected to a use case using a line with no arrowheads.
Includes	<< include >> ◄- - - - - - - - - -	A use case contains a behavior that is common to more than one other use case. The arrow points to the common use case.
Extends	<< extend >> - - - - - - - - - - -►	A different use case handles exceptions from the basic use case. The arrow points from the extended to the basic use case.
Generalizes	———————▷	One UML "thing" is more general than another "thing." The arrow points to the general "thing."

FIGURE 2.13

Four types of behavioral relationships and the lines used to diagram each.

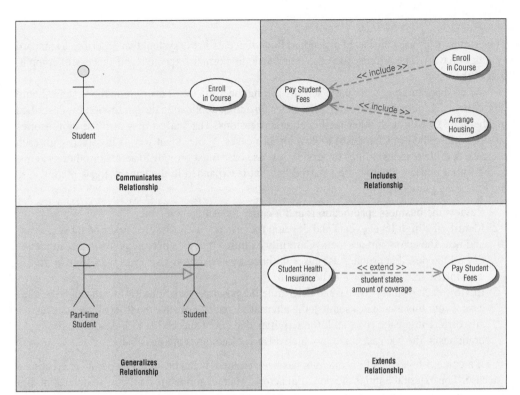

FIGURE 2.14

Some components of use case diagrams showing actors, use cases, and relationships for a student enrollment example.

COMMUNICATES. The behavioral relationship communicates is used to connect an actor to a use case. Remember that the task of a use case is to give some sort of result that is beneficial to the actor in the system. Therefore, it is important to document these relationships between actors and use cases. In our first example, a **Student** communicates with **Enroll in Course**. Examples of some components of a student enrollment example are shown in the use case diagrams in Figure 2.14.

INCLUDES. The includes relationship (also called the uses relationship) describes the situation in which a use case contains behavior that is common to more than one use case. In other words, the common use case is included in the other use cases. A dotted arrow that points to the common use case indicates the includes relationship. An example would be a use case **Pay Student Fees** that is included in **Enroll in Course** and **Arrange Housing** because in both cases, students must pay their fees. **Pay Student Fees** may be used by several use cases. The arrow points toward the common use case.

EXTENDS. The extends relationship describes the situation in which one use case possesses the behavior that allows the new use case to handle a variation or an exception from the basic use case. For example, the extended use case **Student Health Insurance** extends the basic use case **Pay Student Fees.** The arrow goes from the extended to the basic use case.

GENERALIZES. The generalizes relationship implies that one thing is more typical than the other thing. This relationship may exist between two actors or two use cases. For example, a **Part-Time Student** generalizes a **Student.** Similarly, some of the university employees are professors. The arrow points to the general thing.

Developing System Scope

The scope of a system defines its boundaries: what is in scope—or inside the system—and what is out of scope. The project usually has a budget that helps to define scope, as well as start and end times. Actors are always outside the scope of the system. The communicates lines that connect actors to the use cases are the boundaries and define the scope. Since a use case diagram is created early in the system's life cycle, the budget, starting time, and ending time may change as the project progresses; as the analyst learns more about the system, the use case diagrams, use case, and scope may change.

Developing Use Case Diagrams

The primary use case consists of a standard flow of events in the system that describes a standard system behavior. The primary use case represents the normal, expected, and successful completion of the use case.

When diagramming a use case, start by asking the users to list everything the system should do for them. This can be done using interviews, in a joint application design session (as described in Chapter 4), or through other facilitated team sessions. The analyst may also use agile stories sessions (described in Chapter 6) to develop use cases. Write down who is involved with each use case and the responsibilities or services the use case must provide to actors or other systems. In the initial phases, this may be a partial list that is expanded in the later analysis phases. Use the following guidelines:

1. Review the business specifications and identify the actors involved.
2. Identify the high-level events and develop the primary use cases that describe those events and how the actors initiate them. Carefully examine the roles played by the actors to identify all the possible primary use cases initiated by each actor. Use cases with little or no user interaction do not have to be shown.
3. Review each primary use case to determine the possible variations of flow through the use case. From this analysis, establish the alternative paths. Because the flow of events is usually different in each case, look for activities that could succeed or fail. Also look for any branches in the use case logic in which different outcomes are possible.

If a context-level data flow diagram has been created, it can be a starting point for creating a use case. The external entities are potential actors. Then examine the data flow to determine if it would initiate a use case or be produced by a use case.

Figure 2.15 is an example of a use case diagram representing a system used to plan a conference. The actors are the **Conference Chair,** responsible for planning and managing the conference, the conference **Participant, Speakers,** a **Keynote Speaker, Hotel Reservations,** and a **Caterer.** Actors represent the *role* the user plays, and the **Caterer** may be either a hotel employee or an external catering service.

Both the **Conference Chair** and the **Caterer** are involved in planning meals and banquets. The **Conference Chair** is also responsible for arranging speakers. The **Participant** registers for the conference. Notice that the **Reserve Room** use case is involved in an *includes* relationship with the **Arrange Speaker** and **Register for Conference** use cases, since both speakers and participants will need lodging. The **Arrange Language Translation** use case extends the **Register for Conference** use case because not all participants will require language translation services. The **Speaker** actor is a generalization of **Keynote Speaker.**

Developing Use Case Scenarios

Each use case has a description. We will refer to the description as a use case scenario. As mentioned, the primary use case represents the standard flow of events in the system, and alternative paths describe variations to the behavior. Use case scenarios may describe what happens if an item purchased is out of stock, or if a credit card company rejects a customer's requested purchase.

There is no standardized use case scenario format, so each organization is faced with specifying what standards should be included. Often the use cases are documented using a use case document template predetermined by the organization, which makes the use cases easier to read and provides standardized information for each use case in the model.

Use Case Levels

You may want to create use cases for different levels. One method (defined by Alistair Cockburn) uses the following altitude metaphors:

1. White is the highest level, like clouds. This is the enterprise level, and there may be only four to five uses cases at this level for the entire organization. Examples might be to advertise goods, sell goods to customers, manage inventory, manage the supply chain, and optimize shipping.
2. Kite is lower than white but still a high level, providing an overview. The kite use case may be at the business unit or department level and is a summary of goals. Examples

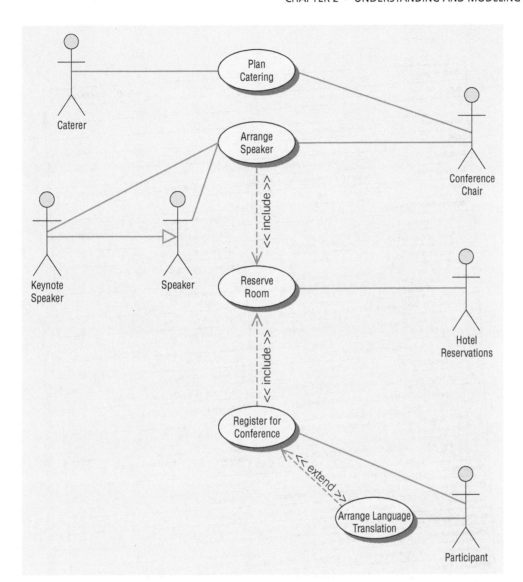

FIGURE 2.15

A use case diagram representing a system used to plan a conference.

would be to register students, or, if working with a travel company, to make an airline, hotel, car, or cruise reservation.

3. Blue is at sea level and is customarily used to depict user goals. This level often has the greatest interest for users and is easiest for a business to understand. It is usually written for a business activity, and each person should be able to do one blue level activity in anywhere from 2 to 20 minutes. Examples are register a continuing student, add a new customer, place an item in a shopping cart, and order checkout.

4. Indigo or fish is a use case that shows lots of detail, often at a functional or subfunctional level. Examples are choose a class, pay academic fees, look up the airport code for a given city, and produce a list of customers after entering a name.

5. Black or clam, like the bottom of the ocean, are the most detailed use cases, at a subfunction level. Examples might be to validate secure logon, add a new field using dynamic HTML, or use Ajax to update a web page in a small way.

A use case scenario example is shown in Figure 2.16. Some of the areas included are optional and may not be used by all organizations. The three main areas are:

1. An area header that contains case identifiers and initiators
2. Steps performed
3. A footer area that contains preconditions, assumptions, questions, and other information

FIGURE 2.16

A use case scenario is divided into three sections: identification and initiation; steps performed; and conditions, assumptions, and questions.

Use case name:	Register for Conference		
Area:	Conference Planning	UniqueID:	Conf RG 003
Actor(s):	Participant		
Stakeholder:	Conference Sponsor, Conference Speakers		
Level:	Blue		
Description:	Allow conference participant to register online for the conference using a secure Web site.		
Triggering Event:	Participant uses Conference Registration Web site, enters userID and password, and clicks the logon button.		
Trigger type:	☐ External ☷ Temporal		

Steps Performed (Main Path)	Information for Steps
1. Participant logs in using the secure Web server.	userID, Password
2. Participant record is read and password is verified.	Participant Record, userID, Password
3. Participant and session information is displayed on the Registration Web page.	Participant Record, Session Record
4. Participant enters information on the Registration Web form and clicks **Submit** button.	Registration Web Form
5. Registration information is validated on the Web server.	Registration Web Form
6. Registration Confirmation page is displayed to confirm registration information.	Confirmation Web Page
7. Credit card is charged for registration fees.	Secure Credit Card Web Page
8. Add Registration Journal record is written.	Confirmation Web Page
9. Registration record is updated on the Registration Master.	Confirmation Web Page, Registration Record
10. Session record is updated for each selected session on the Session Master.	Confirmation Web Page, Session Record
11. Participant record is updated for the participant on the Participant Master.	Confirmation Web Page, Participant Record
12. Successful Registration Confirmation Web page is sent to the participant.	Registration Record Confirmation Number

Preconditions:	Participant has already registered and has created a user account.
Postconditions:	Participant has successfully registered for the conference.
Assumptions:	Participant has a browser and a valid userID and password.
Success Guarantee:	Participant has registered for the conference and is enrolled in all selected sessions.
Minimum Guarantee:	Participant was able to logon.
Requirements Met:	Allow conference participants to be able to register for the conference using a secure Web site.
Outstanding Issues:	How should a rejected credit card be handled?
Priority:	High
Risk:	Medium

The first area, use case identifiers and initiators, orients the reader and contains the use case name and a unique ID; the application area or system that this use case belongs to; the actors involved in the use case; and the stakeholders that have a high level of interest in the use case. Some stakeholders never interact directly with the system, such as the stockholders, the board of directors, or the sales manager. Each primary actor is a stakeholder but not listed in the stakeholder area. The first area should therefore include the level (blue, kite, and so on) and a brief description of what the use case accomplishes.

The header concludes with the initiating (triggering) event—that is, what caused the use case to start—and the type of trigger, either external or temporal. External events are those started by an actor, either a person or another system requesting information, such as an airline reservation system requesting flight information from an airline system. Temporal events are those that are triggered or started by time. Events occur at a specific time, such as sending an email about special offers once a week on a Sunday evening, sending bills on a specific day, or generating government statistics on a specified date every quarter.

The second area of the use case includes the steps performed and the information required for each of the steps. These statements represent the standard flow of events and the steps taken for the successful completion of the use case. It is desirable to write up a use case for the main path and then to write up one for each of the alternative paths separately rather than using IF … THEN … statements. Steps are numbered with an integer. The steps may come from a detailed interview with users or may be derived from agile modeling stories (as described in Chapter 6). These steps should be reviewed with the users for clarification.

The analyst should examine each of the steps and determine the information required for each step. If the analyst cannot determine the information, he or she should schedule a follow-up interview with the user. Some use case descriptions include extensions or alternative scenarios, with the exceptions as additional sections following the standard flow of events. These are numbered with an integer, decimal point, and another integer, such as 3.1, 3.2, 3.3, and so on. These are steps that may or may not be used. Analysts and users can brainstorm what can go wrong with the main path, and may uncover important details and conditions. It is necessary to work with the users to determine what to do when these conditions occur. This helps to detect errors earlier in the life cycle.

Figure 2.17 illustrates how logic and alternative scenarios can be included in the middle section of a use case. In this airline example, notice that step 1 is made up of smaller steps, many of which are preceded by an "if." These are still on the main path but occur only if the condition is met. For example, if there are many airports that serve a city, then all the airports will be displayed. Extensions or alternate scenarios can also appear here. For this airline, other scenarios

FIGURE 2.17

Use cases can include conditional steps as well as extensions or alternative scenarios.

Steps Performed (Main Path)	Information for Steps
1. Enter departing and arriving airports, dates of travel.	Airport Locations
1.1. If an airport code is entered, display matching name, city, country	
1.2. If a city is entered, find all matching cities	
1.3. Customer selects a city	
1.4. If there is more than one airport for the city, display airports	
1.5. Client selects an airport	
1.6. Insert the airport code (3 characters)	
1.7. Display the matching airport country, city, and airport name	
2. Find all matching flights with available seats	
3. Customer selects flight	Flight Information
4. Customer logs on	
5. Customer selects passenger names	Customer Logon
6. Seating chart is displayed showing all available seats	Passenger Records
7. Customer selects seat(s) for each passenger	Plane Number, Seating Chart, Available Seats
8. Display confirmation and credit card page	
9. Credit card verified	
10. Email confirmation sent	
11. Airline reservation made	
Extensions or Alternative Scenarios	
Flight Selection	
1. A list of flights displays	
2. Customer selects a flight	
3. Request is sent to airline	
4. Flight is already full	
Seat Selection	
1. A list of flights displays	
2. Customer selects a flight	
3. Request is sent to airline	
4. Seat reservations are retrieved	
5. Seating chart is displayed	
6. Customer cannot find an acceptable seat	
Meal Selection for International Flights	
1. Customer selects meal from drop-down list	
2. Record is updated with meal selection	Available Airline Meal List
	Customer Meal Record

include flight selection, seat selection, and meal selection. Use cases may even include iterative or looping steps.

The third area of the use case includes:

- Preconditions, or the condition of the system before the use case may be performed, which may be another use case. An example might be, "The viewer has successfully logged into the system," or it might be the successful completion of another use case.
- Postconditions, or the state of the system after the use case has finished, including output people have received, transmissions to other systems, and data that have been created or updated. These relate to the goals or user requirements from a problem definition (described in Chapter 3) or to agile stories (described in Chapter 6).
- Assumptions made that would affect the method of the use case and that could stipulate required technology, such as the minimum technology requirements in a browser or even a specific or higher version of a browser. An assumption might be that cookies or JavaScript are enabled. The analyst must determine what to do if the assumptions are not met. When using Google Maps, JavaScript must be enabled; if it is not enabled, the map will not display. Netflix requires cookies. Good web pages will detect that an assumption has not been met and notify the viewer with a message, including information on how to turn on cookies or JavaScript for different browsers.
- Minimal guarantee is the minimum promised to the users. They may not be happy with this result, and it may be that nothing happens.
- Success guarantee is what would satisfy the users, and it is usually that the goal of the use case has been met.
- Any outstanding issues or questions must be answered before implementation of the use case.
- An optional statement of priority of the use case, which may come from a problem definition or user requirements.
- An optional statement of risk involved in creating the use case.

The "requirements met" area links the use case to user requirements or objectives from a problem definition. Once you develop the use case scenarios, be sure to review your results with the business experts to verify and refine the use cases if needed.

In this particular use case scenario, called **Register for Conference,** the only actor involved is the **Participant.** The overall area is **Conference Planning,** and the use case is triggered by the participant logging on to the **Registration** Web page. The **Steps Performed** area lists the sequence of events that must occur for a successful conference registration. Notice that the information needed to perform each of the steps is listed on the right. This may include web pages and forms, as well as database tables and records.

The **Preconditions** area in the footer section of the use case scenario lists what must occur before the participant can register for a conference. In this example, the participant must have already signed up as a member of the society and have a valid user ID and password. The **Postconditions** area lists what has been accomplished by the use case. The **Assumptions** area lists any basic premises the analyst assumes are fulfilled by the actor beforehand. The **Requirements Met** area shows why this use case is important and necessary for the business area to be successful. **Priority** is an indication of which use cases should be developed first and which may be delayed. **Risk** is a rough assessment of whether there may be problems or difficulties developing the use case. In this case, the risk is medium because the registration use case requires a secure server and is accepting credit card information.

Creating Use Case Descriptions

Use the following four steps to create use case descriptions:

1. Use agile stories, problem definition objectives, user requirements, or a features list as a starting point.
2. Ask about the tasks that must be done to accomplish the transaction. Ask if the use case reads any data or updates any tables.
3. Find out if there are any iterative or looping actions.
4. The use case ends when the customer goal is complete.

- Use cases effectively communicate systems requirements because the diagrams are kept simple.
- Use cases allow people to tell stories.
- Use case stories make sense to nontechnical people.
- Use cases do not depend on a special language.
- Use cases can describe most functional requirements (such as interactions between actors and applications).
- Use cases can describe nonfunctional requirements (such as performance and maintainability) through the use of stereotypes.
- Use cases help analysts define boundaries.
- Use cases can be traceable, allowing analysts to identify links between use cases and other design and documentation tools.

FIGURE 2.18

The main reasons for writing use cases are their effectiveness in communicating with users and their capturing of user stories.

Why Use Case Diagrams Are Helpful

No matter what method you use to develop your system (traditional SDLC methods, agile methods, or object-oriented methods), you will find that use cases are very valuable. The use case diagrams identify all the actors in the problem domain, and a systems analyst can concentrate on what humans want and need to use the system, extend their capabilities, and enjoy their interaction with technology. The main reasons for writing use cases are shown in Figure 2.18.

The actions that need to be completed are also clearly shown on the use case diagram. This not only makes it easy for the analyst to identify processes, but it also aids in communication with other analysts on the team and business executives.

The use case scenario is also worthwhile. Since a lot of the information the users impart to the analyst already takes the form of stories, it is easy to capture the stories on a use case scenario form. The use case scenario always documents the triggering event so that an analyst can always trace the steps that led to other use cases. Since the steps performed are noted, it is possible to employ use case scenarios to write logical processes.

Use case diagrams are becoming popular because of their simplicity and lack of technical detail. They are used to show the scope of a system, along with the major features of the system and the actors who work with those major features. The users see the system and they can react to it and provide feedback. They may also help to determine whether to build or buy the software.

Levels of Management

Management in organizations exists on three broad, horizontal levels: operational control, managerial planning and control (middle management), and strategic management, as shown in Figure 2.19. Each level carries its own responsibilities, and all work toward achieving organizational goals and objectives in their own ways.

FIGURE 2.19

Management in organizations exists on three horizontal levels: operational control, managerial planning and control, and strategic management.

CONSULTING OPPORTUNITY 2.2

Where There's Carbon, There's a Copy

"I don't know what we do with the pink ones yet," Richard Russell admitted. "They're part of a quadruplicate form that rips apart. All I know is that we keep them for the filing clerk, and he files them when he has time."

Richard is a newly hired junior account executive for Carbon, Carbon & Rippy, a brokerage house. You are walking through the steps he takes in making a stock purchase "official" because his boss has asked you to streamline the process whereby stock purchase information is stored in the computer system and retrieved.

After you leave, Richard continues thinking about the pink forms. He tells his clerk, Harry Schultz, "In my two months here, I haven't seen anyone use those. They take up my time and yours, not to mention all the filing space. Let's pitch them."

Richard and Harry proceed to open all the old files kept by Richard's predecessor and throw out the filed pink forms, along with those accumulated but not yet filed. It takes hours, but they make a lot of room. "Definitely worth the time," Richard reassures Harry.

Three weeks later, an assistant to Richard's boss, Carol Vaness, appears. Richard is happy to see a familiar face, greeting her with, "Hi, Carol. What's new?"

"Same old thing," Carol sighs. "Well, I guess it isn't old to you, because you're the newcomer. But I need all those pesky pink forms."

Almost in shock, Richard exchanges looks with Harry, then mumbles, "You're kidding, of course."

Carol looks more serious than Richard ever thought possible, replying, "No joke. I summarize all the pink forms from all the brokers, and then my totals are compared with computerized stock purchase information. It's part of our routine, three-month audit for transaction accuracy. My work depends on yours. Didn't Ms. McCue explain that to you when you started?"

What systems concept did Richard and Harry ignore when tossing out the pink forms? What are the possible ramifications for systems analysts if general systems concepts are ignored?

Operational control forms the bottom tier of three-tiered management. Operations managers make decisions using predetermined rules that have predictable outcomes when implemented correctly. They make decisions that affect implementation in work scheduling, inventory control, shipping, receiving, and control of processes such as production. Operations managers oversee the operating details of the organization.

Middle management forms the second, or intermediate, tier of the three-tiered management system. Middle managers make short-term planning and control decisions about how resources may best be allocated to meet organizational objectives.

Their decisions range all the way from forecasting future resource requirements to solving employee problems that threaten productivity. The decision-making domain of middle managers can usefully be characterized as partly operational and partly strategic, with constant fluctuations.

Strategic management is the third level of three-tiered management control. Strategic managers look outward from the organization to the future, making decisions that will guide middle and operations managers in the months and years ahead.

Strategic managers work in a highly uncertain decision-making environment. Through statements of goals and the determination of strategies and policies to achieve them, strategic managers actually define the organization as a whole. Theirs is the broad picture, wherein the company decides to develop new product lines, divest itself of unprofitable ventures, acquire other compatible companies, or even allow itself to be acquired or merged.

There are sharp contrasts among the decision makers on many dimensions. For instance, strategic managers have multiple decision objectives, whereas operations managers have single ones. It is often difficult for high-level managers to identify problems, but it is easy for operations managers to do so. Strategic managers are faced with semistructured problems, whereas lower-level managers deal mostly with structured problems.

The alternative solutions to a problem facing the strategic managers are often difficult to articulate, but the alternatives that operations managers work with are usually easy to enumerate. Strategic managers most often make one-time decisions, whereas the decisions made by operations managers tend to be repetitive.

CONSULTING OPPORTUNITY 2.3

Pyramid Power

"We really look up to you," says Paul LeGon. As a systems analyst, you have been invited to help Pyramid, Inc., a small, independent book-publishing firm that specializes in paperback books outside the publishing mainstream.

Paul continues, "We deal with what some folks think are fringe topics. You know, pyramid power, end-of-the-world prophecies, and healthier living by thinking of the color pink. Sometimes when people see our books, they just shake their heads and say, 'Tut—uncommon topic.' But we're not slaves to any particular philosophy, and we've been very successful. So much so that because I'm 24, people call me the 'boy king.'" Paul pauses to decipher your reaction.

Paul continues, "I'm at the top as president, and functional areas such as editorial, accounting, production, and marketing are under me."

Paul's assistant, Ceil Toom, who has been listening quietly up to now, barges in with her comments: "The last systems experts that did a project for us recommended the creation of liaison committees of employees between accounting, production, and marketing, so that we could share newly computerized inventory and sales figures across the organization. They claimed that committees such as that would cut down on needless duplication of output, and each functional area would be better integrated with all the rest."

Paul picks up the story, saying, "It was fair—oh, for a while—and the employees shared information, but the reason you're here is that the employees said they didn't have time for committee meetings and were uncomfortable sharing information with people from other departments who were further up the ladder than they were here at Pyramid."

According to Paul and Ceil, what were the effects of installing a management information system at Pyramid, Inc., that required people to share information in ways that were not consistent with their structure? Propose some general ways to resolve this problem so that Pyramid employees can still obtain the sales and inventory figures they need.

Implications for Information Systems Development

Each of the three management levels has different implications for developing information systems. Some of the information requirements for managers are clear-cut, whereas others are fuzzy and overlapping.

Operations managers need internal information that is of a repetitive, low-level nature. They are highly dependent on information that captures current performance, and they are large users of online, real-time information resources. The need of operations managers for past performance information and periodic information is only moderate. They have little use for external information that allows future projections.

On the next management level, middle managers are in need of both short- and longer-term information. Due to the troubleshooting nature of their jobs, middle managers experience extremely high needs for information in real time. To control properly, they also need current information on performance as measured against set standards. Middle managers are highly dependent on internal information. In contrast to operations managers, they have a high need for historical information, along with information that allows for the prediction of future events and simulation of numerous possible scenarios.

Strategic managers differ somewhat from both middle and operations managers in their information requirements. They are highly dependent on information from external sources that supply news of market trends and the strategies of competing corporations. Because the task of managing strategically demands projections into the uncertain future, strategic managers have a high need for information of a predictive nature and information that allows creation of many different what-if scenarios. Strategic managers also exhibit strong needs for periodically reported information as they seek to adapt to fast-moving changes.

Organizational Culture

Organizational culture is an established area of research that has grown remarkably in the past decades. Just as it is appropriate to think of organizations as including many technologies, it is similarly appropriate to see them as hosts to multiple, often competing subcultures.

"**Y**ou seem to have already made a good start at MRE. I'm glad you met Snowden Evans. As you know, you'll be reporting directly to him during your consulting project. As his administrative assistant for the past five years, I can tell you a lot about the company, but remember that there are a number of ways to find out more. You will want to interview users, observe their decision-making settings, and look at archival reports, charts, and diagrams. To do so, you can click on the phone directory to get an appointment with an interviewee, click on the building map to view the layout of the building, or click on the corporate website to see the functional areas and formal hierarchical relationships at MRE.

"Many of the rules of corporate life apply in the MRE HyperCase. You can walk freely in many public areas. If you want to tour a private office, however, you must first book an appointment with one of our employees. Some secure areas are strictly off limits to you as an outsider since you could pose a security risk.

"I don't think you'll find us excessively secretive, however, because you may assume that any employee who grants you an interview will also grant you access to the archival material in his or her files as well as to current work on their desktops or screens.

"Unfortunately, some people in the company never seem to make themselves available to consultants. I suggest you be persistent. There are lots of ways to find out about the people and the systems of MRE. Creativity pays off. You'll notice that the systems consultants who follow their hunches, sharpen their technical skills, and never stop thinking about piecing together the puzzles at MRE are the ones who get the best results.

"Remember to use multiple methods—interviewing, observation, and investigation—to understand what we at MRE are trying to tell you. Sometimes actions, documents, and offices actually speak louder than words!"

There are many ways to graphically depict the system. An analyst should choose among these tools early on to get an overview of the system. These approaches include drawing context-level data flow diagrams, capturing relationships early on with entity-relationship diagrams, and drawing use case diagrams or writing use case scenarios based on user stories. Using these diagrams and techniques at the beginning of analysis can help the analyst define the boundaries and scope of the system and can help bring into focus which people and systems are external to the system being developed.

Entity-relationship diagrams help a systems analyst understand the entities and relationships that comprise an organizational system. E-R diagrams can depict a one-to-one relationship, a one-to-many relationship, a many-to-one relationship, and a many-to-many relationship.

FIGURE 2.HC1

Click on keywords in HyperCase to find out more detail.

The three levels of managerial control are operational, middle management, and strategic. The time horizon of decision making is different for each level.

Organizational cultures and subcultures are important determinants of how people use information and information systems. By grounding information systems in the context of the organization as a larger system, it is possible to realize that numerous factors are important and should be taken into account when ascertaining information requirements and designing and implementing information systems.

HYPERCASE Questions

1. What major organizational change recently took place at MRE? What department(s) was (or were) involved? Why was the change made?
2. What are the goals of the Training and Management Systems Department?
3. Would you categorize MRE as a service industry, a manufacturer, or both? What kind of "products" does MRE "produce"? Suggest how the type of industry MRE is affects the information systems it uses.
4. What type of organizational structure does MRE have? What are the implications of this structure for MRE?
5. Describe in a paragraph the "politics" of the Training and Management Systems Department at MRE. Who is involved, and what are some of the main issues?
6. Draw a use case diagram representing the activities of The Webster Design Group at MRE when developing site and facility master plans. (Use the MRE website to obtain your basic information.)

There is still little agreement on what precisely constitutes an organizational subculture. It is agreed, however, that competing subcultures may be in conflict, attempting to gain adherents to their vision of what the organization should be. Research is in progress to determine the effects of virtual organizations and virtual teams on the creation of subcultures when members do not share a physical workspace but share tasks.

Rather than think about culture as a whole, it is more useful to think about the researchable determinants of subcultures, such as shared verbal and nonverbal symbolism. Verbal symbolism includes shared language used to construct, convey, and preserve subcultural myths, metaphors, visions, and humor. Nonverbal symbolism includes shared artifacts, rites, and ceremonies; clothing of decision makers and workers; the use, placement, and decoration of offices; and rituals for celebrating members' birthdays, promotions, and retirements.

Subcultures coexist within "official" organizational cultures. The officially sanctioned culture may prescribe a dress code, suitable ways to address superiors and coworkers, and proper ways to deal with the outside public. Subcultures may be powerful determinants of information requirements, availability, and use.

Organizational members may belong to one or more subcultures in the organization. Subcultures may exert a powerful influence on member behavior, including sanctions for or against the use of information systems.

Understanding and recognizing predominant organizational subcultures may help a systems analyst overcome the resistance to change that arises when a new information system is installed. For example, the analyst might devise user training to address specific concerns of organizational subcultures. Identifying subcultures may also help in designing decision support systems that are tailored for interaction with specific user groups.

Summary

There are three broad organizational fundamentals to consider when analyzing and designing information systems: the concept of organizations as systems, the various levels of management, and the overall organizational culture.

Organizations are complex systems composed of interrelated and interdependent subsystems. In addition, systems and subsystems are characterized by their internal environments on a continuum from open to closed. An open system allows free passage of resources (people, information, materials) through its boundaries; closed systems do not permit free flow of input or output. Organizations and teams can also be organized virtually, with remote members who are not in the same physical workspace connected electronically. Enterprise resource planning systems are integrated organizational (enterprise) information systems developed with customized, proprietary software that help the flow of information between the functional areas in the organization. They support a systems view of the organization.

Keywords and Phrases

actor	middle management
associative entity	openness
attributive entity	operations management
closedness	organizational boundaries
context-level data flow diagram	organizational culture
crow's foot notation	scope of the system
enterprise resource planning (ERP)	strategic management
enterprise systems	systems
entity (fundamental entity)	use case
entity-relationship (E-R) diagram	use case diagram
environment	use case scenario
feedback	virtual enterprise
four levels of use cases	virtual organization
interdependent	virtual team
interrelatedness	

Review Questions

1. What are the three groups of organizational fundamentals that carry implications for the development of information systems?
2. What is meant by saying that organizational subsystems are interrelated and interdependent?
3. Define the term *organizational boundary.*
4. What are the two main purposes of feedback in organizations?
5. Define *openness* in an organizational environment.
6. Define *closedness* in an organizational environment.
7. What is the difference between a traditional organization and a virtual one?
8. What are the potential benefits and a drawback of a virtual organization?
9. Give an example of how systems analysts could work with users as a virtual team.
10. What are enterprise systems (ERP)?
11. What is the main difference between doing business process analysis for ERP and other types of systems?
12. What problems do analysts often encounter when they try to implement an ERP package?
13. What are the two symbols on a use case diagram, and what do they represent?
14. What is a use case scenario?
15. What are the three main parts of a use case scenario?
16. What are the four steps in creating use case descriptions?
17. What are the five altitude metaphors for describing use case on different levels? What do they represent?
18. What does a process represent on a context-level data flow diagram?
19. What is an entity on a data flow diagram?
20. What is meant by the term *entity-relationship diagram*?
21. What symbols are used to draw E-R diagrams?
22. List the types of E-R diagrams.
23. How do an entity, an associative entity, and an attributive entity differ?
24. List the three broad, horizontal levels of management in organizations.
25. How can understanding organizational subcultures help in the design of information systems?

Problems

1. "It's hard to focus on what we want to achieve. I look at what our real competitors, the convenience stores, are doing and think we should copy that. Then a hundred customers come in, and I listen to each of them, and they say we should keep our little store the same, with friendly clerks and old-fashioned cash registers. Then, when I pick up a copy of *SuperMarket News*, they say that the wave of the future is super grocery stores, with no individual prices marked and UPC scanners replacing clerks. I'm pulled in so many directions I can't really settle on a strategy for our grocery store," admits Geoff Walsham, owner and manager of Jiffy Geoff's Grocery Store.

    In a paragraph, apply the concept of permeable organizational boundaries to analyze Geoff's problem in focusing on organizational objectives.
2. Write seven sentences explaining the right-to-left relationships in Figure 2.8.
3. Draw an entity-relationship diagram of a patient–doctor relationship.
    a. Which of the types of E-R diagrams is it?
    b. In a sentence or two, explain why the patient–doctor relationship is diagrammed in this way.
4. You began drawing E-R diagrams soon after your entry into the health maintenance organization for which you're designing a system. Your team member is skeptical about using E-R diagrams before the design of the database is begun. In a paragraph, persuade your team member that early use of E-R diagrams is worthwhile.
5. Neil is a decision maker for Pepe's Atlantic Sausage Company. Because there are several suppliers of ingredients and their prices fluctuate, he has come up with several different formulations for the various sausages that he makes, depending on the availability of particular ingredients from particular suppliers. He then orders ingredients accordingly twice a week. Even though he cannot predict when ingredients will become available at a particular price, his ordering of supplies can be considered routine.
    a. On what level of management is Neil working? Explain in a paragraph.
    b. What attributes of his job would have to change before you would categorize him as working on a different level of management? List them.

6. Many of the people who work at Pepe's (Problem 5) are extremely dedicated to Pepe's and have devoted their lives to the company. Others feel that the company is behind the times and should use more sophisticated production systems, information systems, and supply chain management to make the company more competitive. Members of a third group feel that what they do is unappreciated. Describe the various subcultures in words. Assign them a name, based on their emotions.

7. Alice in the human resources department at the Cho Manufacturing Company plant is constantly being asked by employees how much is taken out of their paychecks for insurance, taxes, medical, mandatory retirement, and voluntary retirement. "It takes up to a few hours every day," says Alice.

   She would like the company to have a Web system that would allow employees to use a secure logon to view the information. Alice wants the system to interface with health and dental insurance companies to obtain the amount remaining in the employee's account for the year. She would also like to obtain retirement amounts saved, along with investment results. Alice has a high regard for privacy and wants the system to have employees register and give permission to obtain financial amounts from the dental insurance and retirement companies. Draw a use case diagram representing the activities of this employee benefits system.

8. Write up a use case scenario for the use case diagram you constructed for Cho Manufacturing.

9. At what level are you creating your use case for Cho Manufacturing? Choose one of the five altitude metaphors and explain why you chose it.

10. Create a context-level data flow diagram for the employee benefits system in Problem 7. Make any assumptions you need about the data to and from the central process. Do you find this to be better than or not as good at explaining the system to Alice compared to the use case and use case scenarios?

11. Draw a use case and write up a use case scenario for getting two or three email accounts. Think about the steps that are needed to ensure security.

Group Projects

1. Break up into groups of five. Assign one person to act as the website designer, one to write copy for a company's product, one to keep track of customer payments, one to monitor distribution, and one to satisfy customers who have questions about using the product. Then select a simple product (one that does not have too many versions). Good examples are a digital camera, a GPS, a box of candy, or a specialty travel hat (rainproof or sunblocker). Now spend 20 minutes trying to explain to the website designer what to include on the website. Describe in about three paragraphs what experience your group had in coordination. Elaborate on the interrelatedness of subsystems in the organization (your group).

2. In a small group, develop a use case and a use case scenario for making air, hotel, and car reservations for domestic travel.

3. Change your answer in Group Project 2 to include foreign travel. How do the use case and use case scenario change?

4. With your group, draw a context-level data flow diagram of your school's or university's registration system. Label each entity and process. Discuss why there appear to be different ways to draw the diagram. Reach consensus as a group about the best way to draw the diagram and defend your choice in a paragraph. Now, working with the other members of your group, follow the appropriate steps for developing an E-R diagram and create one for your school or university registration system. Make sure your group indicates whether the relationship you depict is one-to-one, one-to-many, many-to-one, or many-to-many.

Selected Bibliography

Bleeker, S. E. "The Virtual Organization." *Futurist,* Vol. 28, No. 2, 1994, pp. 9–14.

Chen, P. "The Entity-Relationship Model—Towards a Unified View of Data." *ACM Transactions on Database Systems,* Vol. 1, March 1976, pp. 9–36.

Ching, C., C. W. Holsapple, and A. B. Whinston. "Toward IT Support for Coordination in Network Organizations." *Information Management,* Vol. 30, No. 4, 1996, pp. 179–199.

Cockburn, A. "Use Case Icons," http://alistair.cockburn.us/Use+case+icons?version=8339&diff= 8339&with=6296. Last accessed March 18, 2009.

Davis, G. B., and M. H. Olson. *Management Information Systems, Conceptual Foundations, Structure, and Development,* 2nd ed. New York: McGraw-Hill, 1985.

Galbraith, J. R. *Organizational Design.* Reading, MA: Addison-Wesley, 1977.

Grabski, S. V., S.A. Leech, and P. J. Schmidt. "A Review of ERP Research: A Future Agenda for Accounting Information Systems." *Journal of Information Systems,* Vol. 25, No. 1, 2011, pp. 37–78.

Kendall, K. E., J. R. Buffington, and J. E. Kendall. "The Relationship of Organizational Subcultures to DSS User Satisfaction." *Human Systems Management,* March 1987, pp. 31–39.

Kulak, D., and E. Guiney. *Use Cases: Requirement in Context,* 2nd ed. Boston: Pearson Education, 2004.

Morris, M. G., and V. Venkatesh. "Job Characteristics and Job Satisfaction: Understanding the Role of Enterprise Resource Planning System Implementation." *MIS Quarterly,* Vol. 34, No. 1, 2010, pp. 143–161.

Warkentin, M., L. Sayeed, and R. Hightower. "Virtual Teams versus Face-to-Face Teams; An Exploratory Study of a Web-Based Conference System." In *Emerging Information Technologies: Improving Decisions, Cooperation, and Infrastructure.* Edited by K. E. Kendall, pp. 241–262. Thousand Oaks, CA: Sage Publications, 1999.

Yager, S. E. "Everything's Coming Up Virtual." www.acm.org/crossroads/xrds4-1/organ.html. Last accessed July 20, 2012.

The CPU Case Episode and accompanying student files are available online at www.pearsonglobaleditions.com/kendall.

CHAPTER 3

Project Management

LEARNING OBJECTIVES

Once you have mastered the material in this chapter you will be able to:

1. Understand how projects are initiated and selected, define a business problem, and determine the feasibility of a proposed project.

2. Evaluate hardware and software alternatives by addressing the trade-offs.

3. Forecast and analyze tangible and intangible costs and benefits.

4. Manage a project by preparing a budget, creating a work breakdown structure, scheduling activities, and controlling the schedule and costs.

5. Build and manage a project team.

6. Write an effective systems proposal, concentrating on both content and design.

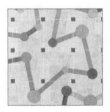

Initiating projects, determining project feasibility, scheduling projects, estimating costs, budgeting, and planning and then managing activities and team members for productivity are all important capabilities for a systems analyst to master. As such, they are considered project management fundamentals.

A systems project begins with problems or with opportunities for improvement in a business that come up as the organization adapts to change. The increasing popularity of ecommerce means that some fundamental changes are occurring as businesses either originate their enterprises on or move their internal operations as well as external relationships to the Internet. Changes that require a systems solution occur in the legal environment as well as in the industry's environment. A systems analyst works with users to create a problem definition that reflects current business systems and concerns. Once a project is suggested, the systems analyst works quickly with decision makers to determine whether the project is feasible. When a project is approved for a full systems study, the project activities are scheduled through the use of tools such as Gantt charts and Program Evaluation and Review Technique (PERT) diagrams so that the project can be completed on time. Part of ensuring the productivity of systems analysis team members is effectively managing their scheduled activities. This chapter is devoted to a discussion of project management fundamentals.

Project Initiation

Systems projects are initiated by many different sources for many reasons. Some of the projects suggested will survive various stages of evaluation to be worked on by you (or you and your team); others will not and should not get that far. Businesspeople suggest systems projects for two broad reasons: (1) because they experience problems that lend themselves to systems solutions and (2) because they recognize opportunities for improvement through upgrading, altering, or installing new systems when they occur. Both situations can arise as an organization adapts to and copes with natural, evolutionary change.

Problems in an Organization

Managers do not like to conceive of their organization as having problems, let alone talk about those problems or share them with someone from outside. Good managers, however, realize that recognizing symptoms of problems or, at a later stage, diagnosing the problems themselves and then confronting them are imperative if the business is to keep functioning at its highest potential.

Problems surface in many different ways. One way of conceptualizing what problems are and how they arise is to think of them as situations in which goals have never been met or are no longer being met. Useful feedback gives information about the gap between actual and intended performance. In this way, feedback spotlights problems.

In some instances, problems that require the services of systems analysts are uncovered because performance measures are not being met. Problems (or symptoms of problems) with processes that are visible in output and that could require the help of a systems analyst include excessive errors and work performed too slowly, incompletely, incorrectly, or not at all. Other symptoms of problems become evident when people do not meet baseline performance goals. Changes in employee behavior such as unusually high absenteeism, high job dissatisfaction, or high worker turnover should alert managers to potential problems. Any of these changes, alone or in combination, might be sufficient reason to request the help of a systems analyst.

Although difficulties such as those just described occur in an organization, feedback on how well the organization is meeting intended goals may come from outside, in the form of complaints or suggestions from customers, vendors, or suppliers, as well as lost or unexpectedly low sales. This feedback from the external environment is extremely important and should not be ignored.

A summary of symptoms of problems and approaches useful in problem detection is provided in Figure 3.1. Notice that checking output, observing or researching employee behavior, and listening to feedback from external sources are all valuable in problem finding. When reacting to accounts of problems in the organization, a systems analyst plays the roles of consultant, supporting expert, and agent of change, as discussed in Chapter 1. As you might expect, roles for the systems analyst shift subtly when projects are initiated because the focus is on opportunities for improvement rather than on the need to solve problems.

FIGURE 3.1

Checking output, observing employee behavior, and listening to feedback are all ways to help the analyst pinpoint systems problems and opportunities.

To Identify Problems	Look for These Specific Signs:
Check output against performance criteria.	• Too many errors • Work completed slowly • Work done incorrectly • Work done incompletely • Work not done at all
Observe behavior of employees.	• High absenteeism • High job dissatisfaction • High job turnover

Listen to external feedback from:
    Vendors and service providers
    Customers.
    Suppliers.
	• Complaints
	• Suggestions for improvement
	• Loss of sales
	• Lower sales

Defining the Problem

Whether using the classical SDLC or an object-oriented approach, an analyst first defines the problems and objectives of a system. These form the foundation of determining what needs to be accomplished by the system. Methods such as Six Sigma (see Chapter 16 for details) start with a problem definition.

A problem definition usually contains some sort of problem statement, summarized in a paragraph or two. This is followed by a series of issues or major independent pieces of the problem. The issues are followed by a series of objectives or goals that match the issues point by point. Issues are the current situation; objectives are the desired situation. The objectives may be very specific or worded using a general statement.

Here are some examples of business questions relating to business objectives:

- What are the purposes of the organization?
- Is the organization a for-profit or nonprofit organization?
- Does the company plan to grow or expand?
- What is the organization's attitude (culture) about technology?
- What is the organization's budget for IT?
- Does the organization's staff have the expertise?

Needless to say, a systems analyst needs to understand how a business works.

Finally, the problem definition contains requirements, the things that must be accomplished, along with the possible solutions and the constraints that limit the development of the system. The requirements section may include security, usability, government requirements, and so on. Constraints often include the word *not*, indicating a limitation, and may contain budget restrictions or time limitations.

The problem definition is produced after completing interviews, observations, and document analysis with the users. The result of gathering this information is a wealth of facts and important opinions in need of summary. The first step in producing the problem definition is to find a number of points that may be included in one issue. Major points can be identified in the interview in a number of ways:

1. Users may identify an issue, a topic, or a theme that is repeated several times, sometimes by different people in several interviews.
2. Users may communicate the same metaphors; such as saying the business is a journey, a war, a game, an organism, a machine, a family, a journey, and so on.
3. Users may tell a story to illustrate a problem that includes a beginning, middle, and an ending, a hero, obstacles to overcome, and a successful (or hoped for) resolution.
4. Users may speak at length on a topic.
5. Users may tell you outright "This is a major problem."
6. Users may communicate importance using body language or may speak emphatically on an issue.
7. The problem may be the first thing the user mentions.

Once the issues have been created, the objectives must be stated. An analyst may have to do a follow-up interview to obtain more precise information about the objectives. After the objectives are stated, the relative importance of the issues or objectives must be determined. If there are not enough funds to develop the complete system, the most critical objectives must be completed first. Users are the best people to identify critical objectives (with the support of analysts) because users are domain experts in their business area and they know how they work best with technologies in the organization.

One technique is to ask the users to assign a weight for each issue or objective in the first draft of the problem definition. This is a subjective judgment by the user, but, if a number of users all assign weights and they are averaged together, the result might reflect the bigger picture. After the weights have been determined, the problem definition issues and objectives are re-sequenced in order of decreasing importance, the most important issues listed first. Software such as that created by Expert Choice (www.expertchoice.com) and other decision support software can assist with weighting and prioritizing objectives.

Besides looking through data and interviewing people, a systems analyst should try to witness the problem firsthand. When looking at the same situation, an employee may view a problem very differently than a system analyst does. This also gives analysts the opportunity to confirm their findings. Using multiple methods strengthens the case for taking appropriate action.

CONSULTING OPPORTUNITY 3.1

The Sweetest Sound I've Ever Sipped

Felix Straw, who represents one of the many U.S. distributors of the European soft drink Sipps, gazes unhappily at a newspaper weather map, which is saturated with dark red, indicating that most of the United States is experiencing an early spring heat wave, with no signs of a letup. Pointing to the paper as he speaks, he tells your systems group, "It's the best thing that could happen to us, or at least it should be. But when we had to place our orders three months ago, we had no idea that this spring monster heat wave was going to devour the country this way!" Nodding his head toward a picture of their European plant hung on the wall, he continues, "We need to be able to tell them when things are hot over here so we can get enough product. Otherwise, we'll miss out every time. This happened two years ago, and it just about killed us.

"Each of us distributors meets with our district managers to do three-month planning. When we agree, we email our orders to European headquarters. They make their own adjustments, bottle the drinks, and then we get our modified orders about 9 to 15 weeks later. But we need ways to tell them what's going on now. Why, we even have some new superstores that are opening up here. They should know we have extra-high demand."

Corky, his assistant, agrees, saying, "Yeah, they should at least look at our past sales around this time of year. Some springs are hot, others are just average."

Straw concurs, saying, "It would be music to my ears, it would be really sweet, if they would work with us to spot trends and changes—and then respond quickly."

Stern's, based in Blackpool, England, is a European beverage maker and the developer and producer of Sipps. Sipps is a sweet, fruit-flavored, nonalcoholic, noncarbonated drink, which is served chilled or with ice, and it is particularly popular when the weather is hot. Sipps has sold briskly in Europe and grown in popularity in the United States since its introduction five years ago, but the company has had a difficult time adequately managing inventory and keeping up with U.S. customer demand, which is affected by seasonal temperature fluctuations. Places with year-round, warm-temperature climates and lots of tourists (such as Florida and California) have large standing orders, but other areas of the country could benefit from a less cumbersome, more responsive order-placing process. Sipps is distributed by a network of local distributors located throughout the United States and Canada.

As one of the systems analysts assigned to work with the U.S. distributors of Sipps, you begin your analysis by listing some of the key symptoms and problems you have identified after studying the information flows, ordering process, and inventory management, and after interviewing Mr. Straw and his assistant. In a paragraph, describe which problems might indicate the need for a systems solution.

Note: This consulting opportunity is loosely based on J. C. Perez, "Heineken's HOPS Software Keeps A-Head on Inventory," *PC Week*, Vol. 14, No. 2, January 13, 1997, pp. 31 and 34.

A PROBLEM DEFINITION EXAMPLE: CATHERINE'S CATERING. Catherine's Catering is a small business that caters meals, receptions, and banquets for business and social occasions such as luncheons and weddings. It was inspired by Catherine's love of cooking and talent for preparing fine meals. At first it was a small company with a handful of employees working on small projects. Catherine met with customers to determine the number of people, the types of meals, and other information necessary to cater an event. As her company's reputation for creating superb food and the quality of the service began to blossom, the number of events started to increase. The building of a new convention center and a prospering business community in the city increased the number of catering events.

Catherine was able to manage the business using spreadsheets and word processing but had difficulty keeping up with the endless phone calls about what types of meals were available, changes to the number of guests attending the event, and the availability of specialty dietary items, such as vegan, vegetarian, low-fat, low-carbohydrate, and so on. Catherine had hired a number of part-time employees to cook and cater the events, and the complexity of scheduling personnel was becoming overwhelming to the new human resources manager. Catherine decided to hire an IT and business consulting company to help her address the problems her catering enterprise was facing.

After performing interviews and observing a number of key staff, the consultants found the following concerns:

1. The master chef ordered supplies (produce, meat, and so on) from suppliers for each event. The suppliers would provide discounts if greater quantities were ordered at a single time for all events occurring in a given time frame.
2. Customers often called to change the number of guests for an event, with some changes made only one or two days before the event was to occur.

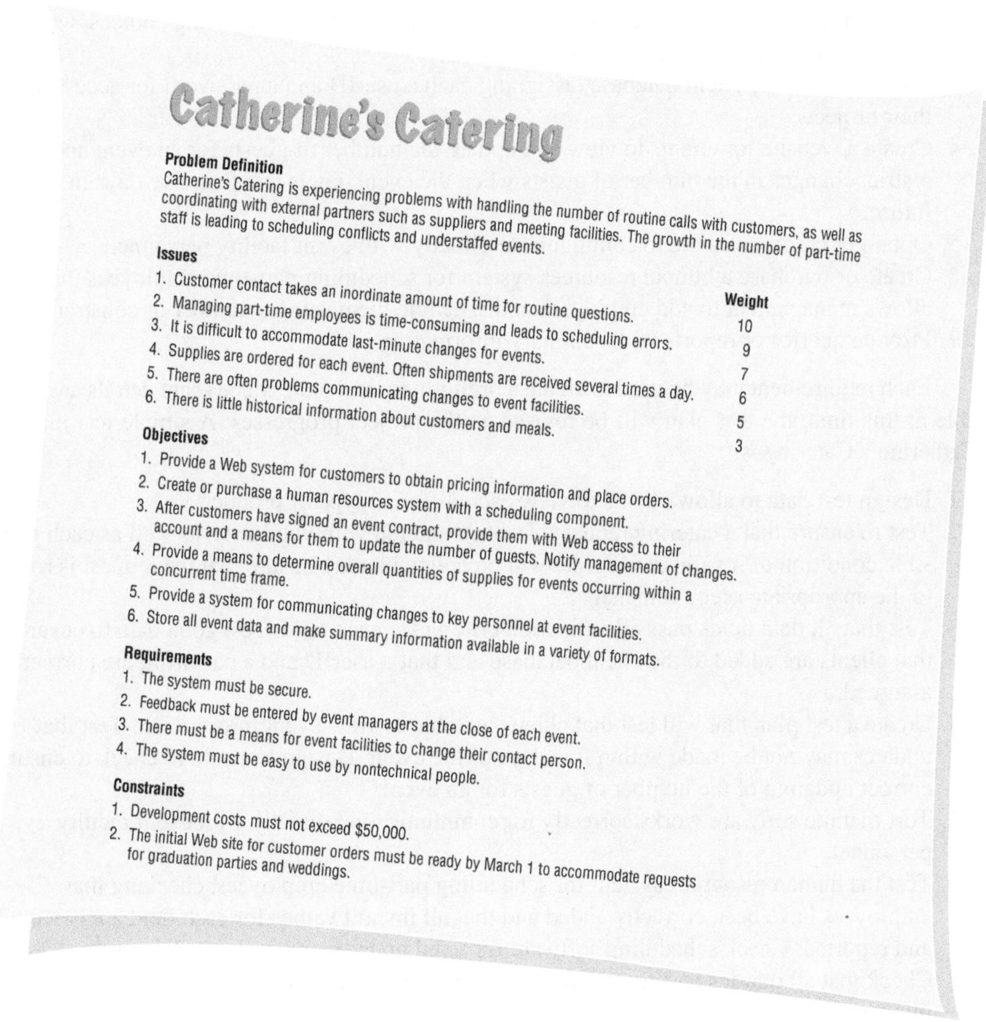

FIGURE 3.2

Problem definition for Catherine's Catering, developed with the help of users.

Catherine's Catering

Problem Definition

Catherine's Catering is experiencing problems with handling the number of routine calls with customers, as well as coordinating with external partners such as suppliers and meeting facilities. The growth in the number of part-time staff is leading to scheduling conflicts and understaffed events.

Issues

	Weight
1. Customer contact takes an inordinate amount of time for routine questions.	10
2. Managing part-time employees is time-consuming and leads to scheduling errors.	9
3. It is difficult to accommodate last-minute changes for events.	7
4. Supplies are ordered for each event. Often shipments are received several times a day.	6
5. There are often problems communicating changes to event facilities.	5
6. There is little historical information about customers and meals.	3

Objectives

1. Provide a Web system for customers to obtain pricing information and place orders.
2. Create or purchase a human resources system with a scheduling component.
3. After customers have signed an event contract, provide them with Web access to their account and a means for them to update the number of guests. Notify management of changes.
4. Provide a means to determine overall quantities of supplies for events occurring within a concurrent time frame.
5. Provide a system for communicating changes to key personnel at event facilities.
6. Store all event data and make summary information available in a variety of formats.

Requirements

1. The system must be secure.
2. Feedback must be entered by event managers at the close of each event.
3. There must be a means for event facilities to change their contact person.
4. The system must be easy to use by nontechnical people.

Constraints

1. Development costs must not exceed $50,000.
2. The initial Web site for customer orders must be ready by March 1 to accommodate requests for graduation parties and weddings.

3. It was too time-consuming for Catherine and her staff to handle each request for catering, with about 60 percent of the calls resulting in a contract.
4. Conflicts in employee schedules were occurring, and some events were understaffed. Complaints about the timeliness of service were becoming more frequent.
5. Catherine does not have any summary information about the number of events and types of meals. It would be helpful to have trend information to help guide customers in their choice of meals.
6. Events that include service for sit-down meals are often held at hotels and other meeting halls. There are problems with scheduling sufficient wait staff and responding to changes in the number of guests.

The problem definition is shown in Figure 3.2. Notice the weights on the right, representing an average of the weights assigned by each employee. Objectives match the issues. Each objective is used to create user requirements.

User requirements are then used to create either use cases and a use case diagram or data flow diagram processes. For each objective there may be one or more user requirements, or several objectives may lead to the creation of one or perhaps no use cases (use cases are not often created for simple reports), or each requirement may lead to the creation of one data flow diagram process. The user requirements for Catherine's Catering are to:

1. Create a dynamic website to allow current and potential clients to view and obtain pricing information for a variety of different products.

2. Allow current and potential clients to submit a request with their catering choices, with the request routed to an account manager.
3. Add clients to the client database, assigning each a userID and a password for access to their projects.
4. Create a website for clients to view and update the number of guests for an event and restrict changes in the number of guests when the event day is less than five days in the future.
5. Obtain or create software to communicate directly with event facility personnel.
6. Create or purchase a human resources system for scheduling part-time employees that allows management to add employees and schedule them using a number of constraints.
7. Provide queries or reports with summary information.

Each requirement may be used to create a preliminary test plan. Since scant details are available at this time, the test plan will be revised as the project progresses. A simple test plan for Catherine's Catering is:

1. Design test data to allow clients to view each different type of product.
2. Test to ensure that a catering request has been entered with valid data, as well as each possible condition of invalid data. (Data will be defined later.) Ensure that the request is routed to the appropriate account manager.
3. Test that all data fields pass all validation criteria for each field. Test good data to ensure that clients are added to the client database and that a userID and a password are correctly assigned.
4. Create a test plan that will test that clients are able to view event information. Test that updates may not be made within five days of the event. Design test data to check to ensure correct updating of the number of guests for an event.
5. Test that the software works correctly for communicating directly with event facility personnel.
6. Test the human resources system for scheduling part-time employees, checking that employees have been correctly added and that all invalid values for each field are detected and reported. Check scheduling software for valid updates and each invalid entry.
7. Check that all queries and reports work correctly and contain the correct summary information.

Selection of Projects

Projects come from many different sources and for many reasons. Not all should be selected for further study. You must be clear in your own mind about the reasons for recommending a systems study on a project that seems to address a problem or could bring about improvement. Consider the motivation that prompts a proposal on the project. You need to be sure that the project under consideration is not being proposed simply to enhance your own political reputation or power, or that of the person or group proposing it, because there is a high probability that such a project will be ill conceived and eventually ill accepted.

As outlined in Chapter 2, prospective projects need to be examined from a systems perspective, considering the impact of the proposed change on the entire organization. Recall that the various subsystems of an organization are interrelated and interdependent, so a change to one subsystem might affect all the others. Even though the decision makers directly involved ultimately set the boundaries for a systems project, a systems project cannot be contemplated or selected in isolation from the rest of the organization.

Beyond these general considerations are five specific questions that need to be asked regarding project selection:

1. Does it have backing from management?
2. Is the timing of the project commitment appropriate?
3. Is it possible to improve attainment of strategic organizational goals?
4. Is it practical in terms of resources for the systems analyst and organization?
5. Is it a worthwhile project compared with other ways the organization could invest resources?

First and foremost is backing from management. Absolutely nothing can be accomplished without the endorsement of the people who will eventually foot the bill. This statement does not mean that you lack influence in directing the project or that people other than management can't be included; however, management backing is essential.

Another important criterion for project selection is timing for you and the organization. Ask yourself and the others who are involved if the business is presently capable of making a time commitment for installation of new systems or improvement to existing ones. You must also be able to commit all or a portion of your time for the duration.

A third criterion is the possibility of improving attainment of strategic organizational goals such as (1) improving corporate profits, (2) supporting the competitive strategy of the organization, (3) improving cooperation with vendors and partners, (4) improving internal operations support so that goods and services are produced efficiently and effectively, (5) improving internal decision support so that decisions are more effective, (6) improving customer service, and (7) improving employee morale. The project should put the organization on target, not deter it from its ultimate goals.

A fourth criterion is selecting a project that is practicable in terms of your resources and capabilities as well as those of the business. Some projects will not fall within your realm of expertise, and you must be able to recognize them.

Finally, you need to come to a basic agreement with the organization about the worthiness of the systems project relative to any other possible project being considered. There are many possibilities for improvements, including (1) speeding up a process, (2) streamlining a process through the elimination of unnecessary or duplicated steps, (3) combining processes, (4) reducing errors in input through changes of forms and display screens, (5) reducing redundant storage, (6) reducing redundant output, and (7) improving integration of systems and subsystems. Remember that when a business commits to one project, it is committing resources that thereby become unavailable for other projects. It is useful to view all possible projects as competing for the business resources of time, money, and people.

Determining Feasibility

Once the number of projects has been narrowed according to the criteria discussed previously, it is still necessary to determine whether the selected projects are feasible. Our definition of *feasibility* goes much deeper than common usage of the term. Systems projects feasibility is assessed in three principal ways: operationally, technically, and economically. A feasibility study is not a full-blown systems study. Rather, a feasibility study is used to gather broad data for the members of management that enables them to make a decision about whether to proceed with a systems study.

Data for a feasibility study can be gathered through interviews, which are covered in detail in Chapter 4. The kind of interview required is directly related to the problem or opportunity being suggested. A systems analyst typically interviews those requesting help and those directly concerned with the decision-making process, typically management. Although it is important to address the correct problem, a systems analyst should not spend too much time doing feasibility studies because many projects will be requested, and only a few can or should be executed. A feasibility study must be highly time compressed, encompassing several activities in a short span of time.

Determining Whether It Is Possible

After an analyst determines reasonable objectives for a project, the analyst needs to determine whether it is possible for the organization and its members to see the project through to completion. Generally, the process of feasibility assessment is effective in screening out projects that are inconsistent with the business's objectives, that are technically impossible, or that have no economic merit.

Although it is painstaking, studying feasibility is worthwhile because it saves businesses and systems analysts time and money. In order for an analyst to recommend further development, a project must show that it is feasible in all three of the following ways: technically, economically, and operationally (see Figure 3.3).

TECHNICAL FEASIBILITY. An analyst must find out whether it is possible to develop a new system given the current technical resources. If not, can the system be upgraded or added to in a manner that fulfills the request under consideration? If existing systems cannot be added to or upgraded, the next question becomes whether there is technology in existence that meets the specifications.

FIGURE 3.3

The three key elements of feasibility are technical, economic, and operational feasibility.

The Three Key Elements of Feasibility

Technical Feasibility
    Add on to present system
    Technology available to meet users' needs

Economic Feasibility
    Systems analysts' time
    Cost of systems study
    Cost of employees' time for study
    Estimated cost of hardware
    Cost of packaged software or software development

Operational Feasibility
    Whether the system will operate when put in service
    Whether the system will be used

At the same time, the analyst can ask whether the organization has staff who are technically proficient to accomplish the objectives. If not, can the organization hire additional programmers, testers, experts, or others who may have different programming skills from theirs, or maybe outsource the project completely? In addition, are there software packages available that can accomplish the objectives, and how significantly does that software need to be customized for the organization?

ECONOMIC FEASIBILITY. Economic feasibility is the second part of resource determination. The basic resources to consider are your time and that of the systems analysis team, the cost of doing a full systems study (including the time of employees you will be working with), the cost of the business employee time, the estimated cost of hardware, and the estimated cost of software, software development, or software customization.

The organization must be able to see the value of the investment it is pondering before committing to an entire systems study. If short-term costs are not overshadowed by long-term gains or if the project produces no immediate reduction in operating costs, the system is not economically feasible and should not proceed further.

OPERATIONAL FEASIBILITY. Suppose for a moment that technical and economic resources are both judged adequate. The systems analyst must now consider the operational feasibility of the requested project. Operational feasibility is dependent on the human resources available for the project and involves projecting whether the system will operate and be used once it is put into service.

If users are virtually wed to the present system, see no problems with it, and generally are not involved in requesting a new system, resistance to implementing the new system will be strong. The new system has a low chance of becoming operational.

Alternatively, if users themselves have expressed a need for a system that is operational more of the time which is more efficient and more accessible, chances are better that the requested system will eventually be used. Much of the art of determining operational feasibility rests with the user interfaces that are chosen, as we see in Chapter 14.

Estimating Workloads

The next step in ascertaining hardware needs is to estimate workloads. Thus, systems analysts formulate numbers that represent both current and projected workloads for the system so that any hardware obtained will be capable of handling current and future workloads.

If estimates are accomplished properly, the business should not have to replace hardware solely due to unforeseen growth in system use. (Other events, however, such as superior technological innovations, may dictate hardware replacement if the business wants to maintain its competitive edge.)

Out of necessity, workloads are sampled rather than actually put through several computer systems. The guidelines given in Chapter 5 can be of use here because in workload sampling, the systems analyst is taking a sample of necessary tasks and the computer resources required to complete them.

Task	Existing System	Proposed System
	Compare performance of distribution warehouses by running the summary program.	Compare performance of distribution warehouses on the Web-based dashboard.
Method	Computer programs are run when needed; processing is done from the workstation.	Updates occur immediately; processing is done online.
Personnel	Distribution manager	Distribution manager
When and how	Daily: Enter shipments on Excel spreadsheet; verify accuracy of spreadsheet manually; and then write files to backup media. Monthly: Run program that summarizes daily records and prints report; get report and make evaluations.	Daily: Enter shipments on the Web-based system using drop-down boxes. Data are automatically backed up to remote location. Monthly: Compare warehouses online using the performance dashboard; print only if needed.
Human time requirements	Daily: 20 minutes Monthly: 30 minutes	Daily: 10 minutes Monthly: 10 minutes
Computer time requirements	Daily: 20 minutes Monthly: 30 minutes	Daily: 10 minutes Monthly: 10 minutes

FIGURE 3.4

Comparisons of workloads between existing and proposed systems.

Figure 3.4 is a comparison of the times required by existing and proposed information systems that are supposed to handle a given workload. Notice that the company is currently using a legacy computer system to prepare a summary of shipments to its distribution warehouses, and a Web-based dashboard is being suggested. The workload comparison looks at when and how each process is done, how much human time is required, and how much computer time is needed. Notice that the newly proposed system should reduce the required human and computer time significantly.

Ascertaining Hardware and Software Needs

Assessing technical feasibility includes evaluating the ability of computer hardware and software to handle workloads adequately. Figure 3.5 shows the steps a systems analyst takes in ascertaining hardware and software needs. First, all current computer hardware the organization owns must be inventoried to discover what is on hand and what is usable.

A systems analyst needs to work with users to determine what hardware will be needed. Hardware determinations can be made only in conjunction with determining human information requirements. Knowledge of the organizational structure and how users interact with technologies in an organizational setting can also be helpful in hardware decisions. Only when systems analysts, users, and management have a good grasp of what kinds of tasks must be accomplished can hardware options be considered.

FIGURE 3.5

Steps in choosing hardware and software.

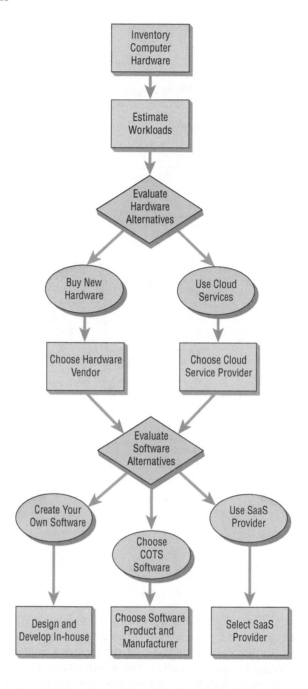

Inventorying Computer Hardware

A systems analyst needs to inventory what computer hardware is already available in the organization. As will become apparent, some of the hardware options involve expanding or recycling current hardware, so it is important to know what is on hand.

If an updated computer hardware inventory is unavailable, the systems analyst needs to set one up quickly and carry through on it. It is important to know the following:

1. The type of equipment, including model number and manufacturer
2. The operation status of the equipment, such as on order, operating, in storage, or in need of repair
3. The estimated age of the equipment
4. The projected life of the equipment
5. The physical location of the equipment
6. The department or person considered responsible for the equipment
7. The financial arrangement for the equipment, such as owned, leased, or rented

Ascertaining the current hardware available will result in a sounder decision-making process when hardware decisions are finally made because much of the guesswork about what exists will have been eliminated. Through your earlier interviews with users, questionnaires surveying them, and research of archival data, you will already know the number of people available for data processing as well as their skills and capabilities. Use this information to project how well the staffing needs for new hardware can be met.

Evaluating Computer Hardware for Purchase

Evaluating new computer hardware is the shared responsibility of management, users, and systems analysts. Although vendors will be supplying details about their particular offerings, analysts need to oversee the evaluation process personally because they will have the best interests of the business at heart. In addition, systems analysts may have to educate users and management about the general advantages and disadvantages of hardware before they can capably evaluate it.

Based on the current inventory of computer equipment and adequate estimates of current and forecasted workloads, the next step in the process is to consider the kinds of equipment available that appear to meet projected needs. Information from vendors on possible systems and system configurations becomes more pertinent at this stage and should be reviewed with management and users.

In addition, workloads can be simulated and run on different systems, including those already used in the organization. This process is referred to as *benchmarking*.

Criteria that the systems analysts and users should use to evaluate performance of different systems hardware include the following:

1. The time required for average transactions (including how long it takes to input data and how long it takes to receive output)
2. The total volume capacity of the system (how much can be processed at the same time before a problem arises)
3. The idle time of the CPU or network
4. The size of the memory provided

Some criteria will be shown in formal demonstrations; some cannot be simulated and must be gleaned from manufacturers' or service providers' specifications. It is important to be clear about the required and desired functions before getting too wrapped up in vendors' claims during demonstrations.

Once functional requirements are known and the products currently available are comprehended and compared with what already exists in the organization, the systems analysts in conjunction with users and management can make decisions about whether obtaining new hardware is necessary.

ADVANTAGES OF BUYING COMPUTER HARDWARE. If the organization purchases its own computers, it will have full control regarding the type of hardware and software. The company will decide when to purchase it and when to replace it. Often it is cheaper in the long run to buy computer equipment outright, but only if the organization buys the appropriate computer equipment. Buying computers often has tax advantages in the United States because of the depreciation rules.

DISADVANTAGES OF BUYING COMPUTER HARDWARE. Purchasing computers presents a problem because the initial cost of buying the equipment is often very high. Phasing in computer equipment over time doesn't always work because of compatibility problems that exist with different models and legacy devices, but buying everything at once means tying up capital or borrowing money. The risk of equipment obsolescence is a serious concern, and getting stuck with equipment that is not useful because someone made the wrong decision is a very large risk. Finally, the organization needs to remember that the full responsibility for the operation and maintenance stays with the company that buys the equipment. This can be a big disadvantage because it ties up people and money.

Renting Time and Space in the Cloud

Instead of purchasing their own equipment, companies can use cloud services. Some of the services available are Web hosting, email hosting, application hosting, backup, storage and

FIGURE 3.6

Comparing the advantages and disadvantages of buying computer resources versus using cloud services.

	Advantages	Disadvantages
Buying computer hardware	• Full control over hardware and software • Often cheaper in long run • Provides tax advantages through depreciation	• Initial cost is high • Risk of obsolescence • Risk of being stuck if choice was wrong • Full responsibility for operation and maintenance
Using cloud services	• Maintenance and upgrades performed by provider • Ability to change software and hardware rapidly • Scalable—can grow quickly • Consistent over multiple platforms • No capital is tied up	• Company doesn't control its own data • Data security is at risk • Reliability risks of the Internet platform • Proprietary APIs and software may make switching providers difficult

processing of databases, archiving, and ecommerce. The advantages and disadvantages to buying hardware versus using cloud services are shown in Figure 3.6.

It is not always easy to estimate what an organization's costs will be per month for using cloud services. However, you can access a cloud service provider such as Rackspace, Datapipe, or GoGrid and examine their pricing information. From Rackspace, for example, you can choose a server with various sizes of RAM and disk space. You will get a price quote of, for example, 24 cents an hour plus a small flat fee per month. GoGrid charges customers for three types of usage: RAM usage measured in RAM hours, transfer usage measured in gigabytes, and storage usage, also measured in gigabytes. For example, a 2GB RAM virtual machine running for one hour uses 2 RAM hours. An organization pays for the time used. It will not need to worry that expensive computers are sitting idle if it subscribes to cloud services.

Cloud services are scalable, which means if the business grows, the organization can easily add computing power without procuring new hardware. While most organizations do not save money by purchasing cloud services, they prize the agility afforded by being able to add or stop cloud services as needed.

Cloud services also allow organizations to handle big data easily. Big data are data sets that are too large to be analyzed through most conventionally sized organizational computing. Products such as Amazon Elastic MapReduce can be valuable in extracting useful information from massive amounts of data.

Three main categories of cloud computing are Software as a Service (SaaS), Infrastructure as a Service (IaaS), and Platform as a Service (PaaS). With SaaS, the cloud vendor sells access to applications (discussed later in this chapter, when we discuss software). With IaaS, also called Hardware as a Service (HaaS), an organization outsources its hardware operations needs to the service provider. Using the PaaS approach, companies use the Internet to rent all manner of hardware, operating systems, storage, and even network capacity.

Decisions on cloud computing can first be addressed on a strategic level. One author states that a business should (1) define a high-level business case that focuses on the high-level benefits of cloud computing for the organization; (2) define core requirements (perhaps making a quick list to start) that support the migration of IT to the cloud, including dimensions such as system performance, security desired in cloud relationships, IT governance, and predicted growth of the enterprise; and (3) define core technologies for the enterprise, starting with a list of cloud computing technologies that are helpful. As an analyst, you may be asked to facilitate any or all of these steps, or your help might be solicited at some point in the process of moving computing services to the cloud.

BENEFITS OF CLOUD COMPUTING. Some of the benefits of cloud computing include less time spent maintaining legacy systems or even performing routine tasks such as maintenance or upgrading of present systems. Using the cloud may make it simpler to acquire IT services and may also make it easier and quicker to separate from or discontinue services that are no longer necessary. Using cloud services makes your applications scalable, which means you can grow them easily by adding more cloud resources. Cloud computing also has the potential to offer consistency across multiple platforms that were previously disjointed or difficult to integrate. Finally, no capital is tied up, and no financing is required.

DRAWBACKS OF CLOUD COMPUTING. As with any other innovation, there are some disadvantages of adopting cloud computing for organizations. Loss of control of data stored in the cloud is perhaps the most striking concern of relying on cloud computing; if the cloud services provider ceased to exist, it is unclear what would happen to the organization's data. Looming almost as large are potential security threats to data that is not stored on premises or even on the organization's own computers. Reliability of the Internet as a platform forms the third main concern for organizations.

Furthermore, something that could dampen an organization's maneuverability when it comes to cloud computing is the prospect that the company could get stuck using application programming interfaces (APIs) that are proprietary with the cloud provider. This might make switching among cloud computing providers more difficult. It is interesting to note that as this text is being written, there are no standards for cloud interoperability, although several are in the works and have been submitted for testing. Some experts believe that companies such as Amazon, Google, and Salesforce are competing to make their cloud APIs standards for cloud service providers.

Evaluation of Vendor Support for Computer Hardware

Several key areas need to be evaluated when weighing the support services vendors offer to businesses. Most hardware vendors offer testing of hardware on delivery and a 90-day warranty covering any factory defects, but you must ascertain what else the vendor has to offer. Cloud service providers offer 30- or 60-day money-back guarantees. Vendors of comparable quality frequently distinguish themselves from others in the range of support services they offer.

A list of key criteria that ought to be checked when evaluating vendor support is provided in Figure 3.7. Support services include routine and preventive maintenance of hardware, specified response time (within six hours, next working day, etc.) in case of emergency equipment breakdowns, loan of equipment in the event that hardware must be permanently replaced or off-site repair is required, and in-house training or off-site group seminars for users. Cloud providers make claims for uptime and promises to get back to you on a question within a certain amount

Vendor Services	Specifics Vendors Typically Offer
Hardware Support	Full line of hardware Quality products Warranty
Software Support	Complete software needs Custom programming Warranty
Installation and Training	Commitment to schedule In-house training Technical assistance
Maintenance	Routine maintenance procedures Specified response time in emergencies Equipment loan while repair is being done
Cloud Services	Web hosting Email hosting Data storage
Software as a Service	Automatic software upgrades Support services Security and antivirus protection

FIGURE 3.7

Guidelines for vendor selection.

of time, such as 24 hours. Peruse the support services documents accompanying the purchase of equipment and remember to involve appropriate legal staff before signing contracts for equipment or services.

Unfortunately, the process of evaluating computer hardware and cloud providers is not as straightforward as simply comparing costs and choosing the least expensive option. Before becoming convinced that buying cheaper compatibles is the way to endow your system with add-on capability, you need to do enough research to feel confident that a hardware vendor or cloud provider is a stable corporate entity.

Understanding the Bring Your Own Device (BYOD) Option

Bring your own device (BYOD) or bring your own technology (BYOT) is a trend appearing in a variety of organizations in different industries. These organizations range in size and have employees of differing skills. In general, BYOD or BYOT means that rather than using a smartphone or tablet issued by the organization, the employee uses his or her own device to access corporate networks, data, and services remotely when working outside the office. The BYOD approach is often presented as a way to keep the organization's hardware costs down and improve morale among employees who, on average, already bring one or more mobile devices to work for personal tasks.

Loading up an employee's mobile device with work software can also extend the workplace into leisure hours. This continual connectivity was initially welcomed as a way to ensure that employees were responsive to the company all of the time, but organizations are now re-examining this 'round–the-clock approach. Now, organizations and unions often curtail the hours that an employee will be emailing or interacting with the corporate office so they are not conducting business during the valuable "off hours" they need to maintain balance between their work and personal lives.

As a systems analyst, you will work to discover through a combination of interviews, surveys, and observation what personal technology is being used by the majority of the group you are designing for. For example, if you observe that all the executives in the decision-making group already bring in iPads every day, you might conclude that designing dashboards for the iPad has some merit. They are devices you want to include in your original design rather than as an after-thought. If your observations show that most executives prefer face-to-face meetings, but they use their handheld devices (such as mobile phones) to check other corporate data and emails during meetings, then you might want to consider supporting corporate email on their personal phones.

BENEFITS AND DRAWBACKS OF BYOD. The benefits of BYOD computing include building employee morale, a potential for lowering the initial cost of organizational IT hardware purchase, facilitating remote, 'round–the-clock access to corporate computer networks regardless of location, and building on a familiar user interface to access corporate computing services, applications, databases, and storage. Security risks posed by untrained users are probably the biggest drawback. Identified threats are loss of the device itself, theft of the device and its data, unauthorized access to corporate networks using personal mobile devices, and so on. Other threats are posed by commonly occurring behavior with mobile devices that may be fine on the private side but not on the corporate side, such as using free Wi-Fi hotspots, using apps such as Dropbox, and others. While these and other behaviors may not meet high levels of corporate security, they are well within the realm of typical behavior for the use of one's own phone or tablet. Users might be lulled into thinking that their mobile device, since it is theirs, is no different once it is put in service to the business.

Creating Custom Software

Analysts and organizations are increasingly faced with a make, buy, or outsource decision when assessing software for information systems projects, particularly when contemplating upgrades to existing or legacy systems. Analysts have three choices: creating their own software, purchasing commercial off-the-shelf (COTS) software, or using software from a Software as a Service (SaaS) provider. Figure 3.8 summarizes the advantages and disadvantages of each of these options.

Several situations call for the creation of original, custom software or software components. The most likely instance is when COTS software does not exist or cannot be identified for the desired application. Alternatively, the software may exist but be unaffordable or not easily purchased or licensed.

	Advantages	Disadvantages
Creating Custom Software	• Specific response to specialized business needs • Innovation may give firm a competitive advantage • In-house staff available to maintain software • Pride of ownership	• May be significantly higher initial cost compared to COTS software or ASP • Necessity of hiring or working with a development team • Ongoing maintenance
Purchasing COTS Packages	• Refined in the commercial world • Increased reliability • Increased functionality • Often lower initial cost • Already in use by other firms • Help and training comes with software	• Programming focused; not business focused • Must live with the existing features • Limited customization • Uncertain financial future of vendor • Less ownership and commitment
Using SaaS	• Organizations that do not specialize in information systems can focus on what they do best (their strategic mission) • There is no need to hire, train, or retain a large IT staff • There is no expenditure of employee time on nonessential IT tasks	• Loss of control of data, systems, IT employees, and schedules • Concern over the financial viability and long-run stability of the SaaS provider • Security, confidentiality, and privacy concerns • Loss of potential strategic corporate advantage regarding innovativeness of applications

FIGURE 3.8

Comparing the advantages and disadvantages of creating custom software, purchasing COTS packages, and outsourcing to a SaaS provider.

Original software should be created when an organization is attempting to gain a competitive advantage through the leveraged use of information systems. This is often the case when an organization is creating ecommerce or other innovative applications where none have existed. An organization also may be a "first mover" in the use of a particular technology or in its particular industry. Organizations that have highly specialized requirements or exist in niche industries can also benefit from creating original software.

The advantages of creating custom software include being able to respond to specialized user and business needs, gaining a competitive advantage by creating innovative software, having in-house staff available to maintain the software, and pride of owning something you have created.

The drawbacks of developing custom software include the potential for a significantly higher initial cost compared to purchasing COTS software or contracting with a SaaS provider, the necessity of hiring or working with a development team, and the fact that you are responsible for the ongoing maintenance because you created the software.

Purchasing COTS Software

COTS software includes such products as the Microsoft Office suite, which includes Word for word processing, Excel for spreadsheets, Access for building databases, and other applications. Other types of COTS software are for organizational-level systems rather than office or personal use. Some authors include popular (but costly) ERP packages such as Oracle and SAP in their examples of COTS software. These packages differ radically in the amount of customization, support, and maintenance required compared to Microsoft Office. COTS software can also refer to software components or objects (also called building blocks) that can be purchased to provide a particular needed functionality in a system.

You should consider using COTS software when you can easily integrate the applications or packages into existing or planned systems and when you have identified no need to immediately or continuously change or customize systems for users. Your forecasts should demonstrate that

Veni, Vidi, Vendi, or, "I Came, I Saw, I Sold"

"It's really some choice. I mean, no single package seems to have everything we want. Some of them come darn close, though," says Roman, an advertising executive for *Empire Magazine* with whom you have been working on a systems project. Recently, the two of you have decided that packaged software would probably suit the advertising department's needs and stem its general decline.

"The last guy's demo we saw—you know, the one who worked for Data Coliseum—really had a well-rounded pitch. And I like their brochure. Full-color printing, on card stock. Classic," Roman asserts. "And what about those people from Vesta Systems? They're really fired up. And their package was easy to use with a minimum of ceremony. Besides, they said they would train all 12 of us, on-site, at no charge. But look at their advertising. They just take things off their printers."

Roman fiddles in his chair as he continues his ad hoc review of software and software vendors. "That one package from Mars, Inc., really sold me all on its own, though. I mean, it had a built-in calendar. And I like the way the menus for the screen displays could all be chosen by Roman numerals. It was easy to follow. And the vendor isn't going to be hard to move on price. I think they're already in a price war."

"Do you want to know my favorite, though?" Roman asks archly. "It's the one put out by Jupiter, Unlimited. I mean, it has everything, doesn't it? It costs a little extra coin, but it does what we need it to do, and the documentation is heavenly. They don't do any training, of course. They think they're above it."

You are already plotting that to answer Roman's burning questions by your March 15 deadline, you need to evaluate the software as well as the vendors, systematically, and then render a decision. Evaluate each vendor and package based on what Roman has said so far. (Assume that you can trust his opinions.) What are Roman's apparent biases when evaluating software and vendors? What further information do you need about each company and its software before you can make a selection? Set up a table to evaluate each vendor. Answer each question in a separate paragraph.

the organization you are designing the system for is unlikely to undergo major changes after the proposed purchase of COTS software, such as a dramatic increase in customers or large physical expansions.

There are some advantages to purchasing COTS software that you should keep in mind as you weigh alternatives. One advantage is that these products have been refined through the process of commercial use and distribution so that often there are additional functionalities offered. Another advantage is that packaged software is typically extensively tested and thus extremely reliable.

Increased functionality is often offered with COTS software because a commercial product is likely to have sister products, add-on features, and upgrades that enhance its attractiveness. Additionally, analysts often find that the initial cost of COTS software is lower than the cost for either in-house software development or the use of a SaaS provider.

Another advantage of purchasing COTS packages is that many other companies use them, so analysts are not experimenting on their clients with one-of-a-kind software applications. Finally, COTS software boasts an advantage in the help and training that accompanies the purchase of the packaged software.

One example of the use of COTS software is from a theatre company in the nonprofit sector, in which organizations (particularly in the performing arts) tend to lag behind their for-profit counterparts in adoption of information communication technologies (ICTs). The theatre company was predictably slow to move to the Web. When the company wanted to create ecommerce applications, it was put in a position of having to hire outside designers to create ecommerce applications for them. In light of the expense and lack of in-house expertise, many nonprofit organizations simply did not move the business portion of their organizations to the Web; instead, they waited for COTS packages, such as PC-based box office software or SaaS, such as online ticketing agencies with automation already in place, to make these services available to patrons. In-house, custom software development was out of the question for most of these groups, which typically have small or nonexistent IT staffs and budgets and minimal internal IT expertise.

There is a downside to the use of COTS software. Because it is not meant to be fully customizable, the theatre company lost its ability to change the software to include key features in its donor database that users relied on. COTS software may also include errors that could expose an organization to liability issues.

There are other disadvantages to consider with the purchase of COTS software, including the fact that packages are programmed rather than being focused on human users working in a business. Additionally, users must live with whatever features exist in the software, whether they are appropriate or not. A disadvantage that grows out of this is the limited customizability of most packaged software. Other disadvantages to purchasing COTS software include the necessity of investigating the financial stability of the software vendor and the diminished sense of ownership and commitment that is inevitable when the software is considered a product rather than a process.

To achieve some perspective on systems being developed, you should recognize that over half of all projects are built from scratch (two-thirds using traditional methods like SDLC and prototyping and one-third using agile or object-oriented technologies). Most of these are developed using an internal systems analysis team. Programmers may be in-house or outsourced.

Fewer than half of all projects are developed from existing applications or components. The great majority are modified, some extensively. Less than 5 percent of software is off-the-shelf software that requires no modifications at all.

Using the Services of a SaaS Provider

Organizations may realize some benefits from taking an entirely different approach to procuring software: outsourcing some of the organization's software needs to a SaaS provider that specializes in IT applications.

There are specific benefits to outsourcing applications to a SaaS provider. For example, organizations that want to retain their strategic focus and do what they're best at may want to outsource the production of information systems applications. Additionally, outsourcing software needs means that the organization doing the outsourcing may be able to sidestep the need to hire, train, and retain a large IT staff. This can result in significant savings. When an organization uses SaaS, there is little or no expenditure of valuable employee time on nonessential IT tasks; these are handled professionally by SaaS.

Hiring a SaaS provider should not be considered a magic formula for addressing software requirements. Some drawbacks to the use of SaaS must be seriously considered. One disadvantage is a general loss of control over corporate data, information systems, IT employees, and even processing and project schedules. Some companies believe that the heart of their business is their information, so even the thought of relinquishing control over it is distressing. Another disadvantage is concern over the financial viability of any SaaS that is chosen. There might also be concerns about the security of the organization's data and records, along with concern about confidentiality of data and client privacy. Finally, when choosing a SaaS provider, there is a potential loss of strategic corporate advantage that might have been gained through the company's deployment of its own innovative applications created by its employees.

Evaluation of Vendor Support for Software and SaaS

Whether you purchase a COTS package or contract for SaaS from a provider, you will be dealing with vendors who may have their own best interests at heart. You must be willing to evaluate software with users and not be unduly influenced by vendors' sales pitches. Specifically, there are six main categories on which to grade software, as shown in Figure 3.9: performance effectiveness, performance efficiency, ease of use, flexibility, quality of documentation, and manufacturer support.

Evaluate packaged software based on a demonstration with test data from the business considering it and an examination of accompanying documentation. Vendors' descriptions alone will not suffice. Vendors typically certify that software is working, but they do not guarantee that it will be error-free in every instance, that it will not crash when users take incorrect actions, or that it will be compatible with all other software the organization is currently running. Obviously, they will not guarantee their packaged software if it is used in conjunction with faulty hardware.

FIGURE 3.9

Guidelines for evaluating software.

Software Requirements	Specific Software Features
Performance Effectiveness	Able to perform all required tasks Able to perform all tasks desired Well-designed display screens Adequate capacity
Performance Efficiency	Fast response time Efficient input Efficient output Efficient storage of data Efficient backup
Ease of use	Satisfactory user interface Help menus available "Read Me" files for last-minute changes Flexible interface Adequate feedback Good error recovery
Flexibility	Options for input Options for output Usable with other software
Quality of Documentation	Good organization Adequate online tutorial Website with FAQ
Manufacturer Support	Technical support hotline Newsletter/email Website with downloadable product updates

Identifying, Forecasting, and Comparing Costs and Benefits

Costs and benefits of a proposed computer system must always be considered together because they are interrelated and often interdependent. Although a systems analyst tries to propose a system that fulfills various information requirements, decisions to continue with the proposed system will be based on a cost-benefit analysis, not on information requirements. In many ways, benefits are measured by costs, which becomes apparent in the next section.

Forecasting

Systems analysts are required to predict certain key variables before submitting a proposal to the client. To some degree, a systems analyst will rely on a what-if analysis, such as, "What if labor costs rise only 5 percent per year for the next three years, rather than 10 percent?" The systems analyst should realize, however, that he or she cannot rely on what-if analysis for everything if the proposal is to be credible, meaningful, and valuable.

A systems analyst has many forecasting models available. The main condition for choosing a model is the availability of historical data. If historical data are unavailable, the analyst must turn to one of the judgment methods: estimates from the sales force, surveys to estimate customer demand, Delphi studies (a consensus forecast developed independently by a group of experts through a series of iterations), creation of scenarios, or historical analogies.

If historical data are available, the next differentiation between classes of techniques involves whether the forecast is conditional or unconditional. Conditional implies that there is an association among variables in the model or that such a causal relationship exists. Common methods in this group include correlation, regression, leading indicators, econometrics, and input/output models.

Unconditional forecasting means the analyst isn't required to find or identify any causal relationships. Consequently, systems analysts find that these methods are low-cost, easy-to-implement

CONSULTING OPPORTUNITY 3.3

We're Off to See the Wizards

Elphaba I. Menzel and Glinda K. Chenoweth are the owners of Emerald City Beautyscapes, a commercial landscaping company. They are trying to decide whether to write their own software, perhaps using Microsoft Access as a basis; adopt a COTS software package such as QuickBooks Pro; or hire a service called Lawn Wizards, Inc., to perform all their bookkeeping functions.

Elphaba turns to Glinda and asks, "Is it possible for us to create a system of our own?"

Glinda replies, "I suppose we could, but it would take forever. We would need to define all our fields, our queries, and our reports. We would need to know who hasn't paid us yet and how long it has been since we last billed them."

"Yes," says Elphaba, "and we would also have to create product descriptions, service descriptions, and codes for everything we sell and provide."

"If that were all we needed, we could probably do it," says Glinda. "But we also need to include a scheduling system. We need to know when we can provide the services to our customers and what to do if we fall behind schedule. Maybe it just isn't worth it."

"Still," reflects Glinda, "my mother used to say 'There's no place like home.' Maybe there's no software like homegrown."

"You see both sides of everything," remarks Elphaba. "But the path you want to take is too long and risky. We need a software package that is ready for us to use now. I hear that there are products they call commercial off-the-shelf software that we can buy and adapt to our lawn service business. I'll investigate." So, Elphaba sets out to look for software that may be suitable.

"I've found something," cries Elphaba. "I found this software called QuickBooks Pro at www.quickbooks.intuit.com, and it looks like we can afford it. There are numerous versions of the software already—one for accounting, one for construction, one for health services. Maybe we can find a package that suits us. If not, it looks like we can customize the generic version of QuickBooks Pro to fit our needs.

"Our system could grow, too. QuickBooks Pro is readily scalable. We can add customers, suppliers, or products easily. I just wanted to plant the idea of buying a ready-made package on you."

"That's interesting," says Glinda, "but I've been doing my own research. Some of our competitors have told me they let a company do all the work for them. The company is called Lawn Wizards. They do landscaping, but they also maintain accounts receivable and scheduling packages."

So off they went to see the Wizards.

Joel Green, the owner and creator of Lawn Wizards, is proud of his software. "I spent a great deal of time working with my suppliers, that is, nurseries, in the area, and we have developed a coding system for everything," he brags. "All the trees, sizes of trees, shrubs, flowers, mulch, and even lawn care tools have numbers.

"I started with a small firm, but when customers realized I paid attention to every little detail, my business blossomed." He adds, "My suppliers love my system because it cuts down on confusion.

"I noticed that my competitors were working with the same suppliers but were getting less preferential treatment because they couldn't communicate about products very effectively. So I decided I would offer my software for hire. I would make money by renting out my software and demand even greater respect from my suppliers. I can even deliver it over the cloud. My end-user license agreement states that I own the software, product codes, and data generated by the system.

"Using my unique Wizards software, I can customize the package a bit for the customer, but essentially all the lawn services in the state will be using my database, codes, and B2B features. I maintain my software. If you could see the software code, it would look just like a manicured lawn."

Now Glinda and Elphaba are even more confused than before. They have three distinct options: create a custom package on their own, buy COTS software such as QuickBooks Pro, or outsource their needs to Lawn Wizards. Help them learn the true secret of (software) happiness by helping them articulate the pros and cons of each of their alternatives. What would you recommend? In two paragraphs, write a recommendation that grows out of your consideration of their specific business situation.

---

alternatives. Included in this group are graphical judgment, moving averages, and analysis of time-series data. Because these methods are simple, reliable, and cost-effective, the remainder of the section focuses on them.

ESTIMATION OF TRENDS. Trends can be estimated in a number of different ways. One way to estimate trends is to use a moving average. This method is useful because some seasonal, cyclical, or random patterns may be smoothed, leaving the trend pattern. The principle behind moving averages is to calculate the arithmetic mean of data from a fixed number of periods; a three-month moving average is simply the average of the past three months. For example, the average sales for January, February, and March are used to predict the sales for April. Then the average sales for February, March, and April are used to predict the sales for May, and so on.

When the results are graphed, it is easily noticeable that the widely fluctuating data are smoothed. The moving average method is useful for its smoothing ability, but it also has many disadvantages. Moving averages are more strongly affected by extreme values than by using graphical judgment or estimating by using other methods, such as least squares. An analyst should learn forecasting well, as it often provides information that is valuable in justifying an entire project.

Identifying Benefits and Costs

Benefits and costs can be either tangible or intangible. Both tangible and intangible benefits and costs must be taken into account when systems are considered.

TANGIBLE BENEFITS. Tangible benefits are advantages that are measurable in dollars that accrue to the organization through the use of the information system. Examples of tangible benefits are an increase in the speed of processing, access to otherwise inaccessible information, access to information on a more timely basis than was possible before, the advantage of the computer's superior calculating power, and a decrease in the amount of employee time needed to complete specific tasks. And there are other tangible benefits. Although measurement is not always easy, tangible benefits can actually be measured in terms of dollars, resources, or time saved.

INTANGIBLE BENEFITS. Some benefits that accrue to an organization from the use of an information system are difficult to measure but are important nonetheless. They are known as intangible benefits.

Intangible benefits include improving the decision-making process, enhancing accuracy, becoming more competitive in customer service, maintaining a good business image, and increasing job satisfaction for employees by eliminating tedious tasks. As you can see from this list, intangible benefits are extremely important and can have far-reaching implications for a business as it relates to people both outside and within the organization.

Although intangible benefits of an information system are important factors that must be considered when deciding whether to proceed with a system, a system built solely for its intangible benefits will not be successful. You must discuss both tangible and intangible benefits in your proposal because presenting both will allow decision makers in the business to make a well-informed decision about the proposed system.

TANGIBLE COSTS. The concepts of tangible and intangible costs present a conceptual parallel to the tangible and intangible benefits discussed already. Tangible costs are costs that a systems analyst and the business's accounting personnel can accurately project.

Included in tangible costs are the cost of equipment such as computers and terminals, the cost of resources, the cost of systems analysts' time, the cost of programmers' time, and other employees' salaries. These costs are usually well established or can be discovered quite easily; they are costs that will require the business to make a cash outlay.

INTANGIBLE COSTS. Intangible costs are difficult to estimate and may not be known. They include losing a competitive edge, losing the reputation for being first with an innovation or the leader in a field, declining company image due to increased customer dissatisfaction, and ineffective decision making due to untimely or inaccessible information. As you can imagine, it is nearly impossible to accurately project a dollar amount for intangible costs. To aid decision makers who want to weigh a proposed system and all its implications, you must include intangible costs even though they are not quantifiable.

Comparing Costs and Benefits

There are many well-known techniques for comparing the costs and benefits of a proposed system. They include break-even analysis, payback, cash-flow analysis, and present value analysis. All these techniques provide straightforward ways of yielding information to decision makers about the worthiness of a proposed system.

BREAK-EVEN ANALYSIS. By comparing costs alone, a systems analyst can use break-even analysis to determine the break-even capacity of a proposed information system. The point at which the total costs of the current system and the proposed system intersect represents the break-even point, the point where it becomes profitable for the business to get the new information system.

Total costs include the costs that recur during operation of a system plus the developmental costs that occur only once (one-time costs of installing a new system)—that is, the tangible costs

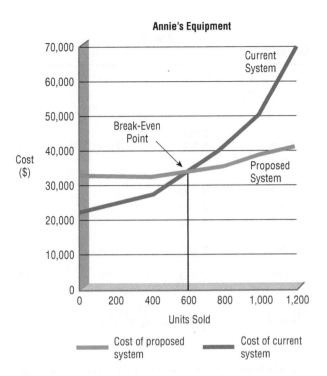

FIGURE 3.10

Break-even analysis for a
proposed inventory system.

that were just discussed. Figure 3.10 is an example of a break-even analysis on a small store that maintains inventory using a manual system. As volume rises, the costs of the manual system rise at an increasing rate. A new computer system would cost a substantial sum up front, but the incremental costs for higher volume would be rather small. The graph shows that the computer system would be cost-effective if the business sold about 600 units per week.

Break-even analysis is useful when a business is growing and volume is a key variable in costs. One disadvantage of break-even analysis is that benefits are assumed to remain the same, regardless of which system is in place. From our study of tangible and intangible benefits, we know that this is clearly not the case.

Break-even analysis can also determine how long it will take for the benefits of the system to pay back the costs of developing it. Figure 3.11 illustrates a system with a payback period of three and a half years.

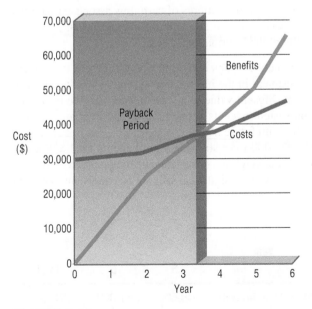

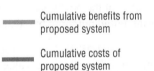

Cumulative benefits from
proposed system

Cumulative costs of
proposed system

Year	Cost ($)	Cumulative Costs ($)	Benefits ($)	Cumulative Benefits ($)
0	30,000	30,000	0	0
1	1,000	31,000	12,000	12,000
2	2,000	33,000	12,000	24,000
3	2,000	35,000	8,000	32,000
4	3,000	38,000	8,000	40,000
5	4,000	42,000	10,000	50,000
6	4,000	46,000	15,000	65,000

FIGURE 3.11

Break-even analysis showing a payback period of three and a half years.

FIGURE 3.12

Cash-flow analysis for a computerized mail-addressing system.

| | Year 1 | | | | Year 2 |
	Quarter 1	Quarter 2	Quarter 3	Quarter 4	Quarter 1
Revenue	$5,000	$20,000	$24,960	$31,270	$39,020
Costs					
Software development	10,000	5,000			
Personnel	8,000	8,400	8,800	9,260	9,700
Training	3,000	6,000			
Equipment lease	4,000	4,000	4,000	4,000	4,000
Supplies	1,000	2,000	2,370	2,990	3,730
Maintenance	0	2,000	2,200	2,420	2,660
Total Costs	26,000	27,400	17,370	18,670	20,090
Cash Flow	−21,000	−7,400	7,590	12,600	18,930
Cumulative Cash Flow	−21,000	−28,400	−20,810	−8,210	10,720

CASH-FLOW ANALYSIS. Cash-flow analysis examines the direction, size, and pattern of the cash flow associated with the proposed information system. If you are proposing the replacement of an old information system with a new one and if the new information system will not be generating any additional cash for the business, only cash outlays are associated with the project. In this case, the new system cannot be justified on the basis of new revenues generated and must be examined closely for other tangible benefits if it is to be pursued further.

Figure 3.12 shows a cash-flow analysis for a small company that is providing a mailing service to other small companies in the city. Revenue projections are that only $5,000 will be generated in the first quarter, but after the second quarter, revenue will grow at a steady rate. Costs will be large in the first two quarters and then level off. Cash-flow analysis is used to determine when a company will begin to make a profit (in this case, it is in the third quarter, with a cash flow of $7,590) and when it will be "out of the red"—that is, when revenue has made up for the initial investment (in this case, in the first quarter of the second year, when accumulated cash flow changes from a negative amount to a positive $10,720).

The proposed system should have increased revenues along with cash outlays. Then the size of the cash flow must be analyzed, along with the patterns of cash flow associated with the purchase of the new system. You must ask when cash outlays and revenues will occur, not only for the initial purchase but also over the life of the information system.

PRESENT VALUE ANALYSIS. Present value analysis helps a systems analyst present to business decision makers the time value of the investment in the information system as well as the cash flow (as discussed in the previous section). Present value is a way to assess all the economic outlays and revenues of the information system over its economic life and to compare costs today with future costs and today's benefits with future benefits.

In Figure 3.13, system costs total $272,000 over six years, and benefits total $280,700. Therefore, we might conclude that benefits outweigh the costs. Benefits only started to surpass costs after the fourth year, however, and dollars in the sixth year will not be equivalent to dollars in the first year.

FIGURE 3.13

Without considering present value, the benefits appear to outweigh the costs.

| | Year | | | | | | |
	1	2	3	4	5	6	Total
Costs	$40,000	42,000	44,100	46,300	48,600	51,000	272,000
Benefits	$25,000	31,200	39,000	48,700	60,800	76,000	280,700

	Year						
	1	**2**	**3**	**4**	**5**	**6**	**Total**
Costs	$40,000	42,000	44,100	46,300	48,600	51,000	
Multiplier	.89	.80	.71	.64	.57	.51	
Present Value of Costs	35,600	33,600	31,311	29,632	27,702	26,010	183,855
Benefits	$25,000	31,200	39,000	48,700	60,800	76,000	
Multiplier	.89	.80	.71	.64	.57	.51	
Present Value of Benefits	22,250	24,960	27,690	31,168	34,656	38,760	179,484

FIGURE 3.14

Taking into account present value, the conclusion is that the costs are greater than the benefits. In calculating the multipliers in this table, the discount rate, i, is assumed to be .12.

For instance, a $1 investment at 7 percent today will be worth $1.07 at the end of the year and will double in approximately 10 years. The present value, therefore, is the cost or benefit measured in today's dollars and depends on the cost of money. The cost of money is the opportunity cost, or the rate that could be obtained if the money invested in the proposed system were invested in another (relatively safe) project.

The present value of $1 at a discount rate of i is calculated by determining the factor:

$$\frac{1}{(1 + i)^n}$$

where n is the number of periods. Then the factor is multiplied by the dollar amount, yielding the present value, as shown in Figure 3.14. In this example, the cost of money—the discount rate—is assumed to be .12 (12 percent) for the entire planning horizon. Multipliers are calculated for each period: $n = 1, n = 2, \ldots, n = 6$. Present values of both costs and benefits are then calculated using these multipliers. When this step is done, the total benefits (measured in today's dollars) are $179,484 and are thus less than the costs (also measured in today's dollars). The conclusion to be drawn is that the proposed system is not worthwhile if present value is considered.

Although this example, which uses present value factors, is useful in explaining the concept, all electronic spreadsheets have a built-in present value function. An analyst can directly compute present value by using this function.

GUIDELINES FOR ANALYSIS. The use of the methods discussed in the preceding subsections depends on the methods employed and accepted in the organization. For general guidelines, however, it is safe to say the following:

1. Use break-even analysis if the project needs to be justified in terms of cost, not benefits, or if benefits do not substantially improve with the proposed system.
2. Use payback when the improved tangible benefits form a convincing argument for the proposed system.
3. Use cash-flow analysis when the project is expensive relative to the size of the company or when the business would be significantly affected by a large drain (even if temporary) on funds.
4. Use present value analysis when the payback period is long or when the cost of borrowing money is high.

Whichever method is chosen, it is important to remember that cost-benefit analysis should be approached systematically, in a way that can be explained and justified to managers, who will eventually decide whether to commit resources to the systems project. Next, we turn to the importance of comparing many systems alternatives.

Managing Time and Activities

The process of analysis and design can become unwieldy, especially when the system being developed is large. To keep the development activities as manageable as possible, you usually use some of the techniques of project management to help get organized.

The Work Breakdown Structure

Systems analysts are responsible for completing projects on time and within budget and for including the features promised. In order to accomplish all three of these goals, a project needs to be broken down into smaller tasks or activities. These tasks together make up a work breakdown structure (WBS).

When defined properly, the tasks that compose a work breakdown structure have special properties:

1. Each task or activity contains one deliverable, or tangible outcome, from the activity.
2. Each task can be assigned to a single individual or a single group.
3. Each task has a responsible person monitoring and controlling performance.

Activities in a work breakdown structure do not need to take the same amount of time or involve the same number of team members. The activities defined must add up to 100 percent of the work, however.

The main method for developing a WBS is decomposition, or starting with large ideas and then breaking them down into manageable activities. This subdivision of ideas into smaller ideas and eventually tasks stops when each task has only one deliverable.

There are different types of work breakdown structures. A WBS can be product oriented. In other words, building a website can be broken down into many parts, with each set of pages having a specific purpose. You could divide a website into its home page, product description pages, a FAQ page, a contact page, and an ecommerce page. Each of these pages could contain activities that you could use in your work breakdown structure.

Another way is to create a process-oriented work breakdown structure. An example of this is shown in Figure 3.15. This type of WBS is typical in systems analysis and design. In this example, we show the development of a website, but rather than show the development of each page, this example emphasizes the importance of each phase in the systems development life cycle.

Time Estimation Techniques

The process of analysis and design can become unwieldy, especially when the system being developed is large. To keep the development activities as manageable as possible, you can employ some of the techniques of project management to help get organized. In this section we discuss keeping a project on schedule by using time management techniques and keeping a project on budget by using cost management and control techniques.

One of the difficult tasks is estimating the time it takes to complete each of the tasks. There are numerous approaches:

1. Relying on experience
2. Using analogies
3. Using three-point estimation
4. Identifying function points
5. Using time estimation software

RELYING ON EXPERIENCE. Experience pays off when it comes to estimating how long an activity will take. If you have previous experience developing software, you will not only know how much time some tasks will likely take, you will also know how much time it will take if something goes wrong. Your experience gives you a most likely estimate as well as a pessimistic estimate.

USING ANALOGIES. If you don't have experience developing a particular piece of software but have worked on other types of projects of any kind, you still may be able to arrive at meaningful estimates. This approach involves identifying a project that is in some ways similar to the one you are about to begin and then describe the analogy. This means you will need to build two models, including two PERT or network diagrams, and compare their similarities. Then you can have confidence when you provide time estimates for the new project.

USING THREE-POINT ESTIMATION. Three-point estimation has been used for many years and is still a valid technique for time estimating. You start by developing three time estimates for

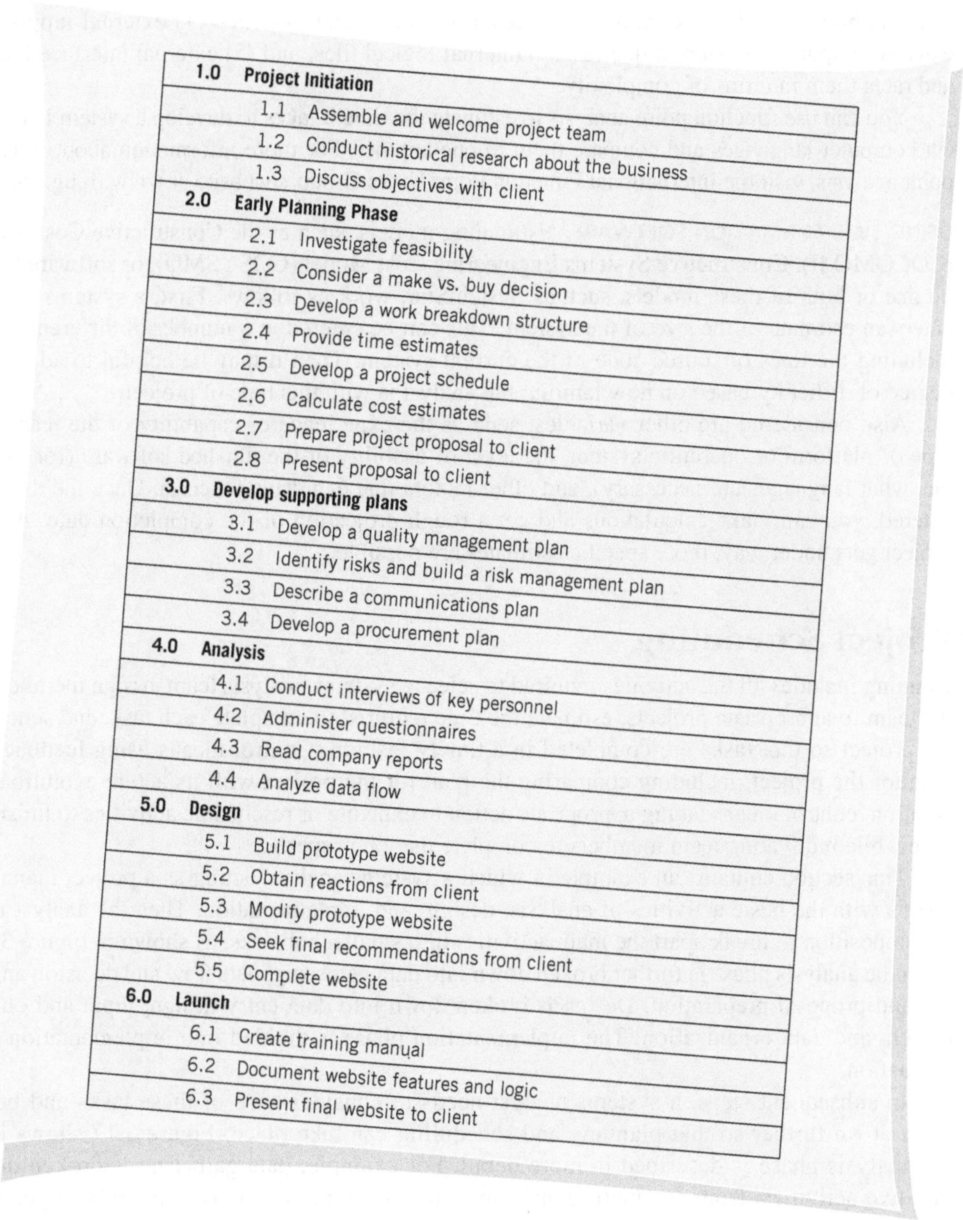

FIGURE 3.15

A sample work breakdown structure (WBS).

1.0	Project Initiation
1.1	Assemble and welcome project team
1.2	Conduct historical research about the business
1.3	Discuss objectives with client
2.0	**Early Planning Phase**
2.1	Investigate feasibility
2.2	Consider a make vs. buy decision
2.3	Develop a work breakdown structure
2.4	Provide time estimates
2.5	Develop a project schedule
2.6	Calculate cost estimates
2.7	Prepare project proposal to client
2.8	Present proposal to client
3.0	**Develop supporting plans**
3.1	Develop a quality management plan
3.2	Identify risks and build a risk management plan
3.3	Describe a communications plan
3.4	Develop a procurement plan
4.0	**Analysis**
4.1	Conduct interviews of key personnel
4.2	Administer questionnaires
4.3	Read company reports
4.4	Analyze data flow
5.0	**Design**
5.1	Build prototype website
5.2	Obtain reactions from client
5.3	Modify prototype website
5.4	Seek final recommendations from client
5.5	Complete website
6.0	**Launch**
6.1	Create training manual
6.2	Document website features and logic
6.3	Present final website to client

completing each task and then apply a simple formula to calculate a weighted average. The formula is:

$$E = (a + 4*m + b)/6$$

where a is the best-case estimate, b is the pessimistic or worst-case time estimate, and m is the most likely time estimate.

Typically the best-case scenario would improve on the time elapsed a little, but the pessimistic or worst-case estimate would imply that something disastrous happens, such as a weather or personnel delay. For example, writing a software module may take 10 days in most cases, but if the project is easier than expected, it might take only 8 days. If, however, the programmer accepts a job elsewhere, it may take 30 days. If you apply the formula, you get this estimate:

$$E = (8 + 4*10 + 30)/6$$
$$= 13 \text{ days}$$

In this case, you will estimate that it will take 13 days to finish the task.

USING FUNCTION POINT ANALYSIS. Another way of estimating the amount of work that needs to be done and how large a staff you need to complete a project is called function point analysis.

This method takes the five main components of a computer system—(1) external inputs, (2) external outputs, (3) external queries, (4) internal logical files, and (5) external interface files—and rates them in terms of complexity.

You can use function point analysis to estimate the time it takes to develop a system in different computer languages and compare them to one another. For more information about function point analysis, visit the International Function Point Users Group's website at www.ifpug.org.

USING TIME ESTIMATION SOFTWARE. Estimating models, such as the Constructive Cost Model (COCOMO II), Constructive Systems Engineering Cost Model (COSYSMO), or software based on one or both of these models, such as SystemStar, work as follows. First, a systems analyst enters an estimate of the size of the system. This can be entered in a number of different ways, including the lines of source code of the current system. Then it may be helpful to adjust the degree of difficulty based on how familiar the analyst is with this type of project.

Also considered are other variables, such as the experience or capability of the team, the type of platform or operating system, the level of usability of the finished software (for example, what languages are necessary), and other factors that can drive up costs. Once the data are entered, you can make calculations and get a rough projection of the completion date. As the project gets under way, more specific estimates are possible.

Project Scheduling

Planning includes all the activities required to select a systems analysis team, assign members of the team to appropriate projects, estimate the time required to complete each task, and schedule the project so that tasks are completed in a timely fashion. Control means using feedback to monitor the project, including comparing the plan for the project with its actual evolution. In addition, control means taking appropriate action to expedite or reschedule activities to finish on time while motivating team members to complete the job properly.

This section contains an example in which a systems analyst, acting as a project manager, begins with the basic activities of analysis, design, and implementation. Then the analyst uses decomposition to break apart the main activities into smaller subtasks, as shown in Figure 3.16. Then the analysis phase is further broken down into data gathering, data flow and decision analysis, and proposal preparation. Design is broken down into data entry design, input and output design, and data organization. The implementation phase is divided into implementation and evaluation.

In subsequent steps, a systems analyst needs to consider each of these tasks and break them down further so that planning and scheduling can take place. Figure 3.17 shows how the analysis phase is described in more detail. For example, data gathering is broken down into five activities, from conducting interviews to observing reactions to the prototype. This particular project requires data flow analysis but not decision analysis, so the systems analyst has written in "analyze data flow" as the single step in the middle phase. Finally, proposal preparation is broken down into three steps: perform cost-benefit analysis, prepare proposal, and present proposal.

FIGURE 3.16

Beginning to plan a project by breaking it into three major activities.

Phase	Activity
Analysis	Data gathering
	Data flow and decision analysis
	Proposal preparation
Design	Data entry design
	Input design
	Output design
	Data organization
Implementation	Implementation
	Evaluation

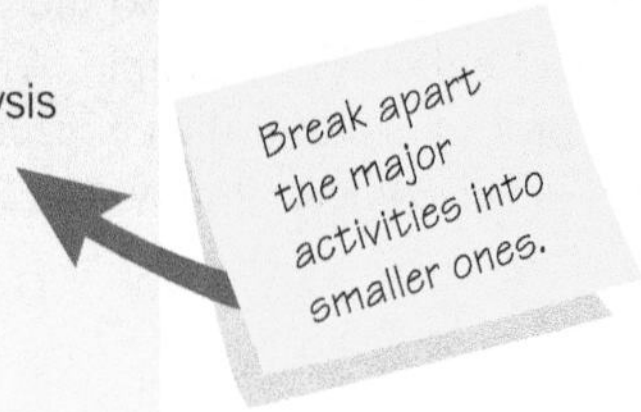

Break apart the major activities into smaller ones.

CONSULTING OPPORTUNITY 3.4

Food for Thought

"We could really make some changes. Shake up some people. Let them know we're with it. Technologically, I mean," said Malcolm Warner, vice president for AllFine Foods, a wholesale dairy products distributor. "That old system should be overhauled. I think we should just tell the staff that it's time to change."

"Yes, but what would we actually be improving?" Kim Han, assistant to the vice president, asks. "I mean, there aren't any substantial problems with the system input or output that I can see."

Malcolm snaps, "Kim, you're purposely not seeing my point. People out there see us as a stodgy firm. A new computer system could help change that. Change the look of our invoices. Send jazzier reports to the food store owners. Get some people excited about us as leaders in wholesale food distributing and computers."

"Well, from what I've seen over the years," Kim replies evenly, "a new system is very disruptive, even when the business really needs it. People dislike change, and if the system is performing the way it should, maybe there are other things we could do to update our image that wouldn't drive everyone nuts in the process. Besides, you're talking big bucks for a new gimmick."

Malcolm says, "I don't think just tossing it around here between the two of us is going to solve anything. Check on it and get back to me. Wouldn't it be wonderful?"

A week later, Kim enters Malcolm's office with several pages of interview notes in hand. "I've talked with most of the people who have extensive contact with the system. They're happy, Malcolm. And they're not just talking through their hats. They know what they're doing."

"I'm sure the managers would like to have a newer system than the guys at Quality Foods," Malcolm replies. "Did you talk to them?"

Kim says, "Yes. They're satisfied."

"And how about the people in systems? Did they say the technology to update our system is out there?" Malcolm inquires insistently.

"Yes. It can be done. That doesn't mean it should be," Kim says firmly.

As the systems analyst for AllFine Foods, how would you assess the feasibility of the systems project Malcolm is proposing? Based on what Kim has said about the managers, users, and systems people, what seems to be the operational feasibility of the proposed project? What about the economic feasibility? What about the technological feasibility? Based on what Kim and Malcolm have discussed, would you recommend that a full-blown systems study be done? Discuss your answer in a paragraph.

The systems analyst, of course, has the option to break down steps further. For instance, the analyst could specify each of the persons to be interviewed. The amount of detail necessary depends on the project, but all critical steps need to appear in the plans.

Sometimes the most difficult part of project planning is the crucial step of estimating the time it takes to complete each task or activity. When quizzed about reasons for lateness on a particular project, project team members cited poor scheduling estimates that hampered the success

Activity	Detailed Activity	Weeks Required
Data gathering	Conduct interviews	3
	Administer questionnaires	4
	Read company reports	4
	Introduce prototype	5
	Observe reactions to prototype	3
Data flow and decision analysis	Analyze data flow	8
Proposal preparation	Perform cost-benefit analysis	3
	Prepare proposal	2
	Present proposal	2

FIGURE 3.17

Refining the planning and scheduling of analysis activities by adding detailed tasks and establishing the time required to complete the tasks.

Break these down further, then estimate time required.

FIGURE 3.18

Using a two-dimensional Gantt chart for planning activities that can be accomplished in parallel.

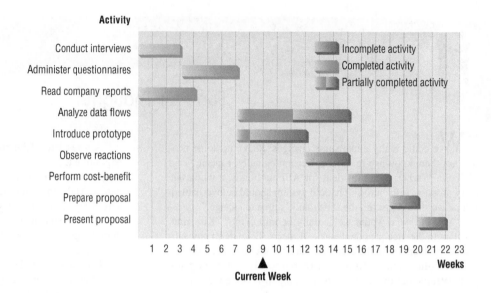

of projects from the outset. There is no substitute for experience in estimating time requirements, and systems analysts who have had the opportunity of an apprenticeship are fortunate in this regard.

Planners have attempted to reduce the inherent uncertainty in determining time estimates by projecting most likely, pessimistic, and optimistic estimates and then using a weighted average formula to determine the expected time an activity will take. This approach offers little more in the way of confidence, however. Perhaps the best strategy for a systems analyst is to adhere to a structured approach in identifying activities and describing these activities in sufficient detail. In this manner, the systems analyst will at least be able to limit unpleasant surprises.

Using Gantt Charts for Project Scheduling

A Gantt chart is a tool that enables you to easily schedule tasks. It is a chart on which bars represent tasks or activities. The length of each bar represents the relative length of the task.

Figure 3.18 is an example of a two-dimensional Gantt chart in which time is indicated on the horizontal dimension, and a description of activities makes up the vertical dimension. In this example, the Gantt chart shows the analysis or information-gathering phase of the project. Notice on the Gantt chart that conducting interviews will take three weeks, administering the questionnaire will take four weeks, and so on. These activities overlap part of the time. In the chart the special symbol ▲ signifies that right now, it is week 9. The bars with color shading represent projects or parts of projects that have been completed, telling us that the systems analyst is behind in introducing prototypes but ahead in analyzing data flows. Action must be taken on introducing prototypes soon so that other activities or even the project itself will not be delayed as a result.

The main advantage of a Gantt chart is its simplicity. Not only is this technique easy to use, but it also lends itself to worthwhile communication with end users. Another advantage of using a Gantt chart is that the bars representing activities or tasks are drawn to scale; that is, the size of the bars indicates the relative length of time it will take to complete each task.

Using PERT Diagrams

PERT is an acronym for Program Evaluation and Review Technique. A program (a synonym for a project) is represented by a network of nodes and arrows that are evaluated to determine the critical activities, improve the schedule if necessary, and review progress once the project is undertaken. PERT was developed in the late 1950s for use in the U.S. Navy's *Polaris* nuclear submarine project. It reportedly saved the U.S. Navy two years of development time.

PERT is useful when activities can be done in parallel rather than in sequence. A systems analyst can benefit from using PERT by applying it to systems projects on a smaller scale, especially when some team members can be working on certain activities at the same time that other team members are working on other tasks.

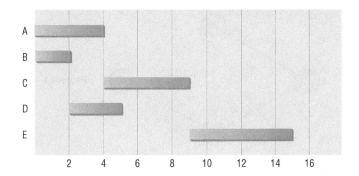

FIGURE 3.19

A Gantt chart compared with a PERT diagram for scheduling activities.

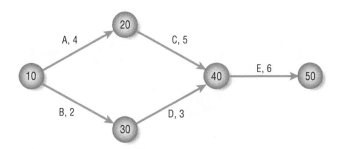

Figure 3.19 compares a simple Gantt chart with a PERT diagram. The activities expressed as bars in the Gantt chart are represented by arrows in the PERT diagram. The length of the arrows has no direct relationship with the activity durations. Circles on the PERT diagram are called events and can be identified by numbers, letters, or any other arbitrary form of designation. The circular nodes are present to (1) recognize that an activity is completed and (2) indicate which activities need to be completed before new activities may be undertaken (which is called *precedence*).

In the PERT example shown here, activity C may not be started until activity A is completed. Precedence is not indicated at all in the Gantt chart, so it is not possible to tell whether activity C is scheduled to start on day 4 on purpose or by coincidence.

A project has a beginning, a middle, and an end; in this example, the beginning is event 10, and the end is event 50. To find the length of the project, each path from beginning to end is identified, and the length of each path is calculated. In this example, path 10–20–40–50 has a length of 15 days, whereas path 10–30–40–50 has a length of 11 days. Even though one person may be working on path 10–20–40–50 and another on path 10–30–40–50, the project is not a race. The project requires that both sets of activities (or paths) be completed; consequently, the project takes 15 days to complete.

The longest path is referred to as the *critical path*. Although the critical path is determined by calculating the longest path, it is defined as the path that will cause the whole project to fall behind if even one day's delay is encountered on that path. Note that if you are delayed one day on path 10–20–40–50, the entire project will take longer, but if you are delayed one day on path 10–30–40–50, the entire project will not suffer. The leeway to fall behind somewhat on noncritical paths is called *slack time*.

Occasionally, PERT diagrams need pseudo-activities, referred to as dummy activities, to preserve the logic of or clarify the diagram. Figure 3.20 shows two PERT diagrams with dummies. Project 1 and project 2 are quite different, and the way the dummy is drawn makes the difference clear. In project 1, activity C can be started only if both A and B are finished because all arrows coming into a node must be completed before leaving the node. In project 2, however, activity C requires only activity B's completion and can therefore be under way while activity A is still taking place.

Project 1 takes 14 days to complete, whereas project 2 takes only 9 days. The dummy in project 1 is necessary, of course, because it indicates a crucial precedence relationship. The dummy in project 2, on the other hand, is not required, and activity A could have been drawn from 10 to 40, and event 20 may be eliminated completely.

FIGURE 3.20

Precedence of activities is
important in determining the
length of a project when using a
PERT diagram.

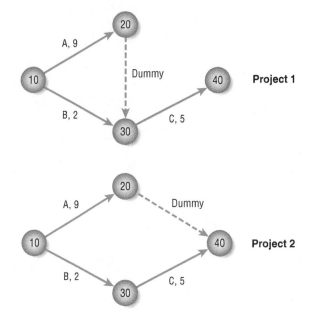

Therefore, there are many reasons for using a PERT diagram over a Gantt chart. The PERT diagram allows:

1. Easy identification of the order of precedence
2. Easy identification of the critical path and thus critical activities
3. Easy determination of slack time

A PERT EXAMPLE. Suppose a systems analyst is trying to set up a realistic schedule for the data-gathering and proposal phases of the systems analysis and design life cycle. The systems analyst looks over the situation and lists activities that need to be accomplished along the way. This list, which appears in Figure 3.21, also shows that some activities must precede other activities. The time estimates were determined as discussed in an earlier section of this chapter.

DRAWING THE PERT DIAGRAM. In constructing the PERT diagram for this example, the analyst looks first at the activities that require no predecessor activities—in this case A (conduct interviews) and C (read company reports). In the example in Figure 3.22, the analyst chose to number the nodes 10, 20, 30, and so on, and he or she drew two arrows out of the beginning node 10. These arrows represent activities A and C and are labeled as such. Nodes numbered 20 and 30 are drawn at the end of these respective arrows. The next step is to look for any activity that requires only A as a predecessor; task B (administer questionnaires) is the only one, so it can be represented by an arrow drawn from node 20 to node 30.

Because activities D (analyze data flow) and E (introduce prototype) require both activities B and C to be finished before they are started, arrows labeled D and E are drawn from node 30, the event that recognizes the completion of both B and C. This process is continued until the entire PERT diagram is completed. Notice that the entire project ends at an event called node 80.

FIGURE 3.21

Listing activities for use in
drawing a PERT diagram.

Activity		Predecessor	Duration
A	Conduct interviews	None	3
B	Administer questionnaires	A	4
C	Read company reports	None	4
D	Analyze data flow	B, C	8
E	Introduce prototype	B, C	5
F	Observe reactions to prototype	E	3
G	Perform cost-benefit analysis	D	3
H	Prepare proposal	F, G	2
I	Present proposal	H	2

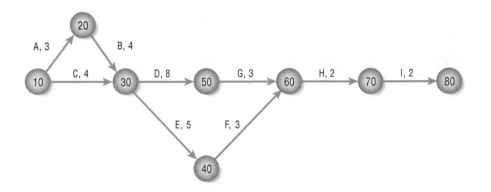

FIGURE 3.22

A completed PERT diagram for the analysis phase of a systems project.

IDENTIFYING THE CRITICAL PATH. Once the PERT diagram is drawn, it is possible to identify the critical path by calculating the sum of the activity times on each path and choosing the longest path. In this example, there are four paths: 10–20–30–50–60–70–80, 10–20–30–40–60–70–80, 10–30–50–60–70–80, and 10–30–40–60–70–80. The longest path is 10–20–30–50–60–70–80, which takes 22 days. It is essential that the systems analyst carefully monitor the activities on the critical path so as to keep the entire project on time or even shorten the project length, if warranted.

Controlling a Project

No matter how well a systems analyst plans a project, things can go wrong. In this section, we'll discuss how to estimate cost and prepare a budget, how to predict and prepare for risk, how to make up time, and how to react when the team falls behind or experiences cost overruns.

Estimating Costs and Preparing the Budget

Keeping a project on time is important, but it is also essential that the cost of a project be managed properly. Once a work breakdown structure is created and a schedule is planned, the analyst needs to:

1. Estimate costs for each activity in the work breakdown structure
2. Prepare a budget for the project and have it approved by the organization or client
3. Manage and control the costs throughout the project

Earlier we discussed cost estimates for equipment and off-the-shelf software. These are resources that we need to complete the project. Now we are concerned with other types of resources needed to complete each task in the work breakdown structure. The main resources for this part of the project are the time of the team members and the type of special equipment and tools needed to finish each of the activities.

Many approaches to cost estimation are available to a systems analyst. They are similar to the estimates used for time estimates. Some are:

1. Basing estimates on similar projects, also called the top-down approach
2. Building bottom-up estimates
3. Using parametric modeling

THE TOP-DOWN APPROACH TO COST ESTIMATION. Previous experience with estimating costs means a great deal, especially if the project you are attempting to estimate resembles a project you have previously worked on. If you've developed a website with similar features, then you can reliably estimate the costs of developing a new website.

Customization is possible. The new website may feature a different number of products for sale, but it is possible to adjust the costs according.

THE BOTTOM-UP APPROACH TO COST ESTIMATION. Often, an analyst is faced with a project that requires something unique, such as development in a different programming language. In this case, the analyst needs to use a bottom-up approach.

A systems analyst can take the work breakdown structure and ask each responsible project team member to estimate the cost involved with completing the activity he or she is responsible for. This method, however, yields estimates that are good or bad, depending on the abilities of

Color-coding helps a project manager sort out similar phases, tasks, and resources. OmniPlan, available for Macs, takes advantage of color-coding to set up a project, identify tasks, identify the critical path, and flag impossible situations.

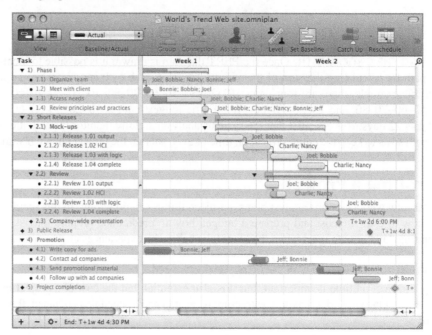

FIGURE 3.MAC

OmniPlan project management software from The Omni Group. (Screenshot from OmniPlan, a registered trademark of the Omni Group. Graphic reprinted with permission.)

each team member. The analyst, acting as a project manager, must review each of these estimates and arrive at a cost estimation that satisfies the team and client. The obvious problem with the bottom-up method is the time it takes to make each estimate along the way.

PARAMETRIC MODELING. This method involves making estimates for each of the many factors, or parameters, that make up a project. For example, you can estimate that it will cost $75 per line of code and $80 per hour for the programmers needed; you can then estimate the lines of code and hours it will take to complete the project. It is often useful to use special parametric modeling software such as COCOMO II, mentioned previously, to model the project.

WHY COST ESTIMATES FAIL. In practice, most systems analysts use a combination of all three of the cost-estimating methods just described. There are two main reasons cost estimates fail. First, an analyst is likely to be overly optimistic. An analyst believes in his or her team members, and an optimistic estimate will show the team completing the project quickly and without errors all the time. An analyst is likely to be optimistic and underestimate the lines of code and effort in general.

Second, an analyst may want to move past the cost estimation process and get on with preparing and presenting the budget and beginning the actual work on the project. The analyst may therefore spend less time on estimates than is needed to properly prepare the estimates. The analyst needs to be as accurate as possible, even though the estimates will be revised as the project continues.

PREPARING THE BUDGET. In the end, a systems analyst needs to prepare a budget even if the estimates are not perfect. The budget is a critical deliverable, and every client wants to see a detailed budget early in the process.

Items	Hours or Units	Cost per Hour or Unit	Subtotals
1. Project Team			
Project manager (systems analyst)	600	$120	$72,000
Project team members	2,400	80	192,000
Outside contractors (testers)	200	20	4,000
2. Hardware resources			
Workstations	5	4000	20,000
iPads	5	900	4,500
3. Software			
Off-the-shelf software	5	400	2,000
Software developed in-house		75,000	75,000
4. Training			
Seminars for team members	5	3,500	17,500
Seminars for trainees	5	1,200	6,000
Trainee hourly costs	1300	20	26,000
Total Project Cost Estimate			$419,000

FIGURE 3.23

Part of a sample budget for a software development project, showing hardware, software, and personnel costs.

When working for a client, a systems analyst needs to conform to the standard process the client uses to develop a budget. Most of the time, the client will have standard forms it uses for this purpose. A sample budget is shown in Figure 3.23.

The budget contains estimates for the hours worked and the rates of pay for each of the internal or outsourced workers. It also contains hardware and software costs and spells out how much of each type of equipment is needed. It also explains the costs involved with training.

Managing Risk

The early discussions you have with management and others requesting a project, along with the feasibility studies you do, are usually the best defenses possible against taking on projects that have a high probability of failure. Your training and experience will improve your ability to judge the worthiness of projects and the motivations that prompt others to request projects. If you are part of an in-house systems analysis team, you must keep current with the political climate of the organization as well as with financial and competitive situations.

It is important, however, to note that systems projects can and do have serious problems. Those that are developed using agile methods are not immune to such troubles. In order to illustrate what can go wrong in a project, a systems analyst may want to draw a fishbone diagram (also called a *cause-and-effect diagram*, or an *Ishikawa diagram*). When you examine Figure 3.24, you will see that it is called a *fishbone diagram* because it resembles the skeleton of a fish.

When using a fishbone diagram, you systematically list all the problems that can possibly occur. In the case of the agile approach, it is useful to organize a fishbone diagram by listing all the resource control variables on the top and all the activities on the bottom. Some problems, such as schedule slips, might be obvious, but others, such as scope creep (the desire to add features after the analyst hears new stories) or developing features with little value, are not as obvious.

You can also learn from the wisdom gained by people involved in earlier project failures. When asked to reflect on why projects had failed, professional programmers cited the setting of impossible or unrealistic dates for completion by management, belief in the myth that simply adding more people to a project would expedite it (even though the original target date on the project was unrealistic), and management behaving unreasonably by forbidding the team to seek professional expertise from outside the group to help solve specific problems.

Remember that you are not alone in the decision to begin a project. Although apprised of your team's recommendations, management will have the final say about whether a proposed project is worthy of further study (that is, further investment of resources). The decision process of your team must be open and stand up to scrutiny from those outside it. The team members should consider that their reputation and standing in the organization are inseparable from the projects they accept.

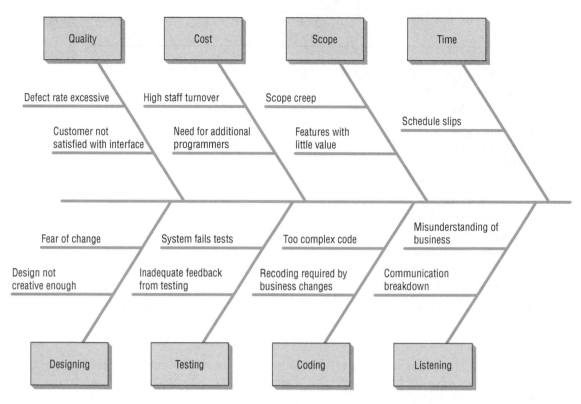

FIGURE 3.24

A fishbone diagram may be used to identify all the things that can go wrong in developing a system.

Managing Time Using Expediting

Speeding up a process is called *expediting*. There are certain instances in which you can benefit from completing a project much faster. One example is getting a bonus if you finish sooner. A second example is realizing that valuable project team resources and team members can be used for some other project if you finish the current project before the due date.

Figure 3.25 is a table that lists each activity in the project and the original estimated time required to complete each of these activities. The third column, labeled "crash time," refers to the absolute minimum time in which an activity can be completed if additional money is funneled to that activity. The final column contains the cost of reducing the activity duration by one week.

Expediting can help reduce the time it takes to complete an entire project, but in order to do so, the expedited activities have to be on the critical path. Which activity is expedited on the critical path depends on the cost, assuming that the activity is not already at its crash time.

FIGURE 3.25

A table showing crash time and the cost of expediting.

Activity	Estimated Duration	Crash Time	Cost/Week
A	3	1	$ 800
B	4	2	500
C	4	2	400
D	8	6	1000
E	5	5	1000
F	3	3	800
G	3	3	800
H	2	2	400
I	2	1	600

Eligible Activities	Activity Chosen	Time for Each Path ㉒ 19 19 16	Cost	Cumulative Cost
A, B, D, or I	B	㉑ 18 19 16	$ 500	$ 500
A, B, D, or I	B	⑳ 17 19 15	500	1000
A, D, or I	I	⑲ 17 18 15	600	1600
A or D	A	⑱ 16 ⑱ 15	800	2400
A and C, or D	D	⑰ 16 ⑰ 15	1000	3400
A and C, or D	D	⑯ ⑯ ⑯ 15	1000	4400
A and C	A and C	⑮ ⑮ ⑮ 14	1200	5600

The maximum number of weeks each activity can be reduced is the difference between the expected time and its crash time. For example, activity B, administering questionnaires, could be reduced from four weeks to two weeks at a cost of $500 per week, but it cannot be reduced less than 2 weeks; activity H, preparing the proposal, cannot be reduced because it is already at its crash time.

The expediting analysis for this example is provided in Figure 3.26. The expediting process takes place one step at a time, until it is impossible to expedite any further. The columns in the table include eligible activities (tasks that are on the critical path and can be reduced by expediting), the activity chosen (because it is the cheapest alternative), the time it currently takes to complete each of the paths, the cost of expediting the chosen activity, and finally the cumulative cost.

In the first step, the critical path is 10–20–30–50–60–70–80, so the eligible activities are A, B, D, and I. Activities G and H are also on the critical path, but they are already at their crash times and are consequently ineligible for expediting. The cheapest alternative is to expedite activity B by one day, which reduces the first path from 22 to 21 weeks and the second from 19 to 18 weeks; the third and fourth paths are not affected by the reduction since activity B is not on either of those paths.

The critical path, and therefore the entire project, is reduced from 22 to 21 weeks (circled on the table). We can repeat this reduction and reduce the project time by another week.

When activity B reaches its crash time, another activity must be chosen. Row 3 in the table shows that activities A, D, and I are eligible, and activity I is the cheapest alternative. Reducing activity I reduces not only the critical path but all paths because it is common to all of them.

In the fourth step, activity A is chosen, reducing paths 1 and 2, but as a result, there are now two critical paths. This implies that any reduction of the project time will take place only if both of the critical paths are reduced at the same time.

We can shorten both paths in the fifth and sixth steps by choosing either a combination of activities A and C (one activity from each of the critical paths) or activity D (an activity common to both critical paths). Reducing activity D by 2 days shortens the paths to 16, 16, 16, and 15 days, respectively, and now there are three critical paths.

Finally, when activity D reaches its crash time, the only available choice is a combination of activities A and C. The minimum project time is therefore 15 weeks, obtainable by reducing activity A by 2 days, activity B by 2 days, activity C by 1 day, activity D by 2 days, and activity I by one day, at a total cost of $5,600.

This example describes all-out expediting to obtain the minimum project time at any cost. But a systems analyst may be faced with a budget. In our example, a budget of $4,000 would result in expediting up to and including step 5. The project would be shortened from 22 to 17 weeks, at a cost of $3,400.

Another possible criterion would be the net amount that could be saved if the project were shortened. Suppose that in the above example, the analyst would save $750 per week, mostly consisting of the opportunities available for the project team to begin new projects sooner. In this case, expediting would take place until step 3, since the incremental cost of step 4 ($800 for expending activity A) would exceed the $750 saved.

Expediting can make or break a successful project. The systems analyst has to remain on top of the situation by managing the project throughout the entire development process. The analyst needs to make sure that the project costs are managed properly.

Controlling Costs Using Earned Value Management

Once a systems analyst has an approved budget, it is imperative that the analyst keeps it updated so the project doesn't exceed the budget as it progresses. The baseline needs to be continually revised, and all the stakeholders need to be informed.

Often changes occur in the middle of a project. The client may request new features, or new technologies may be introduced that will change the way the system is developed. No matter what these changes are, the budget needs to be revised.

One tool available to systems analysts is earned value management (EVM). It is a technique used to help determine progress (or setbacks) on a project and involves project cost, the project schedule, and the performance of the project team. The four key measures in earned value management are:

- *Budget at completion (BAC)* is the total budget for the project, from the beginning until completion. If you are calculating the performance measures for a task, then it is the total budget for the task.
- *Planned value (PV)* is the value of the work that is to be completed on the project (or, alternatively, the work completed on any task). Since the value of the work completed is how much effort and money will be put into it, you can think of planned value (PV) as the budgeted cost of work scheduled.
- *Actual Cost (AC)* is the total cost (direct and indirect) incurred in completing the work on the project (or, alternatively, a task) up to that particular point in time. Another way of referring to this is actual cost of work performed to date.
- *Earned value (EV)* is an estimate of the value of the work performed thus far. Earned value therefore refers to only the work that has been completed to date. We can calculate earned value (EV) as follows:

$$EV = PV * p$$

where p is the percentage of the work completed thus far.

Figure 3.27 shows the cost of developing a website over a five-month period. In this example, the development of our website is well under way. The budget at completion is the planned budget of $18,000, the total estimated cost of developing the website at the end of five months. It is calculated by summing the cost figures in the estimated cost column.

At the end of the first, second, and third months, the actual cost of the project equals the cumulative estimated cost, but in the fourth month, the actual cost is $17,000 versus the $15,000 from the cumulative estimate column. It is obvious that we are going over our budget.

It appears to be worse. We've only completed 50 percent of stage 4 at the end of the fourth month. That means we're falling behind in our schedule as the costs are rising.

Let's see what the damage is. We already know that the budget at completion for the project is $18,000 and that the actual cost at the end of the fourth month is $17,000. Let's determine the planned value and the earned value at this time.

At the end of	Stage	Estimated Cost	Cumulative Estimate	Estimated Duration	Stage Completed	Actual Cost of Stage to Date	Actual Cost of Project to date
Month 1	Stage 1	$6,000	$6,000	1 month	100%	$6,000	$6,000
Month 2	Stage 2	3,000	9,000	1 month	100%	3,000	9,000
Month 3	Stage 3	3,000	12,000	1 month	100%	3,000	12,000
Month 4	Stage 4	3,000	15,000	1 month	50%	5,000	17,000
Month 5	Stage 5	3,000	18,000	1 month	0%	Not yet begun	Not yet begun

FIGURE 3.27

The anticipated cost of a website development project over a five-month period.

Planned value (PV) is simply the cumulative estimate at the end of month 4, or $15,000. In order to calculate the planned value (PV) we need to multiply the earned value (EV) by the percentage of the project that is complete. We see in the table that the first three stages are complete, but the fourth stage is only half finished. To date, this means that:

$$p = (100 + 100 + 100 + 50)/(100 + 100 + 100 + 100)$$
$$= .875$$

So the earned value (EV) is:

$$EV = PV * p$$
$$= \$15,000 * .875$$
$$= \$13,125$$

The four variables—budget at completion (BAC), planned value (PV), actual cost (AC), and earned value (EV)—are the core of earned value management. They can be used to calculate a number of performance measures, including cost variance, schedule variance, the cost performance index, and the schedule performance index:

- *Cost variance (CV)* tells us whether the project or task is costing more than we have planned. If cost variance is negative, our costs are greater than planned. If it is positive, it is costing less than planned. Cost variance is simply:

$$CV = EV - AC$$

- *Schedule variance (SV)* tells us whether the project is taking more time than planned. If so, it will have a negative impact on our costs. Once again, if schedule variance is negative, then it is taking longer than planned to perform the work, and the amount is the cost overage. If schedule variance is positive, then it is taking less time than planned, and the amount indicates how much we are under budget. Schedule variance is obtained by subtracting the planned value from the earned value, as follows:

$$SV = EV - PV$$

- *Cost performance index (CPI)* is a ratio that signifies whether a project is over budget. If the index is less than 1.0, the project or task is over budget. If the ratio is greater than 1.0, the project or task is under budget. Cost performance index can be calculated as follows:

$$CPI = EV / AC$$

- *Schedule performance index (SPI)* is another ratio, this one used to tell whether a project is falling behind schedule. If the schedule performance index is less than 1.0, it implies that the project or task is behind schedule. If the schedule performance index ratio is greater than 1.0, the project is ahead of schedule. Schedule performance index (SPI) can be calculated as follows:

$$CPI = EV / PV$$

In our continuing example of website development, we can now calculate these four performance measures:

Cost variance at the end of the fourth month is:

$$CV = EV - AC$$
$$= \$13,125 - \$17,000$$
$$= (-\$3,875)$$

meaning that we are $3,875 over budget.

Schedule variance is:

$$SV = EV - PV$$
$$= \$13,125 - \$15,000$$
$$= (-\$1,875)$$

implying that we are behind schedule and are over budget.

The cost performance index is:

$$CPI = EV / AC$$
$$= \$13,125 / \$17,000$$
$$= .772$$

or 77.2 percent. This means the team is performing at a rate far below the expected rate of performance. This figure will be used later.

The schedule performance index (SPI) is:

$$SPI = EV / PV$$
$$= \$13,125 / \$15,000$$
$$= .875$$

or 87.5 percent. Once again, the ratio that is lower than 1.0 means that the project is behind schedule.

Finally, there are two key variables that a systems analyst may want to calculate. They tell the analyst how much it will cost to finish the project or task from this point onward. They are estimate to complete (ETC) and estimate at completion (EAC).

Estimate to complete considers how the team has done in the past and estimates how much additional money will be required to complete the project at its current rate of performance. It is calculated by subtracting the earned value (EV) from the budget at completion (BAC). That result is divided by the cost performance index (CPI) calculated earlier, as follows:

$$ETC = (BAC - EV) / CPI$$

Estimate at completion (EAC) represents the revised budget. It is the amount the entire project or task will cost when it is finished. It is calculated by adding the actual cost (AC) to the estimated amount of money it will take to complete the project (ETC). The equation is:

$$EAC = AC + ETC$$

In our ongoing example of the website development project, the estimate to complete the project is:

$$ETC = (\$18,000 - \$13,125)/.772 = \$6,315$$

This means that in order to finish the website, we will need to spend $6,315 more than we have up to this point. The estimate at completion will be:

$$EAC = \$17,000 + \$6,315$$
$$= \$23,315$$

which is significantly more than the $18,000 we originally budgeted for. Figure 3.28 shows how some of these performance measures relate to the original schedule and budget and the revised schedule and budget.

FIGURE 3.28

How the earned value management (EVM) values relate to the planned and actual development of the website.

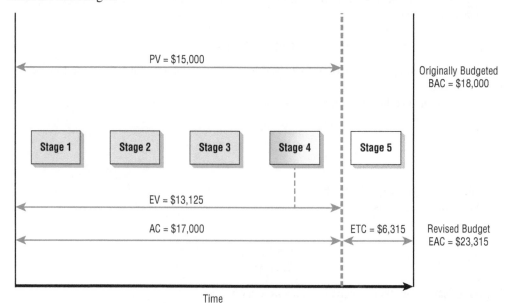

It is important for a systems analyst to keep an eye on cost and address the budgetary implications when unexpected delays or cost overruns occur. An analyst must constantly balance cost, time, and scope (that is, which systems features will be included) when working on any project.

Managing the Project Team

Along with managing time and resources, systems analysts must also manage people. Management is accomplished primarily by communicating accurately to team members who have been selected for their competency and compatibility. Goals for project productivity must be set, and members of systems analysis teams must be motivated to achieve them.

Assembling a Team

Assembling a team is desirable. If a project manager has an opportunity to create a dream team of skilled people to develop a system, who should he or she choose? In general, project managers need to look for others who share their values of teamwork, guided by the desire to deliver a high-quality system on time and on budget. Other desirable team member characteristics include a good work ethic, honesty, competency, readiness to take on leadership based on expertise, motivation, enthusiasm for the project, and trust of teammates.

The project manager needs to know about business principles, but it doesn't hurt to have at least one other person on the team who understands how a business operates. Perhaps this person should be a specialist in the area of the system being developed. When developing an ecommerce site, for example, a team can enlist the help of someone in marketing, and those developing an inventory system can ask a person versed in production and operations to provide expertise.

A team ideally should include two systems analysts. They can help each other, check each other's work, and shift their workloads. There is certainly a need to have people with programming skills on board. Coding is important, but people who know how to conduct walk-throughs, reviews, testing, and documenting systems are important as well. Some people are good at seeing the big picture, while others perform well when tasks are broken down into smaller ones for them. Every team should have both types of individuals.

Beyond the basics just described, a project manager should look for people with both experience and enthusiasm. Experience is especially important when trying to estimate the time required for completing a project. Experience in programming can mean code is developed five times faster than if it is developed by an inexperienced team. A usability expert is also a useful addition to the team.

The team must be motivated. One way to keep a team positively oriented throughout the entire process is to select good people at the outset. Look for enthusiasm, imagination, and an ability to communicate with different kinds of people. These basic attributes hold the potential for success. In addition, superior writers and articulate speakers can present proposals and work directly with customers, so hiring such individuals is a good idea.

Trust is an important part of a team. All members of the project need to act responsibly and agree to do their best and complete their part of the project. People may have different work styles, but they all need to agree to work together toward a common goal.

Communication Strategies for Managing Teams

Teams have their own personalities as a result of combining each individual team member with every other in a way that creates a totally new network of interactions. A way to organize your thinking about teams is to visualize them as always seeking a balance between accomplishing the work at hand and maintaining relationships among team members.

In fact, teams often have two leaders, not just one. Usually one person will emerge who leads members to accomplish tasks, and another person will emerge who is concerned with the social relationships among group members. Both are necessary for the team. These individuals have been labeled by other researchers as, respectively, task leader and socioemotional leader. Every team is subject to tensions that are an outgrowth of seeking a balance between accomplishing tasks and maintaining relationships among team members.

For a team to continue its effectiveness, tensions must be continually resolved. Minimizing or ignoring tensions will lead to ineffectiveness and eventual disintegration of the team. Much of

"I hope everyone you've encountered at MRE has treated you well. Here's a short review of some of the ways you can access our organization through HyperCase. The reception area at MRE contains the key links to the rest of our organization. Perhaps you've already discovered these on your own, but I wanted to remind you of them now because I don't want to get so engrossed in the rest of our organizational problems that I forget to mention them.

"The empty doorway you see is a link to the next room, which we call the East Atrium. You have probably noticed that all open doorways are links to adjacent rooms. Notice the building map displayed in the reception area. You are free to go to public areas such as the canteen, but as you know, you must have an employee escort you into a private office. You cannot go there on your own.

"By now you have probably noticed the two documents and the computer on the small table in the reception area. The little one is the MRE internal phone directory. Just click on an employee name, and if that person is in, he or she will grant you an interview and a tour of the office. I leave you to your own devices in figuring out what the other document is.

"The computer on the table is on and displays the web page for MRE. You should take a look at the corporate site and visit all the links. It tells the story of our company and the people who work here. We're quite proud of it and have gotten positive feedback about it from visitors.

"If you have had a chance to interview a few people and see how our company works, I'm sure you are becoming aware of some of the politics involved. We are also worried, though, about more technical issues, such as what constitutes feasibility for a training project and what does not."

HYPERCASE Questions

1. What criteria do the Training Unit use to judge the feasibility of a new project? List them.
2. List any changes or modifications to these criteria that you would recommend.
3. Snowden Evans has asked you to help prepare a proposal for a new project tracking system for the Training Unit. Briefly discuss the technical, economic, and operational feasibility of each alternative for a proposed project tracking system for the Training Unit.
4. Which option would you recommend? Use evidence from HyperCase to support your decision.

---

the tension release necessary can be gained through skillful use of feedback by all team members. All members, however, need to agree that the way they interact (that is, the process) is important enough to merit some time. Productivity goals for processes are discussed in a later section.

Securing agreement on appropriate member interaction involves creating explicit and implicit team norms (collective expectations, values, and ways of behaving) that guide members in their relationships. A team's norms belong to it and will not necessarily transfer from one team to another. These norms change over time and are better thought of as a team process of interaction than a product.

Norms can be functional or dysfunctional. Just because a particular behavior is a norm for a team does not mean it is helping the team to achieve its goals. For example, an expectation that junior team members should do all project scheduling may be a team norm. By adhering to this norm, the team puts extreme pressure on new members and does not take full advantage of the experience of the team. It is a norm that, if continued, could make team members waste precious resources.

Team members need to make norms explicit and periodically assess whether norms are functional or dysfunctional in helping the team achieve its goals. The overriding expectation for your team must be that change is the norm. Ask yourself whether team norms are helping or hindering the team's progress.

Setting Project Productivity Goals

As you work with your team members on a variety of projects, you or your team leader will acquire acumen for projecting what the team can achieve in a specific amount of time. Using the hints discussed in the earlier section in this chapter on methods for estimating time required and coupling them with experience will enable the team to set worthwhile productivity goals.

Systems analysts are accustomed to thinking about productivity goals for employees who show tangible outputs, such as the number of pairs of blue jeans sewn per hour, the number of entries keyed in per minute, or the number of items scanned per second. As manufacturing

CONSULTING OPPORTUNITY 3.5

Goal Tending

"Here's what I think we can accomplish in the next five weeks," says Hy, the leader of your systems analysis team, as he confidently pulls out a schedule listing each team member's name alongside a list of short-term goals. Just a week ago, your systems analysis team went through an intense meeting on expediting the project schedule for the Kitchener Redwings, an Ontario hockey organization whose management is pressuring you to produce a prototype.

The three other members of the team look at the chart in surprise. Finally, one of the members, Rip, speaks: "I'm in shock. We each have so much to do as it is—and now this."

Hy replies defensively, "We've got to aim high, Rip. They're in the off-season. It's the only time to get them. If we set our goals too low, we won't finish the system prototype, let alone the system itself, before another hockey season passes. The idea is to give the Kitchener Redwings the fighting edge through the use of their new system."

Fiona, another team member, enters the discussion, saying, "Goodness knows their players can't give them that!" She pauses for the customary groan from the assembled group, then continues.

"But seriously, these goals are killers. You could have at least asked us what we thought, Hy. We may even know better than you what's possible."

"This is a pressing problem, not a tea party, Fiona," Hy replies. "Polite polling of team members was out of the question. Something had to be done quickly. So I went ahead with these. I say we submit our schedule to management based on this. We can push back deadlines later if we have to. But this way, they'll know we're committed to accomplishing a lot during the off-season."

As a fourth team member listening to the foregoing exchange, formulate three suggestions that would help Hy improve his approach to goal formation and presentation. How well motivated do you think the team will be if they share Fiona's view of Hy's goals? What are the possible ramifications of supplying management with overly optimistic goals? Write one paragraph devoted to short-term effects and another one discussing the long-term effects of setting unrealistically high goals.

productivity rises, however, it is becoming clear that managerial productivity must keep pace. It is with this aim in mind that productivity goals for the systems analysis team are set.

A team needs to formulate and agree to goals, and these goals should be based on team members' expertise and former performance, as well as the nature of the specific project. Goals will vary somewhat for each project undertaken because sometimes an entire system will be installed, whereas other projects might involve limited modifications to a portion of an existing system.

Motivating Project Team Members

Although motivation is an extremely complex topic, it is a good one to consider, even if briefly, at this point. To oversimplify, recall that people join organizations to provide for some of their basic needs, such as food, clothing, and shelter. All humans, however, also have higher-level needs, including affiliation, control, independence, and creativity. People are motivated to fulfill unmet needs on several levels.

Team members can be motivated, at least partially, through participation in goal setting, as described in the previous section. The very act of setting a challenging but achievable goal and then periodically measuring performance against the goal seems to work in motivating people. Goals act almost as magnets in attracting people to achievement.

Part of the reason goal setting motivates people is that team members know prior to any performance review exactly what is expected of them. The success of goal setting for motivating can also be ascribed to it, affording each team member some autonomy in achieving the goals. Although a goal is predetermined, the means to achieve it may not be. In this instance team members are free to use their own expertise and experience to meet their goals.

Setting goals can also motivate team members by clarifying for them and others what must be done to get results. Team members are also motivated by goals because goals define the level of achievement that is expected of them. This use of goals simplifies the working atmosphere, and it also electrifies it with the possibility that what is expected can indeed be done.

Managing Ecommerce Projects

Many of the approaches and techniques discussed so far in this chapter are transferable to ecommerce project management. You should be cautioned, however, that although there are many similarities, there are also many differences. One difference is that the data used by ecommerce systems are scattered all over an organization. Therefore, you are not just managing data in a self-contained department or even one solitary unit. Hence, many organizational politics can come into play because units often feel protective of the data they generate and do not understand the need to share them across the organization.

Another stark difference is that ecommerce project teams typically need more staff with a variety of skills, including developers, consultants, database experts, and system integrators, from across the organization. Neatly defined, stable project groups that exist within a cohesive IS group or systems development team are the exception rather than the rule. In addition, because so much help may be required initially, ecommerce project managers need to build partnerships externally and internally well ahead of the implementation, perhaps sharing talent across projects to defray costs of ecommerce implementations and to muster the required numbers of people with the necessary expertise. The potential for organizational politics to drive a wedge between team members is very real.

One way to prevent politics from sabotaging a project is for the ecommerce project manager to emphasize the integration of the ecommerce system with the organization's internal systems and in so doing emphasize the organizational aspect embedded in the ecommerce project. As one ecommerce project manager told us, "Designing the front end [what the consumer sees] is the easy part of all this. The real challenge comes from integrating ecommerce strategically into all the organization's systems."

A fourth difference between traditional project management and ecommerce project management is that because the system will be linking with the outside world via the Internet, security is of the utmost importance. Developing and implementing a security plan before the new system is in place is a project in and of itself and must be managed as such.

Creating a Project Charter

Part of the planning process is to agree on what will be done and at what time. Analysts who are external consultants, as well as those who are organization members, need to specify what they will eventually deliver and when they will deliver it. The entire team as well as the client need to be on board. This chapter has elaborated on ways to estimate the delivery date for the completed system and also how to identify organizational goals and assess the feasibility of the proposed system.

The project charter is a written narrative that clarifies the following questions:

1. What does the user expect of the project (that is, what are the objectives)? What will the system do to meet the needs (achieve the objectives)?
2. What is the scope (or what are the boundaries) of the project? (What does the user consider to be beyond the project's reach?)
3. What analysis methods will the analyst use to interact with users in gathering data and developing and testing the system?
4. Who are the key participants? How much time are users willing and able to commit to participating?
5. What are the project deliverables? (What new or updated software, hardware, procedures, and documentation do the users expect to have available for interaction when the project is done?)
6. Who will evaluate the system, and how will they evaluate it? What are the steps in the assessment process? How will the results be communicated, and to whom?
7. What is the estimated project timeline? How often will analysts report project milestones?
8. Who will train the users?
9. Who will maintain the system?

The project charter describes in a written document the expected results of the systems project (deliverables) and the time frame for delivery. It essentially becomes a contract between the chief analyst (or project manager) and the analysis team, with the organizational users requesting the new system.

The Systems Proposal

While the project charter serves the purpose of identifying objects, determining scope, and assigning responsibilities, an analyst still needs to prepare a systems proposal that includes much of the detail about system needs, options, and recommendations. This section covers both the content and style that make up a systems proposal.

What to Include in a Systems Proposal

A written systems proposal has 10 main sections. Each section has a particular function, and eventually the proposal should be arranged in the following order:

1. Cover letter
2. Title page of project
3. Table of contents
4. Executive summary (including recommendations)
5. Outline of systems study with appropriate documentation
6. Detailed results of the systems study
7. Systems alternatives (three or four possible solutions)
8. Systems analysts' recommendations
9. Proposal summary
10. Appendices (assorted documentation, summary of phases, correspondence, and so on)

A cover letter to managers and the IT task force should accompany the systems proposal. It should list the people who did the study and summarize the objectives of the study. Keep the cover letter concise and friendly.

Include on the title page the name of the project, the names of the systems analysis team members, and the date the proposal is submitted. The proposal title must accurately express the content of the proposal, but it can also exhibit some imagination. The table of contents can be useful to readers of long proposals. If the proposal is less than 10 pages long, omit the table of contents.

The executive summary, in 250 to 375 words, provides the who, what, when, where, why, and how of the proposal, just as would the first paragraph in a news story. It should also include the recommendations of the systems analysts and desired management action because some people will only have time to read the summary. It should be written last, after the rest of the proposal is complete.

The outline of the systems study provides information about all the methods used in the study and who or what was studied. Any questionnaires, interviews, sampling of archival data, observation, or prototyping used in the systems study should be discussed in this section.

This detailed results section describes what the systems analyst has found out about human and systems needs through all the methods described in the preceding section. Conclusions about problems workers experience when interacting with technologies and systems that have come to the fore through the study should be noted here. This section should list the problems or suggest opportunities that call forth the alternatives presented in the next section.

In the systems alternatives section of the proposal, present two or three alternative solutions that directly address the aforementioned problems. The alternatives you present should include one that recommends keeping the system the same. Each alternative should be explored separately. Describe the costs and benefits of each situation. Because there are usually trade-offs involved in any solution, be sure to include the advantages and disadvantages of each.

Each alternative must clearly indicate what users and managers must do to implement it. The wording should be as clear as possible, such as, "Buy notebook computers for all middle managers," "Purchase packaged software to support users in managing inventory," or "Modify the existing system through funding in-house programming efforts."

After the systems analysis team has weighed the alternatives, it will have a definite professional opinion about which solution is most workable. The systems analysts' recommendations section expresses the *recommended* solution. Include the reasons supporting the team's recommendation so that it is easy to understand why it is being made. The recommendation should flow logically from the preceding analysis of alternative solutions, and it should clearly relate the human–computer interaction findings to the choice offered.

The proposal summary is a brief statement that mirrors the content of the executive summary. It gives the objectives of the study and the recommended solution. You should once more stress the project's importance and feasibility, along with the value of the recommendations for reaching the users' goals and improving the business. Conclude the proposal on a positive note.

The appendices are the last part of the systems proposal, and they can include any information that you feel may be of interest to specific individuals but that is not essential for understanding the systems study and what is being proposed.

Once the systems proposal is written, carefully select who should receive the report. Personally hand the report to the people you have selected. Your visibility is important for the acceptance and eventual success of the system.

You can propose a special meeting of key decision makers regarding the systems project for the purpose of delivering the results of the systems proposal. Develop a separate oral presentation (illustrated with PowerPoint slides or some other presentation software such as Keynote) that features highlights of your written report. Keep your presentation brief, from 30-40 minutes maximum, with the majority of time devoted to questions. Never read your report aloud. It is preferable to prepare and deliver a dynamic presentation that takes advantage of the fact that you can interact with key decision makers face to face.

Using Figures for Effective Communication

The emphasis so far in this section has been on considering your audience when composing a systems proposal. Tables and graphs as well as words are important in capturing and communicating the basics of the proposed system. Never underestimate good design.

Integrating figures into your proposal helps demonstrate that you are responsive to the different ways people absorb information. Figures in the report supplement written information and must always be interpreted in words; they should never stand alone.

EFFECTIVE USE OF TABLES. Although tables are technically not visual aids, they provide a different way of grouping and presenting analyzed data that an analyst wants to communicate to proposal readers.

Tables use labeled columns and rows to present statistical or alphabetical data in an organized way. Each table must be numbered according to the order in which it appears in the proposal and should be meaningfully titled. Figure 3.29 shows the appropriate layout and labeling for a table.

FIGURE 3.29

Guidelines for creating effective tables.

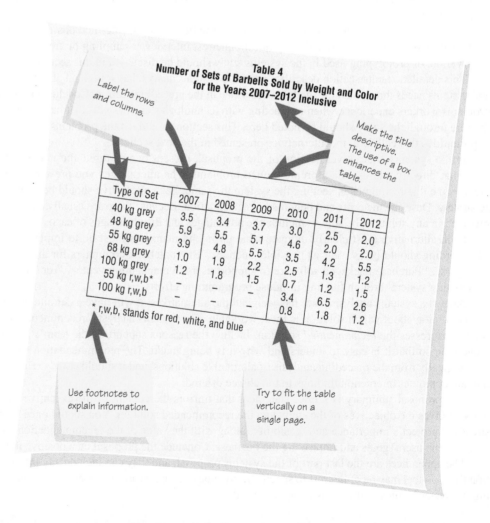

The following are some guidelines for tables:

1. Integrate tables into the body of the proposal. Don't relegate them to the appendices.
2. Try to fit an entire table vertically on a single page, if possible.
3. Number and title a table at the top of the page. Make the title descriptive and meaningful.
4. Label each row and column. Use more than one line for a title, if necessary.
5. Use a boxed table if room permits. Vertically ruled columns will enhance the readability.
6. Use footnotes if necessary to explain detailed information contained in the table.

Several methods for comparing costs and benefits were presented in previous sections. Tabled results of such comparisons should appear in the systems proposal. If a break-even analysis is done, a table illustrating results of the analysis should be included. Payback can be shown in tables that serve as additional support for graphs. A short table comparing computer systems or options might also be included in the systems proposal.

EFFECTIVE USE OF GRAPHS. There are many different kinds of graphs, including line graphs, column graphs, bar charts, pie charts, and area charts. Line graphs, column graphs, and bar charts compare variables, whereas pie charts and area charts illustrate the percentage composition of an entity.

The following guidelines will help you include effective graphs in a proposal (see Figure 3.30):

1. Choose a style of graph that communicates your intended meaning well.
2. Integrate the graph into the body of the proposal.
3. Give the graph a sequential figure number and a meaningful title.
4. Label each axis and any lines, columns, bars, or pieces of the pie on the graph.
5. Include a key to indicate differently colored lines, shaded bars, or crosshatched areas.

Much of the detail that goes into a systems proposal is obtained through interviews, questionnaires, sampling, discovery of other hard data, and observation. These topics are discussed in the next two chapters.

FIGURE 3.30

Guidelines for developing effective line graphs.

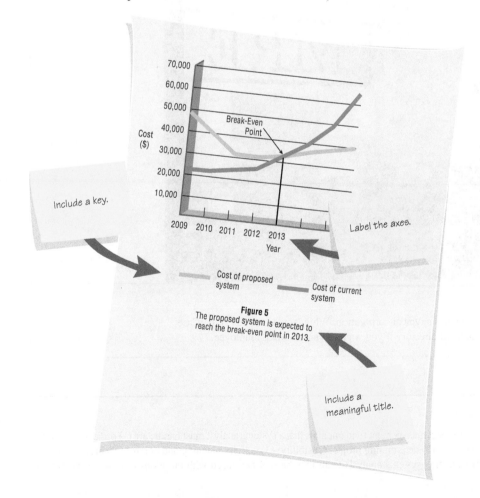

HYPERCASE® EXPERIENCE 3.2

"**S**ometimes the people who have been here for some time are surprised at how much we have actually grown. Yes, I do admit that it isn't easy to keep track of what each person is up to or even what purchases each department has made in the way of hardware and software. We're working on it, though. Snowden would like to see more accountability for computer purchases. He wants to make sure we know what we have, where it is, why we have it, who's using it, and if it's boosting MRE productivity, or, as he so delicately puts it, 'to see whether it's just an expensive toy' that we can live without."

HYPERCASE Questions

1. Complete a computer equipment inventory for the Training and Management Systems Department, describing all the systems you find. *Hint:* Create an inventory form to simplify your task.

2. Using the software evaluation guidelines given in the chapter, do a brief evaluation of GEMS, a software package used by the Training and Management Systems Department employees. In a paragraph, briefly critique this custom software by comparing it with COTS software such as Microsoft Project.

3. List the intangible costs and benefits of GEMS, as reported by employees of MRE.

4. Briefly describe the two alternatives Snowden is considering for the proposed project tracking and reporting system.

5. What organizational and political factors should Snowden consider in proposing his new system at MRE? (In a brief paragraph, discuss three central conflicts.)

FIGURE 3.HC1

The reception room resembles a typical corporation. While you are in this HyperCase screen, click on the small red directory if you want to visit an MRE employee.

Summary

The five major project management fundamentals that a systems analyst must handle are (1) project initiation—defining a problem, (2) determining project feasibility, (3) activity planning and control, (4) project scheduling, and (5) managing systems analysis team members. When faced with questions of how businesses can meet their goals and solve systems problems, an analyst creates a problem definition. A problem definition is a formal statement of the problem, including (1) the issues of the present situation, (2) the objectives for each issue, (3) the requirements that must be included in all proposed systems, and (4) the constraints that limit system development.

Selecting a project is a difficult decision because more projects will be requested than can actually be done. Five important criteria for project selection are (1) that the requested project be backed by management, (2) that it be timed appropriately for a commitment of resources, (3) that it move the business toward attainment of its goals, (4) that it be practical, and (5) that it be important enough to be considered over other possible projects.

If a requested project meets these criteria, a feasibility study of its operational, technical, and economic merits can be done. Through the feasibility study, a systems analyst gathers data to enable management to decide whether to proceed with a full systems study. By inventorying equipment already on hand and on order, systems analysts can better determine whether new, modified, or current computer hardware is to be recommended.

There are advantages and disadvantages to purchasing computer hardware. Alternatively the analyst may help the organization weigh the advantages and disadvantages of renting time and space through use of a cloud service provider. Three main categories of cloud computing are Software as a Service (SaaS), Infrastructure as a Service (IaaS), and Platform as a Service (PaaS). As user sophistication grows, some companies are adopting bring your own device (BYOD) approaches to boost morale of employees and lower the initial cost to the organization of purchasing mobile devices. Security risks from untrained users are likely the biggest threat to organizations adopting a "bring your own device" policy.

Cloud services and other hardware vendors will supply support services such as preventive maintenance and user training that are typically negotiated separately. Software can be created as a custom product, purchased as a commercial off-the-shelf (COTS) software package, or outsourced to a Software as a Service (SaaS) provider.

Preparing a systems proposal means identifying all the costs and benefits of a number of alternatives. A systems analyst has a number of methods available to forecast future costs, benefits, volumes of transactions, and economic variables that affect costs and benefits. Costs and benefits can be tangible (quantifiable) or intangible (nonquantifiable and resistant to direct comparison). A systems analyst can use many methods for analyzing costs and benefits, including break-even analysis, the payback method, and cash-flow analysis.

Project planning includes the estimation of time required for each of the analyst's activities, scheduling the activities, and expediting them if necessary to ensure that a project is completed on time.

In order to complete projects on time, within budget, and with the features that are promised, a project needs to be broken down into smaller tasks or activities. These tasks together make up a work breakdown structure (WBS), usually through decomposition. There are different types of work breakdown structures; they can be product oriented or process oriented. A process-oriented work breakdown structure is typical in systems analysis and design.

One technique available to a systems analyst for scheduling tasks is a Gantt chart, which displays activities as bars on a graph. Another technique, called Program Evaluation and Review Technique (PERT), displays activities as arrows on a network. PERT helps an analyst determine the critical path and slack time, which is the information required for effective project control.

Earned value management (EVM) is a technique used to help the analyst determine progress (or setbacks) on a project and involves project cost, the project schedule, and the performance of the project team. The four key measures in earned value management are: budget at completion (BAC), planned value (PV), actual cost (AC), and earned value (EV). These four measures or variables are the core of earned value management (EVM). They can be used to calculate a number of performance measures, including cost variance (CV), schedule variance (SV), the cost performance index (CPI), and the schedule performance index (SPI).

Creating a project charter that contains user expectations and analyst deliverables is recommended since programmers cited unrealistic management deadlines, adding unneeded personnel to a project that is trying to meet an unrealistic deadline, and not permitting developer teams to seek expert help outside their immediate group as reasons projects had failed. Project failures can usually be avoided by examining the motivations for requested projects, as well as a team's motives for recommending or avoiding a particular project.

A systems analyst must follow these steps when putting together an effective systems proposal: effectively organizing the proposal content, and writing and delivering the proposal in an appropriate business style. To be effective, a proposal should be written in a clear and understandable manner, and its content should be divided into 10 functional sections. Visual considerations are important when putting together a proposal. Face-to face meetings with key decision makers provide a way to orally present the systems proposal in a brief and compelling presentation.

Keywords and Phrases

Actual Cost (AC)
benchmarking
break-even analysis
bring your own data (BYOD)
bring your own technology (BYOT)
budget
Budget at Completion (BAC)
cash-flow analysis
cloud computing
Cost Performance Index (CPI)
Cost Variance (CV)
critical path
Earned Value (EV)
earned value management (EVM)
ecommerce project management
economic feasibility
Estimate at Completion (EAC)
Estimate to Complete (ETC)
expediting
forecasting
function point analysis
Gantt chart
Infrastructure as a Service (IaaS)
Hardware as a Service (HaaS)
intangible benefits
intangible costs

moving average
operational feasibility
payback
Planned Value (PV)
Program Evaluation and Review Technique (PERT)
 diagram
present value
problem definition
productivity goals
project charter
Platform as a Service (PaaS)
Schedule Performance Index (SPI)
Schedule Variance (SV)
socioemotional leader
Software as a Service (SaaS)
systems proposal
tangible benefits
tangible costs
task leader
team motivation
team norms
team process
technical feasibility
vendor support
work breakdown structure (WBS)

Review Questions

1. What are the five major project fundamentals?
2. List three ways to find out about problems or opportunities that might call for a systems solution.
3. List the five criteria for systems project selection.
4. Define *technical feasibility*.
5. Define *economic feasibility*.
6. Define *operational feasibility*.
7. List four criteria for evaluating system hardware.
8. What two main options do organizations have for acquiring or using computer hardware?
9. What does COTS stand for?
10. List five of the many benefits of cloud computing for organizations.
11. List three of the many drawbacks of cloud computing for organizations.
12. What does BYOD stand for?
13. What are the benefits of BYOD to an organization?
14. What are the benefits of BYOD to an employee?
15. What is the biggest drawback of BYOD for an organization?
16. What are three main categories of cloud computing?
17. Define *tangible costs* and *tangible benefits*. Give an example of each one.
18. Define *intangible costs* and *intangible benefits*. Give an example of each one.
19. List four techniques for comparing the costs and benefits of a proposed system.
20. When is break-even analysis useful?
21. What are the three drawbacks of using the payback method?
22. When is cash-flow analysis used?
23. As a general guideline, when should present value analysis be used?
24. What is work breakdown structure (WBS), and when should it be used?
25. What is a Gantt chart?
26. When is a PERT diagram useful for systems projects?
27. List three advantages of using a PERT diagram over a Gantt chart for scheduling systems projects.
28. Define the term *critical path*.

29. How does a project manager assess the risk of things going wrong and take that into consideration when planning the time needed to complete a project?
30. What costs need to be estimated to prepare a budget?
31. Why is preparing a budget so important for a systems analyst who is managing a project?
32. What three instances call for expediting a systems project?
33. What does crash time mean when expediting a project?
34. What is earned value management (EVM)?
35. What are the four key measures in earned value management?
36. For what purposes can a systems analyst use earned value management?
37. List the two types of team leaders.
38. What is meant by a dysfunctional team norm?
39. What is meant by team process?
40. What are three reasons that goal setting seems to motivate systems analysis team members?
41. What are four ways in which ecommerce project management differs from traditional project management?
42. What elements are contained in a project charter?
43. What is a fishbone diagram used for?
44. What are the three steps a systems analyst must follow to put together an effective systems proposal?
45. List the 10 main sections of a systems proposal.

Problems

1. Williwonk's Chocolates of St. Louis makes an assortment of chocolate candy and candy novelties. The company has six in-city stores, five stores in major metropolitan airports, and a small mail-order branch. Williwonk's has a small, computerized information system that tracks inventory in its plant, helps schedule production, and so on, but this system is not tied directly into any of its retail outlets. The mail-order system is handled manually.

    Recently, several Williwonk's stores experienced a rash of complaints from mail-order customers that the candy was spoiled upon arrival, that it did not arrive when promised, or that it never arrived; the company also received several letters complaining that candy in various airports tasted stale. Williwonk's has been selling a new, low-carb, dietetic form of chocolate made with sugar-free artificial sweetener. Sales have been brisk, but there have been problems shipping the wrong type of chocolate to an address for a diabetic person. There were a number of complaints, and Williwonk's sent a number of free boxes of chocolate to ease the situation.

    Management would like to sell products using the Web but only has a few web pages with information about the company and an order form that can be printed. Web ordering does not exist. One of the senior executives would like to sell customized chocolates with the name of a person on each piece. Although the production area has assured management that this could be easily done, there is no method to order customized chocolates.

    Another senior executive has mentioned that Williwonk's has partnered with several European chocolate manufacturers and will be importing chocolate from a variety of countries. At present, this must be done over the phone, by email, or by mail. The executive wants an internal website that will enable employees to order directly from the partner companies. All this has led a number of managers to request trend analysis. Too much inventory results in stale chocolate, while at other times, there is a shortage of a certain kind of chocolate.

    Seasonal and holiday variation trends would help Williwonk's maintain an adequate inventory. The inventory control manager has insisted that all changes must be implemented before the next holiday season. "The time for this to be complete is an absolute due date," remarked Candy, a senior manager. "Make sure that everything works perfectly before the site goes public," she continues. "I don't want any customers receiving the wrong chocolates!" In addition, the order processing manager has mentioned that the system must be secure.

    You have been working for two weeks with Williwonk's on some minor modifications for its inventory information system when you overhear two managers discussing these occurrences. List the possible opportunities or problems among them that might lend themselves to systems projects.
2. Where is most of the feedback on problems with Williwonk's products coming from in Problem 1? How reliable are the sources? Explain in a paragraph.
3. After getting to know Williwonk's management people better, you have approached them with your ideas on possible systems improvements that could address some of the problems or opportunities given in Problem 1.

a. In two paragraphs, provide your suggestions for systems projects. Make any realistic assumptions necessary.

b. Are there any problems or opportunities discussed in Problem 1 that are not suitable? Explain your response.

4. Create a problem definition for Williwonk's, as described in Problem 1. Estimate the weights of importance. Include at least one requirement and one constraint.

5. Create a list of user requirements for the problem definition created in Problem 4.

6. Delicato, Inc., a manufacturer of precise measuring instruments for scientific purposes, has presented you with a list of attributes that its managers think are probably important in selecting a vendor for computer hardware and software, or a cloud service provider. The criteria are listed here, but not in order of importance:

1. Low price
2. Precisely written software for engineering applications
3. Vendor performs routine maintenance on hardware
4. Training for Delicato employees

   a. Critique the list of attributes in a paragraph.

   b. Using its initial input, help Delicato, Inc., draw up a more suitable list of criteria for selecting computer hardware and software vendors to purchase from.

   c. Using its initial input, help Delicato, Inc., draw up a more suitable list of criteria for selecting cloud vendors who could provide HaaS and SaaS.

   d. In a paragraph, state how the list of attributes for selecting a computer hardware vendor to purchase from should differ from the list of attributes for selecting a cloud provider for HaaS.

   e. In a paragraph, state how the list of attributes for selecting a computer software vendor to purchase applications from should differ from the list of attributes for selecting a cloud provider for SaaS.

7. SoftWear Silhouettes is a rapidly growing mail-order house specializing in all-cotton clothing. Management would like to expand sales to the Web by creating an ecommerce site. The company has two full-time systems analysts and one programmer. Company offices are located in a small, isolated New England town, and the employees who handle the traditional mail-order business have little computer training.

   a. Considering the company's situation, draw up a list of software attributes that SoftWear Silhouettes should emphasize in its choice of software to create a website and integrate the mail-order business with business from the website.

   b. Would you recommend COTS software, custom software, or outsourcing to a SaaS provider? State your choice and defend it in a paragraph.

   c. List the variables that contributed to your response in part b.

8. The following is 12 years' demand for Viking Village, a game now available for handhelds and smartphones:

Year	Demand
2001	20,123
2002	18,999
2003	20,900
2004	31,200
2005	38,000
2006	41,200
2007	49,700
2008	46,400
2009	50,200
2010	52,300
2011	49,200
2012	57,600

   a. Graph the demand data for Viking Village.

   b. Determine the linear trend for Viking Village, using a three-year moving average.

9. Do the data for Viking Village appear to have a cyclical variation? Explain.

10. Interglobal Paper Company has asked for your help in comparing its present computer system with a new one its board of directors would like to see implemented. The proposed system costs and present system costs are as follows:

Item	Proposed System Costs	Present System Costs
Year 1		
Equipment lease	$20,000	$11,500
Salaries	30,000	50,000
Overhead	4,000	3,000
Development	30,000	—
Year 2		
Equipment lease	$20,000	$10,500
Salaries	33,000	55,000
Overhead	4,400	3,300
Development	12,000	—
Year 3		
Equipment lease	$20,000	$10,500
Salaries	36,000	60,000
Overhead	4,900	3,600
Development	—	—
Year 4		
Equipment lease	$20,000	$10,500
Salaries	39,000	66,000
Overhead	5,500	4,000
Development	—	—

a. Using break-even analysis, determine the year in which Interglobal Paper will break even.
b. Graph the costs and show the break-even point.

11. The following are system benefits for Interglobal Paper Company (from Problem 10):

Year	Benefits
1	$55,000
2	75,000
3	80,000
4	85,000

a. Use the costs of Interglobal Paper's proposed system from Problem 10 to determine the payback period. (Use the payback method.)
b. Graph the benefits versus the costs and indicate the payback period.

12. Glenn's Electronics, a small company, has set up a computer service. The table that follows shows the revenue expected for the first five months of operation, in addition to the costs for office remodeling and so on. Determine the cash flow and accumulated cash flow for the company. When is Glenn's expected to show a profit?

	July	August	September	October	November
Revenue	$35,000	$36,000	$42,000	$48,000	$57,000
Costs					
Office remodeling	$25,000	$8,000			
Salaries	11,000	12,100	$13,300	$14,600	$16,000
Training	6,000	6,000			
Equipment lease	8,000	8,480	9,000	9,540	10,110
Supplies	3,000	3,150	3,300	3,460	3,630

13. Alamo Foods of San Antonio wants to introduce a new computer system for its perishable products warehouse. The costs and benefits are as follows:

Year	Costs	Benefits
1	$33,000	$21,000
2	34,600	26,200
3	36,300	32,700
4	38,100	40,800
5	40,000	51,000
6	42,000	63,700

FIGURE 3.EX1

Data to help in the organization of a design project for creating an information system that tracks shipments of frozen foods to warehouses.

Description	Task	Must Follow	Time (Weeks)
Draw data flow	A	None	5
Draw decision tree	B	A	4
Revise tree	C	B	10
Write up project	D	C, I	4
Organize data dictionary	E	A	7
Do output prototype	F	None	2
Revise output design	G	F	9
Write use cases	H	None	10
Design database	I	H, E, and G	8

a. Given a discount rate of 8 percent (.08), perform present value analysis on the data for Alamo Foods. (*Hint:* Use this formula:

$$\frac{1}{1(1+i)^n}$$

to find the multipliers for years 1 to 6.)

b. What is your recommendation for Alamo Foods?

14. a. Suppose the discount rate in Problem 13a changes to 13 percent (.13). Perform present value analysis using the new discount rate.

b. What is your recommendation to Alamo Foods now?

c. Explain the difference between Problem 13b and Problem 14b.

15. Solve Problem 13 using an electronic spreadsheet program such as Excel.

16. Use a spreadsheet program to solve Problem 12.

17. Solve Problem 13 using a function for net present value, such as @NPV (*x*, range) in Excel.

18. Brian F. O'Byrne ("F," he says, stands for "frozen") owns a frozen-foods company and wants to develop an information system for tracking shipments to warehouses.

a. Using the data from the table in Figure 3.EX1, draw a Gantt chart to help Brian organize his design project.

b. When is it appropriate to use a Gantt chart? What are the disadvantages of using a Gantt chart? Explain in a paragraph.

19. In addition to using a Gantt chart, you've drawn a PERT diagram for Brian so that you can communicate the necessity of keeping an eye on the critical path. Consult Figure 3.EX2, which was derived from the data from Figure 3.EX1. List all paths and calculate and identify the critical path.

20. Cherry Jones owns a homeopathic medicine company called Faithhealers. She sells vitamins and other relatively nonperishable products for those who want choices regarding alternative medicine. Cherry is developing a new system that would require her staff to be retrained. Given the information in Figure 3.EX3, make a PERT diagram for her and identify the critical path. If Cherry could find a way to save time on the "write use cases" phase, would it help? Why or why not?

21. Using the PERT diagram in Figure 3.EX2 to determine answers:

a. What activities can you expedite to complete the project a week ahead of schedule?

b. Suppose Activity E is the least costly activity to expedite. What happens if you try to expedite the project by more than one week? Explain.

FIGURE 3.EX2

The PERT diagram for Brian's frozen-foods company.

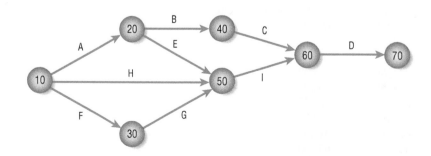

Description	Task	Must Follow	Time (Weeks)
Interview executives	A	None	6
Interview staff in order fulfillment	B	None	3
Design input prototype	C	B	2
Design output prototype	D	A, C	3
Write use cases	E	A, C	4
Record staff reactions to prototypes	F	D	2
Develop system	G	E, F	5
Write up training manual	H	B, G	3
Train staff working in order fulfillment	I	H	2

FIGURE 3.EX3

Tasks to be performed during systems development of an order fulfillment system.

22. You can help Cherry finish her project faster by expediting. Here are the costs.

Activity	Estimated Duration	Crash Time	Cost per Week
A	6	3	600
B	3	2	100
C	2	2	400
D	3	2	400
E	4	2	200
F	2	2	200
G	5	3	300
H	3	2	400
I	2	2	300

If Cherry can keep her expediting expenses to $325 per day, then she would benefit from expediting because she can start a new project earlier than planned. If she goes over the $325 per day limit, it would not be worthwhile.

a. Which three activities cannot be expedited at all because they are at their crash times already?

b. Which two activities cannot be expedited because their expediting cost exceeds the daily limit?

c. Which two activities are unlikely candidates for expediting because they are not on the critical path?

d. Set up a table and show step-by-step which activities should be expedited to shorten the project time. What is the minimum time the project will take if Cherry expedites the project as much as she can and stays within her limits?

e. Why is Cherry's limit exceeded if you try to shorten the project by one day more?

23. Robyn Cucurullo is developing a social networking app for tablet computers. She has the following stages completed so far:

Week or Stage	Estimated Cost of Stage	Stage Completed	Actual Cost of Stage to Date
Week 1	$500	100 %	500
Week 2	400	100 %	500
Week 3	600	100 %	700
Week 4	500	50 %	50
Week 5	400	0	Not yet begun

a. Create a table similar to the one in Figure 3.27

b. For Week 3, calculate the budget at completion (BAC), planned value (PV), actual cost (AC), and earned value (EV) at week 3.

c. For Week 3, calculate the performance measures of cost variance (CV), schedule variance (SV), the cost performance index (CPI), and the schedule performance index (SPI).

d. What can you tell about the *budget* in Week 3 by using these performance measures?

e. What can you tell about the *schedule* in Week 3 by using these performance measures?

f. For Week 3, calculate the estimated time to complete (ETC) and estimate at completion (EAC).

24. Using the data in the table from the previous problem:

a. Create a table similar to the one in Figure 3.27 if you haven't done so already.

b. For Week 4, calculate the budget at completion (BAC), planned value (PV), actual cost (AC), and earned value (EV) at week 3.

    c. For Week 4, calculate the performance measures of cost variance (CV), schedule variance (SV), the cost performance index (CPI), and the schedule performance index (SPI).

    d. What can you say about the *budget* in Week 4 by using these performance measures?

    e. What can you say about the *schedule* in Week 4 by using these performance measures?

    f. For Week 4, calculate the estimated time to complete (ETC) and estimate at completion (EAC).

25. Angus McIndoe wants to modernize his popular restaurant by adapting it more closely to the preferences of his repeat customers. This would involve keeping track of his customers' likes and dislikes. Information such as where they like to sit, what they like to eat, and when they normally arrive at the restaurant are all items of interest to Angus, since he believes that in this way he can better serve his customers. Angus has asked you to develop a system for him that will help make his customers happy while increasing his business.

    You have heard what Angus has to say about his customers. There are certainly more preferences that he can keep track of. Develop a problem definition for Angus, similar to the one developed for Catherine's Catering in this chapter.

26. Recently, two analysts just out of college have joined your systems analysis group at the newly formed company Mega Phone. When talking to you about the group, they mention that some things strike them as odd. One is that group members seem to look up to two group leaders, Bill and Penny, not just one.

    Their observation is that Bill seems pretty relaxed, whereas Penny is always planning and scheduling activities. They have also observed that everyone "just seems to know what to do" when they get into a meeting, even though no instructions are given. Finally, they have remarked on the openness of the group in addressing problems as they arise, instead of letting things get out of hand.

    a. By way of explanation to the new team members, label the types of leaders Bill and Penny appear to be, respectively.

    b. Explain the statement that "everyone just seems to know what to do." What is guiding their behavior?

    c. What concept best describes the openness of the group that the new team members commented on?

27. "I think it's only fair to write up *all* the alternatives you've considered," says Lou Cite, a personnel supervisor for DayGlow Paints. "After all, you've been working on this systems thing for a while now, and I think my boss and everyone else would be interested to see what you've found out." You are talking with Lou as you prepare to put together the final systems proposal that your team will be presenting to upper management.

    a. In a paragraph, explain to Lou why your proposal will not (and should not) contain all the alternatives that your team has considered.

    b. In a paragraph, discuss the sorts of alternatives that should appear in the final systems proposal.

Group Projects

1. The Weil Smile Clinic is a dental practice run by Drs. Bonnie and Jeff, and they need to keep the necessary patient and insurance data safe and secure. They looked into online, cloud-based backup like SOS Online Backup, Spare Backup, MozyPro, and KineticD. Look into the cost of these or other services. Then help Drs. Bonnie and Jeff make a decision. What are the intangible costs and benefits of backing up this way? Should they use a backup service or find some other way? Defend your analysis and recommendations.

2. Explore four or five Voice over IP (VoIP) providers. Make a list of costs, including the setup fee, monthly cost of the basic plan, monthly cost of the unlimited plan, and cost of an adapter or other fees, if required. Then make a list of attributes, such as free in-network calls, international calling, virtual telephone numbers, teleconferencing, support for caller ID, and so on. Explain how a person would use all the quantitative and qualitative information to make an informed decision about which VoIP provider to select. Are any other variables important? Would you recommend any type of software to help compare these services?

3. Make a choice on a VoIP provider based on the analysis in Group Project 2.

4. With your group members, explore project management software such as Microsoft Project. What features are available? Work with your group to list them. Have your group evaluate the usefulness of the software for managing a systems analysis and design team project. In a paragraph, state whether the software you are evaluating facilitates team member communication and management of team activities, time, and resources. Also state which particular features support these aspects of any project. Note whether the software falls short of these criteria in any regard.

5. Draw a fishbone diagram of problems that can possibly occur when constructing a website for a travel company that wants to sell vacations online for the next big travel period (either December or June).

Selected Bibliography

Alter, S. *Information Systems: The Foundation of E-Business,* 4th ed. Upper Saddle River, NJ: Prentice Hall, 2002.

Bales, R. F. *Personality and Interpersonal Behavior.* New York: Holt, Rinehart and Winston, 1970.

Bergstein, B. "IBM Faces the Perils of 'Bring Your Own Device,'" www.technologyreview.com/business/40324/?nlid=nldly&nld=2012-05-24 Last accessed July 28, 2012.

Brownsword, L., C. Sledge, and P. Oberndorf, P. "An Activity Framework for COTS-Based Systems," (CMU/SEI-2000-TR-010), Software Engineering Institute, Carnegie Mellon University website: www.sei.cmu.edu/library/abstracts/reports/00tr010.cfm. Last accessed July 28, 2012.

Construx Software Builders website, www.construx.com. Last accessed July 28, 2012.

Costar website, www.softstarsystems.com. Last accessed July 28, 2012.

Fry, J. "Survey: Cloud Really Is Shaking Up the IT Role—With Some New Job Titles to Prove It," http://datacenterdialog.blogspot.com/2011/03/survey-cloud-really-is-shaking-up-it.html. Last accessed July 28, 2012.

Germonprez, M., J. Kendall, K. Kendall, B. Warner, and L. Mathiassen. "Organizational Participation in Open Communities: Conceptual Framing and Early Findings," http://aisel.aisnet.org/amcis2011_submissions/242. Last accessed July 28, 2012.

Glass, R. "Evolving a New Theory of Project Success." *Communications of the ACM*, Vol. 42, No. 11, 1999, pp. 17–19.

Kaneshige, T. "The BYOD Sea Change Has Already Started," www.cio.com/article/701567/The_BYOD_Sea_Change_Has_Already_Started. Last accessed July 28, 2012.

Levine, D. M., P. R. Ramsey, and M. L. Berenson. *Business Statistics for Quality and Productivity.* Upper Saddle River, NJ: Prentice Hall, 1995.

Linberg, K. R. "Software Perceptions About Software Project Failure: A Case Study." *Journal of Systems and Software*, Vol. 49, Nos. 2 and 3, 1999, pp. 177–192.

Linthicum, D. "3 Easy Steps to Creating Your Cloud Strategy," www.infoworld.com/d/cloud-computing/3-easy-steps-creating-your-cloud-strategy-193593. Last accessed July 28, 2012.

Longstreet Consulting website, www.ifpug.org. Last accessed July 28, 2012.

McBreen, P. *Questioning Extreme Programming.* Boston: Addison-Wesley Co., 2003.

Moyse, I. "Why More Education Is Needed on Cloud," www.cloudcomputing-news.net/news/2012/mar/16/employing-cloud/. Last accessed April 16, 2012.

Moyse, I. "How Consumerisation Is Driving Cloud Acceptance," www.cloudcomputing-news.net/news/2012/mar/13/how-consumerisation-driving-cloud-acceptance. Last accessed July 28, 2012.

Nah, K. S. "Cloud Computing: What CIOs Need to Know About Integration," www.networkworld.com/news/2010/051710-cloud-computing-what-cios-need.html. Last accessed July 28, 2012.

NASA. "NASA Open Source Software," http://ti.arc.nasa.gov/opensource/. Last accessed July 28, 2012.

Peters, M. "How Cloud Computing and Web Services Are Changing the IT Job Market," http://mashable.com/2011/02/26/it-job-market/. Last accessed July 28, 2012.

Schein, E. H. *Process Consultation: Its Role in Organization Development.* Reading, MA: Addison-Wesley, 1969.

Shtub, A., J. F. Bard, and S. Globerson. *Project Management: Processes, Methodologies, and Economics,* 3rd ed. Upper Saddle River, NJ: Pearson, 2005.

Software Product Research website, www.spr.com. Last accessed July 28, 2012.

Stefik, M., G. Foster, D. G. Bobrow, K. Kahn, S. Lanning, and L. Suchman. "Beyond the Chalkboard: Computer Support for Collaboration and Problem Solving in Meetings." *Communications of the ACM*, Vol. 30, No. 1, January 1987, pp. 32–47.

The Linux Foundation website, www.linuxfoundation.org. Last accessed July 23, 2012.

Vigder, M. R., W. M. Gentleman, and J. C. Dean. "Using COTS Software in Systems Development," www.nrc-cnrc.gc.ca/eng/projects/iit/commercial-software.html. Last accessed July 28, 2012.

Walsh, B. "Your Network's Not Ready for E-Commerce," www.networkcomputing.com/922/922colwalsh.html. Last accessed July 28, 2012.

Yang, H. and M. Tate "A Descriptive Literature Review and Classification of Cloud Computing Research," *Communications of the Association for Information Systems*, Vol. 31, Article 2, May 2012, http://aisel.aisnet.org/cais/vol31/iss1/2. Last accessed on July 28, 2012.

The CPU Case Episode and accompanying student files are available online at www.pearsonglobaleditions.com/kendall.

Information Gathering: Interactive Methods

LEARNING OBJECTIVES

Once you have mastered the material in this chapter you will be able to:

1. Recognize the value of using interactive methods for information gathering.

2. Construct interview questions to elicit human information requirements and structure them in a way that is meaningful to users.

3. Understand the purpose of stories and why they are useful in systems analysis.

4. Understand the concept of JAD and when to use it.

5. Write effective questions to survey users about their work.

6. Design and administer effective questionnaires.

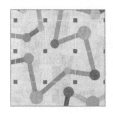

You can use three key interactive methods to elicit human information requirements from organizational members. These three methods are interviewing, joint application design (JAD), and surveying people through questionnaires. Although different in their implementation, these methods have a great deal in common, too. The basis of their shared properties is talking with and listening to people in the organization to understand their interactions with technology through a series of carefully composed questions.

Each of the three interactive methods for information gathering possesses its own established process for you to follow in interacting with users. If followed, these systematic approaches will help ensure proper design and implementation of interviews, JAD workshops, and questionnaires, as well as support insightful analysis of the resulting data. Unobtrusive methods (sampling, investigation, and observing a decision maker's behavior and physical environment) that do not require the same degree of interactivity between analysts and users will be covered in an upcoming chapter. By using interactive methods with unobtrusive methods, you will achieve a more complete portrait of the organization's information requirements.

Interviewing

Before you interview someone, you must, in effect, interview yourself. You need to know your biases and how they will affect your perceptions. Your education, intellect, upbringing, emotions, and ethical framework all serve as powerful filters for what you will be hearing in your interviews.

You need to thoroughly think through an interview before you go. Visualize why you are going, what you will ask, and what will make it a successful interview in your eyes. You must anticipate how to make the interview fulfilling for the individual you interview, as well.

An information-gathering interview is a directed conversation with a specific purpose that uses a question-and-answer format. In the interview, you want to get the opinions of the interviewee and his or her feelings about the current state of the system, organizational and personal goals, and informal procedures for interacting with information technologies.

Above all, seek the opinions of the person you are interviewing. Opinions may be more important and more revealing than facts. For example, imagine asking the owner of a traditional store who has recently added an online store how many customer refunds she typically gives for Web transactions each week. She replies, "About 20 to 25 a week." When you monitor the transactions and discover that the average is only 10.5 per week, you might conclude that the owner is overstating the facts and the problem.

Imagine instead that you ask the owner what her major concerns are and that she replies, "In my opinion, customer returns of goods purchased over the Web are way too high." By seeking opinions rather than facts, you discover a key problem that the owner wants addressed.

In addition to opinions, you should try to capture the feelings of the interviewee. Remember that the interviewee knows the organization better than you do. You can understand the organization's culture more fully by listening to the feelings of the respondent.

Goals are important information that can be gleaned from interviewing. Facts that you obtain from hard data may explain past performance, but goals project the organization's future. Try to find out as many of the organization's goals as possible from interviewing. You may not be able to determine goals through any other data-gathering methods.

The interview is also a valuable opportunity to explore key HCI (human–computer interaction) concerns, including the ergonomic aspects, system usability, how pleasing and enjoyable the system is, and how useful it is in supporting individual tasks.

In the interview, you are setting up a relationship with someone who is probably a stranger to you. You need to build trust and understanding quickly, but at the same time you must maintain control of the interview. You also need to sell the system by providing needed information to your interviewee. Do so by planning for the interview before you go so that conducting it is second nature to you. Fortunately, effective interviewing can be learned. As you practice, you will see yourself improving. Later in the chapter, we discuss joint application design (JAD) (pronounced as "jad," rhymes with *add*), which can serve as an alternative to one-on-one interviewing in certain situations.

Five Steps in Interview Preparation

The five major steps in interview preparation are shown in Figure 4.1. These steps include a range of activities, from gathering basic background material to deciding who to interview.

READING BACKGROUND MATERIAL. It's important to read and understand as much background information about the interviewees and their organization as possible. This material can often be obtained on the corporate website, from a current annual report, a corporate newsletter, or any publications sent out to explain the organization to the public. Check the Internet for any corporate information, such as information from Standard & Poor's.

FIGURE 4.1

Steps a systems analyst follows in planning an interview.

Steps in Planning the Interview
1. Read background material.
2. Establish interviewing objectives.
3. Decide whom to interview.
4. Prepare the interviewee.
5. Decide on question types and structure.

As you read through this material, be particularly sensitive to the language the organizational members use in describing themselves and their organization. What you are trying to do is build up a common vocabulary that will eventually enable you to phrase interview questions in a way that is understandable to your interviewee. Another benefit of researching your organization is to maximize the time you spend in interviews; without such preparation, you may waste time asking general background questions.

ESTABLISHING INTERVIEWING OBJECTIVES. Use the background information you gathered as well as your own experience to establish interview objectives. There should be four to six key areas concerning HCI, information processing, and decision-making behavior about which you will want to ask questions. These areas include HCI concerns (the usefulness and usability of the system; how it fits physical aspects; how it suits a user's cognitive capabilities; whether it is engaging or aesthetically pleasing; and whether using the system is rewarded with desired consequences), information sources, information formats, decision-making frequency, qualities of information, and decision-making style.

DECIDING WHOM TO INTERVIEW. When deciding whom to interview, include key people at all levels who will be affected by the system in some manner. Strive for balance so that as many users' needs are addressed as possible. Your organizational contact will also have some ideas about who should be interviewed.

PREPARING THE INTERVIEWEE. Prepare the person to be interviewed by calling ahead or sending an email message and allowing the interviewee time to think about the interview. If you are doing an in-depth interview, it is permissible to email your questions ahead of time to allow the interviewee time to think over his or her responses. Because there are many objectives to fulfill in the interview (including building trust and observing the workplace), however, interviews should typically be conducted in person and not via email. Interviews should be kept to 45 minutes or an hour at the most. No matter how much your interviewees seem to want to extend the interview beyond this limit, remember that when they spend time with you, they are not doing their work. If interviews go over an hour, it is likely that the interviewees will resent the intrusion, whether or not they articulate their resentment.

DECIDING ON QUESTION TYPES AND STRUCTURE. Write questions to cover the key areas of HCI and decision making that you discovered when you ascertained interview objectives. Proper questioning techniques are the heart of interviewing. Questions have some basic forms you need to know. The two basic question types are open-ended and closed. Each question type can accomplish something a little different from the other, and each has benefits and drawbacks. You need to think about the effect each question type will have.

It is possible to structure your interview in three different patterns: a pyramid structure, a funnel structure, or a diamond structure. Each is appropriate under different conditions and serves a different function, and each one is discussed later in this chapter.

Question Types

There are three main question types that you will use to construct your interview; open-ended questions, closed questions and probes. Each question type is useful in different situations and at special times during the interview. In addition, each type of question will elicit different information from the person you are interviewing.

OPEN-ENDED QUESTIONS. Open-ended questions include those such as "What do you think about putting all the managers on an intranet?" "Please explain how you make a scheduling decision." and "In what ways does the system extend your capability to do tasks that would not be possible otherwise?" Consider the term *open-ended*. "Open" actually describes the interviewee's options for responding. They are open. The response can be two words or two paragraphs. Some examples of open-ended questions are found in Figure 4.2.

The benefits of using open-ended questions are numerous and include the following:

1. Putting the interviewee at ease
2. Allowing the interviewer to pick up on the interviewee's vocabulary, which reflects his or her education, values, attitudes, and beliefs
3. Providing richness of detail

> ### Open-Ended Interview Questions
>
> - What's your opinion of the current state of business-to-business ecommerce in your firm?
> - What are the critical objectives of your department?
> - Once the data are submitted via the Web site, how are they processed?
> - Describe the monitoring process that is available online.
> - What are some of the common data entry errors made in this department?
> - What are the biggest frustrations you've experienced during the transition to ecommerce?

4. Revealing avenues of further questioning that may have otherwise gone untapped
5. Making the interview more interesting for the interviewee
6. Allowing more spontaneity
7. Making phrasing easier for the interviewer
8. Using them in a pinch if the interviewer is caught unprepared

As you can see, there are several advantages to using open-ended questions. There are, however, also many drawbacks:

1. Asking questions that may result in too much irrelevant detail
2. Possibly losing control of the interview
3. Allowing responses that may take too much time for the amount of useful information gained
4. Potentially seeming that the interviewer is unprepared
5. Possibly giving the impression that the interviewer is on a "fishing expedition," with no real objective for the interview

You must carefully consider the implications of using open-ended questions for interviewing.

CLOSED QUESTIONS. The alternative to open-ended questions is found in the other basic question type: closed questions. Such questions are of the basic form "Is it easy to use the current system?" and "How many subordinates do you have?" The possible responses are closed to the interviewee because he or she can only reply with a finite number, such as "None," "One," or "Fifteen." Some examples of closed questions can be found in Figure 4.3.

A closed question limits the responses available to the interviewee. You may be familiar with closed questions through multiple-choice exams in college. You are given a question and five responses, but you are not allowed to write down your own response and get credit for correctly answering the question.

A special kind of closed question is the bipolar question. This type of question limits the interviewee even further by allowing only a choice on either pole, such as yes or no, true or false, or agree or disagree. Examples of bipolar questions can be found in Figure 4.4.

The benefits of using closed questions of either type include the following:

1. Saving time
2. Easily comparing interviews
3. Getting to the point

> ### Closed Interview Questions
>
> - How many times a week is the project repository updated?
> - On average, how many calls does the call center receive monthly?
> - Which of the following sources of information is most valuable to you?
>   - ° Completed customer complaint forms
>   - ° Email complaints from consumers who visit the Web site
>   - ° Face-to-face interaction with customers
>   - ° Returned merchandise
> - List your top two priorities for improving the technology infrastructure.
> - Who receives this input?

Bipolar Interview Questions

- Do you use the Web to provide information to vendors?
- Do you agree or disagree that ecommerce on the Web lacks security?
- Do you want to receive a printout of your account status every month?
- Does your Web site maintain a FAQ page for employees with payroll questions?
- Is this form complete?

FIGURE 4.4

Bipolar interview questions are a special kind of closed question. These examples were selected from different interviews and are not shown in any particular order.

4. Keeping control over the interview
5. Covering lots of ground quickly
6. Getting to relevant data

The drawbacks of using closed questions are substantial, however. They include the following:

1. Boring for the interviewee
2. Failing to obtain rich detail (because the interviewer supplies the frame of reference for the interviewee)
3. Failing to address the main ideas for the preceding reason
4. Failing to build rapport between interviewer and interviewee

Thus, as an interviewer, you must think carefully about the question types you will use.

Both open-ended and closed questions have advantages and drawbacks, as shown in Figure 4.5. Notice that choosing one question type over the other actually involves a trade-off; although an open-ended question affords breadth and depth of reply, responses to open-ended questions are difficult to analyze.

PROBES. A third type of question is the probe, or follow-up. The strongest probe is the simplest—the question "Why?" Other probes are "Please provide an example of a time you did not find the system trustworthy." and "Please elaborate on that for me." Some examples of probing questions can be found in Figure 4.6. The purpose of a probe is to go beyond the initial answer to get more meaning, to clarify, and to draw out and expand on the interviewee's point. Probes may be either open-ended or closed questions.

It is essential to probe. Most beginning interviewers are reticent about probing and consequently accept superficial answers. They are usually grateful that employees have granted interviews and feel somewhat obligated to accept unqualified statements politely.

Arranging Questions in a Logical Sequence

Just as there are two generally recognized ways of reasoning—inductive and deductive—there are two similar ways of organizing your interviews. A third way combines both inductive and deductive patterns.

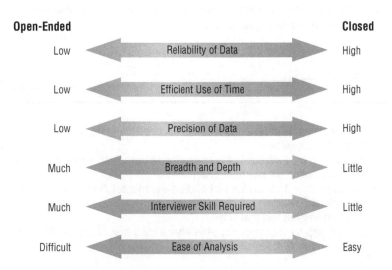

Open-Ended		Closed
Low	Reliability of Data	High
Low	Efficient Use of Time	High
Low	Precision of Data	High
Much	Breadth and Depth	Little
Much	Interviewer Skill Required	Little
Difficult	Ease of Analysis	Easy

FIGURE 4.5

Attributes of open-ended and closed questions.

CONSULTING OPPORTUNITY 4.1

Strengthening Your Question Types

Strongbodies, a large local chain of sports clubs, has experienced phenomenal growth in the past five years. Management would like to refine its decision-making process for purchasing new body-building equipment by using a DSS or a dashboard. Currently, managers listen to customers, attend trade shows, look at advertisements, and put in requests for new equipment purchases based on their subjective perceptions. These are then approved or denied by Harry Mussels.

Harry is the first person you will interview. He is a 37-year-old division manager who runs five area clubs. He travels all over the city to their widespread locations. He keeps an office at the East location, although he is there less than a quarter of the time.

In addition, when Harry is present at a club, he is busy answering business-related phone calls, solving on-the-spot problems presented by managers, and interacting with club members. His time is short, and to compensate for that he has become an extremely well-organized, efficient divisional manager. He cannot grant you a lot of interview time. However, his input is important, and he feels he would be the main beneficiary of the proposed system.

What type of interview question might be most suitable for your interview with Harry? Why is this type most appropriate? How will your choice of question type affect the amount of time you spend in preparation for interviewing Harry? Write 5 to 10 questions of this type. What other techniques might you use to supplement information unavailable through that type of question? Write a paragraph to explain.

USING A PYRAMID STRUCTURE. Inductive organization of interview questions can be visualized as having a pyramid shape. Using this form, the interviewer begins with very detailed, often closed, questions. The interviewer then expands the topics by allowing open-ended questions and more generalized responses, as shown in Figure 4.7.

You should use a pyramid structure if you believe your interviewee needs to warm up to the topic. Using a pyramid structure for question sequencing is also useful when you want an ending determination about the topic. Such is the case in the final question in Figure 4.7, "In general, how do you feel about the security of data versus the importance of Internet access?"

USING A FUNNEL STRUCTURE. In the second kind of structure, the interviewer takes a deductive approach by beginning with generalized, open-ended questions and then narrowing the possible responses by using closed questions. This interview structure can be thought of as funnel shaped, as depicted in Figure 4.8. Using the funnel structure method provides an easy, nonthreatening way to begin an interview. A funnel-shaped question sequence is also useful when the interviewee feels emotional about the topic and needs freedom to express those emotions.

USING A DIAMOND-SHAPED STRUCTURE. Often a combination of the pyramid and funnel structures, resulting in a diamond-shaped interview structure, is best. This structure entails beginning in a very specific way, then examining general issues, and finally coming to a very specific conclusion, as shown in Figure 4.9.

FIGURE 4.6

Probes allow a systems analyst to follow up on questions to get more detailed responses. These examples were selected from different interviews and are not shown in any particular order.

Probes

- Why?
- Give an example of how ecommerce has been integrated into your business processes.
- Please give an illustration of the security problems you are experiencing with your online bill payment system.
- You mentioned both an intranet and an extranet solution. Please give an example of how you think each differs.
- What makes you feel that way?
- Tell me step by step what happens after a customer clicks the "Submit" button on the Web registration form.

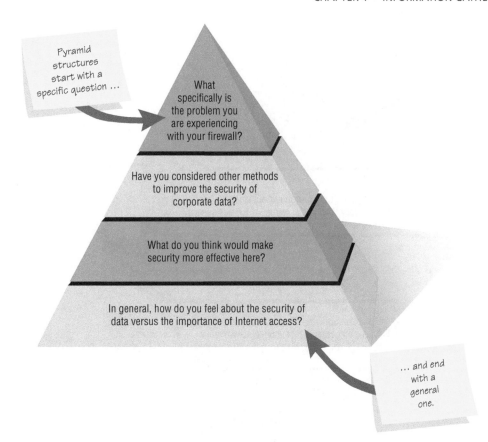

FIGURE 4.7

The pyramid structure for interviewing goes from specific to general questions.

The interviewer begins with easy, closed questions that provide a warm-up to the interview process. In the middle of the interview, the interviewee is asked for opinions on broad topics that obviously have no "right" answer. The interviewer then narrows the questions again to get specific questions answered, thus providing closure for both the interviewee and the interviewer. The diamond structure combines the strengths of the other two approaches but has the disadvantage of taking longer than either other structure.

The end of the interview is a natural place to ask one key question: "Is there anything we haven't touched on that you feel is important for me to know?" Considered a formula question by the interviewee most of the time, the response will often be "No." You are interested

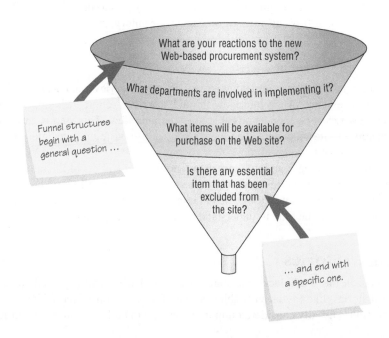

FIGURE 4.8

The funnel structure for interviewing begins with broad questions and then funnels to specific questions.

FIGURE 4.9

The diamond-shaped structure for interviewing combines the pyramid and funnel structures.

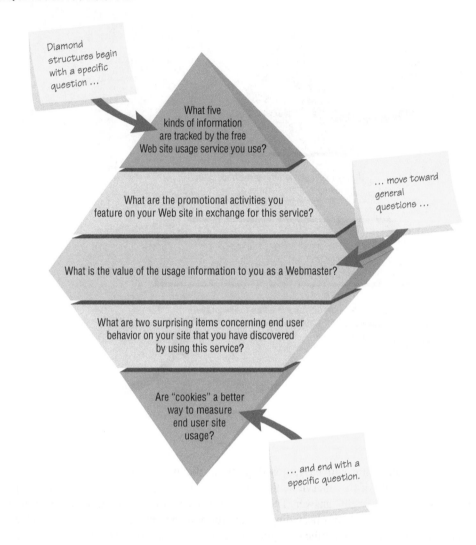

in the other times, when this question opens the proverbial floodgates and much new data are presented, though.

As you conclude the interview, summarize and provide feedback on your overall impressions. Inform the interviewee about the subsequent steps to take and what you and other team members will do next. Ask the interviewee with whom you should talk next. Set up future appointment times for follow-up interviews, thank the interviewee for his or her time, and shake hands.

Writing the Interview Report

Although the interview itself is complete, your work on the interview data is just beginning. You need to capture the essence of the interview through a written report. It is imperative that you write the interview report as soon as possible after the interview. This step is another way you can ensure quality of interview data. The longer you wait to write up your interview, the more suspect the quality of your data becomes.

After this initial summary, go into more detail, noting main points of the interview and your own opinions. Review the interview report with the respondent at a follow-up meeting. This step helps clarify the meaning the interviewee had in mind and lets the interviewee know that you are interested enough to take the time to understand his or her point of view and perceptions.

Listening to Stories

Stories originate in the workplace, and many are shared with, and repeated by, coworkers. Organizational stories, like myths and fables, are used to relay some kind of information. This shared information is usually considered important enough to build an entire story around it, and

CONSULTING OPPORTUNITY 4.2

Skimming the Surface

You are about to leave SureCheck Dairy after a preliminary tour, when another member of your systems analysis team calls you at the dairy to say he cannot make his interview appointment with the plant manager because of illness. The plant manager is extremely busy, and you want to keep his enthusiasm for the project going by doing things as scheduled. You also realize that without the initial interview data, the rest of your data gathering will be slowed. Although you have no interview questions prepared, you make the decision to go ahead and interview the plant manager on the spot.

You have learned that each sales manager at SureCheck is interested in processing their own data and producing their own reports on quantities and kinds of dairy products sold so that they can use that information to better control production of the company's large product line (it includes whole, skim, 2 percent, and 1 percent milk, half-and-half, cottage cheese, yogurt, and frozen novelties). Sales managers are currently sending their sales figures via the Web to corporate headquarters, 600 miles away, and even the production of online reports seems slow. You will base your ad-libbed questions on what you have just found out on the tour.

In the few minutes before your interview begins, decide on a structure for it: funnel, pyramid, or diamond. In a paragraph, justify why you would proceed with the interview structure you have chosen based on the unusual context of this interview. Write a series of questions and organize them in the structure you have chosen.

---

when a story is told and retold, even handed down over time, the story takes on a mythic quality. Isolated stories are welcome when you are looking for facts, while enduring stories capture all aspects of the organization. Enduring stories are the ones a systems analyst should be looking for.

There are four purposes for telling a story: experiential, explanatory, validating, and prescriptive. Experiential stories describe what the business or industry is like. Explanatory stories tell why the organization acted a certain way. Validating stories are used to convince people that the organization made the correct decision. Prescriptive stories tell the listener how to act.

Systems analysts can use storytelling as a complement to other information gathering methods such as interviewing, JAD, and surveys. You need to engage organizational participants by reacting to stories, matching one story to another by recounting it to other participants, and even collaborating with the participant to reframe and understand organizational stories. It is a good way to deeply understand some of the problems associated with information systems use, systems development, systems adoption, and designing for your intended audience.

We have observed that many of the stories that systems analysts have reported on in the past have been mere fragments of stories. The intent or purpose of the story is not understood, and therefore the analyst is not able to fully understand the importance of the story. The systems analyst needs to listen to the story as a whole in order to truly understand its content and purpose.

Joint Application Design

No matter how adept you become as an interviewer, you will inevitably experience situations in which one-on-one interviews do not seem to be as useful as you would like. Personal interviews are time consuming and subject to error, and their data are prone to misinterpretation. IBM developed an alternative approach to interviewing users one by one, called joint application design (JAD). The motivations for using JAD are to cut the time (and hence the cost) required by personal interviews, to improve the quality of the results of information requirements assessment, and to create more user identification with new information systems as a result of the participative processes.

Although JAD can be substituted for personal interviews at any appropriate juncture during the SDLC, it has usually been employed as a technique that allows you, as a systems analyst, to accomplish requirements analysis and to design the user interface jointly with users in a group setting. The many intricacies of this approach can only be learned in a paid seminar demonstrating proprietary methods. We can, however, convey enough information about JAD here to make you aware of some of its benefits and drawbacks in comparison with one-on-one interviews.

Conditions That Support the Use of JAD

The following list of conditions will help you decide when the use of JAD may be fruitful. Consider using joint application design when:

1. User groups are restless and want something new, not a standard solution to a typical problem.
2. The organizational culture supports joint problem-solving behaviors among multiple levels of employees.
3. Analysts forecast that the number of ideas generated via one-on-one interviews will not be as plentiful as the number of ideas possible from an extended group exercise.
4. Organizational workflow permits the absence of key personnel during a two- to four-day block of time.

Who Is Involved?

Joint application design sessions include a variety of participants—analysts, users, executives, and so on—who contribute different backgrounds and skills to the sessions. Your primary concern here is that all project team members are committed to the JAD approach and become involved. Choose an executive sponsor, a senior person who will introduce and conclude the JAD session. Preferably, select an executive from the user group who has some sort of authority over the IS people working on the project. This person will be an important, visible symbol of organizational commitment to the systems project.

At least one IS analyst should be present, but the analyst usually takes a passive role, unlike traditional interviewing in which the analyst controls the interaction. As the project analyst, you should be present during JAD to listen to what users say and what they require. In addition, you will want to give an expert opinion about any disproportionate costs of solutions proposed during the JAD session itself. Without this kind of immediate feedback, unrealistic solutions with excessive costs may creep into the proposal and prove costly to discourage later on.

From eight to a dozen users, of any rank, can be chosen to participate in JAD sessions. Try to select users who can articulate what information they need to perform their jobs as well as what they desire in a new or improved computer system.

The session leader should not be an expert in systems analysis and design but rather someone who has excellent communication skills to facilitate appropriate interactions. Note that you do not want to use a session leader who reports to another person in the group. To avoid this possibility, an organization may want to retain an outside management consultant to serve as session leader. The point is to get a person who can bring the group's attention to important systems issues, satisfactorily negotiate and resolve conflicts, and help group members reach a consensus.

Your JAD session should also include one or two observers who are analysts or technical experts from other functional areas to offer technical explanations and advice to the group during the sessions. In addition, one scribe from the IS department should attend the JAD sessions to formally write down everything that is done.

Where to Hold JAD Meetings

If at all possible, we recommend holding the two- to four-day sessions off-site, away from the organization, in comfortable surroundings. Some groups use executive centers or even group decision support facilities that are available at major universities. The idea is to minimize the daily distractions and responsibilities of the participants' regular work. The room itself should comfortably hold the number of people invited. Minimal presentation support equipment includes two projectors to connect to two laptops, iPads, or PCs, a whiteboard, a flip chart, and easy access to a copier. Group decision support rooms will also provide networked PCs, a projection system, and software written to facilitate group interaction while minimizing unproductive group behaviors.

Schedule your JAD session when all participants can commit to attending. Do not hold the sessions unless everyone who has been invited can actually attend. This rule is critical to the success of the sessions. Ensure that every participant receives an agenda before the meeting, and consider holding an orientation meeting for a half day one week or so before the workshop so that those involved know what is expected of them. Such a premeeting allows you to move rapidly and act confidently once the actual meeting is convened.

HYPERCASE® EXPERIENCE 4.1

"Well, I did warn you that things aren't always smooth here at MRE. By now you've met many of our key employees and are starting to understand the 'lay of the land.' Who would have thought that some innocent decisions about systems, like whether to purchase a COMTEX or Shiroma, would cause such hostility? Well, live and learn, I always say. At least now you'll know what you're up against when you have to start recommending software and hardware!

"It's funny that not all questions are created equal. I myself favor asking open-ended questions, but when I have to answer them, it is not always easy."

HYPERCASE Questions

1. Using the interview questions posed in HyperCase, give five examples of open-ended questions and five examples of closed questions. Explain why your examples are correctly classified as either open-ended or closed question types.
2. List three probing questions that are part of the Daniel Hill interview. In particular, what did you learn by following up on the questions you asked Daniel?
3. List three probing questions that are part of the Snowden Evans interview. In particular, what did you learn by following up on the questions you asked Snowden?

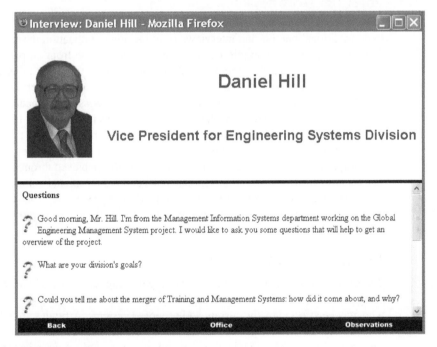

FIGURE 4.HC1

Pointing to a question in HyperCase will reveal an answer.

Accomplishing a Structured Analysis of Project Activities

IBM recommends that the JAD sessions examine these points in the proposed systems project: planning, receiving, receipt processing/tracking, monitoring and assigning, processing, recording, sending, and evaluating. For each topic, the questions who, what, how, where, and why should also be asked and answered. Clearly, ad hoc interactive systems such as decision support systems and other types of systems dependent on decision-maker style (including prototype systems) are not as easily analyzed with the structured approach of JAD.

As the analyst involved with the JAD sessions, you should receive the notes of the scribe and prepare a specifications document based on what happened at the meeting. Systematically present the management objectives as well as the scope and boundaries of the project. Specifics of the system, including details on screen and report layouts, should also be included.

Potential Benefits of Using JAD in Place of Traditional Interviewing

There are four major potential benefits that you, the users, and your systems analysis team should consider when you weigh the possibilities of using joint application design. The first potential benefit is time savings over traditional one-on-one interviews. Some organizations have estimated that JAD sessions have provided a 15 percent time savings over the traditional approach.

Hand-in-hand with time savings is the rapid development possible via JAD. Because user interviews are not accomplished serially over a period of weeks or months, the development can proceed much more quickly.

A third benefit to weigh is the possibility of improved ownership of the information system. Analysts are always striving to involve users in meaningful ways and to encourage users to take early ownership of the systems being designed. Due to its interactive nature and high visibility, JAD helps users become involved early in systems projects and treats their feedback seriously. Working through a JAD session eventually helps reflect user ideas in the final design.

A final benefit of participating in joint application design sessions is the creative development of designs. The interactive character of JAD has a great deal in common with brainstorming techniques that generate new ideas and new combinations of ideas because of the dynamic and stimulating environment. Designs can evolve through facilitated interactions, rather than in relative isolation.

Potential Drawbacks of Using JAD

There are three drawbacks or pitfalls that you should weigh when making a decision on whether to do traditional one-on-one interviews or to use joint application design. The first drawback is that JAD requires the commitment of a large block of time from all participants. Because JAD requires a two- to four-day commitment, it is not possible to do any other activities concurrently or to time-shift any activities, as is typically done in one-on-one interviewing.

A second pitfall occurs if preparation for the JAD sessions is inadequate in any regard or if the follow-up report and documentation of specifications is incomplete. In these instances resulting designs could be less than satisfactory. Many variables need to come together correctly for joint application design to be successful. Conversely, many things can go wrong. The success of designs resulting from JAD sessions is less predictable than that achieved through standard interviews.

Finally, the necessary organizational skills and organizational culture may not be sufficiently developed to enable the concerted effort required to be productive in a JAD setting. In the end you will have to judge whether the organization is truly committed to, and prepared for, this approach.

Using Questionnaires

The use of questionnaires is an information-gathering technique that allows systems analysts to study attitudes, beliefs, behavior, and characteristics of several key people in the organization who may be affected by the current and proposed systems. Attitudes are what people in the organization say they want (in a new system, for instance); beliefs are what people think is actually true; behavior is what organizational members do; and characteristics are properties of people or things.

Responses gained through questionnaires (also called surveys) using closed questions can be quantified. If you are surveying people via email or the Web, you can use software to turn electronic responses directly into data tables for analysis using a spreadsheet application or statistical software packages. Responses to questionnaires using open-ended questions are analyzed and interpreted in other ways. Answers to questions on attitudes and beliefs are sensitive to the wording chosen by the systems analyst.

Through the use of questionnaires, the analyst may be seeking to quantify what was found in interviews. In addition, questionnaires may be used to determine how widespread or limited a sentiment expressed in an interview really is. Conversely, questionnaires can be used to survey a large sample of system users to sense problems or raise important issues before interviews are scheduled.

Throughout this chapter, we compare and contrast questionnaires with interviews. There are many similarities between the two techniques, and perhaps the ideal would be to use them in conjunction with each other, either following up unclear questionnaire responses with an interview or designing the questionnaire based on what is discovered in the interview. Each technique, however, has its own specific functions, and it is not always necessary or desirable to use both.

CONSULTING OPPORTUNITY 4.3

A Systems Analyst, I Presume?

"Know what I think of the work the last systems analyst team did? The proliferation of PDF reports created a jungle on my desktop. To figure out the cost of raw materials to us, I have to cut my way through the overgrowth of data, hacking my path with drill-down capabilities that I have had to create for myself. I delete everything that's irrelevant. Sometimes I rip out the excess vegetation until I see the numbers I need on my screen," says Henry Stanley, accounting supervisor for Zenith Glass Company. As you interview him, he points unhappily to an untidy stack of old printouts sprouting beside his desk. "Those were output from the work of a systems teams who worked here a dozen years ago. I still keep them because you never know when you'll need them as a guide. It's definitely survival of the fittest around here."

Identify the overriding metaphor Henry is using to describe the reports he is receiving and the accessibility of information in them. In a paragraph, describe how this step helps you understand Henry's attitude toward any work proposed by your systems analysis team. In a paragraph, adopt Henry's metaphor and extend it in a more positive sense during your interview with him.

Planning for the Use of Questionnaires

At first glance, using questionnaires may seem to be a quick way to gather massive amounts of data about how users assess the current system, about what problems they are experiencing with their work, and about what people expect from a new or modified system. Although it is true that you can gather a lot of information through questionnaires without spending time in face-to-face interviews, developing a useful questionnaire takes extensive planning time in its own right. When you decide to survey users via email or the Web, you face additional planning considerations concerning confidentiality, authentication of identity, and problems of multiple responses.

You must first decide what you are attempting to gain through using a survey. For instance, if you want to know what percentage of users prefers a FAQ page as a means of learning about new software packages, a questionnaire might be the right technique. If you want an in-depth analysis of a manager's decision-making process, conducting an interview is a better choice.

Here are some guidelines to help you decide whether the use of questionnaires is appropriate. Consider using questionnaires if:

1. The people you need to question are widely dispersed (as in different branches of the same corporation).
2. A large number of people are involved in the systems project, and it is meaningful to know what proportion of a given group (for example, management) approves or disapproves of a particular feature of the proposed system.
3. You are doing an exploratory study and want to gauge overall opinion before the systems project is given any specific direction.
4. You wish to be certain that any problems with the current system are identified and addressed in follow-up interviews.

Once you have determined that you have good cause to use a questionnaire and have pinpointed the objectives to be fulfilled through its use, you can begin formulating questions.

Writing Questions

The biggest difference between the questions used for most interviews and those used on questionnaires is that interviewing permits interaction between the questions and their meanings. In an interview, the analyst has an opportunity to refine a question, define a muddy term, change the course of questioning, respond to a puzzled look, and generally control the context.

Few of these opportunities are possible on a questionnaire. Thus, for an analyst, questions must be transparently clear, the flow of the questionnaire cogent, the respondent's questions anticipated, and the administration of the questionnaire planned in detail. (A respondent is the person who responds to or answers the questionnaire.)

FIGURE 4.10

Open-ended questions used for questionnaires.

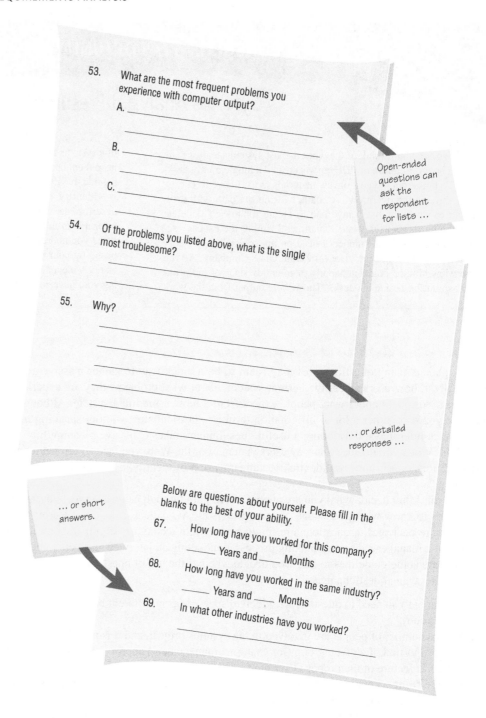

The basic question types used on the questionnaire are open-ended and closed, as discussed for interviewing. Due to the constraints placed on questionnaires, some additional discussion of question types is warranted.

OPEN-ENDED QUESTIONS. Recall that open-ended questions (or statements) are those that leave all possible response options open to the respondent. For example, open-ended questions on a questionnaire might read, "Describe any problems you are currently experiencing with output reports" or "In your opinion, how helpful are the user manuals for the current system's accounting application?"

When you write open-ended questions for a questionnaire, anticipate what kind of response you will get. For instance, if you ask a question such as, "How do you feel about the system?" the responses are apt to be too broad for accurate interpretation or comparison. Therefore, even when you write an open-ended question, it must be narrow enough to guide respondents to answer in a specific way. (Examples of open-ended questions can be found in Figure 4.10.)

FIGURE 4.11

Closed questions on
questionnaires help ensure
responses.

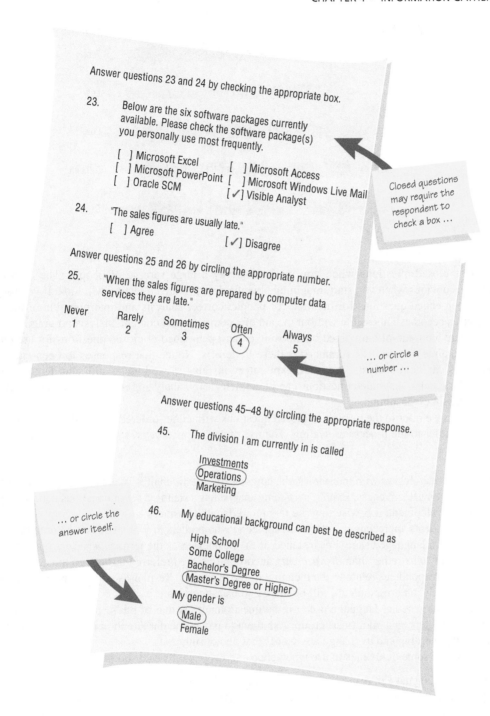

Open-ended questions are particularly well suited to situations in which you want to get at organizational members' opinions about some aspect of the system, whether product or process. In such cases you will want to use open-ended questions when it is impossible to list effectively all the possible responses to the question.

CLOSED QUESTIONS. Recall that closed questions (or statements) are those that limit or close the response options available to the respondent. For example, in Figure 4.11 the statement in question 23 ("Below are the six software packages currently available. Please check the software package(s) you personally use most frequently") is closed. Notice that respondents are not asked why the package is preferred, nor are they asked to select more than one, even if that is a more representative response.

A systems analyst should use closed questions when it is possible to effectively list all the possible responses to the question and when all the listed responses are mutually exclusive, so that choosing one precludes choosing any of the others.

FIGURE 4.12

Trade-offs between the use of open-ended and closed questions on questionnaires.

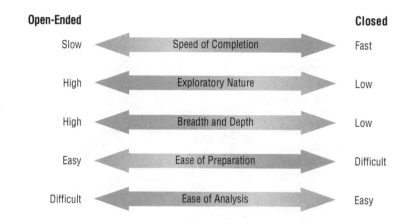

Use closed questions when you want to survey a large sample of people. The reason becomes obvious when you start imagining how the data you are collecting will look. If you use only open-ended questions for hundreds of people, correct analysis and interpretation of their responses becomes impossible without the aid of a computerized content analysis program.

There are trade-offs involved in choosing either open-ended or closed questions for use on questionnaires. Figure 4.12 summarizes these trade-offs. Notice that responses to open-ended questions can help analysts gain rich, exploratory insights as well as breadth and depth on a topic. Although open-ended questions can be written easily, analyzing responses to them is difficult and time-consuming.

When we refer to the writing of closed questions with either ordered or unordered answers, we often refer to the process as scaling. The use of scales in surveys is discussed in detail in a later section.

WORD CHOICE. Just as with interviews, the language of questionnaires is an extremely important aspect of their effectiveness. Even if a systems analyst has a standard set of questions concerning systems development, it is wise to write them to reflect the business's own terminology.

Respondents appreciate the efforts of someone who bothers to write a questionnaire reflecting their own language usage. For instance, if the business uses the term *supervisors* instead of *managers* or *units* rather than *departments*, incorporating the preferred terms into the questionnaire helps respondents relate to the meaning of the questions. Responses will be easier to interpret accurately, and respondents will be more enthusiastic overall.

To check whether language used on the questionnaire is that of the respondents, try some sample questions on a pilot (test) group. Ask them to pay particular attention to the appropriateness of the wording and to change any words that do not ring true.

Here are some guidelines to use when choosing language for your questionnaire:

1. Use the language of respondents whenever possible. Keep wording simple.
2. Be specific rather than vague in your wording. Avoid overly specific questions as well.
3. Keep questions short.
4. Do not patronize respondents by talking down to them through low-level language choices.
5. Avoid bias in wording. Avoiding bias includes avoiding objectionable questions.
6. Target questions to the correct respondents (that is, those who are capable of responding).
7. Ensure that questions are technically accurate before including them.
8. Use software to check whether the reading level is appropriate for the respondents.

Using Scales in Questionnaires

Scaling is the process of assigning numbers or other symbols to an attribute or a characteristic for the purpose of measuring that attribute or characteristic. Scales are often arbitrary and may not be unique. For example, temperature is measured in a number of ways; the two most common are the Fahrenheit scale (where water freezes at 32 degrees and boils at 212 degrees) and the Celsius scale (where freezing occurs at 0 degrees and boiling at 100 degrees).

CONSULTING OPPORTUNITY 4.4

The Unbearable Questionnaire

"I'm going to go into a depression or at least a slump if someone doesn't figure this out soon," say Penny Stox, office manager for Carbon, Carbon, & Rippy, a large brokerage firm. Penny is sitting across a conference table from you and two of her most productive account executives, By Lowe and Sal Hy. You are all mulling over the responses to a questionnaire that has been distributed among the firm's account executives, which is shown in Figure 4.C1.

"We need a crystal ball to understand these," By and Sal call out together.

"Maybe it reflects some sort of optimistic cycle or something," Penny says as she reads more of the responses. "Who designed this gem, anyway?"

"Rich Kleintz," By and Sal call out in unison.

"Well, as you can see, it's not telling us anything!" Penny exclaims.

Penny and her staff are dissatisfied with the responses they have received on the unbearable questionnaire, and they feel that the responses are unrealistic reflections of the amount of information account executives want. In a paragraph, state why these problems are occurring. On a separate sheet, change the scaling of the questions to avoid these problems.

FIGURE 4.C1

A questionnaire developed for the brokerage firm Carbon, Carbon, & Rippy by Rich Kleintz.

Circle the appropriate number for each source of information described.

1. Industry Reports

Less 1   2   About the Same 3   4   More 5

2. Trend Analysis

Less 1   2   About the Same 3   4   More 5

3. Computer-Generated Graphs

Less 1   2   About the Same 3   4   More 5

4. Investment Advisory Services

Less 1   2   About the Same 3   4   More 5

We need to change this questionnaire. -Penny

5. Point and Figure Charts

Less 1   2   About the Same 3   4   More 5

6. Computerized Portfolio Analysis

Less 1   2   About the Same 3   4   More 5

7. Hot Tips

Less 1   2   About the Same 3   4   More 5

MEASUREMENT. Systems analysts commonly use two different forms of measurement scales:

1. Nominal scales
2. Interval scales

Nominal scales are used to classify things. A question such as the following uses a nominal scale:

What type of software do you use the most?

1 = A word processor
2 = A spreadsheet
3 = A database
4 = An email program

Obviously, nominal scales are the weakest forms of measurement. Generally, all an analyst can do with them is obtain totals for each classification.

In interval scales, the intervals between each pair of numbers are equal. Due to this characteristic, mathematical operations can be performed on the questionnaire data, resulting in a more complete analysis. Examples of interval scales are the Fahrenheit and Celsius scales, which measure temperature.

The foregoing example of the information center definitely does not use an interval scale, but by anchoring the scale on either end, the analyst may want to assume that the respondent perceives the intervals to be equal:

How useful is the support given by the Technical Support Group?

Not Useful at All				*Extremely Useful*
1	2	3	4	5

If the systems analyst makes this assumption, more quantitative analysis is possible.

VALIDITY AND RELIABILITY. There are two measures of performance in constructing scales: validity and reliability. A systems analyst should be aware of these concerns.

Validity is the degree to which a question measures what the analyst intends for it to measure. For example, if the purpose of the questionnaire is to determine whether the organization is ready for a major change in computer operations, do the questions measure that?

Reliability measures consistency. If a questionnaire was administered once and then again under the same conditions and if the same results were obtained both times, the instrument is said to have *external consistency*. If the questionnaire contains subparts and these parts have equivalent results, the instrument is said to have *internal consistency*. Both external and internal consistency are important.

CONSTRUCTING SCALES. The actual construction of scales is a serious task. Careless construction of scales can result in the following problems:

1. Leniency
2. Central tendency
3. Halo effect

Leniency is a problem caused by respondents who are easy raters. A systems analyst can avoid the problem of leniency by moving the "average" category to the left (or right) of center.

Central tendency is a problem that occurs when respondents rate everything as average. The analyst can improve the scale (1) by making the differences smaller at the two ends, (2) by adjusting the strength of the descriptors, or (3) by creating a scale with more points.

The halo effect is a problem that arises when the impression formed in one question carries into the next question. For example, if you are rating an employee about whom you have a very favorable impression, you may give a high rating in every category or trait, regardless of whether it is a strong point of the employee's. The solution is to place one trait and several employees on each page rather than one employee and several traits on a page.

Designing Questionnaires

Many of the same principles that are relevant to the design of forms for data input (as covered in Chapter 12) are important in designing questionnaires as well. Although the intent of a questionnaire is to gather information on attitudes, beliefs, behavior, and characteristics whose impact may substantially alter users' work, respondents are not always motivated to respond. Remember that organizational members on the whole tend to receive too many surveys, many of which are often ill conceived and trivial.

A well-designed, relevant questionnaire can help overcome some of the resistance to responding. Here are some rules for designing a good questionnaire:

1. Allow ample white space.
2. Allow ample space to write or type in responses.
3. Make it easy for respondents to clearly mark their answers.
4. Be consistent in style.

When you design questionnaires for the Web, apply the same rules you use when designing paper questionnaires. Most software packages allow you to insert one of the commonly used data entry formats shown in Figure 4.13. Following the four guidelines should help you gain a better response rate to the questionnaire.

QUESTION ORDER. There is no best way to order questions on a questionnaire. Once again, as you order questions, you must think about your objectives in using the questionnaire and then determine the function of each question in helping you to achieve your objectives. It is also important to see the questionnaire through the respondents' eyes. Some guidelines for ordering questions are:

1. Place questions that are important to respondents first.
2. Cluster items of similar content together.
3. Introduce less controversial questions first.

You want respondents to feel as unthreatened by and interested in the questions being asked as possible, without getting overwrought about a particular issue.

Administering Questionnaires

The topic of administering questionnaires centers around two main questions. These are: who in the organization should receive the questionnaire and how should the questionnaire be administered. Both questions are worth considering carefully before a questionnaire is sent. Doing so will help you better visualize and anticipate the data you are trying to collect with the questionnaire, and it will also help you to plan ahead so that appropriate timing is observed in the administering of the questionnaire.

RESPONDENTS. Deciding who will receive the questionnaire is handled in conjunction with the task of setting up objectives for its results. Sampling, which is covered in Chapter 5, helps a systems analyst to determine what sort of representation is necessary and hence what kind of respondents should receive the questionnaire.

FIGURE 4.13

When designing a Web survey, keep in mind that there are different ways to capture responses.

Name	Appearance	Purpose
One-line text box		Used to obtain a small amount of text and limit the answer to a few words
Scrolling text box		Used to obtain one or more paragraphs of text
Check box		Used to obtain a yes-no answer (e.g., Do you wish to be included on the mailing list?)
Radio button		Used to obtain a yes-no or true-false answer
Drop-down menu		Used to obtain more consistent results (Respondent is able to choose the appropriate answer from a predetermined list [e.g., a list of state abbreviations])
Push button	Button	Most often used for an action (e.g., a respondent pushes a button marked "Submit" or "Clear")

150   PART 2 • INFORMATION REQUIREMENTS ANALYSIS

CONSULTING OPPORTUNITY 4.5

Order in the Courts

"I love my work," Tennys says, beginning the interview with a volley. "It's a lot like a game. I keep my eye on the ball and never look back," he continues. Tennyson "Tennys" Courts is a manager for Global Health Spas, Inc., which has popular health and recreation spas worldwide.

"Now that I've finished my MBA, I feel like I'm on top of the world with Global," Tennys says. "I think I can really help this outfit shape up with its computers and health spas."

Tennys is attempting to help your systems group, which is developing a cloud-based system to be used by all 80 outlets. "Can

I bounce this off you?" he asks Terri Towell, a member of your team of systems analysts. "It's a questionnaire I designed for distribution to all spa managers."

Ever the good sport, Terri tells Tennys that she'd love to take a look at the form. But back in the office, Terri puts the ball in your court. Systematically critique Tennys's technique, as depicted in Figure 4.C2, and explain to him point by point what it needs to be a matchless questionnaire with a winning form. Building on your critique, tell Tennys what he should do to rewrite the form as a Web survey instead.

QUESTIONNAIRE FOR ALL MANAGERS OF HEALTH SPAS

URGENTFILL OUT IMMEDIATELY AND RETURN PERSONALLY TO YOUR DIVISION MANAGER. YOUR NEXT PAYCHECK WILL BE WITHHELD UNTIL IT IS CONFIRMED THAT YOU HAVE TURNED THIS IN.

In 10 words or fewer, what complaints have you lodged about the current computer system in the last six months to a year? Are there others who feel the same way in your outlet as you do? Who? List their names and positions.

1.   2.

3.   4.

5.

7.

Terri Please help me improve this form. Tennys

What is the biggest problem you have when communicating your information requirements to headquarters? Describe it briefly.

How much computer downtime did you experience last year?

1 - 2 - 3 - 4 - 5 - 6 - 7 - 8 - 9 - 10 -

Is there any computer equipment you never use?

Description   Serial Number

Do you want it removed? Agree   Neutral   Disagree

In your opinion, what's next as far as computers and Global Health Spas are concerned?

Thanks for filling this out. • • • • • • • • • • • • • • • •

FIGURE 4.C2

A questionnaire developed for managers of Global Health Spas by Tennys Courts.

Recipients are often chosen as representative because of their rank, length of service with the company, job duties, or special interest in the current or proposed system. Be sure to include enough respondents to allow for a reasonable sample in the event that some questionnaires are not returned or some response sheets are incorrectly completed and thus must be discarded.

METHODS OF ADMINISTERING A QUESTIONNAIRE. A systems analyst has several options for administering a questionnaire, and the choice of administration method is often determined by the existing business situation. Options for administering the questionnaire include the following:

1. Convening all concerned respondents at one time.
2. Personally handing out blank questionnaires and taking back completed ones.

HYPERCASE® EXPERIENCE 4.2

"You've probably noticed by now that not everyone enjoys filling out questionnaires at MRE. We seem to get more questionnaires than most other organizations. I think it's because many of the employees, especially those from the old training unit, value the contributions of questionnaire data in our work with clients. When you examine the questionnaire that Snowden distributed, you'll probably want not only to look at the results but also to critique it from a methods standpoint. I always feel strongly that we can improve our internal performance so that eventually we can better serve our clients. The next time we construct a questionnaire, we want to be able to improve three things: the reliability of the data, the validity of the data, and the response rate we get."

HYPERCASE Questions

1. What evidence of questionnaires have you found at MRE? Be specific about what you have found and where.
2. Critique the questionnaire that Snowden circulated. What can be done to improve its reliability, validity, and response rate? Provide three practical suggestions.
3. Write a short questionnaire to follow up on some aspects of the merger between Management Systems and the Training Unit at MRE that are still puzzling you. Be sure to observe all the guidelines for good questionnaire design.
4. Redesign the questionnaire you wrote in Question 3 so that it can be used as a Web survey.

3. Allowing respondents to self-administer the questionnaire at work and drop it in a centrally located box.
4. Mailing questionnaires to employees at branch sites and supplying a deadline, instructions, and return postage.
5. Administering the questionnaire electronically either via email or on the Web.

Each of these five methods has advantages and disadvantages. Most commonly, respondents are allowed to self-administer a questionnaire. Response rates with this method are a little lower than with the other methods because people may forget about the form, lose it, or purposely ignore it. Self-administration, however, allows people to feel that their anonymity is ensured and may result in less guarded answers from some respondents. Both email and Web surveys fall into the category of self-administered questionnaires.

Administering a questionnaire electronically, either via email or posted on the Web, is one way to quickly reach current system users. Costs of duplication are minimized. In addition, responses can be made at the convenience of the respondent and then can be automatically collected and stored electronically. Some software permits respondents to begin answering a survey, save their answers, and return to it for completion if they are interrupted. Reminders to respondents can be easily and inexpensively sent via email, as can notifications to the analyst about when the respondent has opened the email. Some software now turns email data into data tables for use in spreadsheet or statistical analysis software. One of the popular services for creating and administering online surveys used by consultants is SurveyMonkey.com (see www.surveymonkey.com), a Portland, Oregon, company started in 1999 that acquired an email marketing services company called MailChimp to its product offerings. Mailchimp allows organizations to send email newsletters and design, send, and track email campaigns.

Research shows that respondents are willing to answer questions about highly sensitive matters via the Internet. Thus, questions that may be difficult to pose in person regarding systems problems may be acceptable to ask on a Web survey.

Summary

This chapter covers three of the key interactive methods for information gathering that the systems analyst can use: interviewing, joint application design (JAD), and construction of questionnaires. During the process of interviewing analysts, a systems analyst should listen for HCI concerns related to ergonomics, aesthetics, usability, and usefulness, as well as goals, feelings, opinions, and informal procedures in interviews with organizational decision makers. Interviews are planned question-and-answer dialogues between two people. An analyst can use an interview to develop a relationship with a client, to observe the workplace, and to collect data. Interviews should preferably be conducted in person.

The five steps to take in planning an interview are to read background material, establish interviewing objectives, decide who to interview, prepare the interviewee, and decide on question types and structure.

Questions are of two basic types: open ended or closed. Open-ended questions leave open all response options for the interviewee. Closed questions limit the possible options for response. Probes or follow-up questions can be either open ended or closed, but either way, they ask the respondent for a more detailed reply.

Interviews can be structured in three basic ways: pyramid, funnel, or diamond. Pyramid structures begin with detailed, closed questions and broaden to more generalized questions. Funnel structures begin with open-ended, general questions and then funnel down to more specific, closed questions. Diamond-shaped structures combine the strengths of the other two structures, but they take longer to conduct. Trade-offs are involved when deciding how structured to make interview questions and question sequences. Listening to user stories is another way to collect valuable information.

To cut both the time and cost of personal interviews, analysts may want to consider joint application design (JAD) instead. Using JAD, analysts can both analyze human information requirements and design a user interface with users in a group setting. Careful assessment of the particular organizational culture will help an analyst judge whether JAD is suitable.

By using questionnaires (surveys), systems analysts can gather data on HCI concerns, attitudes, beliefs, behavior, and characteristics from key people in the organization. Surveys are useful if people in the organization are widely dispersed, if many people are involved with the systems project, if exploratory work is necessary before recommending alternatives, and if there is a need for problem sensing before interviews are conducted.

Once objectives for the survey are set, the analyst can begin writing either open-ended or closed questions. Ideally, the questions should be simple, specific, short, free of bias, not patronizing, technically accurate, addressed to those who are knowledgeable, and written at an appropriate reading level. The systems analyst may want to use scales either to measure the attitudes or characteristics of respondents or to have respondents act as judges for the subject of the questionnaire. Scaling is the process of assigning numbers or other symbols to an attribute or characteristic.

Consistent control of the questionnaire format and style can result in a better response rate. Web surveys can be designed to encourage consistent responses. In addition, the meaningful ordering and clustering of questions is important for helping respondents understand the questionnaire. Surveys can be administered in a variety of ways, including electronically via email or the Web, or with the analyst present in a group of users. Software is available to automatically tabulate email or Web responses.

Keywords and Phrases

bipolar closed questions
central tendency
closed questions
diamond-shaped structure
funnel structure
halo effect
human–computer interaction (HCI)
informal procedures
interval scale
interviewee feelings
interviewee goals

interviewee opinions
joint application design (JAD)
leniency
nominal scale
open-ended questions
probes
pyramid structure
questionnaire
reliability
storytelling survey respondents
validity

Review Questions

1. What kinds of information should be sought in interviews?
2. List the five steps in interview preparation.
3. Define what is meant by open-ended interview questions. List eight benefits and five drawbacks of using them.
4. When are open-ended questions appropriate for use in interviewing?
5. Define what is meant by closed interview questions. List six benefits and four drawbacks of using them.
6. When are closed questions appropriate for use in interviewing?

7. What is a probing question? What is the purpose of using a probing question in interviews?
8. Define what is meant by *pyramid structure*. When is it useful to employ this structure in interviews?
9. Define what is meant by *funnel structure*. When is it useful to employ this structure in interviews?
10. Define what is meant by *diamond-shaped structure*. When is it useful to employ this structure in interviews?
11. Define *joint application design (JAD)*.
12. List the situations that warrant use of JAD in place of personal organizational interviews.
13. List the potential benefits of using JAD.
14. List the three potential drawbacks of using JAD as an alternative to personal interviews.
15. What kinds of information is a systems analyst seeking through the use of questionnaires or surveys?
16. List four situations in which the use of questionnaires is appropriate.
17. What are the two basic question types used on questionnaires?
18. List two reasons a systems analyst would use a closed question on a questionnaire.
19. List two reasons a systems analyst would use an open-ended question on a questionnaire.
20. What are the seven guidelines for choosing language for a questionnaire?
21. Define what is meant by *scaling*.
22. What are two kinds of information or scales that systems analysts most commonly use?
23. What are nominal scales used for?
24. Give an example of an interval scale.
25. When should an analyst use an interval scale?
26. Define *reliability* as it refers to the construction of scales.
27. Define *validity* as it refers to the construction of scales.
28. List three problems that can occur because of careless construction of scales.
29. What four actions can be taken to ensure that a questionnaire's format is conducive to a good response rate?
30. Which questions should be placed first on a questionnaire?
31. Why should questions on similar topics be clustered together?
32. What is an appropriate placement for controversial questions?
33. List five methods for administering a questionnaire.
34. What considerations are necessary when questionnaires are Web based?

Problems

1. As part of your systems analysis project to update the automated accounting functions for Xanadu Corporation, a manufacturer of digital cameras, you will interview Leo Blum, the chief accountant. Write four to six interview objectives covering his use of information sources, information formats, decision-making frequency, desired qualities of information, and decision-making style.
   a. In a paragraph, write down how you will approach Leo to set up an interview.
   b. State which structure you will choose for this interview. Why?
   c. Leo has four subordinates who also use the system. Would you interview them also? Why or why not?
   d. Would you also try to interview customers (visitors to the website)? Are there better ways to get the opinions of customers? Why or why not?
   e. Write three open-ended questions that you will email to Leo prior to your interview. Write a sentence explaining why it is preferable to conduct an interview in person rather than via email.
2. Here are five questions written by one of your systems analysis team members. Her interviewee is the local manager of LOWCO, an outlet of a national discount chain, who has asked you to work on a management information system to provide inventory information. Review these questions for your team member.
   1. When was the last time you thought seriously about your decision-making process?
   2. Who are the trouble makers in your store—I mean the ones who will show the most resistance to changes in the system that I have proposed?
   3. Are there any decisions you need more information about before you can make them?
   4. You don't have any major problems with the current inventory control system, do you?
   5. Tell me a little about the output you'd like to see.
      a. Rewrite each question so that it is more effective in eliciting information.
      b. Order your questions in either a pyramid, funnel, or diamond-shaped structure and label the questions with the name of the structure you used.
      c. What guidelines can you give your team member for improving her interviewing questions in the future? Make a list of them.

3. Ever since you walked through the door, your interviewee, Max Hugo, has been shuffling papers, looking at his watch, and drumming on his desk with his fingers. Based on what you know about interviews, you guess that Max is nervous because of the other work he needs to do. In a paragraph, describe how you would deal with this situation so that the interview can be accomplished with Max's full attention. (Max cannot reschedule the interview for a different day.)

4. Write a series of six *closed* questions that cover the subject of decision-making style for the manager described in Problem 2.

5. Write a series of six *open-ended* questions that cover the subject of decision-making style for the manager described in Problem 2.

6. Examine the interview structure presented in the sequencing of the following questions:
   1. How long have you been in this position?
   2. What are your key responsibilities?
   3. What reports do you receive?
   4. How do you view the goals of your department?
   5. How would you describe your decision-making process?
   6. How can that process best be supported?
   7. How frequently do you make those decisions?
   8. Who is consulted when you make a decision?
   9. What is the one decision you make that is essential to departmental functioning?
      a. What structure is being used? How can you tell?
      b. Restructure the interview by changing the sequence of the questions (you may omit some, if necessary). Label the reordered questions with the name of the structure you have used.

7. The following is the first interview report filed by one of your systems analysis team members: "In my opinion, the interview went very well. The subject allowed me to talk with him for an hour and a half. He told me the whole history of the business, which was very interesting. The subject also mentioned that things have not changed all that much since he has been with the firm, which is about 16 years. We are meeting again soon to finish the interview because we did not have time to go into the questions I prepared."
   a. In two paragraphs, critique the interview report. What critical information is missing?
   b. What information is extraneous to the interview report?
   c. If what is reported actually occurred, what three suggestions do you have to help your teammate conduct a better interview next time?

8. Cab Wheeler is a newly hired systems analyst with your group. Cab has always felt that questionnaires are a waste. Now that you will be doing a systems project for MegaTrucks, Inc., a national trucking firm with branches and employees in 130 cities, you want to use a questionnaire to elicit some opinions about the current and proposed systems.
   a. Based on what you know about Cab and MegaTrucks, list three persuasive reasons why he should use a survey for this study.
   b. Given your careful arguments, Cab has agreed to use a questionnaire but strongly urges that all questions be open-ended so as not to constrain the respondents. In a paragraph, persuade Cab that closed questions are useful as well. Be sure to point out trade-offs involved with each question type.

9. "Every time we get consultants in here, they pass out some goofy questionnaire that has no meaning to us at all. Why don't they bother to personalize it, at least a little?" asks Ray Dient, head of emergency systems. You are discussing the possibility of beginning a systems project with Pohattan Power Company (PPC) of Far Meltway, New Jersey.
   a. What steps will you follow to customize a standardized questionnaire?
   b. What are the advantages of adapting a questionnaire to a particular organization? What are the disadvantages?

10. A sample question from the draft of the Pohattan Power Company questionnaire reads:
    I have been with the company:
    20–upwards years
    10–15 years upwards
    5–10 years upwards
    less than a year
    Check one that most applies.
    a. What kind of a scale is the question's author using?
    b. What errors have been made in the construction of the question, and what might be the possible responses?
    c. Rewrite the question to achieve clearer results.
    d. Where should the question you've written appear on the questionnaire?

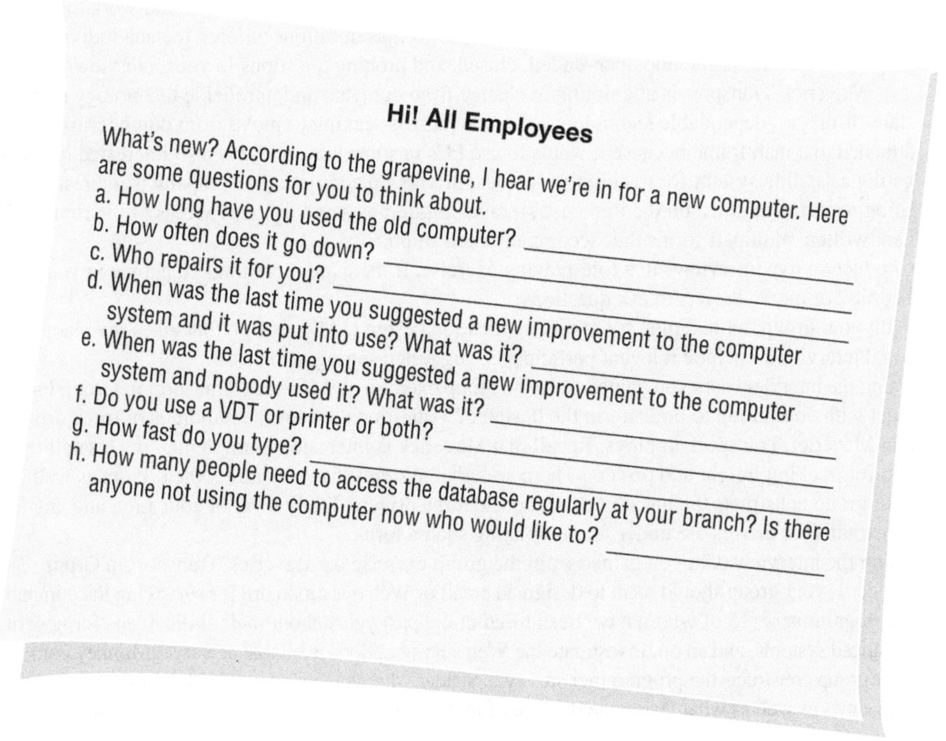

FIGURE 4.EX1

A questionnaire developed by Di Wooly.

Hi! All Employees

What's new? According to the grapevine, I hear we're in for a new computer. Here are some questions for you to think about.
a. How long have you used the old computer? _____
b. How often does it go down? _____
c. Who repairs it for you? _____
d. When was the last time you suggested a new improvement to the computer system and it was put into use? What was it? _____
e. When was the last time you suggested a new improvement to the computer system and nobody used it? What was it? _____
f. Do you use a VDT or printer or both? _____
g. How fast do you type? _____
h. How many people need to access the database regularly at your branch? Is there anyone not using the computer now who would like to? _____

11. Also included on the PPC questionnaire is this question:
    When residential customers call, I always direct them to our website to get an answer.

Sometimes	Never	Always	Usually
1	2	3	4

    a. What type of scale is this intended to be?
    b. Rewrite the question and possible responses to achieve better results.
12. Figure 4.EX1 is a questionnaire designed by an employee of Green Toe Textiles, which specializes in manufacturing men's socks. Di Wooly wrote the questionnaire because, as the office manager at headquarters in Juniper, Tennessee, she is concerned with the proposed purchase and implementation of a new computer system.
    a. Provide a one-sentence critique for each question given.
    b. In a paragraph, critique the layout and style in terms of white space used, room for responses, ease of responding, and so on.
13. Based on what you surmise Ms. Wooly is trying to get through the questionnaire, rewrite and reorder the questions (use both open-ended and closed questions) so that they follow good practice and result in useful information for the systems analysts. Indicate next to each question that you write whether it is open ended or closed and write a sentence indicating why you have written the question this way.
14. Redesign the questionnaire you created for Ms. Wooly in Problem 13 for use on email. Write a paragraph that says what changes were necessary to accommodate email users.
15. Redesign the questionnaire you created for Ms. Wooly in Problem 13 as a Web survey. Write a paragraph that says what changes were necessary to accommodate Web users.

Group Projects

1. With your group members, role-play a series of interviews with various system users at Maverick Transport. Each member of your group should choose one of the following roles: company president, information technology director, dispatcher, customer service agent, or truck driver. Those group members playing roles of Maverick Transport employees should attempt to briefly describe their job responsibilities, goals, and informational needs.

   Remaining group members should play the roles of systems analysts and devise interview questions for each employee. If there are enough people in your group, each analyst may be assigned to interview a different employee. Those playing the roles of systems analysts should work together to develop common questions that they will ask, as well as questions tailored to each individual employee. Be sure to include open-ended, closed, and probing questions in your interviews.

Maverick Transport is attempting to change from outdated and unreliable technology to more state-of-the-art, dependable technology. The company is seeking to move from dumb terminals attached to a mainframe because it wants to use PCs in some way, and it is also interested in investigating a satellite system for tracking freight and drivers. In addition, the company is interested in pursuing ways to cut down on the immense storage requirements and difficult access of the troublesome handwritten, multipart forms that accompany each shipment.

2. Conduct all five interviews in a role-playing exercise. If there are more than 10 people in your group, permit 2 or more analysts to ask questions.
3. With your group, write a plan for a joint application design (JAD) session that takes the place of personal interviews. Include relevant participants, suggested setting, and so on.
4. Using the interview data you gained from the group exercise on Maverick Transport in Group Project 1, meet with your group to brainstorm the design of a questionnaire for the hundreds of truck drivers that Maverick Transport employs. Recall that Maverick is interested in implementing a satellite system for tracking freight and drivers. There are other systems that may affect the drivers as well. As your group constructs the questionnaire, consider the drivers' likely level of education and any time constraints the drivers are under for completing such a form.
5. Using the interview data you gained from the group exercise on Maverick Transport in Group Project 1, your group should meet to design an email or Web questionnaire for surveying the company's 20 programmers (15 of whom have been hired in the past year) about their skills, ideas for new or enhanced systems, and so on. Investigate the Web survey options available at SurveyMonkey.com. As your group constructs the programmer survey, consider what you have learned about users in the other interviews as well as what vision the director of information technology has for the company.

Selected Bibliography

Ackroyd, S., and J. A. Hughes. *Data Collection Context,* 2nd ed. New York: Addison-Wesley, 1992.

Alvesson, M. *Interpreting Interviews,* Thousand Oaks, CA: Sage Publications, Ltd, 2010.

Cash, C. J., and W. B. Stewart, Jr. *Interviewing Principles and Practices,* 13th ed. New York: McGraw-Hill Humanities/Social Sciences/Language, 2010.

Cooper, D. R., and P. S. Schindler. *Business Research Methods,* 11th ed. New York: McGraw-Hill/Irwin, 2010.

Deetz, S. *Transforming Communication, Transforming Business: Building Responsive and Responsible Workplaces.* Cresskill, NJ: Hampton Press, 1995.

Emerick, D., K. Round, and S. Joyce. *Exploring Web Marketing and Project Management.* Upper Saddle River, NJ: Prentice Hall PTR, 2000.

Gane, C. *Rapid System Development.* New York: Rapid System Development, 1987.

Georgia Tech's Graphic, Visualization, and Usability Center. "GVU WWW Survey Through 1998," www.cc.gatech.edu/gvu/user_surveys/survey-1998-10/. Last accessed July 30, 2012.

Hessler, R. M. *Social Research Methods.* New York: West, 1992.

Joint Application Design. GUIDE Publication GPP-147. Chicago: GUIDE International, 1986.

Kendall, J. E. and K. E. Kendall. "Storytelling as a Qualitative Method for IS Research: Heralding the Heroic and Echoing the Mythic." *Australasian Journal of Information Systems,* Vol. 17, No. 2, 2012, pp. 161–187.

Peterson, R. A. *Constructing Effective Questionnaires.* Thousand Oaks, CA: Sage Publications, 1999.

Strauss, J., and R. Frost. *E-Marketing,* 6th ed. Upper Saddle River, NJ: Pearson Prentice Hall, 2012.

Wansink, B., S. Sudman, and N. M. Bradburn. *Asking Questions: The Definitive Guide to Questionnaire Design—For Market Research, Political Polls, and Social and Health Questionnaires,* 2nd ed. New York: Wiley, 2010.

The CPU Case Episode and accompanying student files are available online at www.pearsonglobaleditions.com/kendall.

CHAPTER 5

Information Gathering: Unobtrusive Methods

LEARNING OBJECTIVES

Once you have mastered the material in this chapter you will be able to:

1. Recognize the value of unobtrusive methods of information gathering.
2. Understand the concept of sampling for human information requirements analysis.
3. Construct useful samples of people, documents, and events for determining human information requirements.
4. Create an analyst's playscript to observe decision makers' activities.
5. Apply the STROBE technique to observe and interpret a decision maker's environment and interaction with technologies.

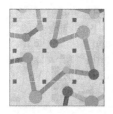

Just by being present in an organization, a systems analyst changes it. However, unobtrusive methods such as sampling, investigation, and observing a decision maker's behavior and interacting with his or her physical environment are less disruptive than other ways of eliciting human information requirements. Unobtrusive methods are considered to be insufficient information-gathering methods when used alone. Rather, they should be used in conjunction with one or many of the interactive methods studied in the previous chapter. This is called a *multiple methods approach*. Using both interactive and unobtrusive methods in approaching an organization is a wise practice that will result in a more complete picture of human information requirements.

Sampling

Sampling is the process of systematically selecting representative elements of a population. When these selected elements are examined closely, it is assumed that the analysis will reveal useful information about the population as a whole.

A systems analyst has to make decisions on two key issues. First, people in the organization have generated many reports, forms, output documents, memos, and websites. Which of these should the systems analyst pay attention to, and which should the systems analyst ignore?

Second, a great many employees can be affected by the proposed information system. Which people should the systems analyst interview, seek information from via questionnaires, or observe in the process of carrying out their decision-making roles?

The Need for Sampling

There are many reasons a systems analyst would want to select either a representative sample of data to examine or representative people to interview, question, or observe. They include:

1. Containing costs
2. Speeding up the data gathering
3. Improving effectiveness
4. Reducing bias

Examining every scrap of paper, talking with everyone, and reading every web page from the organization would be far too costly for the systems analyst. Copying reports, asking employees for valuable time, and duplicating unnecessary surveys would result in much needless expense.

Sampling helps accelerate the process by gathering selected data rather than all data for the entire population. In addition, the systems analyst is spared the burden of analyzing data from the entire population.

Effectiveness in data gathering is an important consideration as well. Sampling can help improve effectiveness if information that is more accurate can be obtained. Such sampling is accomplished, for example, by talking to fewer employees but asking them questions that are more detailed. In addition, if fewer people are interviewed, the systems analyst can afford the time to follow up on missing or incomplete data, thus improving the effectiveness of data gathering.

Finally, data gathering bias can be reduced by sampling. When the systems analyst interviews an executive of the corporation, for example, the executive is involved with the project because this person has already given a certain amount of time to the project and would like it to succeed. When the systems analyst asks for an opinion about a permanent feature of the installed information system, the executive interviewed may provide a biased evaluation because there is little possibility of changing it.

Sampling Design

A systems analyst must follow four steps to design a good sample:

1. Determine the data to be collected or described.
2. Determine the population to be sampled.
3. Choose the type of sample.
4. Decide on the sample size.

These steps are described in detail in the following subsections.

DETERMINING THE DATA TO BE COLLECTED OR DESCRIBED. A systems analyst needs a realistic plan about what will be done with the data once they are collected. If irrelevant data are gathered, then time and money are wasted in the collection, storage, and analysis of useless data.

The duties and responsibilities of the systems analyst at this point are to identify the variables, attributes, and associated data items that need to be gathered in the sample. The objectives of the study must be considered, as well as the type of data-gathering method (for example, investigation, interviews, questionnaires, observation) to be used. The kinds of information sought when using each of these methods are discussed in more detail in this and subsequent chapters.

DETERMINING THE POPULATION TO BE SAMPLED. Next, a systems analyst must determine what the population is. In the case of hard data, the systems analyst needs to decide, for example, if

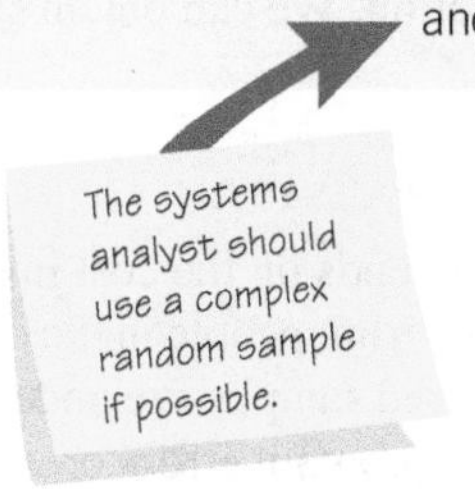

FIGURE 5.1

Four main types of samples a
systems analyst has available.

	Not Based on Probability	**Based on Probability**
Sample elements are selected directly without restrictions	Convenience	Simple random
Sample elements are selected according to specific criteria	Purposive	Complex random (systematic, stratified, and cluster)

The systems analyst should use a complex random sample if possible.

data from the last two months are sufficient, or if an entire year's worth of reports are needed for analysis.

Similarly, when deciding whom to interview, the systems analyst has to determine whether the population should include only one level in the organization or all the levels. Or maybe the analyst should even go outside the system and include the reactions of customers, vendors, suppliers, or competitors. These decisions are explored in more detail in upcoming sections.

CHOOSING THE TYPE OF SAMPLE. A systems analyst can use one of four main types of samples, as pictured in Figure 5.1. They are convenience, purposive, simple random, and complex random. Convenience samples are unrestricted, nonprobability samples. A sample could be called a convenience sample if, for example, the systems analyst posts a notice on the company's intranet asking for everyone interested in working with the new sales performance reports to come to a meeting at 1 P.M. on Tuesday the 12th. Obviously, this sample is the easiest to arrange, but it is also the most unreliable. A purposive sample is based on judgment.

A systems analyst can choose a group of individuals who appear knowledgeable and who are interested in the new information system. Here the systems analyst bases the sample on criteria (knowledge about and interest in the new system), but it is still a nonprobability sample. Thus, purposive sampling is only moderately reliable. If you choose to perform a simple random sample, you need to obtain a numbered list of the population to ensure that each document or person in the population has an equal chance of being selected. This step often is not practical, especially when sampling involves documents and reports. The complex random samples that are most appropriate for a systems analyst are (1) systematic sampling, (2) stratified sampling, and (3) cluster sampling.

In the simplest method of probability sampling, systematic sampling, the systems analyst would, for example, choose to interview every kth person on a list of company employees. This method has certain disadvantages, however. You would not want to use it to select every kth day for a sample because of the potential periodicity problem. Furthermore, a systems analyst would not use this approach if the list were ordered (for example, a list of banks, from the smallest to the largest), because bias would be introduced.

Stratified samples are perhaps the most important to a systems analyst. Stratification is the process of identifying subpopulations, or strata, and then selecting objects or people for sampling in these subpopulations. Stratification is often essential if the systems analyst is to gather data efficiently. For example, if you want to seek opinions from a wide range of employees on different levels of the organization, systematic sampling would select a disproportionate number of employees from the operational control level. A stratified sample would compensate for this. Stratification is also called for when a systems analyst wants to use different methods to collect data from different subgroups. For example, you may want to use a survey to gather data from middle managers, but you might prefer to use personal interviews to gather similar data from executives.

Sometimes a systems analyst must select a group of people or documents to study. This process is referred to as *cluster sampling*. Suppose an organization has 20 help desks scattered across the country. You might want to select 1 or 2 of these help desks, under the assumption that they are typical of the remaining ones.

DECIDING ON THE SAMPLE SIZE. Obviously, if everyone in the population viewed the world the same way or if each of the documents in a population contained exactly the same information as every other document, a sample size of one would be sufficient. Because that is not the case, it is necessary to set a sample size greater than one but less than the size of the population itself.

It is important to remember that the absolute number is more important in sampling than the percentage of the population. We can obtain satisfactory results sampling 20 people in 200 or 20 people in 2,000,000.

The Sample Size Decision

The sample size often depends on the cost involved or the time required by the systems analyst, or even the time available from people in the organization. This subsection gives some guidelines for determining the required sample size under ideal conditions—for example, to determine what percentage of input forms contain errors or what proportion of people to interview.

A systems analyst needs to follow seven steps, some of which involve subjective judgments, to determine the required sample size:

1. Determine the attribute (in this case, the type of errors to look for).
2. Locate the database or reports in which the attribute can be found.
3. Examine the attribute. Estimate p, the proportion of the population having the attribute.
4. Make the subjective decision regarding the acceptable interval estimate, i.
5. Choose the confidence level and look up the confidence coefficient (z value) in a table.
6. Calculate σ_p, the standard error of the proportion, as follows:

$$\sigma_p = \frac{i}{z}$$

7. Determine the necessary sample size, n, using the following formula:

$$n = \frac{p(1-p)}{\sigma_p^2} + 1$$

The first step, of course, is to determine which attribute you will be sampling. Once this is done, you can find out where this data is stored, perhaps in a database, on a form, or in a report.

It is important to estimate p, the proportion of the population having the attribute, so that you set the appropriate sample size. Many textbooks on systems analysis suggest using a heuristic of 0.25 for $p(1-p)$. This value almost always results in a sample size larger than necessary because 0.25 is the maximum value of $p(1-p)$, which occurs only when $p = 0.50$. When $p = 0.10$, as is more often the case, $p(1-p)$ becomes 0.09, resulting in a much smaller sample size.

Steps 4 and 5 are subjective decisions. The acceptable interval estimate of ±0.10 means that you are willing to accept an error of no more than 0.10 in either direction from the actual proportion, p. The confidence level is the desired degree of certainty, such as 95 percent. Once the confidence level is chosen, the confidence coefficient (also called a z value) can be looked up in a table like the one found in this chapter.

Steps 6 and 7 complete the process by taking the parameters found or set in steps 3 through 5 and entering them into two equations to eventually solve the required sample size.

EXAMPLE

The foregoing steps can best be illustrated by an example. Suppose the A. Sembly Company, a large manufacturer of shelving products, asks you to determine what percentage of orders contain errors. You agree to do this job and perform the following steps:

1. Determine that you will be looking for orders that contain mistakes in names, addresses, quantities, or model numbers.
2. Locate copies of order forms from the past six months.
3. Examine some of the order forms and conclude that only about 5 percent (0.05) contain errors.

CONSULTING OPPORTUNITY 5.1

Trapping a Sample

"Real or fake? Fake or real? Who would have thought it, even five years ago?" howls Sam Pelt, a furrier who owns stores in New York; Washington, D.C.; Beverly Hills; and Copenhagen. Sylva Foxx, a systems analyst with her own consulting firm, is talking with Sam for the first time. Currently, P & P, Ltd. (which stands for Pelt and Pelt's son) is using a networked computer that supports package software for a select customer mailing list, accounts payable and accounts receivable, and payroll.

Sam is interested in making some strategic decisions that will ultimately affect the purchasing of goods for his four fur stores. He feels that although the computer might help, other approaches should also be considered.

Sam continues, "I think we should talk to all the customers when they come in the door. Get their opinions. You know, some of them are getting very upset about wearing fur from endangered species. They're very environmentally minded. They prefer fake to real, if they can save a baby animal. Some even like fakes better, calling them 'fun furs.' And I can charge almost the same for a good look-alike.

"It's a very fuzzy proposition, though. If I get too far away from my suppliers of pelts, I may not get what I want when I need it. They see the fake fur people as worms, worse than moths! If I deal with them, the real fur men might not talk to me. They can be animals. On the other hand, I feel strange showing fakes in my stores. All these years, we've prided ourselves on having only the genuine article."

Sam continues, in a nearly seamless monologue, "I want to talk to each and every employee, too."

Sylva glances at him furtively and begins to interrupt, "But that will take months, and purchasing may come apart at the seams unless they know soon what—."

Pelt interrupts, "I don't care how long it takes, if we get the right answers. But they have to be right. Not knowing how to solve this dilemma about fake furs is making me feel like a leopard without its spots."

Sylva talks to Sam Pelt a bit longer and then ends the interview by saying, "I'll talk it all over with the other analysts at the office and let you know what we come up with. I think we can outfox the other furriers if we use software to help us sample opinions, rather than trapping unsuspecting customers into giving an opinion. But I'll let you know what they say. This much is for sure: If we can sample and not talk to everybody before making a decision, every coat you sell will have a silver lining."

As one of the systems analysts who is part of Sylva Foxx's firm, suggest some ways that Sam Pelt can use software to adequately sample the opinions of his customers, store managers, buyers, and any others you feel will be instrumental in making the strategic decision regarding the stocking of fake furs in what has always been a real fur store. Suggest a type of sample for each group and justify it. The constraints you are subject to include the need to act quickly so as to remain competitive, the need to retain a low profile so that competing furriers are unaware of your fact gathering, and the need to keep costs of data gathering to a reasonable level.

4. Make a subjective decision that the acceptable interval estimate will be ±0.02.
5. Choose a confidence level of 95 percent. Look up the confidence coefficient (z value) in Figure 5.2. The z value equals 1.96.
6. Calculate σ_p as follows:

$$\sigma_p = \frac{i}{z} = \frac{0.02}{1.96} = 0.0102$$

7. Determine the necessary sample size, n, as follows:

$$n = \frac{p(1-p)}{\sigma_p^2} + 1 = \frac{0.05(0.95)}{(0.0102)(0.0102)} + 1 = 458$$

The conclusion, then, is to set the sample size at 458. Obviously, a greater confidence level or a smaller acceptable interval estimate would require a larger sample size. If we keep the acceptable interval estimate the same but increase the confidence level to 99 percent (with a z value of 2.58), the necessary sample size becomes 1,827, a figure much larger than the 458 we originally decided to sample.

DETERMINING SAMPLE SIZE WHEN INTERVIEWING. There are no magic formulas to help a systems analyst set the sample size for interviewing. The overriding variable that determines how many people a systems analyst should interview in depth is the time an interview takes. A true in-depth interview and a follow-up interview are very time consuming for both the interviewer and the participant.

FIGURE 5.2

A systems analyst can use a table of area under a normal curve to look up a value once he or she decides on the confidence level.

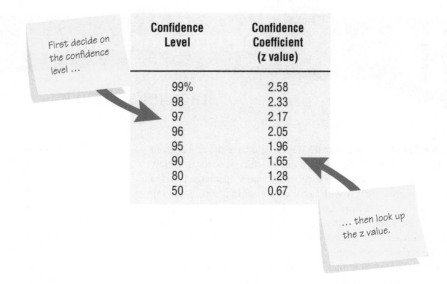

Confidence Level	Confidence Coefficient (z value)
99%	2.58
98	2.33
97	2.17
96	2.05
95	1.96
90	1.65
80	1.28
50	0.67

First decide on the confidence level ...

... then look up the z value.

A good rule of thumb is to interview at least three people at every level of the organization and at least one from each of the organization's functional areas (as described in Chapter 2) who will work directly with a new or updated system. Remember also that it is not necessary to interview more people just because it is a larger organization. If the stratified sample is done properly, a small number of people will adequately represent the entire organization.

Investigation

Investigation is the act of discovery and analysis of data. While investigating evidence in an organization, an analyst acts like Sherlock Holmes, the fabled detective from 221B Baker Street.

As a systems analyst works to understand users, their organization, and its information requirements, it becomes important to examine different types of hard data that offer information unavailable through any other method of data gathering. Hard data reveal where the organization has been and where its members believe it is going. To piece together an accurate picture, the analyst needs to examine both quantitative and qualitative hard data.

Analyzing Quantitative Documents

Many quantitative documents are available for interpretation in any business, and they include reports used for decision making, performance reports, records, and a variety of forms. All these documents have specific purposes and audiences for which they are targeted.

REPORTS USED FOR DECISION MAKING. A systems analyst needs to obtain some of the documents that are used in running the business. These documents are often paper reports regarding the status of inventory, sales, or production. Many of these reports are not complex, but they serve mainly as feedback for quick action. For example, a sales report may summarize the amount sold and the type of sales. In addition, sales reports might include graphical output comparing revenue and income over a set number of periods. Such reports enable the decision maker to spot trends easily.

Production reports include recent costs, current inventory, recent labor, and plant information. In addition to using these key reports, decision makers use many summary reports to provide background information, spot exceptions to normal occurrences, and afford strategic overviews of organizational plans.

PERFORMANCE REPORTS. Most performance reports take the general form of actual versus intended performance. One important function of performance reports is to assess the size of the gap between actual and intended performance. It is also important to be able to determine whether that gap is widening or narrowing as an overall trend in whatever performance is being measured. Figure 5.3 shows a clear improvement in sales performance over two to three months. The analyst will want to note if performance measurement is available and adequate for key organizational areas.

CONSULTING OPPORTUNITY 5.2

A Rose by Any Other
Name . . . Or Quality, Not Quantities

"I think we have everything we need. I've sampled financial statements, sales figures for each branch, waste for each shop—we have it all. With all these numbers, we should be able to figure out how to keep Fields in the green, or at least at the forefront of the flower business. We can even show Seymour Fields himself how his new computer system can make it all happen," says Rod Golden, a junior systems analyst working for a medium-sized consulting group.

The firm, under the supervision of its head systems analyst, Clay Potts, has been working on a systems project for the entire chain of 15 successful florist shops and indoor floral markets called Fields. Each of three Midwestern cities has five Fields outlets.

"Although it's just a budding enterprise now, eventually we want to grow with offshoots to half a dozen states," says Seymour Fields, the owner. "I want to reap the benefits of all the happiness we've sown so far. I think we can do it by playing my hunches about what is the best time to purchase flowers at each European market we buy from, and then we should prune back our purchases.

"Over the past three years, I've written lots of memos to our managers about this plan. They've written some good ones back, too. I think we're ready to stake out some territory on this soon," continues Seymour, painting a rosy picture of Fields's future.

"I agree," says Rod. "When I come back from my analysis of these figures," he says, indicating a large stack of material he has unearthed from Fields field offices, "we'll be able to deliver."

Three weeks later, Rod returns to Clay with wilting confidence. "I don't know what to make of all this. I can't seem to get at what's causing the company's growth, or how it's managed. They've been expanding, but I've been through all the figures, and nothing really seems to make sense yet."

Clay listens empathetically, then says, "You've given me a germ of an idea. What we need is some cross-pollination, a breath of fresh air. We need to dig a little deeper. Did you examine anything but their bottom line?"

Rod looks startled and replies, "No, I—uh—what do you mean?"

How can Clay Potts tactfully explain to Rod Golden that examination of qualitative as well as quantitative documents could be important to delivering an accurate assessment of Fields's potential to be a more fruitful enterprise? In a paragraph, recommend some specific documents that should be read. List the specific steps Rod should follow in evaluating qualitative documents obtained from Fields. Write a paragraph to explain how qualitative documents help in presenting an overall account of Fields's success.

FIGURE 5.3

A performance report showing improvement.

Week	Number of Batches Produced	Number of Batches Rejected	Percentage Rejected	Amount Away from 5% Goal
2/2	245	19	7.8	2.8
2/9	229	19	8.3	3.3
2/16	219	14	6.3	1.3
2/23	252	13	5.2	0.2
3/2	245	13	5.3	0.3
3/9	260	13	5.0	* * *
3/16	275	14	5.1	0.1
3/23	260	13	5.0	* * *
3/30	260	13	5.0	* * *
4/6	244	12	4.9	* * *
4/13	242	11	4.5	* * *
4/20	249	11	4.4	* * *
4/27	249	11	4.4	* * *

* * * indicates met or exceeded the < 5% goal

Performance reports show goals . . .

. . . and trends.

FIGURE 5.4

A manually completed payment record.

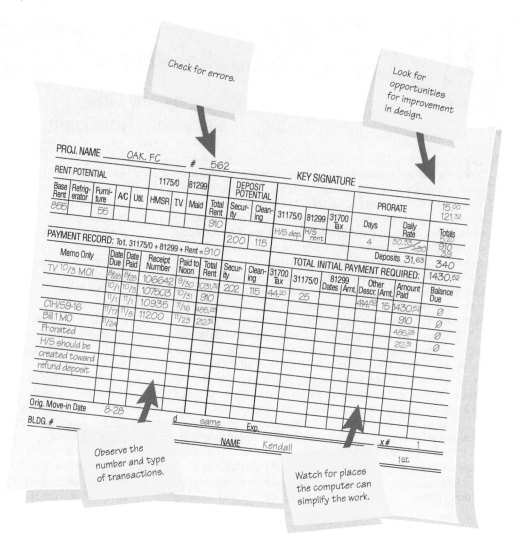

RECORDS. Records provide periodic updates of what is occurring in the business. If the record is updated in a timely fashion by a careful recorder, it can provide much useful information to a systems analyst. Figure 5.4 is a manually completed payment record for an apartment rental. There are several ways that an analyst can inspect a record, many of which are indicative of their usability:

1. Checking for errors in amounts and totals
2. Looking for opportunities for improving the recording form design
3. Observing the number and type of transactions
4. Watching for instances in which the computer can simplify the work (i.e., calculations and other data manipulation)

DATA CAPTURE FORMS. Before you set out to change the information flows in an organization, you need to be able to understand the system that is currently in place. You or one of your team members may want to collect and catalog a blank copy of each form (official or unofficial) that is in use. (Sometimes businesses have a person already charged with forms management, who would be your first source for forms in use.)

Blank forms, along with their instructions for completion and distribution, can be compared with filled-in forms to see if any data items are consistently left blank on the forms; whether the people who are supposed to receive the forms actually do get them; and if they follow standard procedures for using, storing, and discarding them. Remember to print out any Web-based forms that require users to print them. Alternatively, electronic versions that can be submitted via the Web or email can be identified and stored in a database for later inspection.

To proceed when creating a catalog of forms to help you understand the information flow currently in use in the business:

1. Collect examples of all the forms in use, whether officially sanctioned by the business or not (official versus bootleg forms).
2. Note the type of form (whether printed in-house, handwritten, computer-generated in-house, online forms, Web fill-in forms, printed externally and purchased, etc.).
3. Document the intended distribution pattern.
4. Compare the intended distribution pattern with who actually receives the form.

Although this procedure is time-consuming, it is useful. Another approach is to sample data capture forms that have already been completed. Remember to check databases that store consumer data when sampling input from ecommerce transactions. An analyst must keep in mind many particular questions, as illustrated in Figure 5.5. They include the following aspects of HCI relating to usability, aesthetics, and usefulness:

1. Is the form filled out in its entirety? If not, what items have been omitted, and are they consistently omitted? Why?

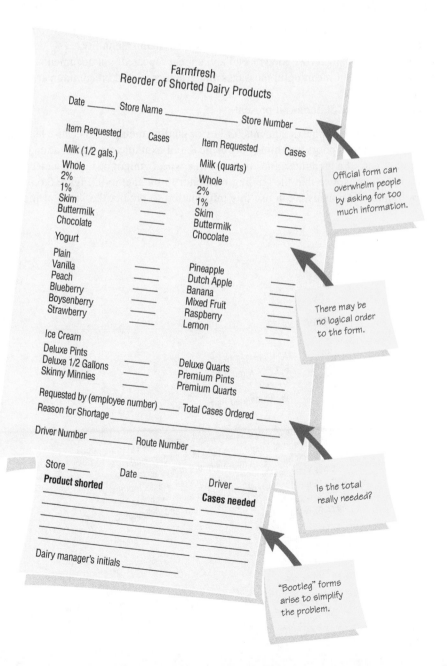

FIGURE 5.5

Questions to ask about official and bootleg forms that are already filled out.

2. Are there forms that are never used? Why? (Check the design and appropriateness of each form for its purported function.)

3. Are all copies of forms circulated to the proper people or filed appropriately? If not, why not? Can people who must access online forms do so?

4. If there is a paper form that is offered as an alternative to a Web-based form, compare the completion rates for both.

5. Are "unofficial" forms being used on a regular basis? (Their use might indicate a problem in standard procedures or may indicate political battles in the organization.)

Analyzing Qualitative Documents

Qualitative documents include email messages, memos, signs on bulletin boards and in work areas, web pages, procedure manuals, and policy handbooks. Many of these documents are rich in details that reveal the expectations for behavior of others that their writers hold and the ways in which users expect to interact with information technologies.

Although many systems analysts are apprehensive about analyzing qualitative documents, they need not be. Several guidelines can help analysts take a systematic approach to this sort of analysis. Many of these relate to the affective, emotional, and motivational aspects of HCI, as well as interpersonal relationships in the organization:

1. Examine documents for key or guiding metaphors.
2. Look for insiders versus outsiders or an "us against them" mentality.
3. List terms that characterize good or evil and appear repeatedly in documents.
4. Look for the use of meaningful messages and graphics posted on common areas or on web pages.
5. Recognize a sense of humor, if present.

It is important to examine documents for key or guiding metaphors because language shapes behavior; thus, the metaphors we employ are critical. For example, an organization that discusses employees as "part of a great machine" or "cogs in a wheel" might be taking a mechanistic view of the organization. Notice that the guiding metaphor in the memo in Figure 5.6 is "We're one big happy family." An analyst can use this information to predict the kinds of metaphors that

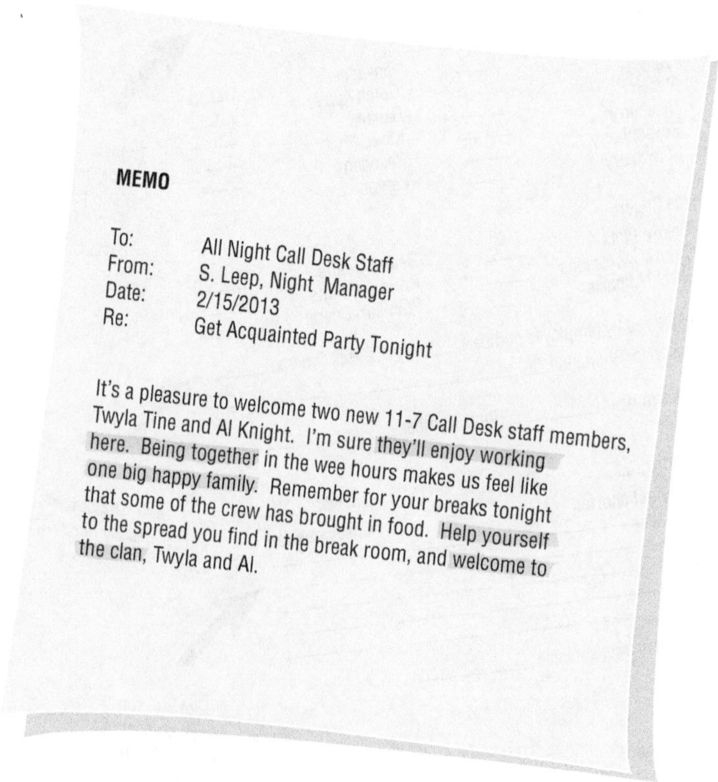

FIGURE 5.6

Analysis of memos provides insight into the metaphors that guide the organization's thinking.

MEMO

To: All Night Call Desk Staff
From: S. Leep, Night Manager
Date: 2/15/2013
Re: Get Acquainted Party Tonight

It's a pleasure to welcome two new 11-7 Call Desk staff members, Twyla Tine and Al Knight. I'm sure they'll enjoy working here. Being together in the wee hours makes us feel like one big happy family. Remember for your breaks tonight that some of the crew has brought in food. Help yourself to the spread you find in the break room, and welcome to the clan, Twyla and Al.

HYPERCASE® EXPERIENCE 5.1

"**W**e're glad you find MRE an interesting place to consult. According to the grapevine, you've been busy exploring the home office. I know, there's so much going on. We find it hard to keep track of everything ourselves. One thing we've made sure of over the years is that we try to use the methods that we believe in. Have you seen any of our reports? How about the data that were collected on one of Snowden's questionnaires? He seems to favor questionnaires over any other method. Some people resent them, but I think you can learn a lot from the results. Some people have been good about cooperating on these projects. Have you met Kathy Blandford yet?"

HYPERCASE Questions

1. Use clues from the case to evaluate the Training Unit's computer experience and its staff's feeling about the PSRS. What do you think the consensus is in the Training Unit toward a computerized project tracking system?
2. What reports and statements are generated by the Training Unit during project development? List each, with a brief description.
3. According to the interview results, what are the problems with the present project tracking system in the Training Unit?
4. Describe the "project management conflict" at MRE. Who is involved? Why is there a conflict?
5. How does the Management Systems Unit keep track of project progress? Briefly describe the method or system.

will be persuasive in the organization. Obviously, if one department is battling another, it may be impossible to gain any cooperation on a systems project until the politics are resolved in a satisfactory manner. Assessing the use of humor provides a quick and accurate barometer of many HCI, interpersonal, and organizational variables, including which subculture a person belongs to and what kind of morale exists.

MEMOS. Along with the five preceding guidelines, an analyst should also consider who sends memos and who receives them. Typically, most information flows downward and horizontally rather than upward in organizations, and extensive email systems mean messages are sent to many work groups and individuals. Memos reveal a lively, continuing dialogue in the organization. Analysis of memo content will provide you with a clear idea of the values, attitudes, and beliefs of organizational members.

SIGNS OR POSTERS ON BULLETIN BOARDS OR IN WORK AREAS. Although signs may seem incidental to what is happening in an organization, they serve as subtle reinforcers of values to those who read them. Slogans posted such as "Quality Is Forever" or "Safety First" give an analyst a sense of the official organizational culture.

CORPORATE WEBSITES. An analyst should view websites used for business-to-consumer (B2C) ecommerce as well as those used for business-to-business (B2B) transactions. Examine the contents for metaphors, humor, use of design features (such as color, graphics, animation, and hyperlinks), and the meaning and clarity of any messages provided. Think about the website from three dimensions: technical, aesthetic, and managerial. Are there discrepancies between the stated goals of the organization and what is presented to the intended viewer? How much customization of the website is available for each user? How much personalization of the website is possible? If you are not designing ecommerce sites for the organization, how does what you see on its website affect the systems you are investigating? Remember to note the level of interactivity of the website or sites, the accessibility of the messages, and the security level.

MANUALS. Other qualitative documents an analyst should examine are organizational manuals, including manuals for computer operating procedures and online manuals. Analyze manuals following the five guidelines spelled out previously. Remember that manuals present the "ideal," the way machines and people are expected to behave. It is important to recall that printed manuals are rarely kept current and are sometimes relegated to shelves, unused.

POLICY HANDBOOKS. The last type of qualitative document we consider is the policy handbook. Although these documents typically cover broad areas of employee and corporate behavior, a systems analyst can be primarily concerned with those that address policies about computer services, use, access, security, and charges. Examining policies allows a systems analyst to gain awareness of the values, attitudes, and beliefs guiding the corporation.

Observing a Decision Maker's Behavior

Observing decision makers, their physical environment, and their interaction with their physical, ergonomic environment is an important unobtrusive method for a systems analyst. Through observing activities of decision makers, an analyst seeks to gain insight about what is actually done, not just what is documented or explained. In addition, through observation of the decision maker, an analyst attempts to see firsthand the relationships that exist between decision makers and other organizational members. Observation of decision makers' interactions with technologies can also reveal important clues regarding HCI concerns, such as how well the system fits with the user.

Observing a Typical Manager's Decision-Making Activities

Managers' workdays have been described as a series of interruptions punctuated by short bursts of work. In other words, pinning down what a manager "does" is a slippery proposition, even under the best of circumstances. A systems analyst can grasp how managers characterize their work by using interactive interviews and questionnaires. Observation, however, allows the analyst to see firsthand how managers gather, process, share, and use information and technology to get work done.

Although it is possible to describe and document how managers make decisions using boxes and arrows, we are primarily describing humans and their activities. Therefore, we suggest that systems analysts use a more humanistic approach to describe what managers do. This method is called the analyst's *playscript*. With this technique, the "actor" is the decision maker who is observed "acting" or making decisions. When you set up a playscript, you list the actor in the left-hand column and all his or her actions in the right-hand column, as shown in Figure 5.7. All activities are recorded with action verbs, so that a decision maker would be described as "talking," "sampling," "corresponding," and "deciding."

Using a playscript is an organized and systematic approach that demands the analyst be able to understand and articulate the action taken by each observed decision maker. This approach eventually assists the systems analyst in determining what information is required for major or frequent decisions made by the observed people. For instance, from the quality assurance manager example in the playscript, it is clear that even though this decision maker is on the middle management level, he or she requires a fair amount of external information to perform the required activities of this specific job.

Observing the Physical Environment

Observing the activities of decision makers is just one way to assess their information requirements. Observing the physical environment where decision makers work also reveals much about their human information requirements. Most often, such observing means systematically examining the offices of decision makers because offices constitute their primary workplace. Decision makers influence, and are in turn influenced by, their physical environments and by their interactions with the technology that takes place there. Many HCI concerns can be identified through structured observation and confirmed with other techniques, such as interviews or questionnaires.

Structured Observation of the Environment (STROBE)

Film critics sometimes use a structured form of criticism called mise-en-scène analysis to systematically assess what is in a single shot of the film. They look at editing, camera angle, set decor, and the actors and their costumes to find out how they are shaping the meaning of the film as intended by the director. Sometimes the film's mise-en-scène will contradict what is said in

FIGURE 5.7

A sample page from an analyst's playscript describing decision making.

Playscript Analysis

Company: Solid Steel Shelving
Analyst: L. Bracket

Scenario: Quality Assurance
Date: 1/3/2013

Decision Maker (Actor)	Information-Related Activity (Script)
Quality Assurance Manager	Asks shop floor supervisor for the day's production report
Shop Floor Supervisor	Prints out daily computerized production report
	Discusses recurring problems in production runs with quality assurance (QA) manager
Quality Assurance Manager	Reads production report
	Compares current report with other reports from the same week
	Inputs data from daily production run into QA model on computer
	Observes onscreen results of QA model
	Calls steel suppliers to discuss deviations from quality standards
Shop Floor Supervisor	Attends meeting on new quality specifications with quality assurance manager and vice president of production
Quality Assurance Manager	Drafts letter to inform suppliers on new quality specifications agreed on in meeting
	Sends draft to vice president via email
Vice President of Production	Reads drafted letter
	Returns corrections and comments via email
Quality Assurance Manager	Reads corrected letter on email
	Rewrites letter to reflect changes

the dialogue. For information requirements analysis, a systems analyst can take on a role similar to that of the film critic. It often is possible to observe the particulars of the surroundings that will confirm or negate the organizational narrative (also called *stories* or *dialogue*) that is found through interviews or questionnaires.

The method *STR*uctured *OB*servation of the *E*nvironment is referred to as STROBE. Successful application of STROBE requires that an analyst explicitly observe seven concrete elements commonly found in offices. The seven observable elements and some key questions that may arise are listed in Figure 5.8. These elements can reveal much about the way a decision maker gathers, processes, stores, and shares information, as well as about the decision maker's credibility in the workplace.

Observable Element	Questions an Analyst Might Investigate
Office location	Who has the corner office? Are the key decision makers dispersed over separate floors?
Desk placement	Does the placement of the desk encourage communication? Does the placement demonstrate power?
Stationary equipment	Does the decision maker prefer to gather and store information personally? Is the storage area large or small?
Props	Is there evidence that the decision maker uses a PC, smartphone, or tablet computer in the office?
External information sources	Does the decision maker get much information from external sources such as trade journals or the Web?
Office lighting and color	Is the lighting set up to do detailed work or more appropriate for casual communication? Are the colors warm and inviting?
Clothing worn by decision makers	Does the decision maker show authority by wearing conservative suits? Are employees required to wear uniforms?

Through the use of STROBE, a systems analyst can gain a better understanding of how managers gather, process, store, and use information. A summary of the characteristics exhibited by decision makers and the corresponding observable elements is shown in Figure 5.9.

Applying STROBE

One way to implement STROBE is through the use of an anecdotal checklist with meaningful shorthand symbols. This approach to STROBE was useful in ascertaining the information requirements for four key decision makers in a franchise clothing store.

In Figure 5.10, the system analysts used five shorthand symbols to evaluate how observation of the STROBE elements compared with the organizational narrative generated through interviews. The five symbols are as follows:

Characteristics of Decision Makers	Corresponding Elements in the Physical Environment
Gathers information informally	Warm, incandescent lighting and colors
Seeks extraorganizational information	Trade journals present in office
Processes data personally	PCs, or tablet computers present in office
Stores information personally	Equipment/files present in office
Exercises power in decision making	Desk placed for power
Exhibits credibility in decision making	Wears authoritative clothing
Shares information with others	Office easily accessible

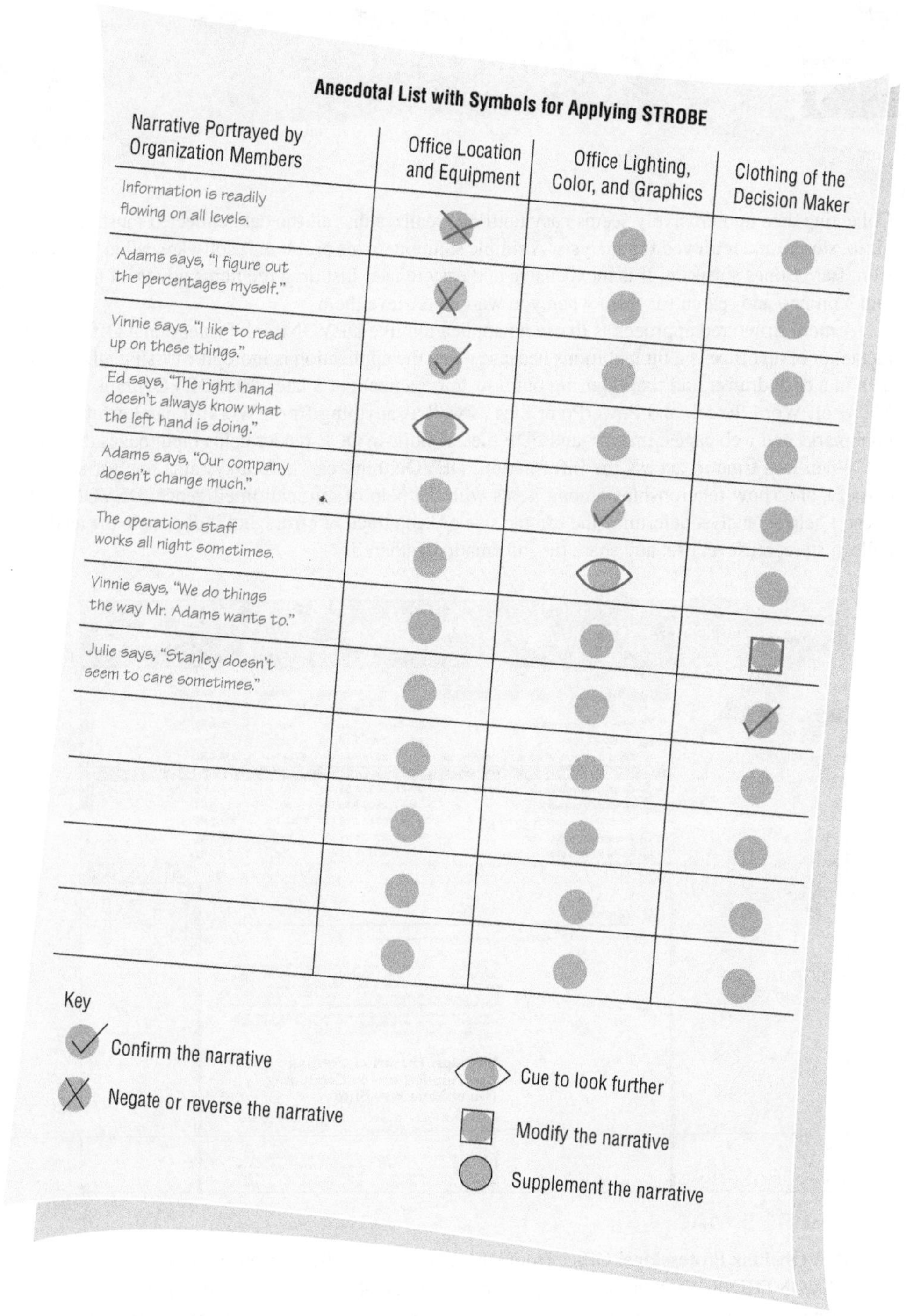

FIGURE 5.10

An anecdotal list with symbols for use in applying STROBE.

Anecdotal List with Symbols for Applying STROBE

1. A check mark means the narrative is confirmed.
2. An "X" means the narrative is reversed.
3. An oval or eye-shaped symbol serves as a cue for the systems analyst to look further.
4. A square means observation of the STROBE elements modifies the narrative.
5. A circle means the narrative is supplemented by what is observed.

When STROBE is implemented in this manner, the analyst first writes down key organizational themes growing out of interviews. Then he or she observes and records the elements of STROBE. The analyst then compares the narrative and observations and uses one of the five appropriate symbols to characterize the relationship. The analyst thus creates a table that first documents and then aids in the analysis of observations.

MAC APPEAL

Collecting data unobtrusively seems easy until one realizes that all the data collected must be organized, stored, and retrieved for analysis. A simple solution to this problem is software called Yojimbo from Bare Bones software. It is inexpensive and easy to use. Just drag the items you want to collect into Yojimbo and search for them when you want to retrieve them.

A more structured approach is to use an application like DEVONthink Professional Office. The metaphor of an office is a bit ambitious because using the application is more like tossing all sorts of data in a desk drawer and then figuring out how to organize it at a later date. DEVONthink accepts Microsoft Word, Excel, and PowerPoint files as well as anything from iWork. It can keep track of bookmarks and web pages, images, and PDF files. A built-in OCR reader helps input pages directly.

When it is time to access the information, DEVONthink can help a systems analyst search, classify, and show relationships among items with the help of artificial intelligence. DEVONthink doesn't help an analyst determine the sample size or keep track of errors, but it does help the analyst collect, store, retrieve, use, and share the information gathered.

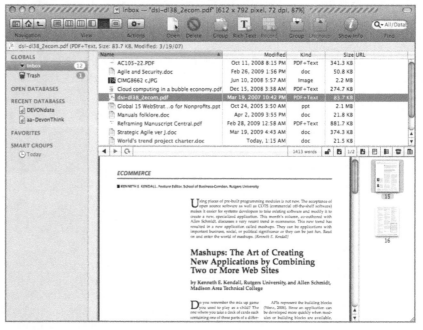

FIGURE 5.MAC

DEVONthink Professional Office from DEVONtechnologies. (Screenshot from DEVONTHINK PRO OFFICE. Copyright © by DEVONtechnologies. Reprinted by permission; Screenshot of "Mashups: The Art of Creating new Applications by Combining Two or More Web Sites" by Kenneth E. Kendall and Allen Schmidt, from DECISION LINE, March 2007. Copyright © 2007 by Decision Sciences Institute. Reprinted by permission.)

Summary

This chapter has covered unobtrusive methods for information gathering, including sampling, investigation of quantitative and qualitative data in current and archived forms, and the observation of the decision maker's activities through the use of the analyst's playscript, as well as observation of the decision maker's physical environment through the use of STROBE.

The process of systematically selecting representative elements of a population is called sampling. The purpose of sampling is to select and study documents such as invoices, sales reports, and memos, or perhaps to select and interview, give surveys to, or observe members of the organization. Sampling can reduce cost, speed data gathering, potentially make the study more effective, and possibly reduce the bias in the study.

HYPERCASE® EXPERIENCE 5.2

"We're proud of our building here in Tennessee. In fact, we used the architectural firm of I. M. Paid to carry the same theme, blending into the local landscape while still reaching out to our clients throughout all the branches. We get lots of people coming through just to admire the building once they catch on to where it is exactly. In fact, by Tennessee standards, we get so many sight-seers that it might as well be the pyramids! Well, you can see for yourself as you go through. The East Atrium is my favorite place: plenty of light, a huge skylight overhead. Yet it has always fascinated me that the building and its furnishings might tell a story quite different from the one its occupants tell.

"Sometimes employees complain that the offices all look the same. The public rooms are spectacular, though. Even the canteen is inviting. Most people can't say that about their cafeterias at work. You'll notice that we all personalize our offices, anyway. So even if the offices were of the 'cookie cutter' kind, their occupants' personalities seem to take over as soon as they have been here a while. What have you seen? Was there anything that surprised you so far?"

HYPERCASE Questions

1. Use STROBE to compare and contrast Evans's and Ketcham's offices. What sort of conclusion about each person's use of information technology can you draw from your observations? How compatible do Evans and Ketcham seem in terms of the systems they use? What other clues to their storage, use, and sharing of information can you discover based on your observations of their offices?
2. Carefully examine Kathy Blandford's office. Use STROBE to confirm, reverse, or negate what you have learned during your interview with her. List anything you found out about Ms. Blandford from observing her office that you did not know from the interview.
3. Carefully examine the contents of the MRE reception area using STROBE. What inferences can you make about the organization? List them. What interview questions would you like to ask, based on your observations of the reception area? Make a list of people you would like to interview and the questions you would ask each of them.

FIGURE 5.HC1

There are hidden clues in HyperCase. Use STROBE to discover them.

A systems analyst must follow four steps in designing a good sample: (1) determine the population itself, (2) determine the type of sample, (3) calculate the sample size, and (4) plan the data that need to be collected or described.

The types of samples useful to a systems analyst are convenience samples, purposive samples, simple random samples, and complex random samples. The last type includes the subcategories of systematic sampling and stratified sampling. There are several guidelines to follow when determining sample size.

Systems analysts need to investigate current and archival data and forms, which reveal where the organization has been and where its members believe it is going. Both quantitative and qualitative documents need to be analyzed. Because documents are persuasive messages, it must be recognized that changing them might well change the organization.

Analysts use observation as an information-gathering technique. Through observation, they gain insight into what is actually done as users interact with information technology. One way to describe how decision makers behave is to use an analyst's playscript that documents each of the major players' activities.

In addition to observing a decision maker's behavior, a systems analyst should observe the decision maker's surroundings for important clues as to how well the system fits the user. One method is Structured Observation of the Environment (STROBE). A systems analyst uses STROBE in the same way that a film critic uses a method called mise-en-scène analysis to analyze a shot in a film.

A systems analyst can observe and interpret several concrete elements in the decision maker's environment. These elements include (1) office location, (2) placement of the decision maker's desk, (3) stationary office equipment, (4) props such as handheld devices and PCs, (5) external information sources such as trade journals and use of the Web, (6) office lighting and color, and (7) clothing worn by the decision maker. A systems analyst can use STROBE to gain a better understanding of how decision makers actually gather, process, store, and share information in order to get their work done.

Keywords and Phrases

analyst's playscript
business-to-business (B2B) ecommerce
business-to-consumer (B2C) ecommerce
cluster sampling
complex random sample
confidence level
convenience sample
corporate websites
external information sources
office lighting and color

purposive sample
sample population
sampling
simple random sample
stratified sampling
STROBE
systematic observation
systematic sampling

Review Questions

1. Define what is meant by *sampling*.
2. List four reasons a systems analyst would want to sample data or select representative people to interview.
3. What are the four steps to follow to design a good sample?
4. List the three approaches to complex random sampling.
5. Define what is meant by *stratification of samples*.
6. What effect on sample size does using a greater confidence level have when sampling attribute data?
7. What is the overriding variable that determines how many people a systems analyst should interview in depth?
8. What information about a decision maker does an analyst seek to gain from observation?
9. List five steps to help an analyst observe a decision maker's typical activities.
10. In the technique known as the analyst's playscript, who is the actor?
11. In an analyst's playscript, what information about managers is recorded in the right-hand column?
12. Noting that the idea of STROBE originally came from the world of film, what does the systems analyst's role resemble?
13. List the seven concrete elements of a decision maker's physical environment that a systems analyst can observe by using STROBE.

Problems

1. Cheryl Stake is concerned that too many forms are being filled out incorrectly. She feels that about 8 percent of all the forms have errors.
   a. How large a sample size should Ms. Stake use to be 99 percent certain she will be within 0.02?
   b. How large a sample size should Ms. Stake use to be 90 percent certain she will be within 0.02?
   c. Explain the difference between parts a and b in words.
   d. Suppose Ms. Stake will accept a confidence level of 95 percent that she will be within 0.02. What will the sample size of forms be now?

2. "I see that you have quite a few papers there. What all do you have in there?" asks Betty Kant, head of the MIS task force that is the liaison group between your systems group and Sawder's Furniture Company. You are shuffling a large bundle of papers as you prepare to leave the building.

   "Well, I've got some financial statements, production reports from the last six months, and some performance reports that Sharon gave me that cover goals and work performance over the last six months," you reply as some of the papers fall to the floor. "Why do you ask?"

   Betty takes the papers from you and puts them on the nearest desk. She answers, "Because you don't need all this junk. You're here to do one thing, and that's talk to us, the users. Bet you can't read one thing in there that'll make a difference."

   a. The only way to convince Betty of the importance of each document is to tell her what you are looking for in each one. In a paragraph, explain what each kind of document contributes to a systems analyst's understanding of the business.

   b. While you are speaking with Betty, you realize that you actually need other quantitative documents as well. List any you are missing.

3. You've sampled the email messages that have been sent to several middle managers of Sawder's Furniture Company, which ships build-your-own particleboard furniture across the country. Here is one that repeats a message found in several other memos:

   > To: Sid, Ernie, Carl
   > From: Imogene
   > Re: Computer/printer supplies
   > Date: November 10, 2012
   > It has come to my attention that I have been waging a war against requests for computer and printer supplies (writable CDs, toner, paper, etc.) that are all out of proportion to what has been negotiated for in the current budget. Because we're all good soldiers here, I hope you will take whatever our supply sergeant says is standard issue. Please, no "midnight requisitioning" to make up for shortages. Thanks for being GI in this regard; it makes the battle easier for us all.

   a. What metaphor(s) is (are) being used? List the predominant metaphor and other phrases that play on that theme.

   b. If you found repeated evidence of this idea in other email messages, what interpretation would you have? Use a paragraph to explain.

   c. In a paragraph, describe how the people in your systems analysis group can use the information from the email messages to shape their systems project for Sawder's.

   d. In interviews with Sid, Ernie, and Carl, there has been no mention of problems with obtaining enough computer and printer supplies. In a paragraph, discuss why such problems may not come up in interviews and discuss the value of examining email messages and other memos in addition to interviewing.

4. "Here's the main policy manual we've put together over the years for system users," says Al Bookbinder, as he blows the dust off the manual and hands it to you. Al is a document keeper for the systems department of Prechter and Gumbel, a large manufacturer of health and beauty aids. "Everything any user of any part of the system needs to know is in what I call the Blue Book. I mean it's chockablock with policies. It's so big, I'm the only one with a complete copy. It costs too much to reproduce it." You thank Al and take the manual with you. When you read through it, you are astonished at what it contains. Most pages begin with a message such as: "This page supersedes page 23.1 in manual Vol. II. Discard previous inserts; do not use."

   a. List your observations about the frequency of use of the Blue Book.

   b. How user friendly are the updates in the manual? Write a sentence explaining your answer.

   c. Write a paragraph commenting on the wisdom of having all-important policies for all systems users in one book.

   d. Suggest a solution that incorporates the use of online policy manuals for some users.

5. "I think I'll be able to remember most everything he does," says Ceci Awll. Ceci is about to interview Biff Welldon, vice president of strategic planning of OK Corral, a steak restaurant chain with 130 locations. "I mean, I've got a good memory. I think it's much more important to listen to what he says than to observe what he does anyway." As one of your systems analysis team members, Ceci has been talking with you about the desirability of writing down her observations of Biff's office and activities during the interview.

   a. In a paragraph, persuade Ceci that listening is not enough in interviews and that observing and recording those observations are also important.

   b. Ceci seems to have accepted your idea that observation is important but still doesn't know what to observe. Make a list of items and behaviors to observe, and in a sentence beside each behavior, indicate what information Ceci should hope to gain through observation of it.

6. "We're a progressive company, always looking to be ahead of the power curve. We'll give anything a whirl if it'll put us ahead of the competition, and that includes every one of us," says I. B. Daring,

an executive with Michigan Manufacturing (2M). You are interviewing him as a preliminary step in a systems project, one in which his subordinates have expressed interest. As you listen to I. B., you look around his office to see that most of the information he has stored on shelves can be classified as internal procedures manuals. In addition, you notice a PC on a back table of I. B.'s office. The display screen is covered with dust, and the manuals stacked beside the PC are still encased in their original shrink-wrap. Even though you know that 2M uses an intranet, no cables are visible going to or from I. B.'s PC. On the wall behind I. B.'s massive mahogany desk you see five framed oil portraits of 2M's founders, all clustered around a gold plaque bearing the corporate slogan, which states, "Make sure you're right, then go ahead."

a. What is the organizational narrative or storyline as portrayed by I. B. Daring? Rephrase it in your own words.

b. List the elements of STROBE that you have observed during your interview with I. B.

c. Next to each element of STROBE that you have observed, write a sentence on how you would interpret it.

d. Construct a table with the organizational story line down the left-hand side of the page and the elements of STROBE across the top. Using the symbols from the "anecdotal list" application of STROBE, indicate the relationship between the organizational story line as portrayed by I. B. and each element you have observed (that is, indicate whether each element of STROBE confirms, reverses, causes you to look further, modifies, or supplements the narrative).

e. Based on your STROBE observations and your interview, state in a paragraph what problems you are able to anticipate in getting a new system approved by I. B. and others. In a sentence or two, discuss how your diagnosis might have been different if you had only talked to I. B. over the phone or had read his written comments on a systems proposal.

Group Projects

1. Assume that your group will serve as a systems analysis and design team for a project designed to computerize or enhance the computerization of all business aspects of a 15-year-old national U.S. trucking firm called Maverick Transport. Maverick is a less-than-a-truckload (LTL) carrier. The people in management work from the philosophy of just-in-time (JIT), in which they have created a partnership that includes the shipper, the receiver, and the carrier (Maverick Transport) for the purpose of transporting and delivering the materials required just in time for their use on the production line. Maverick maintains 626 tractors for hauling freight, and has 45,000 square feet of warehouse space and 21,000 square feet of office space.

a. Along with your group members, develop a list of sources of archival data that should be checked when analyzing the information requirements of Maverick.

b. When this list is complete, devise a sampling scheme that would permit your group to get a clear picture of the company without having to read each document generated in its 15-year history.

2. Arrange to visit a local organization that is expanding or otherwise enhancing its information systems. To allow your group to practice the various observation methods described in this chapter, assign either of these two methods to each team member: (1) developing the analyst's playscript or (2) using STROBE. Many of these strategies can be employed during one-on-one interviews, whereas some require formal organizational meetings. Try to accomplish several objectives during your visit to the organization by scheduling it at an appropriate time, one that permits all team members to try their assigned method of observation. Using multiple methods such as interviewing and observation (often simultaneously) is the only cost-effective way to get a true, timely picture of the organization's information requirements.

3. The members of your group should meet and discuss their findings after completing Group Project 2. Were there any surprises? Did the information garnered through observation confirm, reverse, or negate what was learned in interviews? Were any of the findings from the observational methods in direct conflict with each other? Work with your group to develop a list of ways to address any puzzling information (for example, by doing follow-up interviews).

Selected Bibliography

Cooper, D. R., and P. S. Schindler. *Business Research Methods,* 10th ed. New York: McGraw-Hill/Irwin, 2007.

Edwards, A., and R. Talbot. *The Hard-Pressed Researcher.* New York: Longman, 1994.

Kendall, J. E. "Examining the Relationship Between Computer Cartoons and Factors in Information Systems Use, Success, and Failure: Visual Evidence of Met and Unmet Expectations." *The DATA BASE for Advances in Information Systems,* Vol. 28, No. 2, Spring 1997, pp. 113–126.

Kendall, J. E., and K. E. Kendall. "Metaphors and Methodologies: Living Beyond the Systems Machine." *MIS Quarterly,* Vol. 17, No. 2, June 1993, pp. 149–171.

Kendall J. E., and K. E. Kendall. "Metaphors and Their Meaning for Information Systems Development." *European Journal of Information Systems,* 1994, pp. 37–47.

Kendall, K. E., and J. E. Kendall. "Observing Organizational Environments: A Systematic Approach for Information Analysts." *MIS Quarterly*, Vol. 5, No. 1, 1981, pp. 43–55.

Kendall, K. E., and J. E. Kendall. "STROBE: A Structured Approach to the Observation of the Decision-Making Environment." *Information and Management*, Vol. 7, No. 1, 1984, pp. 1–11.

Kendall, K. E., and J. E. Kendall. "Structured Observation of the Decision-Making Environment: A Validity and Reliability Assessment." *Decision Sciences*, Vol. 15, No. 1, 1984, pp. 107–118.

Markus, M. L., and A. S. Lee. "Special Issue on Intensive Research in Information Systems: Using Qualitative, Interpretive, and Case Methods to Study Information Technology—Second Installment." *MIS Quarterly,* Vol. 24, No. 1, March 2000, p. 1.

Schultze, U. "A Confessional Account of an Ethnography About Knowledge Work." *MIS Quarterly,* Vol. 24, No. 1, March 2000, pp. 3–41.

Shultis, R. L. "'Playscript'—A New Tool Accountants Need." *NAA Bulletin*, Vol. 45, No. 12, August 1964, pp. 3–10.

Webb, E. J., D. T. Campbell, R. D. Schwartz, and L. Sechrest. *Unobtrusive Measures,* 2nd ed. Thousand Oaks, CA: Sage, 1999.

The CPU Case Episode and accompanying student files are available online at www.pearsonglobaleditions.com/kendall.

CHAPTER 6

Agile Modeling and Prototyping

LEARNING OBJECTIVES

Once you have mastered the material in this chapter you will be able to:

1. Understand the roots of agile modeling in prototyping and the four main types of prototyping.
2. Use prototyping for human information requirements gathering.
3. Understand agile modeling and the core practices that differentiate it from other development methodologies.
4. Learn the importance of values critical to agile modeling.
5. Understand how to improve efficiency for users who are knowledge workers using either structured methods or agile modeling.

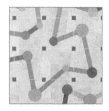 This chapter explores agile modeling, which is a collection of innovative, user-centered approaches to systems development. You will learn the values and principles, activities, resources, practices, processes, and tools associated with agile methodologies. Agile approaches have their roots in prototyping, so this chapter begins with prototyping to provide a proper context for understanding, and then it takes up the agile approach in the last half of the chapter.

Prototyping of information systems is a worthwhile technique for quickly gathering specific information about users' information requirements. Generally speaking, effective prototyping should come early in the SDLC, during the requirements determination phase.

Prototyping is included at this point in the text to underscore its importance as an information-gathering technique. When using prototyping in this way, a systems analyst is seeking initial reactions from users and management to the prototype, user suggestions about changing or cleaning up the prototyped system, possible innovations for it, and revision plans detailing which parts of the system need to be done first or which branches of an organization to prototype next.

Prototyping

As a systems analyst presenting a prototype of an information system, you are keenly interested in the reactions of users and management to the prototype. You want to know in detail how they react to working with the prototype and how good the fit is between their needs and the prototyped features of the system. You gather reactions through observation, interviews, and feedback sheets (possibly questionnaires) designed to elicit each person's opinion about the prototype as he or she interacts with it.

Information gathered in the prototyping phase allows an analyst to set priorities and redirect plans inexpensively, with minimal disruption. Therefore, prototyping and planning go hand-in-hand.

Kinds of Prototypes

The word *prototype* is used in many different ways. Rather than attempt to synthesize all these uses into one definition or try to mandate one correct approach to the somewhat controversial topic of prototyping, we illustrate how each of several conceptions of prototyping may be usefully applied in a particular situation, as shown in Figure 6.1.

PATCHED-UP PROTOTYPE. The first kind of prototyping has to do with constructing a system that works but is patched up or patched together. In engineering, this approach is referred to as breadboarding: creating a patched-together, working model of an (otherwise microscopic) integrated circuit.

An example in information systems is a working model that has all the necessary features but is inefficient. In this instance of prototyping, users can interact with the system, getting accustomed to the interface and types of output available. The retrieval and storage of information may be inefficient, however, because programs were written rapidly, with the objective of being workable rather than efficient.

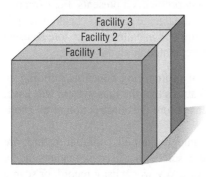

Patched-Up Prototype

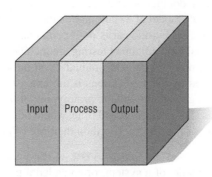

Nonoperational Prototype

FIGURE 6.1

Four kinds of prototypes.

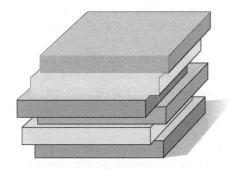

First-of-a-Series Prototype

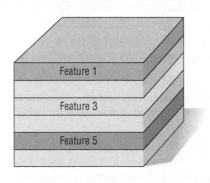

Selected Features Prototype

CONSULTING OPPORTUNITY 6.1

Is Prototyping King?

"As you know, we're an enthusiastic group. We're not a dynasty yet, but we're working on it," Paul LeGon tells you. Paul (introduced in Consulting Opportunity 2.3), at 24 years of age, is the "boy king" of Pyramid, Inc., a small but successful independent book-publishing firm that specializes in paperback books outside of the publishing mainstream. As a systems analyst, you have been hired by Pyramid, Inc., to help develop a computerized warehouse inventory and distribution information system.

"We're hiring lots of workers," Paul continues, as if to convince you of the vastness of Pyramid's undertaking. "And we feel Pyramid is positioned perfectly as far as our markets in the north, south, east, and west are concerned.

"My assistant, Ceil Toom, and I have been slaving away, thinking about the new system. And we've concluded that what we really need is a prototype. As a matter of fact, we've tunneled through a lot of material. Our fascination with the whole idea has really pyramided."

As you formulate a response to Paul, you think back over the few weeks you've worked with Pyramid, Inc. You think that the business problems its information system must resolve are very straightforward. You also know that the people in the company are on a limited budget and cannot afford to spend like kings. Actually, the entire project is quite small.

Ceil, building on what Paul has said, tells you, "We don't mean to be too wrapped up with it, but we feel prototyping represents the new world. And that's where we all want to be. We know we need a prototype. Have we convinced you?"

Based on Paul's and Ceil's enthusiasm for prototyping and what you know about Pyramid's needs, would you support construction of a prototype? Why or why not? Formulate your decision and response in a letter to Paul LeGon and Ceil Toom. Present a justification for your decision based on overall criteria that should be met to justify prototyping.

NONOPERATIONAL PROTOTYPE. The second conception of a prototype is that of a nonworking scale model that is set up to test certain aspects of the design. An example of this approach is a full-scale model of an automobile that is used in wind tunnel tests. The size and shape of the auto are precise, but the car is not operational. In this case, only features of the automobile that are essential to wind tunnel testing are included.

A nonworking scale model of an information system might be produced when the coding required by the applications is too extensive to prototype but when a useful idea of the system can be gained through the prototyping of the input and output only. In this instance, processing, because of undue cost and time, would not be prototyped. Users could still make decisions on the utility of the system, based on their use of prototyped input and output.

FIRST-OF-A-SERIES PROTOTYPE. A third conception of prototyping involves creating a first full-scale model of a system, often called a pilot. An example is prototyping the first airplane of a series and then seeing if it flies before building a second. The prototype is completely operational and is a realization of what the designer hopes will be a series of airplanes with identical features.

This type of prototyping is useful when many installations of the same information system are planned. The full-scale working model allows users to experience realistic interaction with the new system, but it minimizes the cost of overcoming any problems it presents. For example, when a retail grocery chain intends to use electronic data interchange (EDI) to check in suppliers' shipments in a number of outlets, a full-scale model might be installed in one store so users could work through any problems before the system is implemented in all the others.

SELECTED FEATURES PROTOTYPE. A fourth conception of prototyping concerns building an operational model that includes some, but not all, of the features that the final system will have. An analogy would be a new retail shopping mall that opens before the construction of all shops is complete.

When prototyping information systems in this way, some, but not all, essential features are included. For example, users may view a system menu on a screen that lists six features: add a record, update a record, delete a record, search a record for a key word, list a record, or scan a record. In the prototyped system, however, only three of the six may be available for use, so that the user may add a record (feature 1), delete a record (feature 3), and list a record (feature 5).

User feedback can help analysts understand what is working and what isn't. It can also help with suggestions on what features to add next.

When this kind of prototyping is done, the system is accomplished in modules so that if the features that are prototyped are evaluated by users as successful, they can be incorporated into the larger, final system without undertaking immense work in interfacing. Prototypes done in this manner are part of the actual system. They are *not* just mock-ups, as in nonoperational prototyping, considered previously. Unless otherwise mentioned, all further references to prototyping in this chapter refer to the selected-features prototypes.

Prototyping as an Alternative to the SDLC

Some analysts argue that prototyping should be considered as an alternative to the SDLC. Recall that the Systems Development Life Cycle, introduced in Chapter 1, is a logical, systematic approach to follow in the development of information systems.

Complaints about going through the SDLC process center on two interrelated concerns. The first concern is the extended time required to go through the development life cycle. As the investment of analyst time increases, the cost of the delivered system rises proportionately.

The second concern about using the SDLC is that user requirements change over time. During the long interval between the time that user requirements are analyzed and the time that the finished system is delivered, user requirements are evolving. Thus, because of the extended development cycle, the resulting system may be criticized for inadequately addressing current user information requirements.

A corollary of the problem of keeping up with user information requirements is the suggestion that users cannot really know what they do or do not want until they see something tangible. In the traditional SDLC, it often is too late to change an unwanted system once it is delivered.

To overcome these problems, some analysts propose that prototyping be used as an alternative to the SDLC. When prototyping is used in this way, an analyst effectively shortens the time between ascertainment of human information requirements and delivery of a workable system. In addition, using prototyping instead of the traditional SDLC might overcome some of the problems of accurately identifying user information requirements.

Drawbacks to supplanting the SDLC with prototyping include prematurely shaping a system before the problem or opportunity being addressed is thoroughly understood. Also, using prototyping as an alternative may result in producing a system that is accepted by specific groups of users but that is inadequate for overall system needs.

The approach we advocate here is to use prototyping as a part of the traditional SDLC. In this view, prototyping is considered as an additional, specialized method for ascertaining users' information requirements as they interact with prototypes and provide feedback for the analyst.

Developing a Prototype

Prototyping is a superb way to elicit feedback about a proposed system and about how readily it is fulfilling the information needs of its users, as depicted in Figure 6.2. The first step of prototyping is to estimate the costs involved in building a module of the system. If costs of programmers' and analysts' time as well as equipment costs are within the budget, building of the prototype can proceed. Prototyping is an excellent way to facilitate the integration of the information system into the larger system and culture of the organization.

Guidelines for Developing a Prototype

Once the decision to prototype has been made, four main guidelines must be observed when integrating prototyping into the requirements determination phase of the SDLC:

1. Work in manageable modules.
2. Build the prototype rapidly.
3. Modify the prototype in successive iterations.
4. Stress the user interface.

As you can see, the guidelines suggest ways of proceeding with the prototype that are necessarily interrelated. Each guideline is explained in the following subsections.

FIGURE 6.2

Analysts should modify their original screen designs based on user reactions to the prototype.

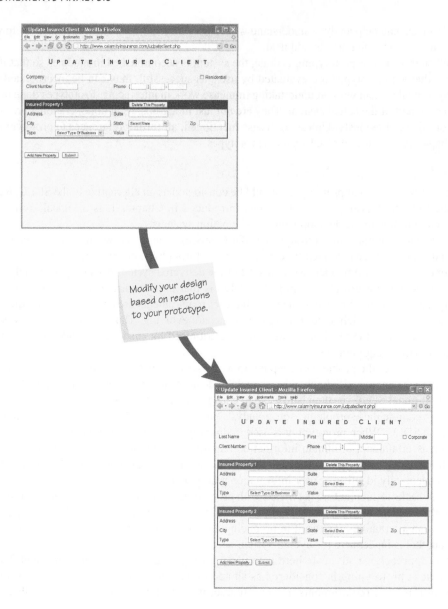

WORKING IN MANAGEABLE MODULES. When prototyping some of the features of a system into a workable model, it is imperative that an analyst work in manageable modules. One distinct advantage of prototyping is that it is not necessary or desirable to build an entire working system for prototype purposes.

A manageable module is one that allows users to interact with its key features but can be built separately from other system modules. Module features that are deemed less important are purposely left out of the initial prototype. As you will see later in this chapter, this is very similar to the agile approach that emphasizes small releases.

BUILDING A PROTOTYPE RAPIDLY. Speed is essential to the successful prototyping of an information system. Recall that one complaint voiced against following the traditional SDLC is that the interval between requirements determination and delivery of a complete system is far too long to address evolving user needs effectively.

Analysts can use prototyping to shorten this gap by using traditional information-gathering techniques to pinpoint salient information requirements, and then quickly make decisions that bring forth a working model. In effect the user sees and uses the system very early in the SDLC instead of waiting for a finished system to gain hands-on experience.

Putting together an operational prototype both rapidly and early in the SDLC allows an analyst to gain valuable insight into how the remainder of the project should go. By showing users very early in the process how parts of the system actually perform, rapid prototyping guards

CONSULTING OPPORTUNITY 6.2

Clearing the Way for Customer Links

World's Trend (see Chapter 7 for a detailed corporate description) is building a website on which to sell clearance merchandise usually sold through the Web and through its catalog operation. As a newly hired Web consultant, Lincoln Cerf finds himself in a very cold, wintry city, fighting his way through several inches of snow to meet with one of the systems team members, Mary Maye, at World's Trend headquarters.

Mary welcomes Lincoln, saying, "At least the weather doesn't seem to affect our Web sales! They're brisk no matter what." Lincoln groans appreciatively at her weak attempt at humor, smiles, and says, "I gathered from your email last week that you are trying to determine the type of information that needs to be displayed on your clearance website."

Mary replies, "Yes, I'm trying to get it organized in the best possible way. Our customers are all so busy. I know photos of all our merchandise can take a long time to appear on the page if a customer is accessing the Web via a slower modem from home." Mary continues, saying, "Linc, I'm not even that concerned about how to design our clearance site at this time. I am worried, though, about how much information we need to include on a page. For example, when items are on clearance, not all colors and sizes are available. Which do you think is better, to include some basic information

and let the customer click a button to ask for more information, or to be as complete as possible on one page? If I use the linking method, then I could fit more items on the screen . . . but it might be too orderly. Customers like the look and feel of a sale in which merchandise is kind of jumbled together."

Linc continues her line of thought, saying, "Yeah, I wonder how customers want the information organized. Have you actually watched them use the Web? I mean, do they look for shoes when they buy a suit? If so, should shoes appear on the suit page or be linked in some way?"

Mary comments, "Those are my questions, too. Then I wonder if we should just try this approach for men's clothes first, before we implement it for women's clothing. What if men's and women's approaches to shopping on the Web are different?"

As a third member of the World's Trend website development group, respond in a brief written report to Lincoln and Mary about whether you should use a prototype to elicit recommendations from potential customers about the proposed website. What type of prototype is appropriate? Consider each form of prototype and explain why each type would apply (or would not apply) to this problem. Devote a paragraph to each explanation.

---

against overcommitting resources to a project that may eventually become unworkable. In addition, agile modeling also builds on quick turnaround times.

MODIFYING A PROTOTYPE. A third guideline for developing a prototype is that its construction must support modifications. Making a prototype modifiable means creating it in modules that are not highly interdependent. If this guideline is observed, less resistance is encountered when modifications in the prototype are necessary.

The prototype is generally modified several times, going through several iterations. Changes in the prototype should move the system closer to what users say is important. Each modification necessitates another evaluation by users.

A prototype is not a finished system. Entering the prototyping phase with the idea that the prototype will require modification is a helpful attitude that demonstrates to users how necessary their feedback is if the system is to improve.

STRESSING THE USER INTERFACE. The user interface for a prototype (and eventually the system) is very important. What you are trying to achieve with a prototype is to get users to further articulate their information requirements, so they must be able to interact easily with the system's prototype. They should be able to see how the prototype will enable them to accomplish their tasks. For many users, the interface is the system. It should not be a stumbling block.

Although many aspects of the system will remain undeveloped in the prototype, the user interface must be well developed enough to enable users to pick up the system quickly and not be put off. Online, interactive systems using GUI interfaces are ideally suited to prototypes. Chapter 14 describes in detail the considerations that are important in designing HCI.

Disadvantages of Prototyping

As with any other information-gathering technique, there are several disadvantages to prototyping. The first is that it can be quite difficult to manage prototyping as a project in the larger systems effort. The second disadvantage is that users and analysts may adopt a prototype as a completed system when it is in fact inadequate and was never intended to serve as a finished system. Analysts need to work to ensure that communication with users is clear regarding the timetable for interacting with and improving the prototype.

An analyst needs to weigh these disadvantages against the known advantages when deciding whether to prototype, when to prototype, and how much of the system to prototype.

Advantages of Prototyping

Prototyping is not necessary or appropriate in every systems project, as we have seen. The advantages, however, should also be given consideration when deciding whether to prototype. The three major advantages of prototyping are the potential for changing the system early in its development, the opportunity to stop development on a system that is not working, and the possibility of developing a system that more closely addresses users' needs and expectations.

Successful prototyping depends on early and frequent user feedback, which analysts can use to modify the system and make it more responsive to actual needs. As with any other systems effort, early changes are less expensive than changes made late in the project's development. In the later part of the chapter, you will see how the agile approach to development uses an extreme form of prototyping that requires an on-site customer to provide feedback during all iterations.

Prototyping Using COTS Software

Sometimes the quickest way to prototype is through the modular installation of COTS software. Although the concept of COTS software can be easily grasped by looking at familiar and relatively inexpensive packages such as the Microsoft Office products, some COTS software is elaborate and expensive but highly useful.

The Users' Role in Prototyping

The users' role in prototyping can be summed up in two words: honest involvement. Without user involvement, there is little reason to prototype. The precise behaviors necessary for interacting with a prototype can vary, but it is clear that the user is pivotal to the prototyping process. Realizing the importance of the user to the success of the process, the members of a systems analysis team must encourage and welcome input and guard against their own natural resistance to changing the prototype.

There are three main ways a user can be of help in prototyping:

1. Experimenting with the prototype
2. Giving open reactions to the prototype
3. Suggesting additions to or deletions from the prototype

Users should be free to experiment with the prototype. In contrast to a mere list of systems features, a prototype allows users the reality of hands-on interaction. Mounting a prototype on an interactive website is one way to facilitate this interaction.

Another aspect of the users' role in prototyping requires that they give open reactions to the prototype. Analysts need to be present at least part of the time when experimentation is occurring. They can then observe users' interactions with the system, and they are bound to see interactions they never planned. A filled-in form for observing user experimentation with the prototype is shown in Figure 6.3. Some of the variables you should observe include user reactions to the prototype, user suggestions for changing or expanding the prototype, user innovations for using the system in completely new ways, and any revision plans for the prototype that aid in setting priorities.

A third aspect of the users' role in prototyping is their willingness to suggest additions to or deletions from the features being tried. The analyst's role is to elicit such suggestions by assuring users that the feedback they provide is taken seriously, by observing users as they interact with the system, and by conducting short, specific interviews with users concerning their experiences with the prototype. Although users will be asked to articulate suggestions and innovations for the prototype, in the end it is the analyst's responsibility to weigh this feedback and translate it into workable changes where necessary. To facilitate the prototyping process, the analyst must clearly communicate the purposes of prototyping to users, along with the idea that prototyping is valuable only when users are meaningfully involved.

FIGURE 6.3

An important step in prototyping
is to properly record user
reactions, user suggestions,
innovations, and revision plans.

Prototype Evaluation Form				
Observer Name Michael Cerveris				
System or Project Name			Date 1/06/2013	
Cloud Computing Data Center		Company or Location		
Program Name or Number Prev. Maint.		Aquarius Water Filters		
		Version	1	
	User 1	**User 2**	**User 3**	**User 4**
User Name	Andy H.	Pam H.		
Period Observed	1/06/2013	1/06/2013		
User Reactions	Generally favorable, got excited about project	Excellent!		
User Suggestions	Add the date when maintenance was performed.	Place a form number on top for reference. Place word WEEKLY in title.		
Innovations				
Revision Plans	Modify on 1/08/2013 Review with Andy and Pam.			

Agile Modeling

Agile methods are a collection of innovative, user-centered approaches to systems development. You will learn the values and principles, activities, resources, practices, processes, and tools associated with agile methodologies in the upcoming sections. Agile methods can be credited with many successful systems development projects and, in numerous cases, even credited with rescuing companies from a failing system that was designed using a structured methodology.

Values and Principles of Agile Modeling

The agile approach is not based just on results. It is based on values, principles, and practices. Essential to agile programming are stated values and principles that create the context for collaboration among programmers and customers. In order to be agile analysts, you must adhere to the following values and principles as developed by Beck (2000) in his work on agile modeling that he called "extreme programming," or "XP."

FOUR VALUES OF AGILE MODELING. There are four values that create an environment in which both developers and businesses can be adequately served. Because there is often tension between what developers do in the short term and what is commercially desirable in the long term, it is important that you knowingly espouse values that will form a basis for acting together on a software project. The four values are communication, simplicity, feedback, and courage, as shown in Figure 6.4.

Let's begin with communication. Every human endeavor is fraught with possibilities for miscommunication. Systems projects that require constant updating and technical design are especially prone to such errors. Add to this tight project deadlines, specialized jargon, and the stereotype that programmers would prefer to talk to machines rather than people, and you have the potential for some serious communication problems. Projects can be delayed, the wrong problem can be solved, programmers may be punished for even bringing up problems to managers, people may leave or join the project in midstream without proper updates, and so the litany goes.

FIGURE 6.4

Values are crucial to the agile
approach.

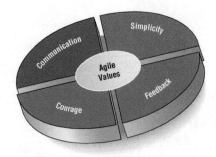

Typical agile practices such as pair programming (where two programmers collaborate, described later in the chapter), tasks estimation, and unit testing rely heavily on good communication. Problems are fixed rapidly, holes are closed, and weak thinking is quickly strengthened through interaction with others on the team.

A second value of the agile approach is simplicity. When we are working on a software development project, our first inclination is to become overwhelmed with the complexity and bigness of the task. However, you cannot run until you know how to walk, nor walk until you know how to stand. Simplicity for software development means beginning with the simplest possible thing we can do.

The agile value simplicity asks us to do the simplest thing today, with the understanding that it might have to be changed a little tomorrow. This requires a clear focus on the goals of the project and really is a basic value.

Feedback is the third basic value that is important when taking an extreme programming approach. When you think of feedback in this context, it is good to consider that feedback is wrapped up with the concept of time. Good, concrete feedback that is useful to the programmer, analyst, and customer can occur within seconds, minutes, days, weeks, or months, depending on what is needed, who is communicating, and what will be done with the feedback. A fellow programmer may hand you a test case that breaks the code you wrote only hours before, but that feedback is almost priceless in terms of being able to change what is not working before it is accepted and further embedded in the system.

Feedback occurs when customers create functional tests for all the stories that programmers have subsequently implemented. (See more on user stories later in this chapter.) Critical feedback about the schedule comes from customers who compare the goal of the plan to the progress that has been made. Feedback helps programmers make adjustments and lets the business start experiencing very early on what the new system will be like once it is fully functional.

Courage is the fourth value in agile programming. Courage has to do with the level of trust and comfort that must exist in a development team. It means not being afraid to throw out an afternoon or a day of programming and begin again if all is not right. It means being able to stay in touch with one's instincts (and test results) concerning what is working and what is not.

Courage also means responding to concrete feedback, acting on your teammates' hunches when they believe that they have a simpler, better way to accomplish a goal. Courage is a high-risk, high-reward value that encourages experimentation that can take the team to its goal more rapidly, in an innovative way. Courage means that you and your teammates trust each other and your customers enough to act in ways that will continuously improve what is being done on the project, even if they require throwing out code, rethinking solutions, or further simplifying approaches. Courage also implies that you, as a systems analyst, eagerly apply the practices of the agile approach.

Analysts can best reflect all four values through an attitude of humility. Historically, computer software was developed by experts who often thought they knew how to run a business better than the local customers who were the true domain experts. Computer experts were often referred to as "gurus." Some of the gurus displayed large egos and insisted on their infallibility, even when customers did not believe it. Many gurus lacked the virtue of humility.

However, maintaining a humble attitude during systems development is critical. You must continually embrace the idea that if the user is expressing a difficulty, then that difficulty must be addressed. It cannot be ignored. Agile modelers are systems analysts who make suggestions, voice opinions, but never insist that they are right 100 percent of the time. Agile modelers

CONSULTING OPPORTUNITY 6.3

To Hatch a Fish

"**J**ust be a little patient. I think we need to add a few more features before we turn it over to them. Otherwise, this whole prototype will sink, not swim," says Sam Monroe, a member of your systems analysis team. All four members of the team are sitting together in a hurriedly called meeting, and they are discussing the prototype that they are developing for an information system to help managers monitor and control water temperature, number of fish released, and other factors at a large, commercial fish hatchery.

"They've got plenty to do already. Why, the system began with four features and we're already up to nine. I feel like we're swimming upstream on this one. They don't need all that. They don't even want it," argues Belle Uga, a second member of the systems analysis team. "I don't mean to carp, but just give them the basics. We've got enough to tackle as it is."

"I think Monroe is more on target," volunteers Wally Ide, a third member of the team, baiting Belle a little. "We have to show them our very best, even if it means being a few weeks later in hatching our prototype than we promised."

"Okay," Belle says warily, "but I want the two of you to tell the managers at the hatchery why we aren't delivering the prototype. I don't want to. And I'm not sure they'll let you off the hook that easily."

Monroe replies, "Well, I guess we could, but we probably shouldn't make a big deal out of being later than we wanted. I don't want to rock the boat."

Wally chimes in, "Yeah. Why point out our mistakes to everyone? Besides, when they see the prototype, they'll forget any complaints they had. They'll love it."

Belle finds a memo in her notebook from their last meeting with the hatchery managers and reads it aloud. "Agenda for meeting of September 22. 'Prototyping—the importance of rapid development, putting together the user analyst team, getting quick feedback for modification. . . .'" Belle's voice trails off, omitting the last few agenda items. In the wake of her comments, Monroe and Wally look unhappily at each other.

Monroe speaks first: "I guess we did try to get everyone primed for receiving a prototype quickly and to be involved from day one." Noting your silence up until now, Monroe continues, "But still waters run deep. What do you think we should do next?" he asks you.

As the fourth member of the systems analysis team, what actions do you think should be taken? In a one- or two-paragraph email message to your teammates, answer the following questions: Should more features be added to the hatchery system prototype before giving it to the hatchery managers to experiment with? How important is the rapid development of the prototype? What are the trade-offs involved in adding more features to the prototype versus getting a more basic prototype to the client when it was promised? Complete your message with a recommendation.

---

possess the self-confidence to allow their customers to question, critique, and sometimes complain about the system under development. Analysts learn from their customers, who have been in business a long time.

THE BASIC PRINCIPLES OF AGILE MODELING. In a perfect world, customers and a software development team would see eye to eye, and communication would not be necessary. We would all be in agreement at all times. But we know that the ideal world doesn't exist. So how can we bring our software development projects closer to the ideal? Part of why this will not happen is that so far we are trying to operate on a vague system of shared values. They're a good beginning, but they are really not operationalized to the point at which we can measure our success in any meaningful way. So we work to derive the basic principles that can help us check whether what we are doing in our software project is actually measuring up to the values that we share.

Agile principles are the reflections and specifications of agile values. They serve as guidelines for developers to follow when developing systems. They also serve to set agile methodologies apart from the more traditional plan-driven methodologies such as SDLC as well as object-oriented methodologies.

Agile principles were first described by Beck and have evolved ever since. These principles can be expressed in a series of sayings such as:

1. Satisfy the customer through delivery of working software.
2. Embrace change, even if introduced late in development.
3. Continue to deliver functioning software incrementally and frequently.
4. Encourage customers and analysts to work together daily.
5. Trust motivated individuals to get the job done.

6. Promote face-to-face conversation.
7. Concentrate on getting software to work.
8. Encourage continuous, regular, and sustainable development.
9. Adopt agility with attention to mindful design.
10. Support self-organizing teams.
11. Provide rapid feedback.
12. Encourage quality.
13. Review and adjust behavior occasionally.
14. Adopt simplicity.

Often you will hear agile developers communicate their point through sayings like these or even simpler phrases, such as "model with a purpose," "software is your primary goal," and "travel light," a way of saying a little documentation is good enough. Listen to these carefully. These sayings (some call them proverbs) are further discussed in Chapter 16. Catchy phrases are easy to understand, easy to memorize, and easy to repeat. They are very effective.

Activities, Resources, and Practices of Agile Modeling

Agile modeling involves a number of activities that need to be completed sometime during the agile development process. This section discusses these activities, the resources, and the practices that are unique to the agile approach.

FOUR BASIC ACTIVITIES OF AGILE DEVELOPMENT. Agile methods use four basic activities of development: coding, testing, listening, and designing. An agile analyst needs to identify the amount of effort that will go into each activity and balance that with the resources needed to complete the project.

Coding is designated as the one activity that it is not possible to do without. One author states that the most valuable thing that we receive from code is "learning." The process is basically this: Have a thought, code it, test it, and see whether the thought was a logical one. Code can also be used to communicate ideas that would otherwise remain fuzzy or unshaped. When I see your code, I may get a new thought. Source code is the basis for a living system. It is essential for development.

Testing is the second basic activity of development. The agile approach views automated tests as critical. The agile approach advocates writing tests to check coding, functionality, performance, and conformance. Agile modeling relies on automated tests, and large libraries of tests exist for most programming languages. These tests need to be updated as necessary during the progress of the project.

There are both long-term and short-term reasons for testing. Testing in the short term provides you with extreme confidence in what you are building. If tests run perfectly you can continue on with renewed confidence. In the long term, testing keeps a system alive and allows you to make changes longer than would be possible if no tests were written or run.

The third basic activity of development is listening. In Chapter 4, we learned about the importance of listening during interviews. In the agile approach, listening is done in the extreme. Developers use active listening to hear their programming partner. In agile modeling, there is less reliance on formal, written communication, and so listening becomes a paramount skill.

A developer also uses active listening with the customer. Developers assume that they know nothing about the business they are helping, and so they must listen carefully to businesspeople to get answers to their questions. A developer needs to come to an understanding of what effective listening is. If you don't listen, you will not know what you should code or what you should test.

The fourth basic activity in development is designing, which is a way of creating a structure to organize all the logic in the system. Designing is evolutionary, and so systems that are designed using the agile approach are conceptualized as evolving, always being designed.

Good design is often simple. Design should allow flexibility as well. Designing well permits you to make extensions to the system by making changes only in one place. Effective design locates logic near the data on which it will be operating. Above all, design should be useful to all those who will need it as the development effort proceeds, including customers as well as programmers.

FOUR RESOURCE CONTROL VARIABLES OF AGILE MODELING. Completing all the activities in a project on time within all the constraints is admirable, but, as you have probably realized by now, in order to accomplish this, project management is crucial. Managing a project doesn't mean simply getting all the tasks and resources together. It also means that the analyst is faced with a number of trade-offs. Sometimes cost may be predetermined, and at other junctures time may be the most important factor. These resource control variables (time, cost, quality, and scope) are discussed next.

TIME. You need to allow enough time to complete a project. Time, however, is split into many separate pieces. You need time to listen to the customers, time to design, time to code, and time to test.

One of our friends is an owner of a Chinese restaurant. Recently, he found himself short-staffed as one of the members of his reliable crew returned to Hong Kong to get married. The owner placed himself in the kitchen so the food was served on time, but he stopped greeting his customers out front in the usual way. He sacrificed the listening activity to achieve another activity, but in this case he found out it was hurting his business. Customers wanted the attention.

It is the same in systems development. You can create quality software but fail to listen. You can design a perfect system but not allow enough time to test it. Time is difficult to manage. If you find yourself running short of time, what do you do?

The agile approach challenges the notion that more time will give you the results you want. Perhaps the customer would prefer that you finish on time rather than extend the deadline to add another feature. Customers, we often find, are happy if some of the functionality is up and running on time. Our experience shows that often a customer is 80 percent satisfied with the first 20 percent of the functionality. This means that when you complete the final 80 percent of the project, the customer may be only slightly happier than he or she was after you completed the first 20 percent. The message here is be careful not to extend your deadline. The agile approach insists on finishing on time.

COST. Cost is the second variable we can consider adjusting. Suppose that the activities of coding, designing, testing, and listening are weighing down the project, and the resources we put into time, scope, and quality are not sufficient, even with a normal amount devoted to cost, to balance the project. Essentially we might be required to contribute more resources that require money to balance the project.

The easiest way to increase spending (and hence costs) is to hire more people. This may appear to be the perfect solution. If we hire more programmers, we'll finish faster. Right? Not necessarily. Picture hiring two people to repair a roof and increase that number to four. Soon the people are bumping into one another. Furthermore, they need to ask each other what still needs to be done. And if there's a lightning storm, no one will be working. Going from two to four doesn't mean it will take half the time. Consider the required increase in communication and other intangible costs when you are considering hiring more people. Remember that when new people join a team, they do not know the project or the team. They will slow down the original members because the original members must devote time to getting new members up to speed.

Overtime doesn't help much either. It increases the cost, but productivity doesn't always follow. Tired programmers are less effective than alert programmers. Tired programmers take a long time to complete a task, and they also make mistakes that are even more time-consuming to fix.

Is there anything else we can spend our money on to facilitate finishing a project? Perhaps. In other chapters, you will read about a variety of tools that support analysts and programmers. These tools are often wise investments. Analysts, for example, use graphical packages such as Microsoft Visio to communicate ideas about a project to others, and CASE tools such as Visible Analyst also help speed up projects.

Even new hardware could be a worthwhile expenditure. Laptops and smartphones improve productivity away from the office. Larger visual displays, Bluetooth-enabled keyboards and mice and more powerful graphics cards can also increase productivity.

QUALITY. The third resource control variable is quality. If ideal systems are perfect, why is so much effort placed in maintaining systems? Are we already practicing agile development by sacrificing quality in software development? In Chapter 16 we will see the importance of quality and methods (such as TQM and Six Sigma) which help ensure that software quality is high.

The agile philosophy, however, allows an analyst to adjust the quality resource and perhaps put less effort into maintaining quality than otherwise would be expected. Quality can be adjusted both internally and externally. Internal quality involves testing software for factors such as functionality (Does a program do what it is supposed to do?) and conformance (Does the software meet certain conformance standards, and is it maintainable?). It usually doesn't pay to tinker with internal quality.

That leaves us with external quality, or how the customer perceives the system. The customer is interested in performance. Some of the questions a customer may ask are: Does the program act reliably (or do software bugs still exist)? Is the output effective? Does the output reach me on time? Does the software run effortlessly? Is the user interface easy to understand and use?

The extreme philosophy of agile development allows some of the external quality issues to be sacrificed. In order for a system to be released on time, the customer may have to contend with some software bugs. If we want to meet our deadline, the user interface may not be perfect. We can make it better in a follow-up version.

Commercial off-the-shelf software manufacturers do sacrifice quality, and it is debatable whether this is a good approach. Developers may be using extreme programming as one of their agile practices, so don't be surprised when your PC software applications (not to mention your operating system and Web browser) are updated often.

SCOPE. Finally, there is scope. In the agile approach, scope is determined by listening to customers and getting them to write down their stories. Then the stories are examined to see how much can be done in a given time to satisfy the customer. Stories should be brief and easy to grasp. Stories will be described in more detail later in this chapter, but here is a brief example showing four short stories from an online air travel system. Each story is shown in bold type:

Display alternative flights.

> *Prepare a list of the five cheapest flights.*

Offer cheaper alternatives.

> *Suggest to customers that they travel on other days, make weekend stays, take special promotions, or use alternate airports.*

Purchase a ticket.

> *Allow the customer to purchase a ticket directly using a credit card (check validity).*

Allow the customer to choose his or her seat.

> *Direct the customer to a visual display of the airplane and ask the customer to select a seat.*

Ideally, the analyst would be able to determine how much time and money is needed to complete each of these stories and would be able to set the level of quality for them as well. It is obvious that this system must not sacrifice quality, or credit card purchases may be invalid or customers may show up at the airport without reservations.

Once again, agile practices allow extreme measures, so in order to maintain quality, manage cost, and complete a project on time, an agile analyst may want to adjust the scope of the project. This can be accomplished by agreeing with the customer that one or more of the stories can be delayed until the next version of the software. For example, maybe the functionality of allowing customers to choose their own seats can be put off for another time.

In summary, an agile analyst can control any of the four resource variables of time, cost, quality, and scope. Agility calls for extreme measures and places a great deal of importance on completing a project on time. In doing so, sacrifices must be made and the agile analyst will find out that the trade-offs available involve difficult decisions.

FOUR CORE AGILE PRACTICES. Four core practices markedly distinguish the agile approach from other approaches: short releases, the 40-hour workweek, hosting an on-site customer, and using pair programming:

1. Short releases means that the development team compresses the time between releases of the product. Rather than release a full-blown version in a year, using the short release

CONSULTING OPPORTUNITY 6.4

This Prototype Is All Wet

"It can be changed. It's not a finished product, remember," affirms Sandy Beach, a systems analyst for RainFall, a large manufacturer of fiberglass bathtub and shower enclosures for bathrooms. Beach is anxiously reassuring Will Lather, a production scheduler for RainFall, who is poring over the first hard-copy output produced for him by the prototype of the new information system.

"Well, it's okay," Lather says quietly. "I wouldn't want to bother you with anything. Let's see, . . . yes, *here* they are," he says as he finally locates the monthly report summarizing raw materials purchased, raw materials used, and raw materials in inventory.

Lather continues paging through the unwieldy computer printout. "This will be fine." Pausing at a report, he remarks, "I'll just have Miss Fawcett copy this part for the people in Accounting." Turning a few more pages, he says, "And the guy in Quality Assurance should really see this column of figures, although the rest of it isn't of much interest to him. I'll circle it and make a copy of it for him. Maybe I should phone part of this in to the warehouse, too."

As Sandy prepares to leave, Lather bundles up the pages of the reports, commenting, "The new system will be a big help. I'll make sure everybody knows about it. Anything will be better than the 'old monster' anyway. I'm glad we've got something new."

Sandy leaves Will Lather's office feeling a little lost at sea. Thinking it over, he starts wondering why Accounting, Quality Assurance, and the warehouse aren't getting what Will thinks they should. Sandy phones a few people, and he confirms that what Lather has told him is true. They need the reports, and they're not getting them.

Later in the week Sandy approaches Lather about rerouting the output as well as changing some of the features of the system. These modifications would allow Lather to get onscreen answers regarding what-if scenarios about changes in the prices suppliers are charging or changes in the quality rating of the raw materials available from suppliers (or both), as well as allow him to see what would happen if a shipment were late.

Lather is visibly upset with Sandy's suggestions for altering the prototype and its output. "Oh, don't do it on my account. It's okay really. I don't mind taking the responsibility for routing information to people. I'm always showering them with stuff anyway. Really, this is working pretty well. I would hate to have you take it away from us at this point. Let's just leave it in place."

Sandy is pleased that Lather seems so satisfied with the prototyped output, but he is concerned about Lather's unwillingness to change the prototype, because he has been encouraging users to think of it as an evolving product, not a finished one.

Write a brief report to Sandy listing changes to the prototype prompted by Lather's reactions. In a paragraph, discuss ways that Sandy can calm Lather's fears about having the prototype "taken away." Discuss in a paragraph some actions that can be taken *before* a prototype is tried out to prepare users for its evolutionary nature.

---

practice, they will shorten the release time by tackling the most important features first, release that system or product, and then improve it later.

2. Forty-hour workweek means that agile development teams purposely endorse a cultural core practice in which the team works intensely together during a typical 40-hour workweek. As a corollary to this practice, the culture reinforces the idea that working overtime for more than a week in a row is very bad for the health of the project and the developers. This core practice attempts to motivate team members to work intensely at the job, and then to take time off so that when they return they are relaxed and less stressed. This helps team members spot problems more readily, and prevents costly errors and omissions due to ineffectual performance or burnout.

3. On-site customer means that a user who is an expert in the business aspect of the systems development work is on-site during the development process. This person is integral to the process, writes user stories, communicates to team members, helps prioritize and balance the long-term business needs, and makes decisions about which feature should be tackled first.

4. Pair programming is an important core practice. It means that you work with another programmer of your own choosing. You both do coding, you both run tests. Often the senior person will take the coding lead initially, but as the junior person becomes involved, whoever has the clear vision of the goal will typically do the coding for the moment. When you ask another person to work with you, the protocol of pair programming says he or she is obligated to consent. Working with another programmer helps you clarify your thinking. Pairs change frequently, especially during the exploration stage of the development

FIGURE 6.5

The core practices are interrelated with agile modeling's resources, activities, and values.

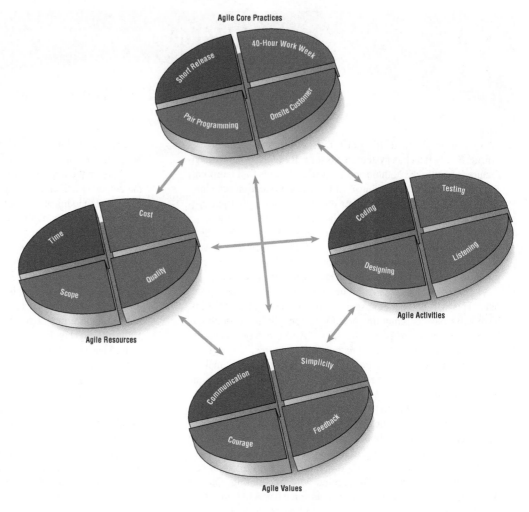

process. Pair programming saves time, cuts down on sloppy thinking, sparks creativity, and is a fun way to program.

How core agile practices interrelate with and support agile development activities, resources, and values is shown in Figure 6.5.

The Agile Development Process

Modeling is a keyword in agile methods. Agile modeling seizes on the opportunity to create models. These can be logical models such as drawings of systems, or mock-ups such as the prototypes described earlier in this chapter. A typical agile modeling process would go something like this:

1. Listen for user stories from the customer.
2. Draw a logical workflow model to gain an appreciation for the business decisions represented in the user story.
3. Create new user stories based on the logical model.
4. Develop some display prototypes. In doing so, show the customers what sort of interface they will have.
5. Using feedback from the prototypes and the logical workflow diagrams, develop the system until you create a physical data model.

Agile is the other keyword in agile modeling. Agile implies maneuverability. Today's systems, especially those that are Web based, pose twin demands: getting software released as soon as possible and continually improving the software to add new features. A systems analyst needs to have the ability and methods to create dynamic, context-sensitive, scalable, and evolutionary applications. Agile modeling is therefore a change-embracing method.

WRITING USER STORIES. Even though the title of this section is "Writing User Stories," the emphasis in the creation of user stories is on spoken interaction between developers and users, not written communication. In user stories, a developer is seeking first and foremost to identify valuable business user requirements. Users will typically engage in conversations every day with the developers about the meaning of the user stories they have written. These frequent conversations are purposeful interactions that have as their goal the prevention of misunderstandings or misinterpretations of user requirements. Therefore, user stories serve as reminders to the developers that they must hold conversations devoted to those requirements.

The following is an example of a series of stories written for an ecommerce application for an online merchant of books, CDs, and other media products. The stories give a fairly complete picture of what is needed at each of the stages in the purchase process, but the stories are very short and easy to comprehend. The point here is to get all the needs and concerns of the online store out in the open. Although there is not enough of a story to begin programming, an agile developer might begin to see the overall picture clearly enough to begin estimating what it will take to complete the project. The stories are as follows:

Welcome the customer.

If the customer has been at this site before using this same computer, welcome the customer back to the online store.

Show specials on homepage.

Show any recent books or other products that have recently been introduced. If the customer is identified, tailor the recommendations to that specific customer.

Search for desired product.

Include an effective search engine that will locate the specific product and similar products.

Show matching titles and availability.

Display the results of the search on a new web page.

Allow customer to ask for greater detail.

Offer the customer more product details, such as sample pages in a book, more photos of a product, or to play a partial track from a CD.

Display reviews of the product.

Share the comments that other customers have about the product.

Place a product into a shopping cart.

Make it easy for the customer to click on a button that places the product into a shopping cart of intended purchases.

Keep purchase history on file.

Keep details about the customer and his or her purchases in a cookie on the customer's computer. Also keep credit card information for faster checkout.

Suggest other books that are similar.

Include photos of other books that have similar themes or were written by the same authors.

Proceed to checkout.

Confirm the identity of the customer.

Review the purchases.

Allow the customer to review the purchases.

Continue shopping.

Offer the customer a chance to make further purchases at the same time.

Apply shortcut methods for faster checkout.

If the identity of the customer is known and the delivery address matches, speed up the transaction by accepting the credit card on file and the remainder of the customer's preferences, such as shipping method.

Add names and shipping addresses.

If the purchase is a gift, allow the customer to enter the name and address of the recipient.

Need or Opportunity:	Apply shortcut methods for faster checkout.					
Story:	If the identity of the customer is known and the delivery address matches, speed up the transaction by accepting the credit card on file and the rest of the customer's preferences such as shipping method.					
		Well Below	Below Average	Average	Above Average	Well Above
Activities:	Coding			✓		
	Testing			✓		
	Listening			✓		
	Designing			✓		
Resources:	Time				✓	
	Cost				✓	
	Quality		✓			
	Scope				✓	
				✓		

FIGURE 6.6

User stories can be recorded on cards. The user story should be brief enough for an analyst to determine what systems features are needed.

Offer options for shipping.

Allow the customer to choose a shipping method based on cost.

Complete the transaction.

Finish the transaction. Ask for credit card confirmation if the shipping address is different from the customer's address on file.

As you can easily see, there is no shortage of stories. An agile analyst needs to choose a few stories, complete the programming, and release the product. Once this is done, more stories are selected and a new version is released until all the stories are included in the system (or the analyst and customer agree that a particular story lacks merit or is not pressing and so need not be included).

An example of a user story as it might appear to an agile developer is shown in Figure 6.6. On paper cards (or electronically), an analyst might first identify the need or opportunity and then follow it with a brief story description. The analyst might take the opportunity to begin thinking broadly about the activities that need to be completed as well as the resources it will take to finish the project. In this example from the online merchant, the analyst indicates that the designing activity will take above-average effort, and the time and quality resources are required to rise above average. Notice that the analyst is not trying to be more precise than currently possible on this estimate, but it is still a useful exercise.

SCRUM. Another agile approach is named scrum. The word *scrum* is taken from a starting position in rugby in which the rugby teams form a huddle and fight for possession of the ball. Scrum is really about teamwork, similar to what is needed in playing a game of rugby.

Just as rugby teams will come to a game with an overall strategy, development teams begin a project with a high-level plan that can be changed on the fly as the "game" progresses. Systems development team members realize that the success of the project is most important, and their individual success is secondary. The project leader has some, but not much, influence on the detail. Rather, the tactical game is left up to the team members, just as if they were on the field. The systems team works within a strict time frame (30 days for development), just as a rugby team would play in a strict time constraint of a game.

We can describe the components of the scrum methodology as:

1. Product backlog, in which a list is derived from product specifications.
2. Sprint backlog, a dynamically changing list of tasks to be completed in the next sprint.
3. Sprint, a 30-day period in which the development team transforms the backlog into software that can be demonstrated.

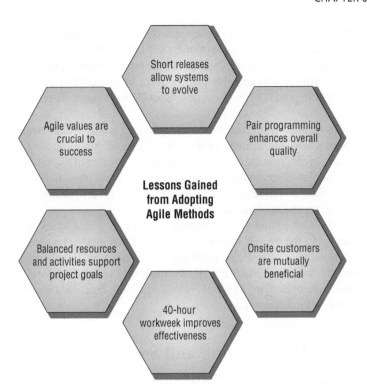

FIGURE 6.7

There are six vital lessons that can be drawn from the agile approach to systems.

4. Daily scrum, a brief meeting in which communication is the number-one rule. Team members need to explain what they did since the last meeting, whether they encountered any obstacles, and what they plan to do before the next daily scrum.
5. Demo, working software that can be demonstrated to the customer.

Scrum is indeed a high-intensity methodology. It is just one of the approaches that adopts the philosophy of agile modeling.

Lessons Learned from Agile Modeling

Often posed as an alternative way to develop systems, the agile approach seeks to address common complaints arising about the traditional SDLC approach (for example, being too time-consuming, focusing on data rather than on humans, being too costly) by being rapid, iterative, flexible, and participative in responding to changing human information requirements, business conditions, and environments.

Several agile development projects have been chronicled in books, articles, and on websites. Many of them were successes, some have been failures, but we can learn a great deal from studying them, as well as the agile values, principles, and core practices. Following are the six major lessons we draw from our examination of agile modeling. Figure 6.7 depicts the six lessons.

The first lesson is that short releases allow systems to evolve. Product updates are made often, and changes are incorporated quickly. In this way the system is permitted to grow and expand in ways that the customer finds useful. Through the use of short releases, the development team compresses the time between releases of their product, improving the product later as the dynamic situation demands.

The second lesson is that pair programming enhances overall quality. Although pair programming is controversial, it clearly fosters other positive activities necessary in systems development such as good communication, identifying with the customer, focusing on the most valuable aspects of the project first, testing all code as it is developed, and integrating the new code after it successfully passes its tests.

The third lesson is that on-site customers are mutually beneficial to the business and the agile development team. Customers serve as a ready reference and reality check, and the focus of the system design will always be maintained via their presence: customers become more like developers and developers empathize more fully with customers.

The fourth lesson we take from the agile approach is that the 40-hour workweek improves effectiveness. Even the hardest-hitting developers are susceptible to errors and burnout if they work too hard for too long a period. When the development team is together, however, every moment counts. Working at a sustainable pace is much more desirable for the life of the project, the life of the system, and the life of the developer! We all know the parable of the hare and the tortoise.

The fifth lesson we draw from taking the agile approach is that balanced resources and activities support project goals. Managing a project doesn't mean simply getting all resources and tasks together. It also means that the analyst is faced with a number of trade-offs. Sometimes cost may be predetermined, at other junctures time may be the most important factor. The resource control variables of time, cost, quality, and scope need to be properly balanced with the activities of coding, designing, testing, and listening.

The last lesson we take from agile modeling approaches is that agile values are crucial to success. It is essential to the overall success of the project that analysts wholeheartedly embrace the values of communication, simplicity, feedback, and courage in all the work that they do. This type of personal and team commitment enables the analyst to succeed where others, who possess similar technical competencies but who lack values, will fail. True dedication to these values is fundamental to successful development.

Comparing Agile Modeling and Structured Methods

As you have seen, agile methods are developed quickly, they reportedly work, and users are customers who are directly involved. While it is true that projects developed by using agile methods often require tweaking to work properly, agile developers admit that tweaking is part of the process. The agile approach implies many short releases, with features added along the way.

Improving Efficiency in Knowledge Work: SDLC Versus Agile

Researchers Davis and Naumann (1999) developed a list of seven strategies that can improve the efficiency of knowledge work: reducing interface time and errors, reducing process learning time and dual processing losses, reducing time and effort to structure tasks and format outputs, reducing nonproductive expansion of work, reducing data and knowledge search and storage time and costs, reducing communication and coordination time and costs, and reducing losses from human information overload. These researchers believe this is important because, based on their study of a group of programmers, they claim that the best programmers are 5 to 10 times more productive than the worst ones. They further point out this ratio is only 2:1 for workers in clerical or physical tasks. Their suggestion is that software can help improve many situations.

We use the standard, traditional systems development approach of structured methods to compare and contrast how structured approaches versus agile methods would implement the seven strategies proposed to improve the efficiency of knowledge workers.

While adopting more software may indeed improve performance, it is reasonable to suggest that changing an approach or a methodology may also improve performance. Consequently, we will examine each aspect of knowledge work productivity through lenses from both structured and agile methodologies. Figure 6.8 lists the original seven strategies for productivity improvement and explains what methods are used to improve the efficiency of systems development for both structured and agile methodologies.

In the upcoming sections we will compare and contrast structured approaches with the agile approach. An overarching observation about the agile methodology is that it is a human-oriented approach that permits people to create nuanced solutions that are impossible to create through formal specifications of process.

REDUCING THE INTERFACE TIME AND ERRORS. Systems analysts and programmers need to analyze, design, and develop systems using knowledge work tools that range from Microsoft Office to sophisticated and costly CASE tools. They also need to document as they develop systems. It is important that analysts and programmers are capable of understanding the interface they use. They need to know how to classify, code, store, and write about the data they gather.

Strategies for Improving Efficiency in Knowledge Work	Implementation Using Structured Methodologies	Implementation Using Agile Methodologies
Reduce interface time and errors	Adopting organizational standards for coding, naming, etc.; using forms	Adopting pair programming
Reduce process learning time and dual processing losses	Managing when updates are released so the user does not have to learn and use software at the same time	Ad hoc prototyping and rapid development
Reduce time and effort to structure tasks and format outputs	Using CASE tools and diagrams; using code written by other programmers	Encouraging short releases
Reduce nonproductive expansion of work	Managing project; establishing deadlines	Limiting scope in each release
Reduce data and knowledge search and storage time and costs	Using structured data gathering techniques, such as interviews, observation, sampling	Allowing for an onsite customer
Reduce communication and coordination time and costs	Separating projects into smaller tasks; establishing barriers	Timeboxing
Reduce losses from human information overload	Applying filtering techniques to shield analysts and programmers	Sticking to a 40-hour workweek

FIGURE 6.8

How Davis and Naumann's strategies for improving efficiency can be implemented using two different development approaches.

Systems developers also need to quickly access a program, enter the required information, and retrieve it when it is needed again.

Structured approaches encourage adopting standards for everything. Rules set forth include items such as "Google Chrome will be the default on office desktops for browsing rather than Firefox." They may be more detailed instructions to ensure clean data, such as "Always use M for Male and F for Female," thereby ensuring that analysts do not unthinkingly choose codes of their own, such as 0 for Male and 1 for Female. These rules then become part of the data repository. Forms are also useful, requiring all personnel to document their procedures so that another programmer might be able to take over if necessary.

In an agile approach, forms and procedures work well too, but another element is added. The additional practice of pair programming ensures that one programmer will check the work of another, thereby reducing the number of errors. Pair programming means that ownership of the design or software itself is shared in a partnership. Both partners (typically one a programmer, often a senior one) will say they chose a programming partner who desired to have a quality product that is error free. Since two people work on the same design and code, interface time is not an issue; it is an integral part of the process. Davis and Naumann have noted that programmers are quite emotional when the topic of pair programming is broached.

REDUCING THE PROCESS LEARNING TIME AND DUAL PROCESSING LOSSES. Analysts and programmers learn specific techniques and software languages required for the completion of a current project. Inefficiencies often result when some analysts and programmers already know the products used while others still need to learn them. Unfortunately, companies often ask systems developers to learn new applications at the same time they are using these apps to build the system. This on-the-job training slows down the entire systems development project considerably.

A traditional, structured project requires more learning. If CASE tools were used, an analyst may need to learn the proprietary CASE tools used in the organization. The same applies to the use of a specific computer language. Documentation is also a concern.

Using an agile philosophy, the ability to launch projects without using CASE tools and detailed documentation allows analysts and programmers to spend most of their time on system development rather than on learning specific tools.

REDUCING THE TIME AND EFFORT TO STRUCTURE TASKS AND FORMAT OUTPUTS. Whenever a project is started, a developer needs to determine the boundaries. In other words, the developers

MAC APPEAL

Just as agile methodologies are an alternative to the SDLC, OmniFocus is an alternative to Microsoft Project and other Gantt chart or PERT diagram approaches.

A casual observer might think that agile methods are unstructured because systems are built without detailed specifics and documentation. A student of agile methods realizes that there is actually quite a bit of structure in the agile approach. Principles include sticking to the 40-hour workweek and coordination through pair programming. An analyst who adopts agile techniques needs a way to set goals, stay within budget, set priorities for features, and find a way to get things done.

OmniFocus is based on an alternative task management system by David Allen, called Getting Things Done. The overriding principle is to free your mind from remembering things, so that you can concentrate on completing them. An analyst using this system goes through five actions: collect, process, organize, review, and do.

A systems analyst using OmniFocus collects items from his or her Web browser, address book or calendar, and most other applications on a Mac. The analyst can categorize the data or assign it to a larger project. OmniFocus contains a planning mode so the analyst can see which task is part of a larger project and a context mode that organizes the tasks so the analyst knows all the tasks that must be done either by phone, by browsing the Web, or by using email. OmniFocus is also available as an iPhone app.

FIGURE 6.MAC

OmniFocus from The Omni Group. (Screenshot form OmniFocus, a registered trademark used under license. Graphic reprinted with permission of the Omni Group.)

need to know what the deliverable will be and how they will go about organizing the project so they can complete all the necessary tasks.

A traditional approach would include using CASE tools, drawing diagrams (such as E-R diagrams and data flow diagrams), using project management software (such as Microsoft Project), writing detailed job descriptions, using and reusing forms and templates, and reusing code written by other programmers.

Systems development using an agile approach addresses the need to structure tasks by scheduling short releases. The agile philosophy suggests that system developers create a series of deadlines for many releases of the system. The first releases would possess fewer features, but, with each new release, additional features would be added.

REDUCING THE NONPRODUCTIVE EXPANSION OF WORK. Parkinson's law states that "work expands so as to fill the time available for its completion." If there are no specified deadlines, it is possible that knowledge work will continue to expand.

With traditional structured methodologies, deadlines at first seem far in the future. Analysts may use project management techniques to try to schedule the activities, but there is a built-in bias to extend earlier tasks longer than they need to be and then try to shorten tasks later on in the development. Analysts and programmers are less concerned about distant deadlines than approaching ones.

Once again, the agile approach stresses short releases. Releases can be delivered at the time promised, minus some of the features originally promised. Making all deadlines imminent pushes a realistic expectation for (at least partial) completion to the fore.

REDUCING THE DATA AND KNOWLEDGE SEARCH AND STORAGE TIME AND COSTS. System developers need to gather information about an organization, goals, priorities, and details about current information systems before they can proceed to develop a new system. Data-gathering methods include interviewing, administration of questionnaires, observation, and investigation by examining reports and memos.

Structured methodologies encourage structured data-gathering methods. Structured techniques would normally be used to structure interviews and design the interview process. Questionnaires would be developed in a structured way, and structured observational techniques such as STROBE would encourage an analyst to specifically observe key elements and form conclusions based on the observations of the physical environment. A sampling plan would be determined quantitatively, in order for the systems analyst to select reports and memos to examine.

Knowledge searches are less structured in an agile modeling environment. The practice of having an on-site customer greatly enhances access to information. The on-site customer is present to answer questions about the organization itself, its goals, the priorities of organizational members and customers, and whatever knowledge is necessary about existing information systems. As the project continues, the picture of customer requirements becomes clearer. This approach seems relatively painless because, when the system developers want to know something, they can just ask. The downside, however, is that the on-site representative may make up information if it is unknown or unavailable or evade telling the truth for some ulterior purpose.

REDUCING COMMUNICATION AND COORDINATION TIME AND COSTS.
Communication between analysts and users, as well as among analysts themselves, is at the heart of developing systems. Poor communication is certainly the root of multiple development problems. We know that communication increases when more people join a project. When two people work on a project, there is one opportunity for a one-to-one conversation; when three people are involved, there are three possibilities; when four are involved, there are six possibilities, and so on. Inexperienced team members need time to get up to speed, and they can slow down a project even though they are meant to help expedite it.

Traditional structured development encourages the separation of big tasks into smaller tasks. This allows more tightly knit groups and decreases the time spent communicating. Another approach involves setting up barriers. For example, customers may not be given access to programmers. This is a common practice in many industries. However, increased efficiency often means decreased effectiveness, and it has been noted that dividing up groups and setting up barriers often introduces errors.

Agile methods, on the other hand, limit time instead of tasks. Timeboxing is used in agile methodologies to encourage completion of activities in shorter periods. Timeboxing is simply setting a time limit of one or two weeks to complete a feature or module. The agile method scrum puts a premium on time, while the developers communicate effectively as a team. Since communication is one of the four values of the agile philosophy, communication costs tend to increase rather than decrease.

REDUCING LOSSES FROM HUMAN INFORMATION OVERLOAD. We have long known that people do not react well in information-overload situations. When telephones were an emerging technology, switchboard operators manually connected calls between two parties. It was demonstrated that this system would work until an information overload occurred, at which point the entire system broke down. When too many calls came in, the overwhelmed switchboard operator would simply stop working and give up completely on connecting callers. An analogous overload situation can occur anytime to anyone, including systems analysts and programmers.

A traditional approach would be to try to filter information to shield analysts and programmers from customer complaints. This approach allows developers to continue working on the problem without the interference and subjectivity that would normally occur.

Using an agile philosophy, analysts and programmers are expected to stick to a 40-hour workweek. Some might view this as a questionable practice. How will all the work ever get done? The agile philosophy states that quality work is usually done during a routine schedule, and it is only when overtime is added that problems of poor quality design and programming enter the scene. By sticking to a 40-hour workweek schedule, agile methodology claims that you will eventually come out ahead.

Risks Inherent in Organizational Innovation

In consultation with users, analysts must consider the risks that organizations face when adopting new methodologies. Clearly, this is part of a larger question of when is the appropriate time to upgrade human skills, adopt new organizational processes, and institute internal change.

In a larger sense, these are questions of a strategic dimension for organizational leadership. Specifically, we consider the case of the systems analysis team adopting agile methods in light of the risks to the organization and the eventual successful outcome for the systems development team and their clients. Figure 6.9 shows many of the variables that need to be considered when assessing the risk of adopting organizational innovation.

ORGANIZATIONAL CULTURE. A key consideration in organizational innovation is the overall culture of the organization and how the culture of the development team fits within it. A conservative organizational culture with many stable features that does not seek to innovate may be an inappropriate or even inhospitable context for the adoption of agile methodologies by a systems development group. Analysts and other developers must use caution in introducing new techniques into this type of setting since their success is far from assured, and long-standing development team members or other organizational members may be threatened by new ways of working that depart from customary, dependable approaches with proven results.

FIGURE 6.9

Adopting new information systems involves balancing several risks.

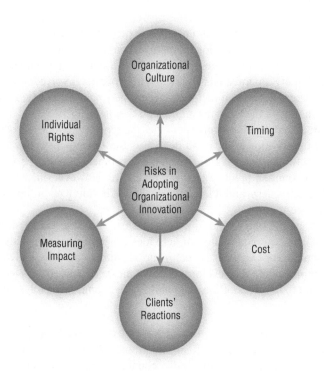

Conversely, an organization that is dependent on innovation to retain its cutting edge in its industry might be the organization most welcoming toward agile innovations in systems development methods. In this instance, the culture of the organization is already permeated with the understanding of the critical nature of many of the core principles of agile development methodologies. From the strategic level downward, the company's members have internalized the need for rapid feedback, dynamic responses to changing environments in real time, dependence on the customer for guidance and participation in problem solving, and so on.

Located between these extremes are organizations that do not rely on innovation as a key strategic strength (in other words, they are not dependent on research and development of new products or services to remain afloat) but that might still wish to adopt innovative practices in small units or groups. Indeed, such small, innovative centers or kernels might eventually drive the growth or competitive advantage of this type of organization.

TIMING. Organizations must ask and answer the question of when is the best time to innovate with the adoption of new systems development methodologies, when all other projects and factors (internally and externally) are taken into account. Organizations must consider the entire panoply of projects in which they are investing, look ahead at project deadlines, schedule the upgrading of physical plants, and absorb key industry and economic forecasts.

COST. Another risk to the adoption of agile methodologies for organizations is the cost involved in education and training of systems analysts and programmers in the new approach. This can involve either costly off-site seminars and courses or hiring consultants to work with current staff on-site. Further, opportunity costs are involved when systems developers are necessarily diverted (albeit temporarily) from ongoing projects to learn new skills. Education in itself can be costly, but an additional burden is recognized when analysts cannot earn income during their training period.

CLIENTS' REACTIONS. When clients (whether they are internal or external) are involved as users or initiators of information systems development efforts, reactions to the use of new methods entailed by the agile approach are also a key consideration. Some clients react with joy once the benefits of timeliness and involvement are described. Others do not want to be used for systems "experiments" with uncertain outcomes. The client–analyst relationship must be resilient enough to absorb and adapt to changes in expected behaviors. For example, the on-site presence of a client during development is a major commitment that should be thoroughly understood and agreed upon by those adopting agile methods.

MEASURING IMPACT. Another consideration for organizations adopting agile methodologies is how to certify and measure that the new methods are going to facilitate successful systems development. The strengths and weaknesses of traditional structured methods used to develop information systems are well known.

While there is ample anecdotal evidence that agile methodologies are superior for development under some conditions, their history is short-lived and not yet empirically supported. Therefore, the adoption of agile methodologies carries with it the risk that systems created with them will not be successful or will not adequately interface with legacy systems. Measuring the impact of the use of agile methodologies has begun, but organizations need to be vigilant in proposing impact measurements in tandem with the adoption of new methods.

THE INDIVIDUAL RIGHTS OF PROGRAMMERS/ANALYSTS. Successful systems developers (analysts and programmers) exercise creativity in their approach to their work, and they deserve the right to work in the most fruitful configuration possible. It is possible that the working requirements of new agile methods (for example, pair programming) encroach upon some basic rights of creative people to work alone or in groups as the design work dictates. There is no "one best way" to design a system, a module, an interface, a form, or a web page. In the case of systems developers, creativity, subjectivity, and the right to achieve design objectives through numerous individual paths need to be balanced against the organizational adoption of innovative approaches such as agile methodologies.

As you can see, adopting organizational innovations poses many risks to an organization as well as to individuals. We examined risks to an organization as a whole as well as risks to an individual systems analyst who is involved in the organization's desire to innovate.

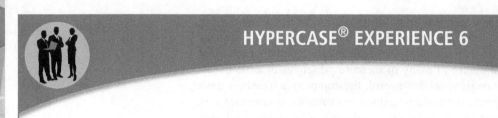

HYPERCASE® EXPERIENCE 6

"Thank goodness it's the time of year when everything is new. I love spring; it's the most exhilarating time here at MRE. The trees are so green, with leaves in so many different shades. So many new projects to do, too; so many new clients to meet. We have a new intern, too. Anna Mae Silver. Sometimes the newest employee is the most eager to help. Call on her if you need more answers."

"All the newness reminds me of prototyping. Or what I know about prototyping, anyway. It's something new and fresh, a quick way to find out what's happening.

"I believe that we have a few prototypes already started. Sometimes our new on-site customer, Tessa Silverstone, gets involved by helping create user stories on which to build the proto-types. But the best thing about prototypes is that they can change. I don't know anyone who's really been satisfied with a first pass at a prototype. But it's fun to be involved with something that's happening fast, and something that will change."

HYPERCASE Questions

1. Make a list of the user stories Tessa Silverstone shared as examples.
2. Locate the prototype currently proposed for use in one of MRE's departments. Suggest a few modifications that would make this prototype even more responsive to the unit's needs.
3. Using a word processor, construct a nonoperational prototype for a Training Department Project Reporting System. Include features brought up by the user stories you found. *Hint:* See sample screens in Chapters 11 and 12 to help you in your design.

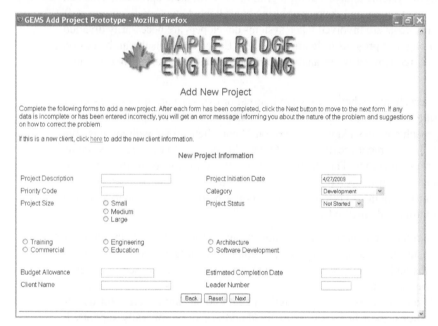

FIGURE 6.HC1

One of the many prototype screens found in HyperCase.

Summary

Prototyping is an information-gathering technique useful for supplementing the traditional SDLC; however, both agile methods and human–computer interaction share roots in prototyping. When systems analysts use prototyping, they are seeking user reactions, suggestions, innovations, and revision plans to make improvements to the prototype, and thereby modify system plans with a minimum of expense and disruption. The four major guidelines for developing a prototype are to (1) work in manageable modules, (2) build the prototype rapidly, (3) modify the prototype, and (4) stress the user interface.

Although prototyping is not always necessary or desirable, it should be noted that there are three main, interrelated advantages to using it: (1) the potential for changing the system early in its development, (2) the opportunity to stop development on a system that is not working, and (3) the possibility of developing a system that more closely addresses users' needs and expectations. Users play a distinct role in the prototyping process, and systems analysts must work systematically to elicit and evaluate users' reactions to a prototype.

Agile modeling is a software development approach that defines an overall plan quickly, develops and releases software quickly, and then continuously revises software to add additional features. The values of the agile approach that are shared by the customer as well as the development team are communication, simplicity, feedback, and courage. Agile activities include coding, testing, listening, and designing. Resources available include time, cost, quality, and scope.

Agile core practices distinguish agile methods, including a type of agile method called extreme programming (XP), from other systems development processes. The four core practices of the agile approach are (1) short releases, (2) 40-hour workweek, (3) on-site customer, and (4) pair programming. The agile development process includes choosing a task that is directly related to a customer-desired feature based on user stories, choosing a programming partner, selecting and writing appropriate test cases, writing the code, running the test cases, debugging it until all test cases run, implementing it with the existing design, and integrating it into what currently exists.

Later in this chapter we compared how SDLC and agile approaches handle improving knowledge work efficiency differently. We then discussed several inherent dangers to organizations adopting innovative approaches, including an incompatible organizational culture, poor timing of the project, cost of training systems analysts, unfavorable client reactions to new behavioral expectations, difficulties in measuring the impact, and the possible compromise of the individual creative rights of programmers and analysts.

Keywords and Phrases

40-hour workweek
agile modeling
agile principles
agile values
assume simplicity
embracing change
extreme programming (XP)
first-of-a-series prototype
implementation
incremental change
modifying the prototype
nonoperational prototype

on-site customer
pair programming
patched-up prototype
prototype
scrum methodology
selected-features prototype
short release
stressing the user interface
user involvement with prototyping
user stories
working in manageable modules

Review Questions

1. What four kinds of information is an analyst seeking through prototyping?
2. What is meant by the term *patched-up prototype*?
3. Define a prototype that is a nonworking scale model.
4. Give an example of a prototype that is a first full-scale model.
5. Define what is meant by a prototype that is a model with some, but not all, essential features.
6. List the advantages and disadvantages of using prototyping to *replace* the traditional SDLC.
7. Describe how prototyping can be used to augment the traditional SDLC.
8. What are the criteria for deciding whether a system should be prototyped?
9. List four guidelines an analyst should observe in developing a prototype.
10. What are the two main problems identified with prototyping?
11. List the three main advantages in using prototyping.
12. How can a prototype mounted on an interactive website facilitate the prototyping process? Answer in a paragraph.
13. What are three ways that a user can be of help in the prototyping process?
14. What are the four values that must be shared by the development team and business customers when taking an agile approach?
15. What are agile principles? Give five examples.
16. What are the four core practices of the agile approach?
17. Name the four resource control variables used in the agile approach.
18. Outline the typical steps in an agile development episode.
19. What is a user story? Is it primarily written or spoken? State your choice, then defend your answer with an example.

20. List software tools that can aid the developer in doing a variety of tests of code
21. What is scrum?
22. Name the seven strategies for improving efficiency in knowledge work.
23. Identify six risks in adopting organizational innovation.

Problems

1. As part of a larger systems project, Clone Bank of Clone, Colorado, wants your help in setting up a new monthly reporting form for its checking and savings account customers. The president and vice presidents are very attuned to what customers in the community are saying. They think that their customers want a checking account summary that looks like the one offered by the other three banks in town. They are unwilling, however, to commit to that form without a formal summary of customer feedback that supports their decision. Feedback will not be used to change the prototype form in any way. They want you to send a prototype of one form to one group and to send the old form to another group.
   a. In a paragraph, discuss why it probably is not worthwhile to prototype the new form under these circumstances.
   b. In a second paragraph, discuss a situation under which it would be advisable to prototype a new form.
2. C. N. Itall has been a systems analyst for Tun-L-Vision Corporation for many years. When you came on board as part of the systems analysis team and suggested prototyping as part of the SDLC for a current project, C. N. said, "Sure, but you can't pay any attention to what users say. They have no idea what they want. I'll prototype, but I'm not 'observing' any users."
   a. As tactfully as possible, so as not to upset C. N. Itall, make a list of the reasons that support the importance of observing user reactions, suggestions, and innovations in the prototyping process.
   b. In a paragraph, describe what might happen if part of a system is prototyped and no user feedback about it is incorporated into the successive system.
3. "Every time I think I've captured user information requirements, they've already changed. It's like trying to hit a moving target. Half the time, I don't think they even know what they want themselves," exclaims Flo Chart, a systems analyst for 2 Good 2 Be True, a company that surveys product use for the marketing divisions of several manufacturing companies.
   a. In a paragraph, explain to Flo Chart how prototyping can help her to better define users' information requirements.
   b. In a paragraph, comment on Flo's observation: "Half the time, I don't think they even know what they want themselves." Be sure to explain how prototyping can actually help users better understand and articulate their own information requirements.
   c. Suggest how an interactive website featuring a prototype might address Flo's concerns about capturing user information requirements. Use a paragraph.
4. Harold, a district manager for the multioutlet chain Sprocket's Gifts, thinks that building a prototype can mean only one thing: a nonworking scale model. He also believes that this way is too cumbersome to prototype information systems and thus is reluctant to do so.
   a. Briefly (in two or three paragraphs) compare and contrast the other three kinds of prototyping that are possible so that Harold has an understanding of what prototyping can mean.
   b. Harold has an option of implementing one system, trying it, and then having it installed in five other Sprocket locations if it is successful. Name a type of prototyping that would fit well with this approach, and in a paragraph defend your choice.
5. "I've got the idea of the century!" proclaims Bea Kwicke, a new systems analyst with your systems group. "Let's skip all this SDLC garbage and just prototype everything. Our projects will go a lot more quickly, we'll save time and money, and all the users will feel as if we're paying attention to them instead of going away for months on end and not talking to them."
   a. List the reasons you (as a member of the same team as Bea) would give Bea to dissuade her from trying to scrap the SDLC and prototype every project.
   b. Bea is pretty disappointed with what you have said. To encourage her, use a paragraph to explain the situations you think would lend themselves to prototyping.
6. The following remark was overheard at a meeting between managers and a systems analysis team at the Fence-Me-In fencing company: "You told us the prototype would be finished three weeks ago. We're still waiting for it!"
   a. In a paragraph, comment on the importance of rapid delivery of a portion of a prototyped information system.
   b. List three elements of the prototyping process that must be controlled to ensure prompt delivery of the prototype.
   c. What are some elements of the prototyping process that are difficult to manage? List them.

7. Prepare a list of activities for a systems development team for an online travel agent that is setting up a website for customers. Now suppose you are running out of time. Describe some of your options. Describe what you will trade off to get the website released in time.

8. Given the situation for Williwonk's chocolates (Problem 1 in Chapter 3), which of the four agile modeling resource variables may be adjusted?

9. Examine the collection of user stories from the online merchant shown earlier in the chapter. The online media store would now like to have you add some features to its website. Following the format shown earlier in this chapter in Figure 6.7, write a user story for the features listed below:
   a. Include pop-up ads.
   b. Offer to share the details of the customer's purchases with his or her friends.
   c. Extend offer to purchase other items.

10. Go to the Android website, at www.palmgear.com. Explore the website and write up a dozen brief user stories for improving the website.

11. Go to the iTunes website and write up a dozen brief user stories for improving the website.

12. Using the stories you wrote for Problem 9, walk through the five stages of the agile development process and describe what happens at each one of the stages.

Group Projects

1. Divide your group into two smaller subgroups. Have Group 1 follow the processes specified in this chapter for creating prototypes. Using a CASE tool or a word processor, Group 1 should devise two nonworking prototype screens using the information collected in the interviews with Maverick Transport employees accomplished in the group exercise in Chapter 4. Make any assumptions necessary to create two screens for truck dispatchers. Group 2 (playing the roles of dispatchers) should react to the prototype screens and provide feedback about desired additions and deletions.

2. The members of Group 1 should revise the prototype screens based on the user comments they received. Those in Group 2 should respond with comments about how well their initial concerns were addressed with the refined prototypes.

3. As a united group, write a paragraph discussing your experiences with prototyping for ascertaining information requirements.

4. Within your united group, assign some of the roles that people take on in agile development. Make sure that one person is an on-site customer and at least two people are programmers. Assign other roles, as you see fit. Simulate the systems development situation discussed in Problem 7, or have the person acting as the on-site customer choose an ecommerce business with which he or she is familiar. Assume that the customer wants to add some functionality to his or her website. Role-play a scenario showing what each person would do if this was being approached through agile methods. Write a paragraph that discusses the constraints that each person faces in enacting his or her role.

Selected Bibliography

Alavi, M. "An Assessment of the Prototyping Approach to Information Systems Development." *Communications of the ACM,* Vol. 27, No. 6, June 1984, pp. 556–563.

Avison, D., and D. N. Wilson. "Controls for Effective Prototyping." *Journal of Management Systems,* Vol. 3, No. 1, 1991.

Beck, K. *Extreme Programming Explained: Embrace Change.* Boston: Addison-Wesley Publishing Co., 2000.

Beck, K., and M. Fowler. *Planning Extreme Programming.* Boston: Addison-Wesley Publishing Co., 2001.

Cockburn, A. *Agile Software Development.* Boston: Addison-Wesley Publishing Co., 2002.

Davis, G. B., and J. D. Naumann. "Knowledge Work Productivity." In *Emerging Information Technologies: Improving Decisions, Cooperation, and Infrastructure.* Edited by K. E. Kendall, pp. 343–357. Thousand Oaks, CA: Sage, 1999.

Davis, G. B., and M. H. Olson. *Management Information Systems: Conceptual Foundations, Structure, and Development,* 2nd ed. New York: McGraw-Hill, 1985.

Fitzgerald, B., and G. Hartnett. "A Study of the Use of Agile Methods Within Intel." In *Business Agility & IT Diffusion.* Edited by L. Matthiassen, J. Pries-Heje, and J. DeGross, pp. 187–202. Proc Conference, Atlanta, May 2005. New York: Springer, 2005.

Ghione, J. "A Web Developer's Guide to Rapid Application Development Tools and Techniques." *Netscape World,* June 1997.

Gremillion, L. L., and P. Pyburn. "Breaking the Systems Development Bottleneck." *Harvard Business Review,* March–April 1983, pp. 130–137.

Harrison, T. S. "Techniques and Issues in Rapid Prototyping." *Journal of Systems Management,* Vol. 36, No. 6, June 1985, pp. 8–13.

Kendall, J. E., and K. E. Kendall. "Agile Methodologies and the Lone Systems Analyst: When Individual Creativity and Organizational Goals Collide in the Global IT Environment." *Journal of Individual Employment Rights,* Vol. 11, No. 4, 2004–2005, pp. 333–347.

Kendall, J. E., K. E. Kendall, and S. Kong. "Improving Quality Through the Use of Agile Methods in Systems Development: People and Values in the Quest for Quality." In *Measuring Information Systems Delivery Quality.* Edited by E. W. Duggan and H. Reichgelt, pp. 201–222. Hershey, PA: Idea Group Publishing, 2006.

Kong, S., J.E. Kendall, and K. E. Kendall. "Project Contexts and Use of Agile Software Development Methodology in Practice: A Case Study, *Journal of Academy of Business and Economics.*" Vol.12, No. 2, 2012, pp. 1–15.

McBreen, P. *Questioning Extreme Programming.* Boston: Addison-Wesley Publishing Co., 2003.

Naumann, J. D., and A. M. Jenkins. "Prototyping: The New Paradigm for Systems Development." *MIS Quarterly,* September 1982, pp. 29–44.

The CPU Case Episode and accompanying student files are available online at www.pearsonglobaleditions.com/kendall.

CHAPTER 7

Using Data Flow Diagrams

LEARNING OBJECTIVES

Once you have mastered the material in this chapter you will be able to:

1. Comprehend the importance of using logical and physical data flow diagrams (DFDs) to graphically depict data movement for humans and systems in an organization.

2. Create, use, and explode logical DFDs to capture and analyze the current system through parent and child levels.

3. Develop and explode logical DFDs that illustrate the proposed system.

4. Produce physical DFDs based on logical DFDs you have developed.

5. Understand and apply the concept of partitioning of physical DFDs.

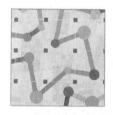

A systems analyst needs to make use of the conceptual freedom afforded by data flow diagrams, which graphically characterize data processes and flows in a business system. In their original state, data flow diagrams depict the broadest possible overview of system inputs, processes, and outputs, which correspond to those of the general systems model discussed in Chapter 2. A series of layered data flow diagrams may be used to represent and analyze detailed procedures in the larger system.

The Data Flow Approach to Human Requirements Determination

When systems analysts attempt to understand the information requirements of users, they must be able to conceptualize how data move through the organization, the processes or transformation that the data undergo, and what the outputs are. Although interviews and the investigation of hard data provide a verbal narrative of the system, a visual depiction can crystallize this information for users and analysts in a useful way.

Through a structured analysis technique called data flow diagrams (DFDs), a systems analyst can put together a graphical representation of data processes throughout an organization. By using combinations of only four symbols, a systems analyst can create a pictorial depiction of processes that will eventually provide solid system documentation.

Advantages of the Data Flow Approach

The data flow approach has four chief advantages over narrative explanations of the way data move through the system:

1. Freedom from committing to the technical implementation of the system too early.
2. Further understanding of the interrelatedness of systems and subsystems.
3. Communicating current system knowledge to users through data flow diagrams.
4. Analysis of a proposed system to determine if the necessary data and processes have been defined.

Perhaps the biggest advantage lies in the conceptual freedom found in the use of the four symbols (covered in the upcoming subsection on DFD conventions). (You will recognize three of the symbols from Chapter 2.) None of the symbols specify the physical aspects of implementation. DFDs emphasize the processing of data or the transforming of data as they move through a variety of processes. In logical DFDs, there is no distinction between manual or automated processes. Neither are the processes graphically depicted in chronological order. Rather, processes are eventually grouped together if further analysis dictates that it makes sense to do so. Manual processes are put together, and automated processes can also be paired with each other. This concept, called *partitioning,* is taken up in a later section.

Conventions Used in Data Flow Diagrams

Four basic symbols are used to chart data movement on data flow diagrams: a double square, an arrow, a rectangle with rounded corners, and an open-ended rectangle (closed on the left side and open ended on the right), as shown in Figure 7.1. An entire system and numerous subsystems can be depicted graphically with these four symbols in combination.

The double square is used to depict an external entity (another department, a business, a person, or a machine) that can send data to or receive data from the system. The external entity, or just entity, is also called a source or destination of data, and it is considered to be external to the system being described. Each entity is labeled with an appropriate name. Although it interacts with the system, it is considered as being outside the boundaries of the system. An entity should be named with a noun. The same entity may be used more than once on a given data flow diagram to avoid crossing data flow lines.

The arrow shows movement of data from one point to another, with the head of the arrow pointing toward the data's destination. Data flows occurring simultaneously can be depicted doing just that through the use of parallel arrows. Because an arrow represents data about a person, place, or thing, it too should be described with a noun.

A rectangle with rounded corners is used to show the occurrence of a transforming process. Processes always denote a change in or transformation of data; hence, the data flow leaving a process is *always* labeled differently than the one entering it. Processes represent work being performed in the system and should be named using one of the following formats. A clear name makes it easier to understand what the process is accomplishing:

1. When naming a high-level process, assign the process the name of the whole system. An example is INVENTORY CONTROL SYSTEM.

Symbol	Meaning	Example

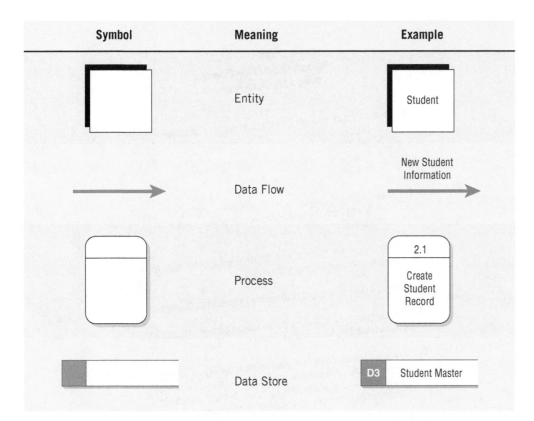

FIGURE 7.1

The four basic symbols used in data flow diagrams, their meanings, and examples.

2. When naming a major subsystem, use a name such as INVENTORY REPORTING SUBSYSTEM or INTERNET CUSTOMER FULFILLMENT SYSTEM.
3. When naming detailed processes, use a verb–adjective–noun combination. The verb describes the type of activity, such as COMPUTE, VERIFY, PREPARE, PRINT, or ADD. The noun indicates what the major outcome of the process is, such as REPORT or RECORD. The adjective illustrates which specific output, such as BACKORDERED or INVENTORY, is produced. Examples of complete process names are COMPUTE SALES TAX, VERIFY CUSTOMER ACCOUNT STATUS, PREPARE SHIPPING INVOICE, PRINT BACK-ORDERED REPORT, SEND CUSTOMER EMAIL CONFIRMATION, VERIFY CREDIT CARD BALANCE, and ADD INVENTORY RECORD.

A process must also be given a unique identifying number that indicates its level in the diagram. This organization is discussed later in this chapter. Several data flows may go into and out of each process. Examine processes with only a single flow in and out for missing data flows.

The last basic symbol used in data flow diagrams is an open-ended rectangle, which represents a data store. The rectangle is drawn with two parallel lines that are closed by a short line on the left side and are open ended on the right. These symbols are drawn only wide enough to allow identifying lettering between the parallel lines. In logical data flow diagrams, the type of physical storage is not specified. At this point the data store symbol is simply showing a depository for data that allows examination, addition, and retrieval of data.

The data store may represent a manual store, such as a filing cabinet, or a computerized file or database. Because data stores represent a person, place, or thing, they are named with a noun. Temporary data stores, such as scratch paper or a temporary computer file, are not included on the data flow diagram. Give each data store a unique reference number, such as D1, D2, D3, and so on.

Developing Data Flow Diagrams

Data flow diagrams can and should be drawn systematically. Figure 7.2 summarizes the steps involved in successfully completing data flow diagrams. First, a systems analyst needs to conceptualize data flows from a top-down perspective.

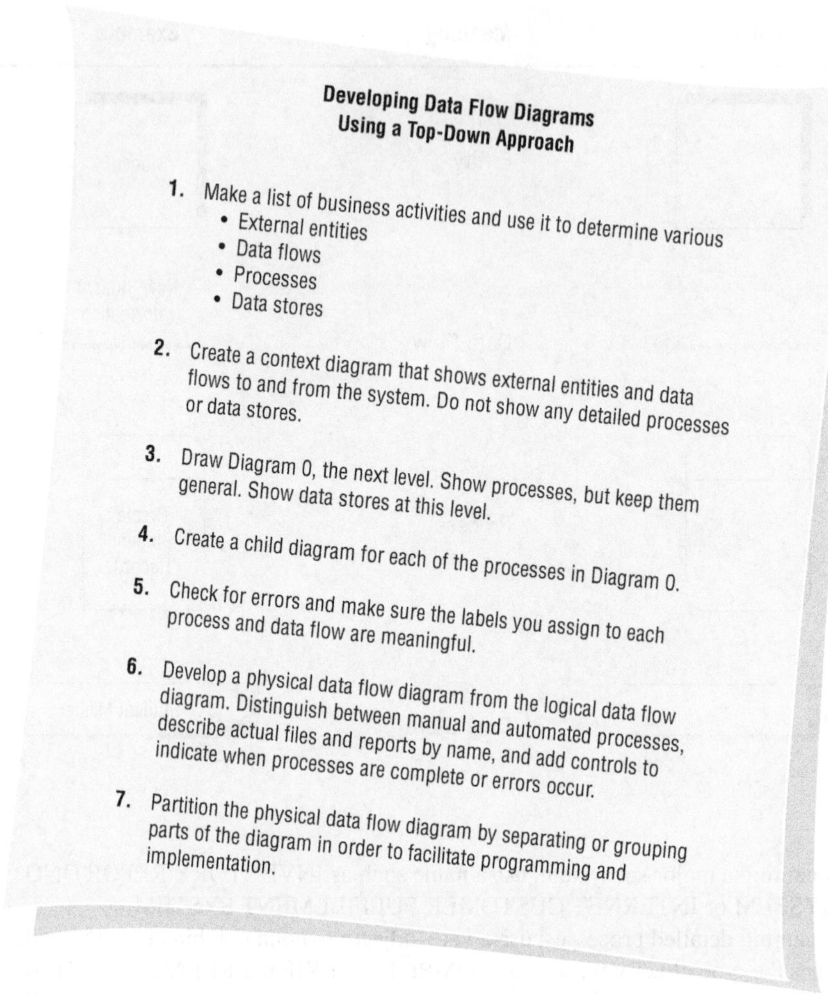

FIGURE 7.2

Steps in developing data flow
diagrams.

**Developing Data Flow Diagrams
Using a Top-Down Approach**

1. Make a list of business activities and use it to determine various
   - External entities
   - Data flows
   - Processes
   - Data stores

2. Create a context diagram that shows external entities and data
   flows to and from the system. Do not show any detailed processes
   or data stores.

3. Draw Diagram 0, the next level. Show processes, but keep them
   general. Show data stores at this level.

4. Create a child diagram for each of the processes in Diagram 0.

5. Check for errors and make sure the labels you assign to each
   process and data flow are meaningful.

6. Develop a physical data flow diagram from the logical data flow
   diagram. Distinguish between manual and automated processes,
   describe actual files and reports by name, and add controls to
   indicate when processes are complete or errors occur.

7. Partition the physical data flow diagram by separating or grouping
   parts of the diagram in order to facilitate programming and
   implementation.

To begin a data flow diagram, collapse the organization's system narrative (or story) into a list with the four categories of external entity, data flow, process, and data store. This list in turn helps determine the boundaries of the system you will be describing. Once a basic list of data elements has been compiled, begin drawing a context diagram.

Here are a few basic rules to follow:

1. A data flow diagram must have at least one process, and it must not have any freestanding objects or objects connected to themselves.
2. A process must receive at least one data flow coming into the process and create at least one data flow leaving from the process.
3. A data store should be connected to at least one process.
4. External entities should not be connected to each other. Although they communicate independently, that communication is not part of the system we design using DFDs.

Creating the Context Diagram

With a top-down approach to diagramming data movement, DFDs move from general to specific. The first diagram helps a systems analyst grasp basic data movement, but its general nature limits its usefulness. The initial context diagram should be an overview, including basic inputs, the general system, and outputs. This diagram will be the most general one, providing a bird's-eye view of data movement in the system and the broadest possible conceptualization of the system.

The context diagram is the highest level in a data flow diagram and contains only one process, representing the entire system. The process is given the number zero. All external entities are shown on the context diagram, as well as major data flow to and from them. The diagram does not contain any data stores and is fairly simple to create, once the external entities and the data flow to and from them are known to analysts.

Drawing Diagram 0 (The Next Level)

More detail than the context diagram permits is achievable by "exploding the diagrams." Inputs and outputs specified in the first diagram remain constant in all subsequent diagrams. The rest of the original diagram, however, is exploded into close-ups involving three to nine processes and showing data stores and new lower-level data flows. The effect is that of taking a magnifying glass to view the original data flow diagram. Each exploded diagram should use only a single sheet of paper. By exploding DFDs into subprocesses, the systems analyst begins to fill in the details about data movement. The handling of exceptions is ignored for the first two or three levels of data flow diagramming.

Diagram 0 is the explosion of the context diagram and may include up to nine processes. Including more processes at this level will result in a cluttered diagram that is difficult to understand. Each process is numbered with an integer, generally starting from the upper left-hand corner of the diagram and working toward the lower right-hand corner. The major data stores of the system (representing master files) and all external entities are included on Diagram 0. Figure 7.3 schematically illustrates both the context diagram and Diagram 0.

Because a data flow diagram is two-dimensional (rather than linear), you may start at any point and work forward or backward through the diagram. If you are unsure of what you would include at any point, take a different external entity, process, or data store, and then start drawing the flow from it. You may:

1. Start with the data flow from an entity on the input side. Ask questions such as: "What happens to the data entering the system?" "Is it stored?" "Is it input for several processes?"

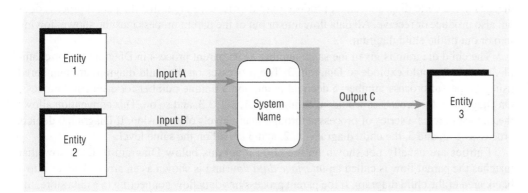

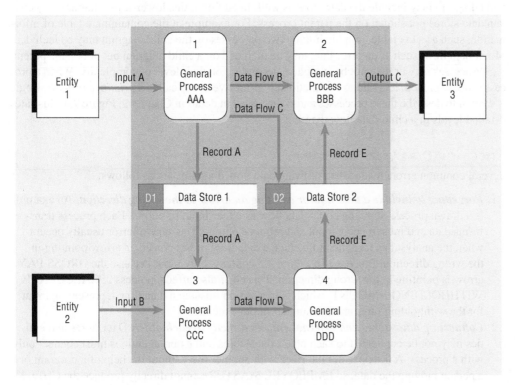

FIGURE 7.3

Context diagrams (above) can be "exploded" into Diagram 0 (below). Note the greater detail in Diagram 0.

2. Work backward from an output data flow. Examine the output fields on a document or screen. (This approach is easier if prototypes have been created.) For each field on the output, ask: "Where does it come from?" or "Is it calculated or stored on a file?" For example, when the output is a PAYCHECK, the EMPLOYEE NAME and ADDRESS would be located on an EMPLOYEE file, the HOURS WORKED would be on a TIME RECORD, and the GROSS PAY and DEDUCTIONS would be calculated. Each file and record would be connected to the process that produces the paycheck.

3. Examine the data flow to or from a data store. Ask: "What processes put data into the store?" or "What processes use the data?" Note that a data store used in the system you are working on may be produced by a different system. Thus, from your vantage point, there may not be any data flow into the data store.

4. Analyze a well-defined process. Look at what input data the process needs and what output it produces. Then connect the input and output to the appropriate data stores and entities.

5. Take note of any fuzzy areas where you are unsure of what should be included or what input or output is required. Awareness of problem areas will help you formulate a list of questions for follow-up interviews with key users.

Creating Child Diagrams (More Detailed Levels)

Each process on Diagram 0 may in turn be exploded to create a more detailed child diagram. The process on Diagram 0 that is exploded is called the *parent process,* and the diagram that results is called the *child diagram.* The primary rule for creating child diagrams, vertical balancing, dictates that a child diagram cannot produce output or receive input that the parent process does not also produce or receive. All data flow into or out of the parent process must be shown flowing into or out of the child diagram.

The child diagram is given the same number as its parent process in Diagram 0. For example, process 3 would explode to Diagram 3. The processes on the child diagram are numbered using the parent process number, a decimal point, and a unique number for each child process. On Diagram 3, the processes would be numbered 3.1, 3.2, 3.3, and so on. This convention allows the analyst to trace a series of processes through many levels of explosion. If Diagram 0 depicts processes 1, 2, and 3, the child diagrams 1, 2, and 3 are all on the same level.

Entities are usually not shown on the child diagrams below Diagram 0. Data flow that matches the parent flow is called an *interface data flow* and is shown as an arrow from or into a blank area of the child diagram. If the parent process has data flow connecting to a data store, the child diagram may include the data store as well. In addition, this lower-level diagram may contain data stores not shown on the parent process. For example, a file containing a table of information, such as a tax table, or a file linking two processes on the child diagram may be included. Minor data flow, such as an error line, may be included on a child diagram but not on the parent.

Processes may or may not be exploded, depending on their level of complexity. When a process is not exploded, it is said to be functionally primitive and is called a *primitive process.* Logic is written to describe these processes and is discussed in detail in Chapter 9. Figure 7.4 illustrates detailed levels in a child data flow diagram.

Checking Diagrams for Errors

Several common errors made when drawing data flow diagrams are as follows:

1. ***Forgetting to include a data flow or pointing an arrow in the wrong direction.*** An example is a drawn process showing all its data flow as either input or output. Each process transforms data and must receive input and produce output. This type of error usually occurs when the analyst has forgotten to include a data flow or has placed an arrow pointing in the wrong direction. Process 1 in Figure 7.5 contains only input because the GROSS PAY arrow is pointing in the wrong direction. This error also affects process 2, CALCULATE WITHHOLDING AMOUNT, which is in addition missing a data flow representing input for the withholding rates and the number of dependents.

2. ***Connecting data stores and external entities directly to each other.*** Data stores and entities may not be connected to each other; data stores and external entities must connect only with a process. A file does not interface with another file without the help of a program or a person moving the data, so EMPLOYEE MASTER cannot directly produce the CHECK

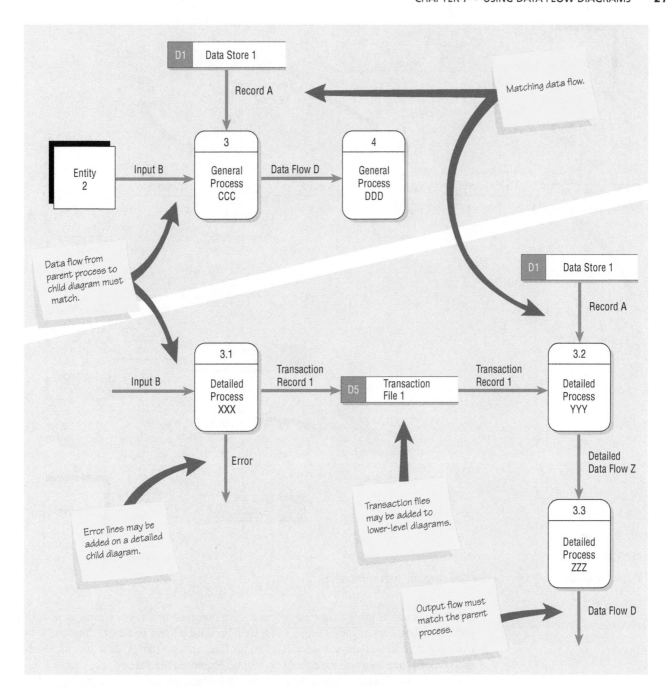

FIGURE 7.4

Differences between the parent diagram (above) and the child diagram (below).

RECONCILIATION file. External entities do not directly work with files. For example, you would not want a customer rummaging around in the customer master file. Thus, the EMPLOYEE does not create the EMPLOYEE TIME FILE. Two external entities directly connected indicate that they wish to communicate with each other. This connection is not included on the data flow diagram unless the system is facilitating the communication. Producing a report is an instance of this sort of communication. A process must still be interposed between the entities to produce the report, however.

3. ***Incorrectly labeling processes or data flow.*** Inspect the data flow diagram to ensure that each object or data flow is properly labeled. A process should indicate the system name or use the verb–adjective–noun format. Each data flow should be described with a noun.

4. ***Including more than nine processes on a data flow diagram.*** Having too many processes creates a cluttered diagram that is confusing to read and hinders rather than enhances

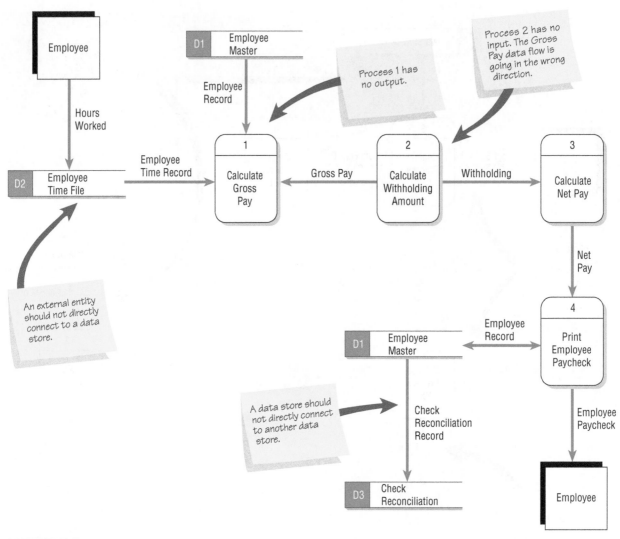

FIGURE 7.5

Typical errors that can occur in a data flow diagram (payroll example).

communication. If more than nine processes are involved in a system, group some of the processes that work together into a subsystem and place them in a child diagram.

5. *Omitting data flow.* Examine your diagram for linear flow—that is, data flow in which each process has only one input and one output. Except in the case of very detailed child data flow diagrams, linear data flow is somewhat rare. Its presence usually indicates that the diagram has missing data flow. For instance, the process CALCULATE WITHHOLDING AMOUNT needs the number of dependents that an employee has and the WITHHOLDING RATES as input. In addition, NET PAY cannot be calculated solely from the WITHHOLDING, and the EMPLOYEE PAYCHECK cannot be created from the NET PAY alone; it also needs to include an EMPLOYEE NAME, as well as the current and year-to-date payroll and WITHHOLDING AMOUNT figures.

6. *Creating unbalanced decomposition (or explosion) in child diagrams.* Each child diagram should have the same input and output data flow as the parent process. An exception to this rule is minor output, such as error lines, which are included only on the child diagram. The data flow diagram in Figure 7.6 is correctly drawn. Note that although the data flow is not linear, you can clearly follow a path directly from the source entity to the destination entity.

Logical and Physical Data Flow Diagrams

Data flow diagrams are categorized as either logical or physical. A logical data flow diagram focuses on the business and how the business operates. It is not concerned with how the system will be constructed. Instead, it describes the business events that take place and the data required

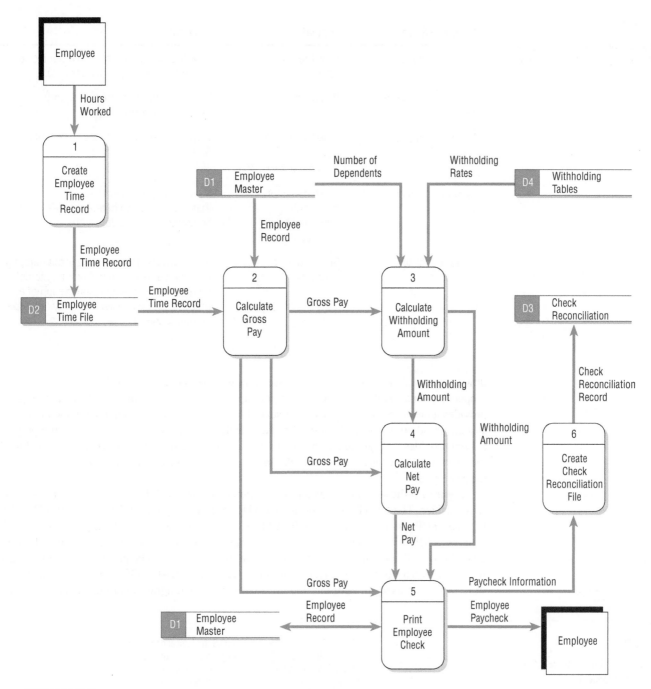

FIGURE 7.6

The correct data flow diagram for the payroll example.

and produced by each event. Conversely, a physical data flow diagram shows how the system will be implemented, including the hardware, software, files, and people involved in the system. The chart shown in Figure 7.7 contrasts the features of logical and physical models. Notice that the logical model reflects the business, whereas the physical model depicts the system.

Ideally, systems are developed by analyzing the current system (the current logical DFD) and then adding features that the new system should include (the proposed logical DFD). Finally, the best methods for implementing the new system should be developed (the physical DFD). This progression is shown in Figure 7.8.

Developing a logical data flow diagram for the current system affords a clear understanding of how the current system operates, and thus a good starting point for developing the logical model of the current system. One argument in favor of taking the time to construct the logical data flow diagram of the current system is that it can be used to create the logical data

FIGURE 7.7

Features common to both logical
and physical data flow diagrams.

Design Feature	Logical	Physical
What the model depicts	How the business operates.	How the system will be implemented (or how the current system operates).
What the processes represent	Business activities.	Programs, program modules, and manual procedures.
What the data stores represent	Collections of data regardless of how the data are stored.	Physical files and databases, manual files.
Type of data stores	Show data stores representing permanent data collections.	Master files, transition files. Any processes that operate at two different times must be connected by a data store.
System controls	Show business controls.	Show controls for validating input data, for obtaining a record (record found status), for ensuring successful completion of a process, and for system security (example: journal records).

flow diagram of the new system. Processes that will be unnecessary in the new system may be dropped, and new features, activities, output, input, and stored data may be added. This approach provides a means of ensuring that the essential features of the old system are retained in the new system. In addition, using the logical model for the current system as a basis for the proposed system provides for a gradual transition to the design of the new system. After the logical model for the new system has been developed, it may be used to create a physical data flow diagram for the new system.

Figure 7.9 shows a logical data flow diagram and a physical data flow diagram for a grocery store cashier. The CUSTOMER brings the ITEMS to the register; PRICES for all ITEMS are LOOKED UP and then totaled; next, PAYMENT is given to the cashier; finally, the CUSTOMER is given a RECEIPT. The logical data flow diagram illustrates the processes involved without going into detail about the physical implementation of activities. The physical data flow diagram shows that a bar code—the universal product code (UPC) BAR CODE found on most grocery store items—is used. In addition, the physical data flow diagram mentions manual processes

FIGURE 7.8

The progression of models from
logical to physical.

Derive the logical data flow diagram for the current system by examining the physical data flow diagram and isolating unique business activities.

Create the logical data flow diagram for the new system by adding the input, output, and processes required in the new system to the logical data flow diagram for the current system.

Derive the physical data flow diagram by examining processes on the new logical diagram. Determine where the user interfaces should exist, the nature of the processes, and necessary data stores.

Logical Data Flow Diagram

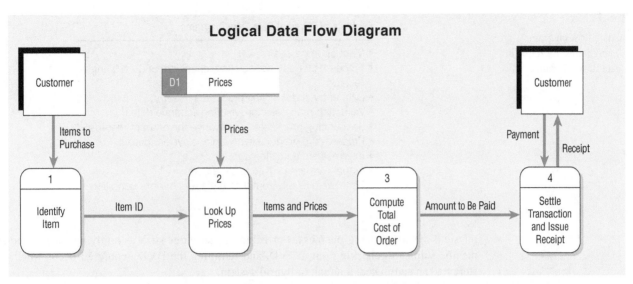

Physical Data Flow Diagram

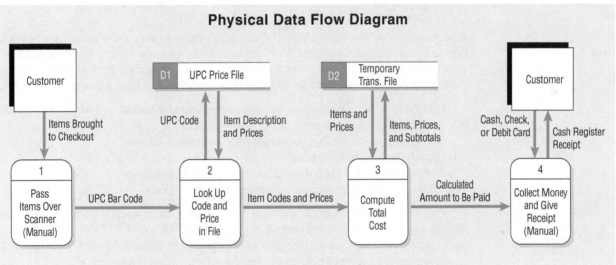

FIGURE 7.9

The physical data flow diagram (below) shows certain details not found on the logical data flow diagram (above).

such as scanning, explains that a temporary file is used to keep a subtotal of items, and indicates that the PAYMENT could be made by CASH, CHECK, or DEBIT CARD. Finally, it refers to the receipt by its name, CASH REGISTER RECEIPT.

Developing Logical Data Flow Diagrams

Before you construct a physical data flow diagram you need to first construct a logical data flow diagram for the current system. There are a number of advantages to using a logical model, including:

1. Better communication with users
2. More stable systems
3. Better understanding of the business by analysts
4. Flexibility and maintenance
5. Elimination of redundancies and easier creation of the physical model

A logical model is easiest to use when communicating with users of the system because it is centered on business activities. Users will thus be familiar with the essential activities and many of the human information requirements of each activity.

Systems formed using a logical data flow diagram are often relatively stable because they are based on business events and not on a particular technology or method of implementation. Logical data flow diagrams represent features of a system that would exist no matter what the

FIGURE 7.10

Physical data flow diagrams contain many items not found in logical data flow diagrams.

Contents of Physical Data Flow Diagrams

- Manual processes
- Processes for adding, deleting, changing, and updating records
- Data entry and verifying processes
- Validation processes for ensuring accurate data input
- Sequencing processes to rearrange the order of records
- Processes to produce every unique system output
- Intermediate data stores
- Actual file names used to store data
- Controls to signify completion of tasks or error conditions

physical means of doing business. For example, activities such as applying for a video store membership card, checking out a DVD, and returning the DVD, would all occur whether the store had an automated, manual, or hybrid system.

Developing Physical Data Flow Diagrams

After you develop the logical model of a new system, you may use it to create a physical data flow diagram. The physical data flow diagram shows how the system will be constructed and usually contains most, if not all, of the elements found in Figure 7.10. Just as logical data flow diagrams have certain advantages, physical data flow diagrams have advantages, including:

1. Clarifying which processes are performed by humans (manual) and which are automated
2. Describing processes in more detail than logical DFDs
3. Sequencing processes that have to be done in a particular order
4. Identifying temporary data stores
5. Specifying actual names of files, database tables, and printouts
6. Adding controls to ensure the processes are done properly

Physical data flow diagrams are often more complex than logical data flow diagrams simply because of the many data stores present in a system. The acronym CRUD is often used for *create*, *read*, *update*, and *delete*, the activities that must be present in a system for each master file. A CRUD matrix is a tool to represent where each of these processes occurs in a system. Figure 7.11 is a CRUD

FIGURE 7.11

A CRUD matrix for an Internet storefront. This tool can be used to represent where each of four processes (create, read, update, and delete) occurs within a system.

Activity	Customer	Item	Order	Order Detail
Customer Logon	R			
Item Inquiry		R		
Item Selection		R	C	C
Order Checkout	U	U	U	R
Add Account	C			
Add Item		C		
Close Customer Account	D			
Remove Obsolete Item		D		
Change Customer Demographics	RU			
Change Customer Order	RU	RU	RU	CRUD
Order Inquiry	R	R	R	R

matrix for an Internet storefront. Notice that some of the processes include more than one activity. Data entry processes such as keying and verifying are also part of physical data flow diagrams.

Physical data flow diagrams also have intermediate data stores, often a transaction file or a temporary database table. Intermediate data stores often consist of transaction files used to store data between processes. Because most processes that require access to a given set of data are unlikely to execute at the same instant in time, transaction files must hold the data from one process to the next. An easily understood example of this concept is found in the everyday experiences of grocery shopping, meal preparation, and eating. The activities are:

1. Selecting items from shelves
2. Checking out and paying the bill
3. Transporting the groceries home
4. Preparing a meal
5. Eating the meal

Each of these five activities would be represented by a separate process on a physical data flow diagram, and each one occurs at a different time. For example, you would not typically transport the groceries home and eat them at the same time. Therefore, a "transaction data store" is required to link each task. When you are selecting items, the transaction data store is the shopping cart. After the next process (checking out), the cart is unnecessary. The transaction data store linking checking out and transporting the groceries home is the shopping bag (cheaper than letting you take the cart home!). Storing the groceries in bags once they are home would be inefficient, so cupboards and a refrigerator are used as a transaction data store between the activity of transporting the goods home and preparing the meal. Finally, a plate, bowl, and cup constitute the link between preparing and eating the meal.

Timing information may also be included. For example, a physical DFD may indicate that an edit program must be run before an update program. Updates must be performed before producing a summary report, or an order must be entered on a website before the amount charged to a credit card may be verified with the financial institution. Note that because of such considerations, a physical data flow diagram may appear more linear than a logical model.

You create the physical data flow diagram for a system by analyzing its output and input. When creating a physical data flow diagram, input data flow from an external entity is sometimes called a *trigger* because it starts the activities of a process, and output data flow to an external entity is sometimes called a *response* because it is sent as the result of some activity. You need to determine which data fields or elements need to be keyed. These fields are called *base elements* and must be stored in a file. Elements that are not keyed but are rather the result of a calculation or logical operation are called *derived elements*.

Sometimes it is not clear how many processes to place in one diagram and when to create a child diagram. One suggestion is to examine each process and count the number of data flows entering and leaving it. If the total is greater than four, the process is a good candidate for a child diagram. Physical data flow diagrams are illustrated later in this chapter.

EVENT MODELING AND DATA FLOW DIAGRAMS. A practical approach to creating physical data flow diagrams is to create a simple data flow diagram fragment for each unique system event. Events cause the system to do something and act as a trigger to the system. Triggers start activities and processes, which in turn use data or produce output. An example of an event is a customer reserving a flight on the Web. As each Web form is submitted, processes are activated, such as validating and storing the data and formatting and displaying the next web page.

Events are usually summarized in an event response table. An example of an event response table for an Internet storefront business is illustrated in Figure 7.12. A data flow diagram fragment is represented by a row in the table. Each DFD fragment is a single process on a data flow diagram. All the fragments are then combined to form Diagram 0. The trigger and response columns become the input and output data flows, and the activity becomes the process. The analyst must determine the data stores required for the process by examining the input and output data flows. Figure 7.13 illustrates data flow diagrams for the first three rows of the event response table.

The advantage of building data flow diagrams based on events is that the users are familiar with the events that take place in their business area and know how the events drive other activities.

Event	Source	Trigger	Activity	Response	Destination
Customer logs on	Customer	Customer number and password	Find customer record and verify password. Send Welcome Web page.	Welcome Web page	Customer
Customer browses items at Web storefront	Customer	Item information	Find item price and quantity available. Send Item Response Web page.	Item Response Web page	Customer
Customer places item into shopping basket at Web storefront	Customer	Item purchase (item number and quantity)	Store data on Order Detail Record. Calculate shipping cost using shipping tables. Update customer total. Update item quantity on hand.	Items Purchased Web page	Customer
Customer checks out	Customer	Clicks "Check Out" button on Web page	Display Customer Order Web page.	Verification Web page	
Obtain customer payment	Customer	Credit card information	Verify credit card amount with credit card company. Send.	Credit card data Customer feedback	Credit card company Customer
Send customer email		Temporal, hourly	Send customer an email confirming shipment.		Customer

FIGURE 7.12

An event response table for an Internet storefront.

USE CASES AND DATA FLOW DIAGRAMS. In Chapter 2, we introduced the concept of a *use case*. We use this notion of a use case in creating data flow diagrams. A use case summarizes an event and has a similar format to process specifications (described in Chapter 9). Each use case defines one activity and its trigger, input, and output. Figure 7.14 illustrates a use case for Process 3, Add Customer Item.

This approach allows an analyst to work with users to understand the nature of the processes and activities and then create a single data flow diagram fragment. When creating use cases, you should first make an initial attempt to define the use cases without going into detail. This step provides an overview of the system and leads to the creation of Diagram 0. Then you can decide what the names should be and provide a brief description of each activity. List the activities, inputs, and outputs for each one.

You should make sure to document the steps used in each use case. These should be in the form of business rules that list or explain the human and system activities completed for each use case. If at all possible, you should list them in the sequence that they would normally be executed. Next, you can determine the data used by each step. This step is easier if a data dictionary has been completed. Finally, you need to ask the users to review and suggest modifications of the use cases. It is important that the use cases be written clearly. (See Chapter 10 for a further discussion of UML, use cases, and use case diagrams.)

Partitioning Data Flow Diagrams

Partitioning is the process of examining a data flow diagram and determining how it should be divided into collections of manual procedures and collections of computer programs. You should analyze each process to determine whether it should be a manual or automated procedure. Then you can group automated procedures into a series of computer programs. It can be helpful to

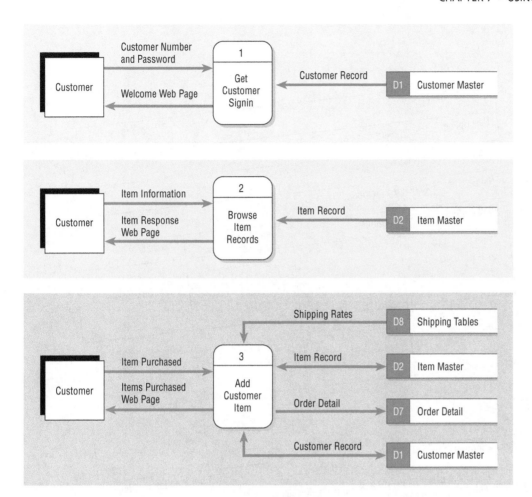

FIGURE 7.13

Data flow diagrams for the first three rows of the Internet storefront event response table.

draw a dashed line around a process or group of processes that should be placed into a single computer program.

There are six reasons for partitioning data flow diagrams:

1. ***Different user groups.*** Are the processes performed by several different user groups, often at different physical locations in the company? If so, they should be partitioned into different computer programs. An example is the need to process customer returns and customer payments in a department store. Both processes involve obtaining financial information that is used to adjust customer accounts (subtracting from the amount the customer owes), but the two processes are performed by different people at different locations. Each group needs a different screen for recording the particulars of the transaction, either a credit screen or a payment screen.

2. ***Timing.*** It is important to examine the timing of the processes. If two processes execute at different times, they cannot be grouped into one program. Timing issues may also involve how much data is presented at one time on a web page. If an ecommerce site has rather lengthy web pages for ordering items or making an airline reservation, the web pages may be partitioned into separate programs that format and present the data.

3. ***Similar tasks.*** If two processes perform similar tasks, they may be grouped into one computer program.

4. ***Efficiency.*** Several processes may be combined into one program for efficient processing. For example, if a series of reports needs to use the same large input files, producing them together may save considerable computer run time.

5. ***Consistency of data.*** Processes may be combined into one program for consistency of data. For example, a credit card company may take a "snapshot" and produce a variety of reports at the same time just so figures are consistent.

6. ***Security.*** Processes may be partitioned into different programs for security reasons. A dashed line may be placed around web pages that are on a secure server to separate them from those

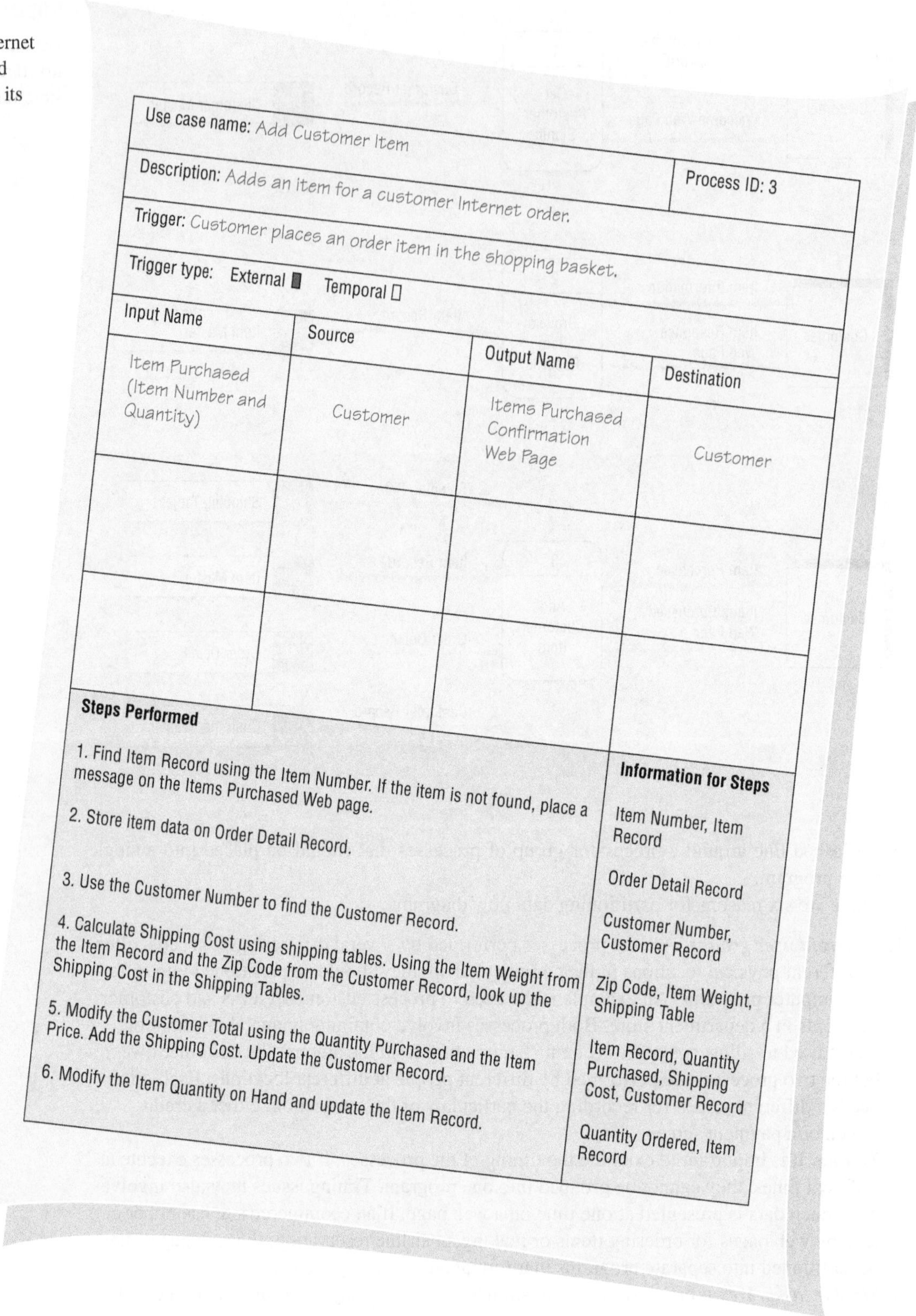

FIGURE 7.14

A use case form for the Internet storefront describes the Add Customer Item activity and its triggers, input, and output.

Use case name: Add Customer Item			
Description: Adds an item for a customer Internet order.			Process ID: 3
Trigger: Customer places an order item in the shopping basket.			
Trigger type: External ■ Temporal ☐			
Input Name	Source	Output Name	Destination
Item Purchased (Item Number and Quantity)	Customer	Items Purchased Confirmation Web Page	Customer

Steps Performed

Steps	Information for Steps
1. Find Item Record using the Item Number. If the item is not found, place a message on the Items Purchased Web page.	Item Number, Item Record
2. Store item data on Order Detail Record.	Order Detail Record
3. Use the Customer Number to find the Customer Record.	Customer Number, Customer Record
4. Calculate Shipping Cost using shipping tables. Using the Item Weight from the Item Record and the Zip Code from the Customer Record, look up the Shipping Cost in the Shipping Tables.	Zip Code, Item Weight, Shipping Table
5. Modify the Customer Total using the Quantity Purchased and the Item Price. Add the Shipping Cost. Update the Customer Record.	Item Record, Quantity Purchased, Shipping Cost, Customer Record
6. Modify the Item Quantity on Hand and update the Item Record.	Quantity Ordered, Item Record

web pages on a server that is not secured. A web page that is used for obtaining the user's identification and password is usually partitioned from order entry or other business pages.

A Data Flow Diagram Example

The following example is intended to illustrate the development of a data flow diagram by selectively looking at each of the components explored earlier in this chapter. This example, called "World's Trend Catalog Division," will also be used to illustrate concepts covered in Chapters 8 and 9.

FIGURE 7.15

A summary of business activities for World's Trend Catalog Division.

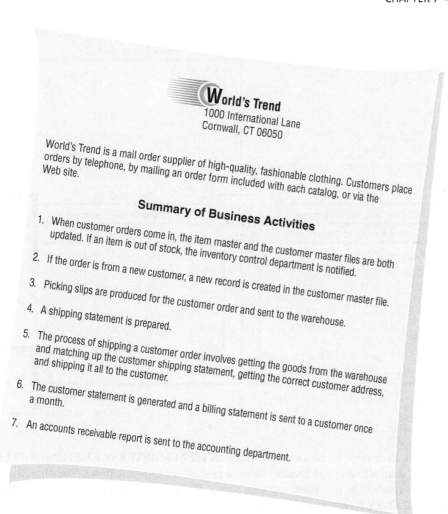

World's Trend
1000 International Lane
Cornwall, CT 06050

World's Trend is a mail order supplier of high-quality, fashionable clothing. Customers place orders by telephone, by mailing an order form included with each catalog, or via the Web site.

Summary of Business Activities

1. When customer orders come in, the item master and the customer master files are both updated. If an item is out of stock, the inventory control department is notified.

2. If the order is from a new customer, a new record is created in the customer master file.

3. Picking slips are produced for the customer order and sent to the warehouse.

4. A shipping statement is prepared.

5. The process of shipping a customer order involves getting the goods from the warehouse and matching up the customer shipping statement, getting the correct customer address, and shipping it all to the customer.

6. The customer statement is generated and a billing statement is sent to a customer once a month.

7. An accounts receivable report is sent to the accounting department.

Developing the List of Business Activities

A list of business activities for World's Trend can be found in Figure 7.15. You could develop this list using information obtained through interacting with people in interviews, through investigation, and through observation. The list can be used to identify external entities such as CUSTOMER, ACCOUNTING, and WAREHOUSE as well as data flows such as ACCOUNTS RECEIVABLE REPORT and CUSTOMER BILLING STATEMENT. Later (when developing level 0 and child diagrams), the list can be used to define processes, data flows, and data stores.

Creating a Context-Level Data Flow Diagram

Once the list of activities is developed, you can create a context-level data flow diagram as shown in Figure 7.16. This diagram shows the ORDER PROCESSING SYSTEM in the middle (no processes are described in detail in the context-level diagram) and five external entities (the two separate entities both called CUSTOMER are really one and the same). The data flows that come from and go to the external entities are shown as well (for example, CUSTOMER ORDER and ORDER PICKING LIST).

Drawing Diagram 0

Next, you need to go back to the activity list and make a new list of as many processes and data stores as you can find. You can add more later, but start making the list now. If you think you have enough information, draw a level 0 diagram such as the one shown in Figure 7.17. Call this Diagram 0 and keep the processes general so as not to overcomplicate the diagram. Later, you can add detail. When you are finished drawing the seven processes, draw data flows between them and to the external entities (the same external entities shown in the context-level diagram).

FIGURE 7.16

A context-level data flow diagram
for the order processing system at
World's Trend.

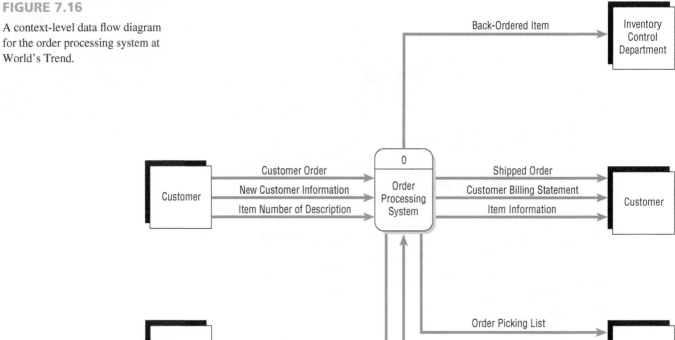

If you think there need to be data stores such as ITEM MASTER or CUSTOMER MASTER, you should draw those in and connect them to processes using data flows. You should also take the time to number the processes and data stores. Pay particular attention to making the labels meaningful. Check for errors and correct them before moving on.

Creating a Child Diagram

At this point, you should try to draw a child diagram (sometimes also called a level 1 diagram) such as the one in Figure 7.18. Child diagram processes are more detailed, illustrating the logic required to produce the output. You should number your child diagrams Diagram 1, Diagram 2, and so on, in accordance with the number you assigned to each process in the level 0 diagram.

When you draw a child diagram, you should make a list of subprocesses first. A process such as ADD CUSTOMER ORDER can have subprocesses (in this case, there are seven). Connect these subprocesses to one another and also to data stores when appropriate. Subprocesses do not have to be connected to external entities, because we can always refer to the parent (or level 0) data flow diagram to identify these entities. Label the subprocesses 1.1, 1.2, 1.3, and so on. Take the time to check for errors and make sure the labels make sense.

Creating a Physical Data Flow Diagram from the Logical DFD

If you want to go beyond the logical model and draw a physical model as well, look at Figure 7.19, which is an example of a physical data flow child diagram of process 3, PRODUCE PICKING SLIPS. Physical DFDs give you the opportunity to identify processes for scanning bar codes, displaying screens, locating records, and creating and updating files. The sequence of activities is important in physical DFDs because the emphasis is on how the system will work and in what order events happen.

When you label a physical model, take care to describe the process in great detail. For example, Subprocess 3.3 in a logical model could simply be SORT ORDER ITEM, but in the physical model, a better label is SORT ORDER ITEM BY LOCATION WITHIN CUSTOMER. When you write a label for a data store, refer to the actual file or database, such as CUSTOMER MASTER FILE or SORTED ORDER ITEM FILE. When you describe data flows, describe the actual form, report, or screen. For example, when you print a slip for order picking, call the data flow ORDER PICKING SLIP.

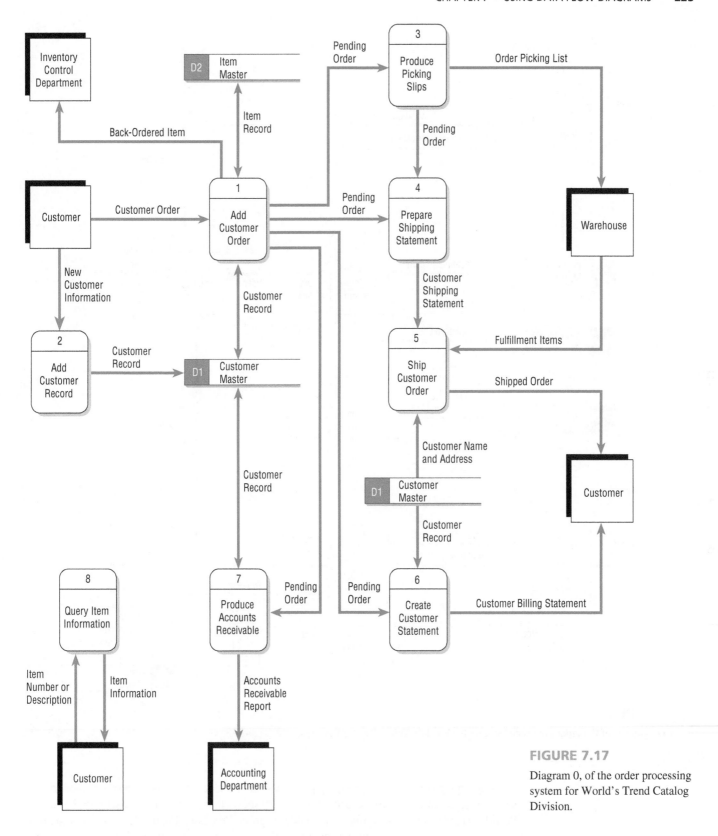

FIGURE 7.17

Diagram 0, of the order processing system for World's Trend Catalog Division.

Partitioning the Physical DFD

Finally, you suggest partitioning of the physical data flow diagram through combining or separating the processes. As stated earlier, there are many reasons for partitioning: identifying distinct processes for different user groups, separating processes that need to be performed at different

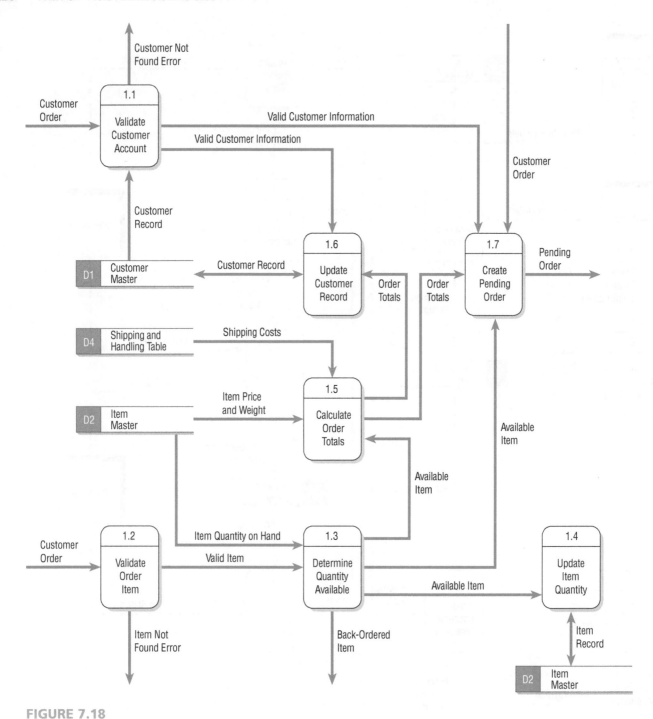

Diagram 1, of the order processing system for World's Trend Catalog Division.

times, grouping similar tasks, grouping processes for efficiency, combining processes for consistency, or separating them for security. Figure 7.20 shows that partitioning is useful in the case of World's Trend Catalog Division. You would first group Processes 1 and 2 because it would make sense to add new customers at the same time their first order was placed. You would then put Processes 3 and 4 in two separate partitions because these must be done at different times from each other and thus cannot be grouped into a single program.

The process of developing a data flow diagram is now complete from the top down. You first drew a companion physical data flow diagram to accompany the logical data flow diagram and then partitioned the data flow diagram by grouping or separating the processes. The World's Trend example is used again in Chapters 8 and 9.

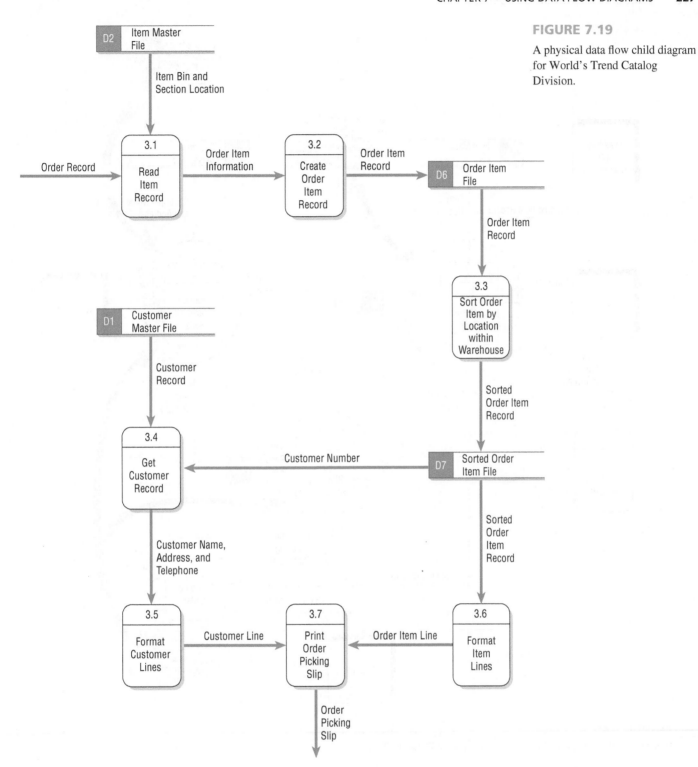

FIGURE 7.19

A physical data flow child diagram for World's Trend Catalog Division.

Partitioning Websites

Partitioning is a very useful principle when designing a website. Website designers who use forms to collect data may find it appropriate to divide a website into a series of web pages, which will improve the way humans use the site, the speed of processing, and the ease of maintaining the site. Each time data must be obtained from a data store or an external partner, a website designer might consider creating a unique Web form and DFD process to validate and process the data.

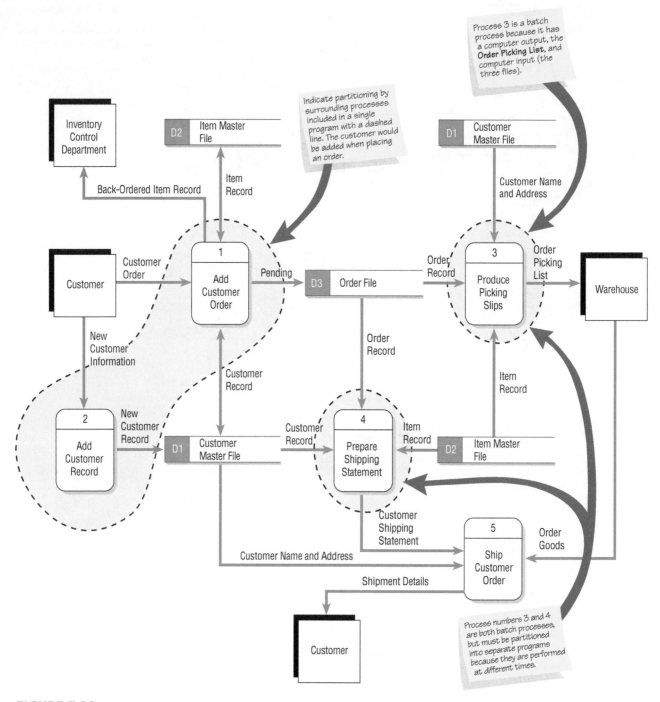

FIGURE 7.20

Partitioning the data flow diagram (showing part of Diagram 0).

A Web developer may also use Ajax, sending a request to the server and obtaining a small amount of data or an XML document returned to the same page. Ajax may be used to avoid creating too many small pages that contain only a few extra or changed Web form elements. However, the analyst should create several pages when needed. One consideration is when a large amount of data needs to be obtained from the server, such as a list of all the flights that match starting and destination airports for specific travel days. When accessing different database tables on the same database, the data that is obtained may contain fields from different database tables and may be passed to one process. However, if different databases are involved, an analyst may decide to use separate web pages. When user input is required, the analyst may

FIGURE 7.21

Partitioning is important for Web-based systems, as this physical data flow diagram of an online ticket purchasing system demonstrates.

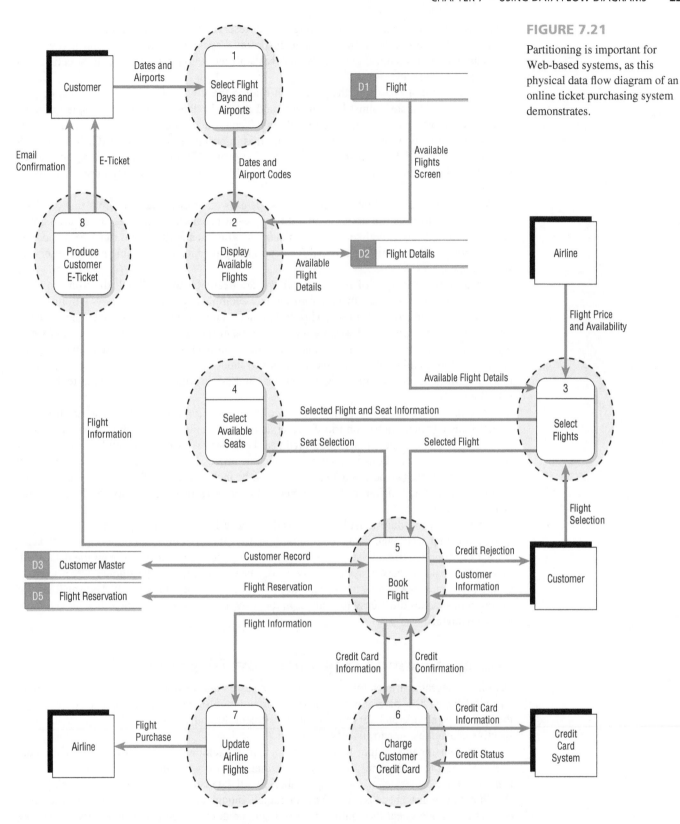

either use separate web pages or use Ajax to facilitate a change in a drop-down list or to change a small amount of data.

A good example of partitioning can be seen in the development of a Web-based travel booking site. To simplify, we will only look at the airline booking portion of the website, shown in the data flow diagram in Figure 7.21. Notice that the Web designer has chosen to create several processes and unique partitions in making a flight reservation. Process 1 receives and validates the dates and airports entered by the customer (or travel agent acting for a customer). The selection data is used to obtain flight details and create a transaction data store of flight details that match the flight request.

It is advisable to partition the process of finding the flight information as a separate process because a data store must be searched, and the flight details are used to display a series of successive web pages with matching flights. Then, once a customer chooses a flight, the information must be sent to a selected airline. It is important to have the FLIGHT DETAILS transaction file available to display each web page of new flights because redoing the search may take a large amount of time that is unacceptable to a human user trying to complete a transaction.

The selection of available flights (process 2) uses an internal database, but this database does not have information about availability of seats, because the airlines are receiving reservations from many travel service organizations. This means that there must be a separate process and small program partitioned for determining if seats are available and for reserving specific seats.

Because there is a lot of user input, forms are designed to handle all the user requests. Having separate forms means that the forms are less complex, and therefore users will find them more attractive and easier to fill out. This design meets both the usability and usefulness criteria important when designing websites for human–computer interaction. It also means that processing will take place more quickly because once the flight is chosen, the next step involving the choice of seats should not require the customer to input or even see the flight details again at this time. Most airline websites now use pop-up windows in which customers point to their seat selection.

Another reason for partitioning is to keep the transaction secure. Once the seat has been selected, the customer must confirm the reservation and supply credit card information. This is done using a secure connection, and the credit card company is involved in validating the amount of purchase. The secure connection means a separate process must be used. Once the credit card has been confirmed, two additional processes must be included, one to format and send an email confirmation and an e-ticket to the customer and another to send notification of the flight purchase to the airline.

The entire procedure must be partitioned into a series of interacting processes, each with a corresponding web page or interaction with an external system. Each time a new data store is used to obtain additional data, a process must be included to format or obtain the data. Each time an external company or system is involved, a process needs to be partitioned into a separate program. When processes or forms need to be revised, it is not a major task. The small size of the programs makes them easy to change. In this way, the website is secure, efficient, and more easily maintained.

Communicating Using Data Flow Diagrams

Data flow diagrams are useful throughout the analysis and design process. You use original, unexploded data flow diagrams early when ascertaining information requirements. At this stage, they can help provide an overview of data movement through the system, lending a visual perspective unavailable in narrative data.

A systems analyst might be quite competent at sketching through the logic of the data stream for data flow diagrams, but to make the diagrams truly communicative to users and other members of the project team, meaningful labels for all data components are also required. Labels should not be generic because then they do not tell enough about the situation at hand. All general systems models bear the configuration of input, process, and output, so labels for a data flow diagram need to be more specific than that.

Finally, remember that data flow diagrams are used to document the system. Assume that data flow diagrams will be around longer than the people who drew them, which is, of course, always true if an external consultant is drawing them. Data flow diagrams can be used for documenting high or low levels of analysis and helping to substantiate the logic underlying the data flows of the organizations.

CONSULTING OPPORTUNITY 7.1

There's No Business Like Flow Business

The phone at Merman's Costume Rentals rings, and Annie Oaklea, head of costume inventory, picks it up and answers a query by saying, "Let me take a look at my inventory cards. Sorry, it looks as if there are only two male bear suits in inventory, with extra growly expressions at that. We've had a great run on bears. When do you need them? Perhaps one will be returned. No, can't do it, sorry. Would you like these two sent, regardless? The name of your establishment? Manhattan Theatre Company? London branch? Right. Delightful company! I see by our account card that you've rented from us before. And how long will you be needing the costumes?"

Figure 7.C1 is a data flow diagram that sets the stage for processing of costume rentals from Merman's. It shows rentals such as the one Annie is doing for Manhattan Theatre Company.

After conversing for another few moments about shop policy on alterations, Annie concludes her conversation by saying, "You are very lucky to get the bears on such short notice. I've got another company reserving them for the first week in July. I'll put you down for

the bear suits, and they'll be taken to you directly by our courier. As always, prompt return will save enormous trouble for us all."

Merman's costume rental enterprise is located in London's world-famous West End theatre district. When a theatre or television production company lacks the resources (either time or expertise) to construct a costume in its own shop, the cry goes up to "Ring up Merman's!" and it proceeds to rent what it needs with a minimum of fuss.

Merman's shop (more aptly visualized as a warehouse) goes on for three floors full of costume racks, holding thousands of costumes hung together by historical period, then grouped as to whether they are for men or women, and then by costume size.[1] Most theatre companies are able to locate precisely what they need through Annie's capable assistance.

Now tailor-make the *rental return* portion of the data flow diagram given earlier. Remember that timely returns are critical for keeping the spotlight on costumes rented from Merman's.

FIGURE 7.C1

A data flow diagram for Merman's Costume Rentals.

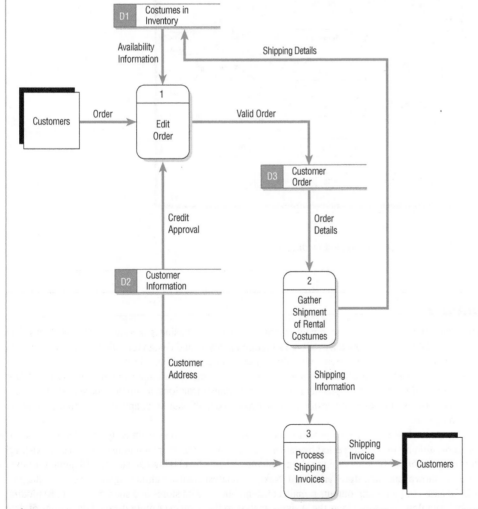

[1]Western Costume Company in Hollywood, California, is said to have more than 1 million costumes, worth about $40 million.

HYPERCASE® EXPERIENCE 7

"Y ou take a very interesting approach to the problems we have here at MRE. I've seen you sketching diagrams of our operation almost since the day you walked in the door. I'm actually getting used to seeing you doodling away now. What did you call those? Oh, yes. Context-level diagrams. And flow networks? Oh, no. Data flow diagrams. That's it, isn't it?"

HYPERCASE Questions

1. Find the data flow diagrams already drawn in MRE. Make a list of those you found and add a column to show where in the organization you found them.
2. Draw a context-level diagram to model the Training Unit Project Development process, one that is based on interviews with relevant Training Unit staff. Then draw a level 0 diagram that details the process.

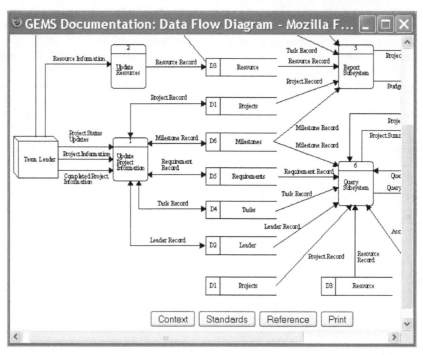

FIGURE 7.HC1

In HyperCase you can click on elements in a data flow diagram.

Summary

To better understand the logical movement of data throughout a business, a systems analyst draws data flow diagrams (DFDs). Data flow diagrams are structured analysis and design tools that allow the analyst to comprehend the system and subsystems visually as a set of interrelated data flows.

Graphical representations of data movement storage and transformation are drawn with the use of four symbols: a rounded rectangle to depict data processing or transformations, a double square to show an outside data entity (the source or receiver of data), an arrow to depict data flow, and an open-ended rectangle to show a data store.

A systems analyst extracts data processes, sources, stores, and flows from early organizational narratives or stories told by users or revealed by data and uses a top-down approach to first draw a context-level data flow diagram of the system within the larger picture. Then a level 0 logical data flow diagram is drawn. Processes are shown, and data stores are added. Next, the analyst creates a child diagram for each of the processes in Diagram 0. Inputs and outputs remain constant, but the data stores and sources change. Exploding the original data flow diagram allows the systems analyst to focus on ever more detailed depictions of data movement in the system. The analyst then develops a physical data flow diagram from the logical data flow diagram, partitioning it to facilitate programming. Each process is analyzed to determine whether it should be a manual or automated procedure.

Six considerations for partitioning data flow diagrams are whether processes are performed by different user groups, whether processes execute at the same times, whether processes perform similar tasks, whether batch processes can be combined for efficient processing, whether processes may be combined into one program for consistency of data, and whether processes may be partitioned into different programs for security reasons.

Keywords and Phrases

Ajax
base element
child diagram
context-level data flow diagram
data flow diagram (DFD)
data flow diagram fragment
data store
derived element
event modeling
event response table
event trigger
exploding
external entity (source or destination)
functionally primitive
interface data flow

level 0 diagram
logical model
online process
parent process
partitioning
physical data store
physical model
primitive process
top-down approach
transaction data store
transforming process
Unified Modeling Language (UML)
use case
vertical balancing

Review Questions

1. What is one of the main methods available for an analyst to use when analyzing data-oriented systems?
2. What are the four advantages of using a data flow approach over narrative explanations of data movement?
3. What are the four data items that can be symbolized on a data flow diagram?
4. What is a context-level data flow diagram? Contrast it to a level 0 DFD.
5. Define the top-down approach as it relates to drawing data flow diagrams.
6. Describe what "exploding" data flow diagrams means.
7. What are the trade-offs involved in deciding how far data streams should be exploded?
8. Why is labeling data flow diagrams so important? What can effective labels on data flow diagrams accomplish for those who are unfamiliar with the system?
9. What is the difference between a logical data flow diagram and a physical data flow diagram?
10. List three reasons for creating a logical data flow diagram.
11. List five characteristics of a physical data flow diagram that a logical data flow diagram does not have.
12. When are transaction files required in a system's design?
13. How can an event table be used to create a data flow diagram?
14. List the major sections of a use case.
15. How can a use case be used to create a data flow diagram?
16. What is partitioning, and how is it used?
17. How can an analyst determine when a user interface is required?
18. List three ways of determining partitioning in a data flow diagram.
19. List three ways to use completed data flow diagrams.

Problems

1. Up to this point, you seem to have had excellent rapport with Kevin Cahoon, the owner of a musical instrument manufacturing company. When you showed him a set of data flow diagrams you drew, he wasn't able to see how the system you were proposing was described in the diagrams.
   a. In a paragraph, write down in general terms how to explain a data flow diagram to a user. Be sure to include a list of symbols and what they mean.

b. It takes some effort to educate users about data flow diagrams. Is it worthwhile to share DFDs with users? Why or why not? Defend your response in a paragraph.

c. Compare data flow diagrams to use cases and use case scenarios. What do data flow diagrams show that use case diagrams have a difficult time trying to explain?

2. Your latest project is to combine two systems used by Producers Financial. Angie Schworer's loan application system is fairly new, but has no documentation. Scott Wittman's loan management system is older and needs much revision, and the records are coded and kept independently of the other system. The loan application system accepts applications, processes them, and recommends loans for approval. The loan management system takes loans that have been approved and follows them through their final disposition (paid, sold, or defaulted). Draw a context diagram and a level 1 data flow diagram that show what an idealized combined system would look like.

3. One common experience that students in every college and university share is enrolling in a college course.

a. Draw a level 1 data flow diagram of data movement for enrollment in a college course. Use a single sheet and label each data item clearly.

b. Explode one of the processes in your original data flow diagram into subprocesses, adding data flows and data stores.

c. List the parts of the enrollment process that are "hidden" to the outside observer and about which you have had to make assumptions to complete a second-level diagram.

4. Figure 7.EX1 is a level 1 data flow diagram of data movement in a Niagara Falls tour agency called Marilyn's Tours. Read it over, checking for any inaccuracies.

a. List and number the errors that you have found in the diagram.

b. Redraw and label the data flow diagram of Marilyn's so that it is correct. Be sure that your new diagram employs symbols properly so as to cut down on repetitions and duplications where possible.

5. Perfect Pizza wants to install a system to record orders for pizza and chicken wings. When regular customers call Perfect Pizza on the phone, they are asked their phone number. When the number is typed into a computer, the name, address, and last order date is automatically brought up on the screen. Once the order is taken, the total, including tax and delivery, is calculated. Then the order is given to the cook. A receipt is printed. Occasionally, special offers (coupons) are printed so the customer can get a discount. Drivers who make deliveries give customers a copy of the receipt and a coupon (if any). Weekly totals are kept for comparison with last year's performance. Write a summary of business activities for taking an order at Perfect Pizza.

6. Draw a context-level data flow diagram for Perfect Pizza (Problem 5).

7. Explode the context-level diagram in Problem 6 showing all the major processes. Call this Diagram 0. It should be a logical data flow diagram.

FIGURE 7.EX1

A hand-sketched data flow diagram for Marilyn's Tours.

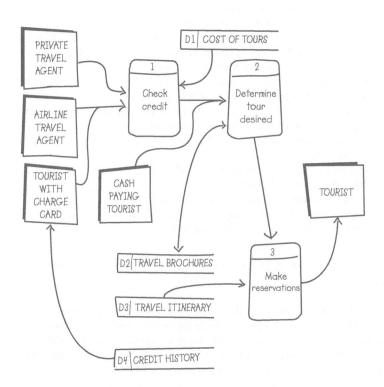

8. Draw a logical child diagram for Diagram 0 in Problem 7 for the process that adds a new customer if he or she is not currently in the database (that is, has never ordered from Perfect Pizza before).

9. Draw a physical data flow diagram for Problem 7.

10. Draw a physical data flow diagram for Problem 8.

11. Partition the physical data flow diagram in Problem 7, grouping and separating processes as you deem appropriate. Explain why you partitioned the data flow diagram in this manner. (Remember that you do not have to partition the entire diagram, only the parts that make sense to partition.)

12. a. Draw a logical child diagram for Process 6 in Figure 7.17.
    b. Draw a physical child diagram for Process 6 in Figure 7.17.

13. Draw a physical data flow diagram for Process 1.1 in Figure 7.18.

14. Create a context diagram for a real estate agent trying to create a system that matches buyers with potential houses.

15. Draw a logical data flow diagram showing general processes for Problem 14. Call it Diagram 0.

16. Create a context-level diagram for billing in a dental office. External entities include the patients and insurance companies.

17. Draw a logical data flow diagram showing general processes for Problem 16. Call it Diagram 0.

18. Create an event response table for the activities listed for World's Trend order processing system.

19. Create a use case for the list of seven processes for the World's Trend order processing system.

20. Create a CRUD matrix for the files of World's Trend.

21. Use the principles of partitioning to determine which of the processes in Problem 18 should be included in separate programs.

22. Create a physical data flow child diagram for the following situation: The local running club holds meetings once a month, with informative speakers, door prizes like pre-paid entry fees for races, designer running gear, and so on, and sessions for special interest groups (people who want to coach, people who want to run marathons, those who want to get fit, and so on). A laptop computer is taken to the meetings and is used to add the names of new members to the group. The diagram represents an online process and is the child of Process 1, ADD NEW MEMBERS. The following tasks are included:
    a. Key the new member information.
    b. Validate the information. Errors are displayed on the screen.
    c. When all the information is valid, a confirmation screen is displayed. The operator visually confirms that the data are correct and either accepts the transaction or cancels it.
    d. Accepted transactions add new members to the MEMBERSHIP MASTER file, which is stored on the laptop hard drive.
    e. Accepted transactions are written to a MEMBERSHIP JOURNAL file, which is stored on a second hard drive.

Group Projects

1. Meet with your group to develop a context-level data flow diagram for Maverick Transport (first introduced in Chapter 4). Use any data you have subsequently generated with your group about Maverick Transport. (*Hint:* Concentrate on one of the company's functional areas rather than try to model the entire organization.)

2. Using the context-level diagram developed in Group Project Problem1, develop with your group a level 0 logical data flow diagram for Maverick Transport. Make any assumptions necessary to draw it. List them.

3. With your group, choose one key process and explode it into a logical child diagram. Make any assumptions necessary to draw it. List follow-up questions and suggest other methods to get more information about processes that are still unclear to you.

4. Use the work your group has done to date to create a physical data flow diagram of a portion of the new system you are proposing for Maverick Transport.

Selected Bibliography

Ambler, S. W., and L. L. Constantine (Eds.). *The Unified Process Inception Phase: Best Practices for Implementing the UP.* Lawrence, KS: CMP Books, 2000.

Gane, C., and T. Sarson. *Structured Systems Analysis and Design Tools and Techniques.* Englewood Cliffs, NJ: Prentice Hall, 1979.

Hoffer, J. A., M. Prescott, and H. Topi. *Modern Database Management,* 9th ed. Upper Saddle River: Prentice Hall, 2009.

Kotonya, G., and I. Sommerville. *Requirements Engineering: Processes and Techniques.* New York: John Wiley & Sons, 1999.

Lucas, H. *Information Systems Concepts for Management,* 3rd ed. New York: McGraw-Hill, 1986.

Martin, J. *Strategic Data-Planning Methodologies.* Englewood Cliffs, NJ: Prentice Hall, 1982.

Thayer, R. H., M. Dorfman, and D. Garr. *Software Engineering: Vol. 1: The Development Process,* 2nd ed. New York: Wiley-IEEE Computer Society Press, 2002.

The CPU Case Episode and accompanying student files are available online at www.pearsonglobaleditions.com/kendall.

CHAPTER 8

Analyzing Systems Using Data Dictionaries

LEARNING OBJECTIVES

Once you have mastered the material in this chapter you will be able to:

1. Understand how analysts use data dictionaries for analyzing data-oriented systems.

2. Understand the concept of a repository for analysts' project information and the role of CASE tools in creating them.

3. Create data dictionary entries for data processes, stores, flows, structures, and logical and physical elements of the systems being studied, based on DFDs.

4. Recognize the functions of data dictionaries in helping users update and maintain information systems.

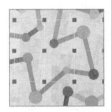

When successive levels of data flow diagrams are complete, systems analysts use them to help catalog the data processes, flows, stores, structures, and elements in a data dictionary. Of particular importance are the names used to characterize data items. When given an opportunity to name components of data-oriented systems, a systems analyst needs to work at making names meaningful but exclusive of other existing data component names. This chapter covers the data dictionary, which is another tool that aids in the analysis of data-oriented systems.

The Data Dictionary

A data dictionary is a specialized application of the kinds of dictionaries used as references in everyday life. A data dictionary is a reference work of data about data (that is, *metadata*). Systems analysts compile data dictionaries to guide them through analysis and design. A data dictionary is a document that collects and coordinates specific data terms, and it confirms what each term means to different people in the organization. The data flow diagrams covered in Chapter 7 are an excellent starting point for collecting data dictionary entries.

One important reason for maintaining a data dictionary is to keep clean data. This means that data must be consistent. If you store data about a man's sex as "M" in one record, "Male" in a second record, and as the number "1" in a third record, the data are not clean. Keeping a data dictionary will help in this regard.

Automated data dictionaries (part of the CASE tools mentioned earlier) are valuable for their capacity to cross-reference data items, thereby allowing necessary program changes to all programs that share a common element. This feature supplants changing programs on a haphazard basis, and it prevents waiting until the program won't run because a change has not been implemented across all programs sharing the updated item. Clearly, automated data dictionaries are important for large systems that produce several thousand data elements requiring cataloging and cross-referencing.

Need for Understanding the Data Dictionary

Many database management systems now come equipped with an automated data dictionary. These dictionaries can be either elaborate or simple. Some computerized data dictionaries automatically catalog data items when programming is done; others simply provide a template to prompt the person filling in the dictionary to do so in a uniform manner for every entry.

Despite the existence of automated data dictionaries, a systems analyst should understand what data compose a data dictionary, the conventions used in data dictionaries, and how a data dictionary is developed. Understanding the process of compiling a data dictionary can aid a systems analyst in conceptualizing the system and how it works. The upcoming sections allow the systems analyst to see the rationale behind what exists in automated data dictionaries.

In addition to providing documentation and eliminating redundancy, a data dictionary may be used to:

1. Validate the data flow diagram for completeness and accuracy.
2. Provide a starting point for developing screens and reports.
3. Determine the contents of data stored in files.
4. Develop the logic for data flow diagram processes.
5. Create XML (Extensible Markup Language).

The Data Repository

Whereas a data dictionary contains information about data and procedures, a larger collection of project information is called a *repository*. One of the benefits of using a CASE tool to develop the data dictionary is the ability to develop a repository, or a shared collection of project information and team contributions. The repository may contain the following:

1. Information about the data maintained by the system, including data flows, data stores, record structures, elements, entities, and messages
2. Procedural logic and use cases
3. Screen and report design
4. Data relationships, such as how one data structure is linked to another
5. Project requirements and final system deliverables
6. Project management information, such as delivery schedules, achievements, issues that need resolving, and project users

The data dictionary is created by examining and describing the contents of the data flows, data stores, and processes, as illustrated in Figure 8.1. Each data store and data flow should be defined and then expanded to include the details of the elements it contains. The logic of each process should be described using the data flowing into or out of the process. Omissions and other design errors should be noted and resolved.

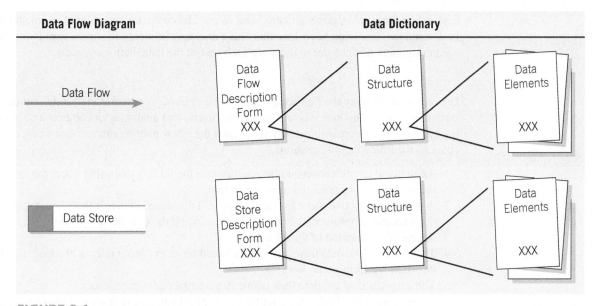

FIGURE 8.1

How data dictionaries relate to data flow diagrams.

The four data dictionary categories—data flows, data structures, data elements, and data stores—should be developed to promote understanding of the data of the system. Procedural logic is presented in Chapter 9, entities are discussed in Chapter 13, and messages and use cases are presented in Chapters 2 and 10.

To illustrate how data dictionary entries are created, we use an example for World's Trend Catalog Division. This company sells clothing and other items by mail order, using a toll-free phone order system (or faxing the mail order form), and via the Internet, using customized Web forms. Regardless of the origin of the order, the underlying data captured by the system are the same for all three methods.

The World's Trend order form shown in Figure 8.2 gives some clues about what to enter into a data dictionary. First, you need to capture and store the name, address, and telephone number of the person placing the order. Then you need to address the details of the order: the item

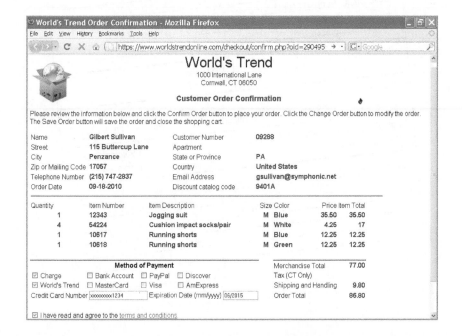

FIGURE 8.2

An online order form from World's Trend Catalog Division.

description, size, color, price, quantity, and so on. The customer's method of payment must also be determined. Once you have done this, these data may be stored for future use. This example is used throughout this chapter to illustrate each part of the data dictionary.

Defining the Data Flows

Data flows are usually the first components to be defined. A systems analyst determines system inputs and outputs by interviewing, observing users, and analyzing documents and other existing systems. The information captured for each data flow may be summarized using a form that contains the following information:

1. An optional identification number. Sometimes the ID is coded using a scheme to identify the system and the application in the system.
2. A unique descriptive name for the data flow. This name is the text that should appear on the diagram and be referenced in all descriptions using the data flow.
3. A general description of the data flow.
4. The source of the data flow. The source could be an external entity, a process, or a data flow coming from a data store.
5. The destination of the data flow (same items listed under the source).
6. An indication of whether the data flow is a record entering or leaving a file or a record containing a report, form, or screen. If the data flow contains data that are used between processes, it is designated as *internal.*
7. The name of the data structure describing the elements found in this data flow. For a simple data flow, it could be one or several elements.
8. The volume per unit of time. The data could be records per day or any other unit of time.
9. An area for further comments and notations about the data flow.

Once again, we can use our World's Trend Catalog Division example from Chapter 7 to illustrate a completed form. Figure 8.3 is an example of the data flow description representing the screen used to add a new CUSTOMER ORDER and to update the customer and item files. Notice that

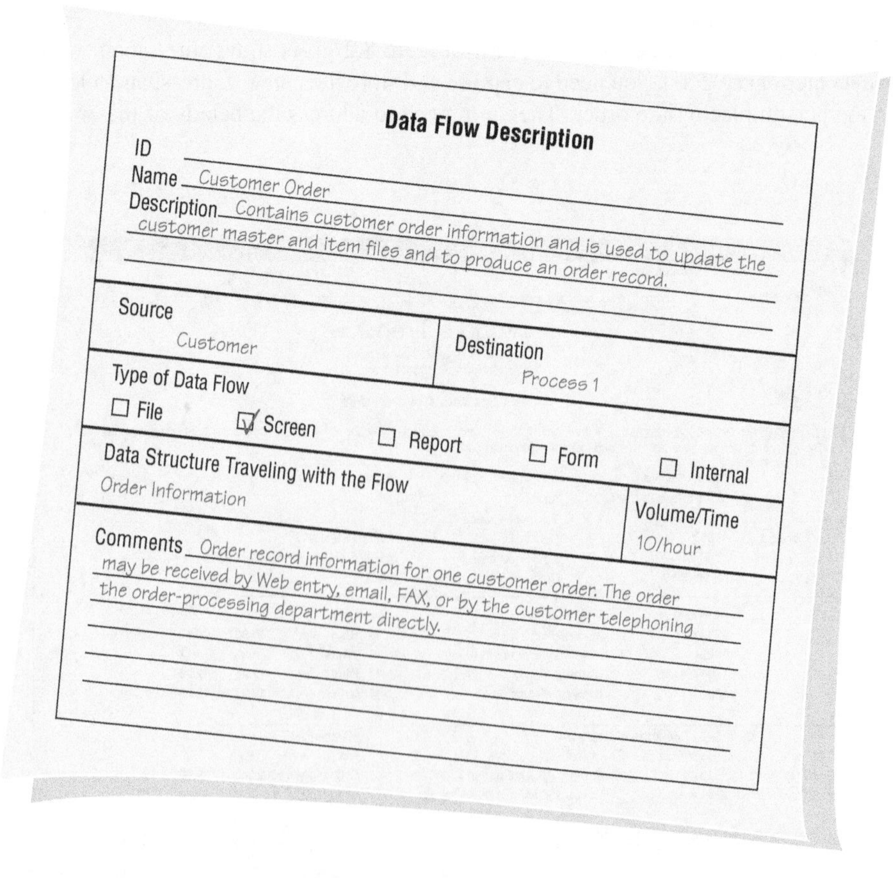
Data Flow Description
ID
Name Customer Order
Description Contains customer order information and is used to update the customer master and item files and to produce an order record.
Source
Customer
Destination
Process 1
Type of Data Flow
☐ File ☑ Screen ☐ Report ☐ Form ☐ Internal
Data Structure Traveling with the Flow
Order Information
Volume/Time
10/hour
Comments Order record information for one customer order. The order may be received by Web entry, email, FAX, or by the customer telephoning the order-processing department directly.

FIGURE 8.3

An example of a data flow description from World's Trend Catalog Division.

the external entity CUSTOMER is the source and that PROCESS 1 is the destination, providing linkage back to the data flow diagram. The checked box for "Screen" indicates that the flow represents an input screen. It could be any screen, such as a web page, graphical user interface (GUI), mobile phone, or perhaps a mainframe screen. The detailed description of the data flow could appear on this form, or it could be represented as a data structure.

Data flows for all inputs and outputs should be described first because they usually represent the human interface, followed by the intermediate data flows and the data flows to and from data stores. The detail of each data flow is described using elements, sometimes called fields; a data structure; or a group of elements.

A simple data flow may be described using a single element, such as a customer number used by an inquiry program to find the matching customer record.

Describing Data Structures

Data structures are usually described using algebraic notation. This method allows an analyst to produce a view of the elements that make up the data structure, along with information about those elements. For instance, the analyst will denote whether there are many of the same elements in the data structure (a repeating group) or whether two elements may exist mutually exclusive of each other. The algebraic notation uses the following symbols:

1. An equal sign (=) means "is composed of."
2. A plus sign (+) means "and."
3. Braces { } indicate repetitive elements, also called repeating groups or tables. There may be one repeating element or several in a group. The repeating group may have conditions, such as a fixed number of repetitions, or upper and lower limits for the number of repetitions.
4. Brackets [ ] are used for an either/or situation. Either one element or another may be present, but not both. The elements listed between the brackets are mutually exclusive.
5. Parentheses ( ) are used for an optional element. Optional elements may be left blank on entry screens and may contain spaces or zeros for numeric fields in file structures.

Figure 8.4 is an example of the data structure for adding a customer order at World's Trend Catalog Division. Each NEW CUSTOMER screen consists of the entries found on the right side of the equal signs. Some of the entries are elements, but others, such as CUSTOMER NAME, ADDRESS, and TELEPHONE, are groups of elements or structural records. For example, CUSTOMER NAME is made up of FIRST NAME, MIDDLE INITIAL, and LAST NAME. Each structural record must be further defined until the entire set is broken down into its component elements. Notice that following the definition for the CUSTOMER ORDER screen are definitions for each structural record. Even a field as simple as the TELEPHONE NUMBER is defined as a structure so that the area code may be processed individually.

Structural records and elements that are used in many different systems are given a nonsystem-specific name, such as street, city, and zip, that does not reflect the functional area in which they are used. This method allows the analyst to define these records once and use them in many different applications. For example, a city may be a customer city, a supplier city, or an employee city. Notice the use of parentheses to indicate that (MIDDLE INITIAL), (APARTMENT), and (ZIP EXPANSION) are optional ORDER information (but not more than one). Indicate the OR condition by enclosing the options in square brackets and separating them with the symbol.

Logical and Physical Data Structures

When data structures are first defined, only the data elements that the user would see, such as a name, address, and balance due, are included. This stage is the logical design, showing what data the business needs for its day-to-day operations. As we learned from HCI, it is important that the logical design accurately reflect the mental model of how the user views the system. Using the logical design as a basis, the analyst then designs the physical data structures, which include additional elements necessary for implementing the system. The following are examples of physical design elements:

1. Key fields used to locate records in a database table. An example is an item number, which is not required for a business to function but is necessary for identifying and locating computer records.

FIGURE 8.4

Data structure example for adding a customer order at World's Trend Catalog Division.

Customer Order = Customer Number +
Customer Name +
Address +
Telephone +
Catalog Number +
Order Date +
{Available Order Items} +
Merchandise Total +
(Tax) +
Shipping and Handling +
Order Total +
Method of Payment +
(Credit Card Type) +
(Credit Card Number) +
(Expiration Date)

Customer Name = First Name +
(Middle Initial) +
Last Name

Address = Street +
(Apartment) +
City +
State +
Zip +
(Zip Expansion) +
(Country)

Telephone = Area Code +
Local Number

Available Order Items = Quantity Ordered +
Item Number +
Item Description +
Size +
Color +
Price +
Item Total

Method of Payment = [Check ¦ Charge ¦ Money Order]

Credit Card Type = [World's Trend ¦ American Express ¦ MasterCard ¦ Visa]

2. Codes to identify the status of master records, such as whether an employee is active (currently employed) or inactive. Such codes can be maintained on files that produce tax information.
3. Transaction codes are used to identify types of records when a file contains different record types. An example is a credit file containing records for returned items as well as records of payments.
4. Repeating group entries containing a count of how many items are in the group.
5. Limits on the number of items in a repeated group.
6. A password used by a customer accessing a secure website.

Figure 8.5 is an example of the data structure for a CUSTOMER BILLING STATEMENT, one showing that the ORDER LINE is both a repeating item and a structural record. The ORDER LINE limits are from 1 to 5, indicating that the customer may order from one to five items on this screen. Additional items would appear on subsequent orders.

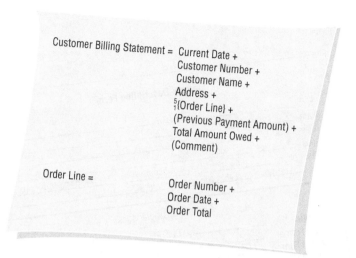

FIGURE 8.5

Physical elements added to a data structure.

The repeating group notation may have several other formats. If the group repeats a fixed number of times, that number is placed next to the opening brace, as in 12 {Monthly Sales}, where there are always 12 months in the year. If no number is indicated, the group repeats indefinitely. An example is a table containing an indefinite number of records, such as Customer Master Table = {Customer Records}.

The number of entries in repeating groups may also depend on a condition, such as an entry on the Customer Master Record for each item ordered. This condition could be stored in the data dictionary as {Items Purchased} 5, where 5 is the number of items.

Data Elements

Each data element should be defined once in the data dictionary and may also be entered previously on an element description form, such as the one illustrated in Figure 8.6. Characteristics commonly included on the element description form are the following:

1. Element ID. This optional entry allows the analyst to build automated data dictionary entries.
2. The name of the element. The name should be descriptive, unique, and based on what the element is commonly called in most programs or by the major user of the element.
3. Aliases, which are synonyms or other names for the element. Aliases are names used by different users in different systems. For example, a CUSTOMER NUMBER may also be called a RECEIVABLE ACCOUNT NUMBER or a CLIENT NUMBER.
4. A short description of the element.
5. Whether the element is base or derived. A base element is one that is initially keyed into the system, such as a customer name, address, or city. Base elements must be stored in files. Derived elements are created by processes as the result of a calculation or a series of decision-making statements.
6. The length of an element. Some elements have standard lengths. In the United States, for example, lengths for state name abbreviations, zip codes, and telephone numbers are all standard. For other elements, the lengths may vary, and the analyst and user community must jointly decide the final length, based on the following considerations:
   a. Numeric amount lengths should be determined by figuring the largest number the amount will probably contain and then allowing reasonable room for expansion. Lengths designated for totals should be large enough to accommodate the sum of the numbers accumulated in them.
   b. Name and address fields may be given lengths based on the following table. For example, a last name field of 11 characters will accommodate 98 percent of the last names in the United States.
   c. For other fields, it is often useful to examine or sample historical data found in the organization to determine a suitable field length. If the element is too small, the data that need to be entered will be truncated. The analyst must decide how that will affect the

FIGURE 8.6

An element description form example from World's Trend Catalog Division.

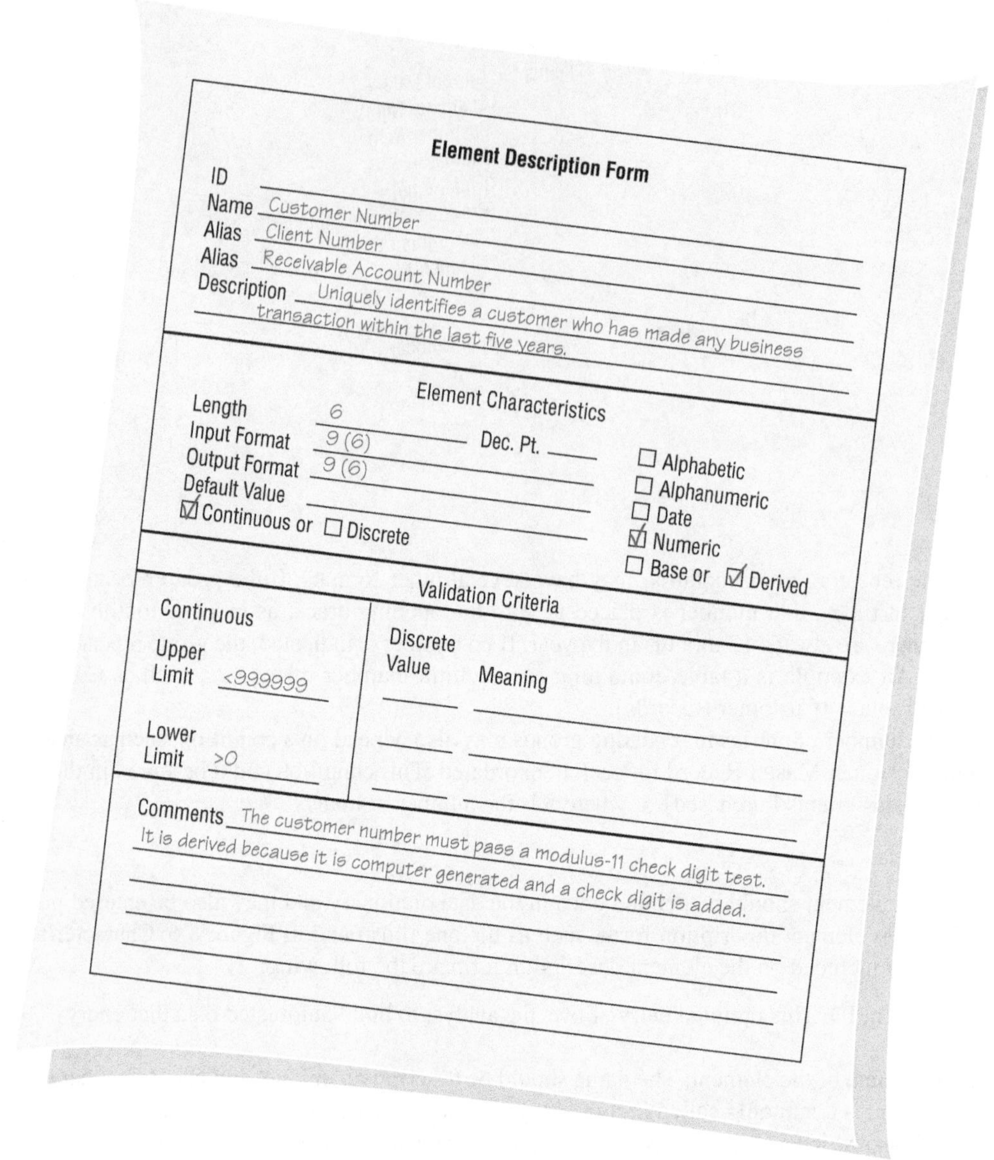

Element Description Form

ID _____

Name _Customer Number_

Alias _Client Number_

Alias _Receivable Account Number_

Description _Uniquely identifies a customer who has made any business transaction within the last five years._

Element Characteristics

Length ___6___

Input Format ___9 (6)___ Dec. Pt. _____

Output Format ___9 (6)___

Default Value _____

☑ Continuous or ☐ Discrete

☐ Alphabetic
☐ Alphanumeric
☐ Date
☑ Numeric
☐ Base or ☑ Derived

Validation Criteria

Continuous

Upper Limit ___<999999___

Lower Limit ___>0___

Discrete
Value _____ Meaning _____

Comments _The customer number must pass a modulus-11 check digit test. It is derived because it is computer generated and a check digit is added._

system outputs. For example, if a customer's last name is truncated, mail would usually still be delivered; if an email address is truncated, however, it will be returned as not found. Common lengths of fields are shown below.

Field	Length	Percentage of Data That Will Fit (U.S.)
Last Name	11	98
First Name	18	95
Company Name	20	95
Street	18	90
City	17	99

7. The type of data—numeric, date, alphabetic, varchar, or character, which is sometimes called alphanumeric or text data. Varchar data may contain any number of characters, up to a limit set by the database software. When using varchar, specifying the length is optional. Several of these formats are shown in Figure 8.7. Character fields may contain a mixture of letters, numbers, and special characters. If the element is a date, its format—for example, MMDDYYYY—must be determined. If the element is numeric, its storage type should be determined.

Data Type	Meaning
Bit	A value of 1 or 0, a true/false value
Char, varchar, text	Any alphanumeric character
Datetime, smalldatetime	Alphanumeric data, several formats
Decimal, numeric	Numeric data that are accurate to the least significant digit; can contain a whole and decimal portion
Float, real	Floating-point values that contain an approximate decimal value
Int, smallint, tinyint	Only integer (whole digit) data
Currency, money, smallmoney	Monetary numbers accurate to four decimal places
Binary, varbinary, image	Binary strings (sound, pictures, video)
Cursor, timestamp, uniqueidentifier	A value that is always unique within a database
Autonumber	A number that is always incremented by one when a record is added to a database table

FIGURE 8.7

Some examples of data formats used in PC systems.

Personal computer formats, such as currency, number, or scientific, depend on how the data will be used. Number formats are further defined as integer, long integer, single precision, double precision, and so on. There are many other types of formats used with PC systems. Unicode is a standardized coding system for defining graphic symbols, such as Chinese or Japanese characters. Unicode is described in greater detail in Chapter 15. There are three standard formats for mainframe computers: zoned decimal, packed decimal, and binary. The zoned decimal format is used for printing and displaying data. The packed decimal format is commonly used to save space on file layouts and for elements that require a high level of arithmetic to be performed on them. The binary format is suitable for the same purposes as the packed decimal format but is less commonly used.

8. Input and output formats should be included, using special coding symbols to indicate how the data should be presented. These symbols and their uses are illustrated in Figure 8.8. Each symbol represents one character or digit. If the same character repeats several times, the character followed by a number in parentheses indicating how many times the character repeats is substituted for the group. For example, XXXXXXXX would be represented as X(8).

9. Validation criteria for ensuring that accurate data are captured by the system. Elements are either discrete, meaning they have certain fixed values, or continuous, with a smooth range of values. Here are common editing criteria:

a. A range of values is suitable for elements that contain continuous data. For example, in the United States a student grade point average may be from 0.00 through 4.00. If there is only an upper or lower bound to the data, a limit is used instead of a range.

b. A list of values is indicated if the data are discrete. Examples are codes representing the colors of items for sale in World's Trend's catalog.

Formatting Character	Meaning
X	May enter or display/print any character
9	Enter or display only numbers
Z	Display leading zeros as spaces
,	Insert commas into a numeric display
.	Insert a period into a numeric display
/	Insert slashes into a numeric display
-	Insert a hyphen into a numeric display
V	Indicate a decimal position (when the decimal point is not included)

FIGURE 8.8

Format character codes.

   c. A table of codes is suitable if the list of values is extensive (for example, state abbreviations, telephone country codes, or U.S. telephone area codes.)

   d. For key or index elements, a check digit is often included.

10. Any default value the element may have. The default value is displayed on entry screens and is used to reduce the amount of keying that the operator may have to do. Usually, several fields in each system have default values. When using GUI lists or drop-down lists, the default value is the one currently selected and highlighted. When using radio buttons, the option for the default value is selected, and when using check boxes, the default value (either "yes" or "no") determines whether the check box will have an initial check in it.

11. An additional comment or remarks area. This might be used to indicate the format of the date, special validation that is required, the check digit method used (explained in Chapter 15), and so on.

Data element descriptions such as CUSTOMER NUMBER may be called CLIENT NUMBER elsewhere in the system (and perhaps old code written with this alias needs to be updated).

    Another kind of data element is an alphabetic element. At World's Trend Catalog Division, codes are used to describe colors—for example, BL for blue, WH for white, and GR for green. When this element is implemented, a table will be needed for users to look up the meanings of these codes. (Coding is discussed further in Chapter 15.)

Data Stores

All base elements must be stored in the system. Derived elements, such as the employee year-to-date gross pay, may also be stored in the system. Data stores are created for each different data entity being stored. That is, when data flow base elements are grouped together to form a structural record, a data store is created for each unique structural record.

    Because a given data flow may only show part of the collective data that a structural record contains, you may have to examine many different data flow structures to arrive at a complete data store description.

    Figure 8.9 is a typical form used to describe a data store. The information included on the form is as follows:

1. The data store ID. The ID is often a mandatory entry to prevent the analyst from storing redundant information. An example would be D1 for the CUSTOMER MASTER.
2. The data store name, which is descriptive and unique.
3. An alias for the table, such as CLIENT MASTER for the CUSTOMER MASTER.
4. A short description of the data store.
5. The file type, either computer or manual.
6. The format designates whether the file is a database table or whether it has the format of a simple flat file. (File formats are detailed in Chapter 13.)
7. The maximum and average number of records on the file as well as the growth per year. This information helps the analyst to predict the amount of disk space required for the application and is necessary for hardware acquisition planning.
8. The file or data set name specifies the file name, if known. In the initial design stages, this item may be left blank. An electronic form produced using Visible Analyst is shown in Figure 8.10. This example shows that the CUSTOMER MASTER is stored on a computer in the form of a database with a maximum number of 45,000 records. (Records and the keys used to sort the database are explained in Chapter 13.)
9. The data structure should use a name found in the data dictionary, providing a link to the elements for this data store. Alternatively, the data elements could be described on the data store description form or on the CASE tool screen for the data store. Primary and secondary keys must be elements (or a combination of elements) found in the data structure. In the example, the CUSTOMER NUMBER is the primary key and should be unique. The CUSTOMER NAME, ZIP, and YEAR-TO-DATE AMOUNT PURCHASED are secondary keys used to control record sequencing on reports and to locate records directly. (Keys are discussed in Chapter 13.) Comments are used for information that does not fit into any of the above categories. They may include update or backup timing, security, or other considerations.

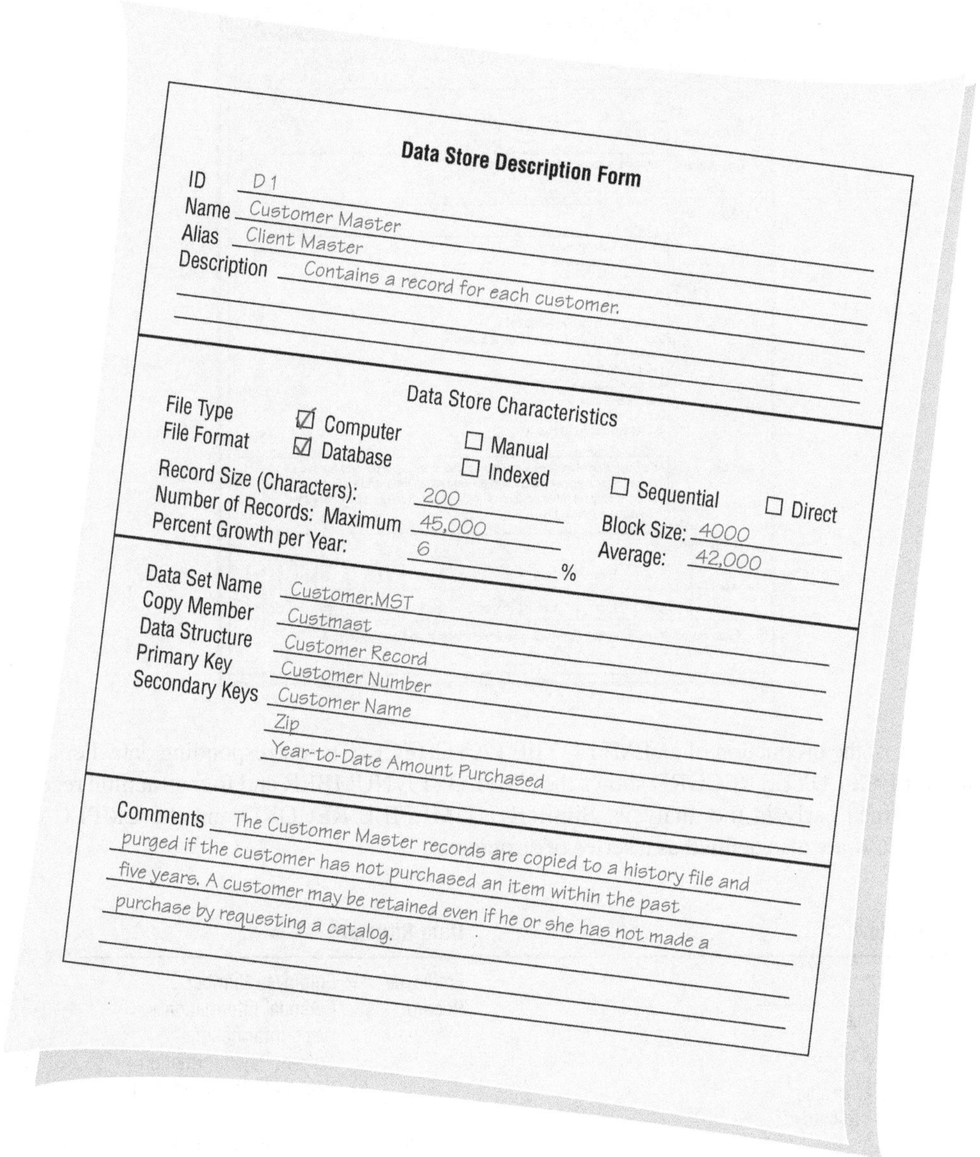

FIGURE 8.9

An example of a data store form description for World's Trend Catalog Division.

Data Store Description Form

ID   D 1
Name   Customer Master
Alias   Client Master
Description   Contains a record for each customer.

Data Store Characteristics

File Type   ☑ Computer
File Format   ☑ Database   ☐ Manual
                      ☐ Indexed
Record Size (Characters):   200   ☐ Sequential   ☐ Direct
Number of Records: Maximum 45,000   Block Size: 4000
Percent Growth per Year: 6   Average: 42,000
                                    %

Data Set Name   Customer.MST
Copy Member   Custmast
Data Structure Customer Record
Primary Key   Customer Number
Secondary Keys Customer Name
                Zip
                Year-to-Date Amount Purchased

Comments   The Customer Master records are copied to a history file and purged if the customer has not purchased an item within the past five years. A customer may be retained even if he or she has not made a purchase by requesting a catalog.

Creating a Data Dictionary

Data dictionary entries may be created after the data flow diagram has been completed, or they may be constructed as the data flow diagram is being developed. The use of algebraic notation and structural records allows an analyst to develop the data dictionary and the data flow diagrams using a top-down approach. For instance, the analyst may create a Diagram 0 data flow after the first few interviews and, at the same time, make the preliminary data dictionary entries. Typically, these entries consist of the data flow names found on the data flow diagram and their corresponding data structures.

After conducting several additional interviews with users to learn the details of the system and the ways they interact with it, the analyst will expand the data flow diagram and create the child diagrams. The data dictionary is then modified to include the new structural records and elements gleaned from further interviews, observation, and document analysis.

Each level of a data flow diagram should use data appropriate for the level. Diagram 0 should include only forms, screens, reports, and records. As child diagrams are created, the data flow into and out of the processes becomes more and more detailed, including structural records and elements.

Figure 8.11 illustrates a portion of two data flow diagram levels and corresponding data dictionary entries for producing an employee paycheck. Process 5, found on Diagram 0, is an

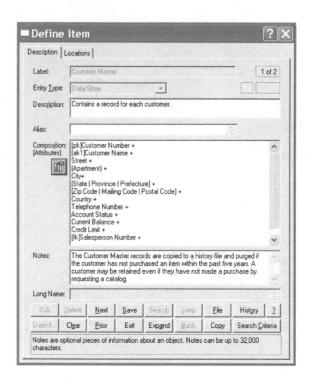

overview of the production of an EMPLOYEE PAYCHECK. The corresponding data dictionary
entry for EMPLOYEE RECORD shows the EMPLOYEE NUMBER and four structural records,
data obtained early in the analysis. Similarly, TIMEFILE RECORD and the EMPLOYEE
PAYCHECK are also defined as a series of structures.

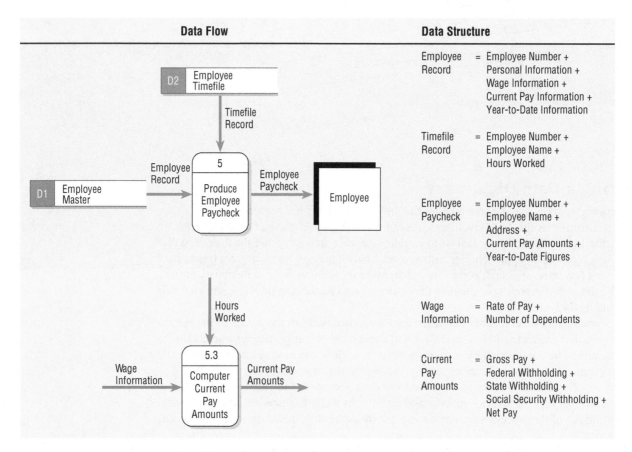

FIGURE 8.11

Two data flow diagrams and corresponding data dictionary entries for producing an employee paycheck.

CONSULTING OPPORTUNITY 8.1

Want to Make It Big in the Theatre?
Improve Your Diction(ary)!

As you enter the door of Merman's, Annie Oaklea greets you warmly, saying, "I'm delighted with the work you have done on the data flow diagrams. I would like you to keep playing the role of systems analyst for Merman's and see if you can eventually get a new information system for our costume inventory sewn up. Unfortunately, some of the terms you're using don't come off very well in the language of Shakespeare. Bit of a translation problem, I suspect."

Clinging to Annie's initial praise, you are undaunted by her exit line. You determine that a data dictionary based on the rental and return data flow diagrams would make a big hit.

Begin by writing entries manually in as much detail as possible. Prepare two data flow entries, two data store entries, one data structure entry, and four data element entries using the formats in this chapter. Portraying interrelated data items with preciseness will result in rave reviews. (Refer to Consulting Opportunity 7.1.)

It is important that the data flow names on the child data flow diagram be contained as elements or structural records in the data flow on the parent process. Returning to the example, WAGE INFORMATION (input into process 5.3, COMPUTE CURRENT PAY AMOUNTS) is a structural record contained in the EMPLOYEE RECORD (input to process 5). Similarly, GROSS PAY (output from process 5.3.4, a lower-level process not shown in the figure) is contained in the structural record CURRENT PAY AMOUNTS (output from the parent process 5.3, COMPUTE CURRENT PAY AMOUNTS).

Analyzing Input and Output

An important step in creating a data dictionary is to identify and categorize system input and output data flow. Input and output analysis forms contain the following commonly included fields:

1. A descriptive name for the input or output. If the data flow is on a logical diagram, the name should identify what the data are (for example, CUSTOMER INFORMATION). If the analyst is working on the physical design or if the user has explicitly stated the nature of the input or output, however, the name should include that information regarding the format. Examples are CUSTOMER BILLING STATEMENT and CUSTOMER DETAILS INQUIRY.
2. The user contact responsible for further details clarification, design feedback, and final approval.
3. Whether the data is input or output.
4. The format of the data flow. In the logical design stage, the format may be undetermined.
5. Elements indicating the sequence of the data on a report or screen (perhaps in columns).
6. A list of elements, including their names, lengths, and whether they are base or derived, and their editing criteria.

Once the form has been completed, each element should be analyzed to determine whether the element repeats, whether it is optional, or whether it is mutually exclusive of another element. Elements that fall into a group or that regularly combine with several other elements in many structures should be placed together in a structural record.

These considerations can be seen in the completed Input and Output Analysis Form for World's Trend Catalog Division (see Figure 8.12). In this example of a CUSTOMER BILLING STATEMENT, the CUSTOMER FIRST NAME, CUSTOMER LAST NAME, and CUSTOMER MIDDLE INITIAL should be grouped together in a structural record.

Developing Data Stores

Another activity in creating a data dictionary is developing data stores. Up to now, we have determined what data need to flow from one process to another. This information is described in data

FIGURE 8.12

An example of an input/output analysis form for World's Trend Catalog Division.

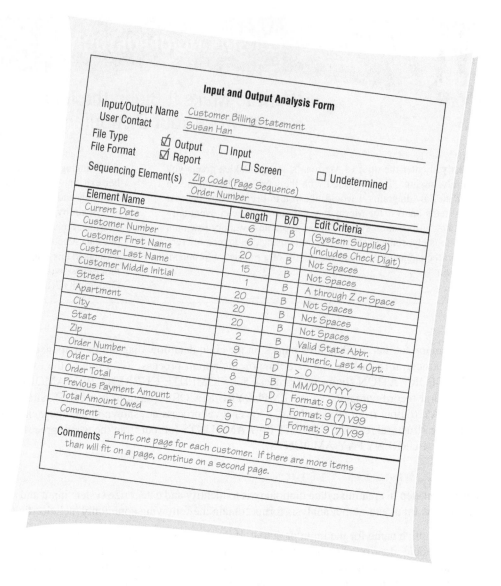

Input and Output Analysis Form

Input/Output Name _Customer Billing Statement_
User Contact _Susan Han_

File Type ☑ Output ☐ Input
File Format ☑ Report ☐ Screen ☐ Undetermined

Sequencing Element(s) _Zip Code (Page Sequence)_
Order Number

Element Name	Length	B/D	Edit Criteria
Current Date	6	B	(System Supplied)
Customer Number	6	D	(Includes Check Digit)
Customer First Name	20	B	Not Spaces
Customer Last Name	15	B	Not Spaces
Customer Middle Initial	1	B	A through Z or Space
Street	20	B	Not Spaces
Apartment	20	B	Not Spaces
City	20	B	Not Spaces
State	2	B	Valid State Abbr.
Zip	9	B	Numeric, Last 4 Opt.
Order Number	6	D	> 0
Order Date	8	B	MM/DD/YYYY
Order Total	9	D	Format: 9 (7) V99
Previous Payment Amount	5	D	Format: 9 (7) V99
Total Amount Owed	9	D	Format: 9 (7) V99
Comment	60	B	

Comments _Print one page for each customer. If there are more items_
than will fit on a page, continue on a second page.

structures. The information, however, may be stored in numerous places, and in each place the data store may be different. Whereas data flows represent data in motion, data stores represent data at rest.

For example, when an order arrives at World's Trend (see Figure 8.13), it contains mostly temporary information—that is, the information needed to fill that particular order—but some information might be stored permanently. Examples of the latter include information about customers (so catalogs can be sent to them) and information about items (because these items will appear on many other customers' orders).

Data stores contain information of a permanent or semipermanent (temporary) nature. An ITEM NUMBER, DESCRIPTION, and ITEM COST are examples of information that is relatively permanent. So is the TAX RATE. When the ITEM COST is multiplied by the TAX RATE, however, the TAX CHARGED is calculated (or derived). Derived values do not have to be stored in a data store.

When data stores are created for only one report or screen, we refer to them as "user views" because they represent the way that the user wants to see the information.

Using a Data Dictionary

An ideal data dictionary is automated, interactive, online, and evolutionary. As a systems analyst learns about an organization's systems, he or she adds data items to the data dictionary. On the other hand, the data dictionary is not an end in itself and must never become so. To avoid

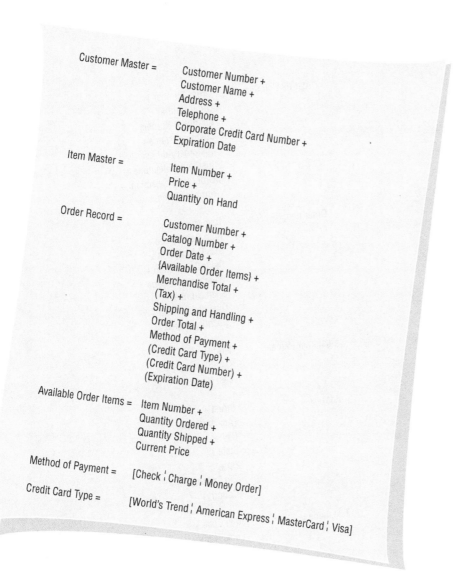

FIGURE 8.13

Data stores derived from a pending order at World's Trend Catalog Division.

becoming sidetracked with the building of a complete data dictionary, a systems analyst should view it as an activity that parallels systems analysis and design.

To have maximum power, the data dictionary should be tied into a number of systems programs so that when an item is updated or deleted from the data dictionary, it is automatically updated or deleted from the database. The data dictionary becomes simply a historical curiosity if it is not kept current.

The data dictionary may be used to create screens, reports, and forms. For example, examine the data structure for the World's Trend ORDER PICKING SLIP in Figure 8.14. Because the necessary elements and their lengths have been defined, the process of creating physical documents consists of arranging the elements in a pleasing and functional way using design guidelines and common sense.

Repeating groups become columns, and structural records are grouped together on the screen, report, or form. The report layout for the World's Trend ORDER PICKING SLIP is shown in Figure 8.15. Notice that FIRST NAME and LAST NAME are grouped together in NAME and that QUANTITY (PICKED and ORDERED), SECTION, SHELF NUMBER, ITEM NUMBER, ITEM DESCRIPTION, SIZE, and COLOR form a series of columns because they are the repeating elements.

The data structure and elements for a data store are commonly used to generate corresponding computer language source code, which is then incorporated into computer programs. The

FIGURE 8.14

Data structure for an order picking slip at World's Trend Catalog Division.

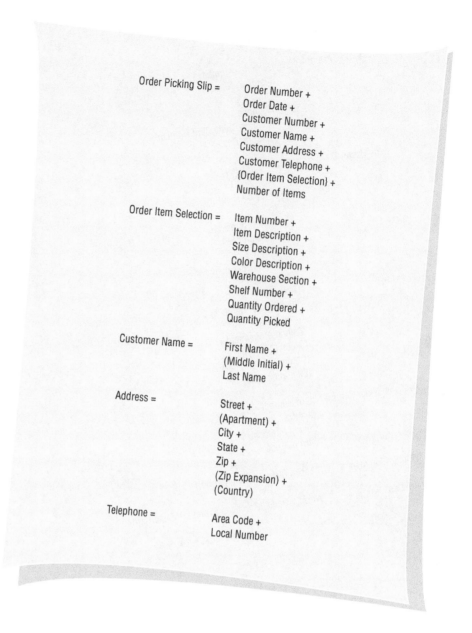

Order Picking Slip =

    Order Number +
    Order Date +
    Customer Number +
    Customer Name +
    Customer Address +
    Customer Telephone +
    {Order Item Selection} +
    Number of Items

Order Item Selection =

    Item Number +
    Item Description +
    Size Description +
    Color Description +
    Warehouse Section +
    Shelf Number +
    Quantity Ordered +
    Quantity Picked

Customer Name =

    First Name +
    (Middle Initial) +
    Last Name

Address =

    Street +
    (Apartment) +
    City +
    State +
    Zip +
    (Zip Expansion) +
    (Country)

Telephone =

    Area Code +
    Local Number

data dictionary may be used in conjunction with a data flow diagram to analyze the system design, detecting flaws and areas that need clarification. Some considerations are:

1. All base elements on an output data flow must be present on an input data flow to the process producing the output. Base elements are keyed and should never be created by a process.
2. A derived element must be created by a process and should be output from at least one process into which it is not input.
3. The elements that are present in a data flow coming into or going out of a data store must be contained in the data store.

If begun early, a data dictionary can save many hours of time in the analysis and design phases. The data dictionary is the one common source in the organization for answering questions and settling disputes about any aspect of data definition. An up-to-date data dictionary can serve as an excellent reference for maintenance efforts on unfamiliar systems. Automated data dictionaries can serve as references for both people and programs.

Using Data Dictionaries to Create XML

Extensible Markup Language (XML) is a language that can be used to exchange data between businesses or between systems within a business. It is similar to HTML, the markup language

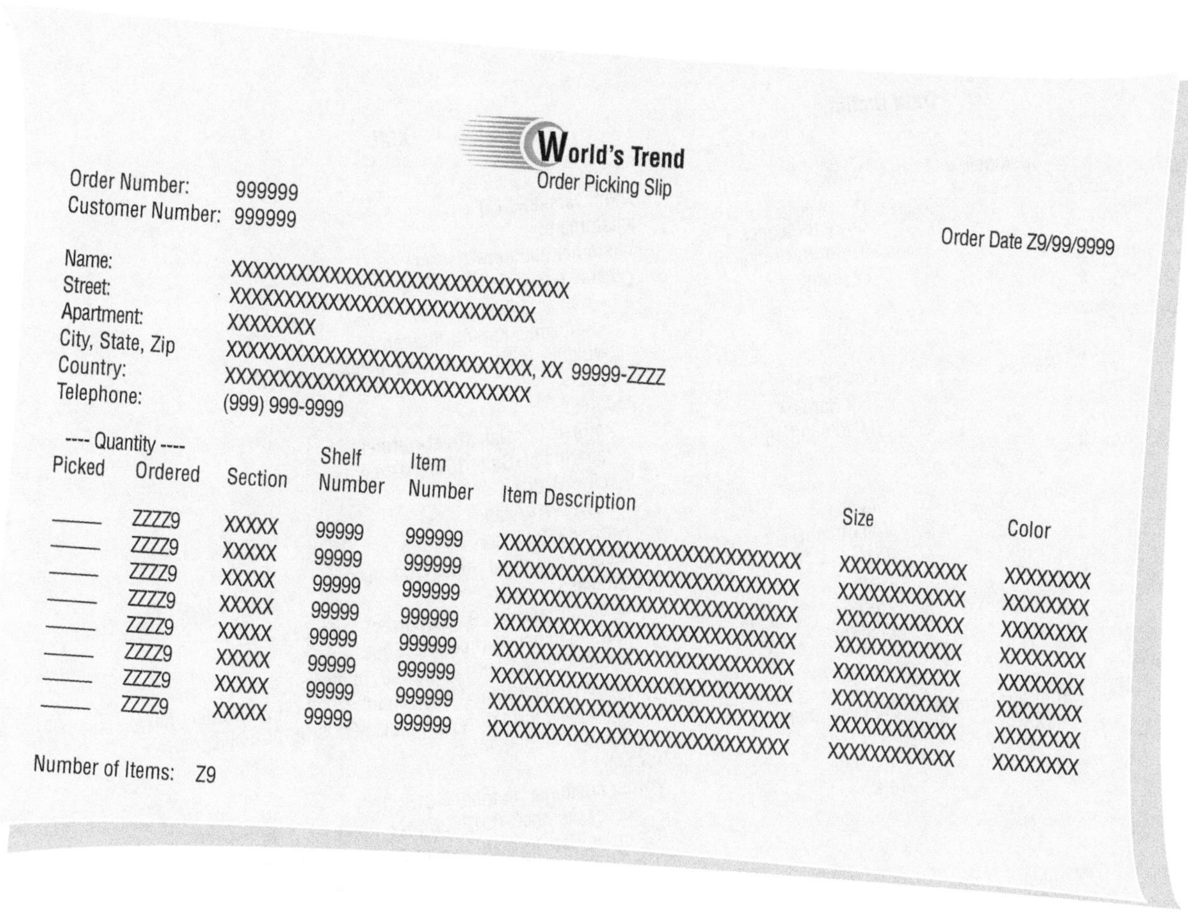

FIGURE 8.15

Order picking slip created from the data dictionary.

used to create web pages, but is more powerful. HTML is concerned primarily with formatting a document; XML addresses the problem of sharing data when users have different computer systems and software or different database management systems (for example, one company using Oracle and another using IBM's DB2). If everyone used the same software or database management system, there would be little need for XML.

Once an XML document has been created, the data may be transformed into a number of different output formats and displayed in many different ways, including printed output, web pages, output for a handheld device, and portable document format (PDF) files. Thus, the document's data content is separated from the output format. The XML content is defined once as data and then transformed as many times as necessary.

The advantage of using an XML document is that the analyst may select only the data that an internal department or external partner needs to have in order to function. This helps to ensure the confidentiality of data. For example, a shipping company may receive only the customer name, the address, the item number, and the quantity to ship, but *not* credit card information or other financial data. This efficient approach also cuts down on information overload.

Using XML, therefore, is a way to define, sort, filter, and translate data into a universal data language that can be used by anyone. XML may be created from databases, a form, or software programs, or it may be keyed directly into a document, a text editor, or an XML entry program.

The data dictionary is an ideal starting point for developing XML content. The key to using XML is creating a standard definition of the data. This is accomplished by using a set of tags or data names that are included before and after each data element or structure. The tags become the metadata, or data about the data. Data may be further subdivided

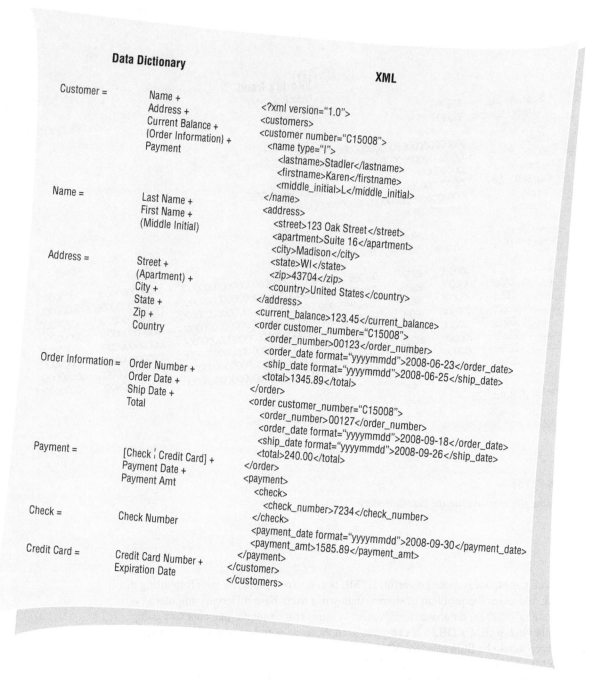

FIGURE 8.16

Using a data dictionary entry to develop XML content. The XML document mirrors the data dictionary structure.

into smaller elements and structures until all elements are defined. XML elements may also include attributes, additional data included within the tag that describe something about the XML element.

Figure 8.16 illustrates a data dictionary containing customer, order, and payment information. The overall collection of customers is included in what is called the root element: customers. An XML document may contain only one root element, so it is often the plural of the data contained in the XML document. Each customer may place many orders. The structure is defined in the two left columns, and the XML code appears on the right. CUSTOMER, as you can see, consists of a NAME, ADDRESS, CURRENT BALANCE, multiple ORDER INFORMATION entries, and a PAYMENT. Some of these structures are further subdivided.

The XML document tends to mirror the data dictionary structure. The first entry (other than an XML line identifying the document) is <customers>, which defines the entire collection of customer information. The less than (<) and greater than (>) symbols are used to identify tag names (similar to HTML). The last line of the XML document is a closing tag, </customers>, signifying the end of the customer information.

Customer is defined first and contains an attribute, the customer number. There is often a discussion about whether data should be stored as an element or an attribute. In this case, they are stored as an attribute.

The name tag, <name>, is defined next because it is the first entry in the data dictionary. NAME is a structure consisting of LAST NAME, FIRST NAME, and an optional MIDDLE INITIAL. In the XML document, this structure starts with <name> and is followed by <lastname>, <firstname>, and <middle_initial>. Because spaces are not allowed in XML tag names, an underscore is typically used to separate words. The closing </name> tag signifies the end of the group of elements. Using a structure such as name saves time and coding if the transformation displays the full name. Each of the child elements will be on one line separated by a space. Name also contains an attribute, either I for individual or C for corporation.

Indentation is used to show which structures contain elements. Note that <address> is similar to <customer>, but when we get to <order_information> there is a big difference.

There are multiple entries for <order_information>, each containing an <order_number>, <order_date>, <shipping_date>, and <total>. Because the payment is made either by check or credit card, only one of these may be present. In our example, payment is by check. The dates have an attribute called format that indicates whether the date appears as month, day, year; year, month, day; or day, month, year. If a credit card is used to make a payment, a TYPE attribute contains either an M, V, A, D, or O, indicating the type of credit card (MasterCard, Visa, and so on).

XML Document Type Definitions

Often the element structure of XML content is defined using a document type definition (DTD). A DTD is used to determine whether the XML document content is valid—that is, whether it conforms to the order and type of data that must be present in the document. The DTD is easy to create and well supported by standard software. Once the DTD has been completed, it may be used to validate the XML document using standard XML tools. The DTD is easier to create if a data dictionary has been completed, since the analyst has worked with users and made decisions on the structure of the data.

Figure 8.17 illustrates the document type definition for the Customer XML document. Keywords, such as !DOCTYPE, indicating the start of the DTD, must be in capital letters. !ELEMENT describes an element, and !ATTLIST describes an attribute, listing the element name followed by the attribute name. An element that has the keyword #PCDATA, for parsed character data, is a primitive element, not further defined. An element that has a series of other elements within parentheses means that they are child elements and must be in the order listed. The statement <!ELEMENT name (lastname, firstname, middle_initial?)> means that the name must have the last name followed by the first name followed by the middle initial.

The question mark after "middle_initial" means that the element is optional and may be left out of the document for a particular customer. A plus sign means that there are one or more repeatable elements. Customers must contain at least one customer tag but could contain many customer tags. An asterisk means that there is zero or more of the elements. Each customer may have zero to many orders. A vertical bar separates two or more child elements that are mutually exclusive. Payment contains either check or credit card as options.

The attribute list definition for a customer number contains a keyword ID (in uppercase letters). This means that the attribute number must appear only once in the XML document as an attribute for an element with an ID. That it is somewhat similar to a primary key. The difference is that, if the document had several different elements, each with an ID attribute, the given ID (C15008 in this example) could appear only once. An ID must start with a letter or an underscore and cannot be solely a number. The reason behind putting the customer number as an ID is to ensure that it is not repeated in a longer document. The keyword #REQUIRED means that the attribute must be present. A keyword of #IMPLIED means that the attribute is optional. A document may also have an IDREF attribute, which links one element with another that is an ID. The ORDER tag has a customer_number attribute defined as an IDREF, and the value C15008 must

FIGURE 8.17

A document type definition
for the customer XML
document.

```
< !DOCTYPE   customers   [
< !ELEMENT   customers                        (customer) + >
< !ELEMENT   customer                         (name, address, current_balance, order*) >
< !ATTLIST   customer number                  ID #REQUIRED>
< !ELEMENT   name                             (lastname, firstname, middle_initial?) >
< !ATTLIST   name type                        (I|C) #REQUIRED>
< !ELEMENT   lastname                         (#PCDATA) >
< !ELEMENT   firstname                        (#PCDATA) >
< !ELEMENT   middle_initial                   (#PCDATA) >
< !ELEMENT   address                          (street, apartment?, city, state, zip, country) >
< !ELEMENT   street                           (#PCDATA) >
< !ELEMENT   apartment                        (#PCDATA) >
< !ELEMENT   city                             (#PCDATA) >
< !ELEMENT   state                            (#PCDATA) >
< !ELEMENT   zip                              (#PCDATA) >
< !ELEMENT   country                          (#PCDATA) >
< !ELEMENT   current_balance                  (#PCDATA) >
< !ELEMENT   order                            (order_number, order_date, ship_date, total) >
< !ATTLIST   order customer_number            IDREF #REQUIRED>
< !ELEMENT   order_number                     (#PCDATA) >
< !ELEMENT   order_date                       (#PCDATA) >
< !ATTLIST   order_date format                (mmddyyyy|yyyymmdd|ddmmyyyy) #REQUIRED>
< !ELEMENT   payment                          (check|credit_card) >
< !ELEMENT   check                            (check_number) >
< !ELEMENT   credit_card                      (credit_card_number, expiration_date) >
< !ATTLIST   credit_card type                 (M|V|A|D|O) #REQUIRED>
< !ELEMENT   credit_card_number               (#PCDATA) >
< !ELEMENT   expiration_date                  (#PCDATA) >
< !ELEMENT   payment_date                     (#PCDATA) >
< !ATTLIST   payment_date format              (mmddyyyy|yyyymmdd|ddmmyyyy) #REQUIRED>
< !ELEMENT   payment_amt                      (#PCDATA) >
             ]>
```

be present in an ID somewhere in the document. An attribute list containing values in parentheses means that the attribute must contain one of the values. A DTD definition <!ATTLIST credit_card type (M|V|A|D|O) #REQUIRED> means that the credit card type must be either an M, V, A, D, or O.

XML Schemas

A schema is another, more precise way to define the content of an XML document. Schemas may include the exact number of times an element may occur as well as the type of data within elements, such as character or numeric values, including the length of the element, limits on the data, and the number of places to the left and right of a decimal number.

A data dictionary is an excellent starting point for developing an XML document and a document type of definition or schema. The advantage of using XML to define data is that, in the XML format, data are stored in a pure text format and not dependent on any proprietary software. The document may be easily validated and transformed into many different output formats.

Industry groups or organizations may be involved in defining an industry-specific XML structure so that all involved parties understand what the data mean. This is very important when an element name may have several meanings. An example is "state," which may mean a postal state abbreviation or the state of an order or account. Examples of industry-specific XML document type definitions and schemas may be found at www.xml.org.

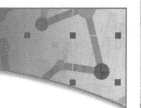

HYPERCASE® EXPERIENCE 8

"You're really doing very well. Snowden says you've given him all sorts of new ideas for running the new department. That's saying quite a lot, when you consider that he has a lot of his own ideas. By now I hope you've had a chance to speak with everyone you would like to: certainly Snowden himself, Tom Ketcham, Daniel Hill, and Mr. Hyatt.

"Mr. Hyatt is an elusive soul, isn't he? I guess I didn't meet him until well into my third year. I hope you get to find out about him much sooner. Oh, but when you do get to see him, he cuts quite a figure, doesn't he? And those crazy airplanes. I've almost been conked on the head by one in the parking lot. But how can you get angry, when it's The Boss who's flying it? He's also got a secret—or should I say private—oriental garden off his office suite. No, you'll never see it on the building plans. You have to get to know him very well before he'll show you that, but I would wager it's the only one like it in Tennessee and maybe in the whole United States. He fell in love with the wonderful gardens he saw in Southeast Asia as a young man. It goes deeper than that, however.

Mr. Hyatt knows the value of contemplation and meditation. If he has an opinion, you can be sure it has been well thought through."

HYPERCASE Questions

1. Briefly list the data elements that you have found on three different reports produced at MRE.
2. Based on your interviews with Snowden Evans and others, list the data elements that you believe you should add to the Management Unit's project reporting systems to better capture important data on project status, project deadlines, and budget estimates.
3. Create a data dictionary entry for a new data store and a new data flow that you are suggesting based on your response to Question 2.
4. Suggest a list of new data elements that might be helpful to Jimmy Hyatt but are clearly not being made available to him currently.

FIGURE 8.HC1

In HyperCase, you can look at the data dictionary kept at MRE.

Summary

Using a top-down approach, a systems analyst uses data flow diagrams to begin compiling a data dictionary, which is a reference work that contains data about data—or metadata—on all data processes, stores, flows, structures, and logical and physical elements in the system being studied. One way to begin is by including all data items from data flow diagrams.

A larger collection of project information is called a repository. CASE tools permit the analyst to create and share a repository with team mates that may include information about data flows, stores, record structures, and elements; about procedural logic screen and report design; and about data relationships. A repository can also contain information about project requirements and final system deliverables; and about project management information.

Each entry in a data dictionary contains the item name, an English description, aliases, related data elements, the range, the length, encoding, and necessary editing information. A data dictionary is useful in all phases of analysis, design, and ultimately documentation because it is the authoritative source on how a data element is used and defined by users in the system. Many large systems feature computerized data dictionaries that cross-reference all programs in the database using a particular data element. The data dictionary can also be used to create XML that enables businesses with different systems, software, or database management systems to exchange data.

Keywords and Phrases

base element	packed decimal
binary format	physical data structure
data dictionary	repeating group
data element	repeating item
data structure	repository
derived element	schema
document type definition (DTD)	structural record
Extensible Markup Language (XML)	system deliverables
ID	varchar
IDREF	zoned decimal

Review Questions

1. Define the term *data dictionary*. Define *metadata*.
2. What are four reasons for compiling a complete data dictionary?
3. What information is contained in the data repository?
4. What is a structural record?
5. List the eight specific categories that each entry in the data dictionary should contain. Briefly give the definition of each category.
6. What are the basic differences among data dictionary entries prepared for data stores, data structures, and data elements?
7. Why are structural records used?
8. What is the difference between logical and physical data structures?
9. Describe the difference between base and derived elements.
10. How do the data dictionary entries relate to levels in a set of data flow diagrams?
11. List the four steps to take in compiling a data dictionary.
12. Why shouldn't compiling the data dictionary be viewed as an end in itself?
13. What are the main benefits of using a data dictionary?
14. What does Extensible Markup Language (XML) describe?
15. What is a document type definition?
16. How does a document type definition help to ensure that an XML document contains all necessary elements?
17. When should attributes be used in an XML document?
18. What does an ID attribute ensure?
19. What does an IDREF attribute validate?

Problems

1. Based on Figure 7.EX1 in Chapter 7, Joe, one of your systems analysis team members, made the following entry for the data dictionary used by Marilyn's Tours:

    DATA ELEMENT = TOURIST* * * * PAYMENT
    ALIAS = TOURIST PAY
    CHARACTERS = 12–24
    RANGE = $5.00–$1,000
    VARIABLES = $5.00, $10.00, $15.00 up to $1,000, and anything in between in dollars and cents.

TO CALCULATE = TOTAL COST OF ALL TOURS, ANY APPLICABLE N.Y. STATE TAX, minus any RESERVATION DEPOSITS made.

   a. Is this truly a data element? Why or why not?

   b. Rewrite the data dictionary entry for TOURIST PAYMENT, reclassifying it if necessary. Use the proper form for the classification you choose.

2. Sue Kong, the systems analyst, has made significant progress in understanding the data movement at Shanghai Megabank. To share what she has done with other members of her team as well as the head of regional operations, she is composing a data dictionary.

   a. Write an entry in Sue's data dictionary for three of the data flows in regional banking. Be as complete as possible.

   b. Write an entry in Sue's data dictionary for three of the data stores in regional banking. Be as complete as possible.

3. Jorge Alvarez, the manager of the bookstore that your systems analysis team has been working with to build a computerized inventory system, thinks that one of your team members is making a nuisance of himself by asking him extremely detailed questions about data items used in the system. For example, he asks, "Jorge, how much space, in characters, does the listing of an ISBN take?"

   a. What are the problems created by going directly to the manager with questions concerning data dictionary entries? Use a paragraph to list the problems you can see with your team member's approach.

   b. In a paragraph, explain to your team member how he can better gather information for the data dictionary.

4. Michael Bush owns a store that specializes in travel gear and clothes. Manufacturers have their own coding, but there are many manufacturers. Set up data elements for six different travel hats from three different suppliers.

5. Michael (from Problem 4) also assembles packages of camping kits. Each kit is a group of separate products that are sold as a package. Each package (called a PRODUCT) is built using many parts, which vary from product to product. Interviews with the head parts clerk have resulted in a list of elements for the PRODUCT PART web page, showing which parts are used in the manufacture of each product. A prototype of the PRODUCT-PART web page is illustrated in Figure 8.EX1. Create a data structure dictionary entry for the PRODUCT-PART.

6. Analyze the elements found on the PRODUCT-PART web page and create the data structure for the PRODUCT MASTER and the PART MASTER data stores.

7. Which of the elements on the PRODUCT-PART web page are derived elements?

8. The Pacific Holiday Company arranges cruise vacations of varying lengths at several locations. When customers call to check on the availability of a cruise, a CRUISE AVAILABILITY INQUIRY, illustrated in Figure 8.EX2, is used to supply them with information. Create the data dictionary structure for the CRUISE AVAILABILITY INQUIRY.

9. List the master files that would be necessary to implement the CRUISE AVAILABILITY INQUIRY.

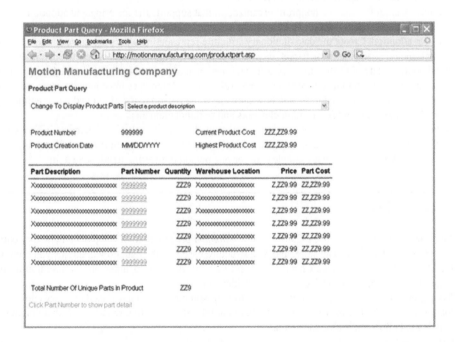

FIGURE 8.EX1

A prototype of the PRODUCT-PART web page.

FIGURE 8.EX2

A screen prototype to show cruise availability.

```
MM/DD/YYYY                              CRUISE AVAILABILITY              HH:MM
ENTER STARTING DATE Z9-ZZZ-9999


 - - - - - - - - - - - - - - - - - - - - - - - - - - - - - - - - - - - - - - -


CRUISE INFORMATION:
CRUISE SHIP                XXXXXXXXXXXXXXXXXX
LOCATION                   XXXXXXXXXXXXXXXXXX
STARTING DATE              Z9-ZZZ-9999              ENDING DATE  Z9-ZZZ-9999
NUMBER OF DAYS             ZZ9
COST                       ZZ,ZZZ.99
DISCOUNTS ACCEPTED         XXXXXXXXXXXX    XXXXXXXXXXXX    XXXXXXXXXXXX
OPENINGS REMAINING  ZZZZ9
```

10. The following ports of call are available for the Pacific Holiday Company:

Apia	Nuku Hiva	Auckland
Pago Pago	Papeete	Wellington
Bora Bora	Raiatea	Christ Church
Moorea	Napier	Dunedin

Create the PORT OF CALL element. Examine the data to determine the length and format of the element.

11. Raúl Esparza, the ecommerce manager for Moonlight Mugs, a company that sells customized coffee mugs, would like to send information to another company that maintains the warehouse and provides shipping services. Order information is obtained from a secure website, including customer number, name and address, telephone number, email address, product number and quantity, as well as credit card information. There may be several different products shipped on one order. The shipping company handles items for other small businesses as well. Define an XML document that will include only the information that the shipping company needs to ship goods to the customer.

12. Once the order in Problem 11 has been shipped, the shipping company sends information back to Moonlight Mugs, including the customer name and address, shipper tracking number, data shipped, quantity ordered, quantity shipped, and quantity backordered. Define an XML document that will include the information sent to Moonlight Mugs.

13. Create a document type definition for Problem 11.

14. Western Animal Rescue is a nonprofit organization that supports the fostering and adoption of animals, such as cats, dogs, and birds. People can register to adopt animals. Others register and add animals for adoption. Create the data dictionary structure representing a person registering to adopt an animal. Include name, address (street, city, state or province, zip or postal code), telephone number, email address, date of birth, current pets (type, breed, age of pet), and references. Each person may have multiple pets and must have at least three references. References must include name, address, telephone number, email address, and how they know the person registering to adopt an animal. Be sure to include notation for repeating elements and optional elements.

15. Define the length, the type of data, and the validation criteria for each of the elements in Problem 14.

16. List the data stores that would be required to implement the person registering in Problem 14.

17. Create an XML document with sample data for one person registering to adopt an animal.

Group Projects

1. Meet with your group and use a CASE tool or a manual procedure to develop data dictionary entries for a process, data flow, data store, and data structure based on the data flow diagrams you completed for Maverick Transport in the Chapter 7 group exercises. As a group, agree on any assumptions necessary to make complete entries for each data element.

2. Your group should develop a list of methods to help you make complete data dictionary entries for this exercise as well as for future projects. For example, study existing reports, base them on new or existing data flow diagrams, and so on.

Selected Bibliography

Baskerville, R., and J. Pries-Heje. "Short Cycle Time Systems Development." *Information Systems Journal,* Vol. 14, 2004, pp. 237–264.

Conboy, K., and B. Fitzgerald. "Toward a Conceptual Framework of Agile Methods: A Study of Agility in Different Disciplines." *WISER '04,* November 5, 2004, Newport Beach, CA, pp. 37–44.

Davis, G. B., and M. H. Olson. *Management Information Systems, Conceptual Foundations, Structure, and Development,* 2nd ed. New York: McGraw-Hill, 1985.

Gane, C., and T. Sarson. *Structured Systems Analysis and Design Tools and Techniques.* Englewood Cliffs, NJ: Prentice Hall, 1979.

Hoffer, J. A., R. Venkataraman, and H. Topi. *Modern Database Management,* 10th ed. Upper Saddle River, NJ: Prentice Hall, 2010.

Kelly, B., N. Sheppard, J. Delasalle, M. Dewey, O. Stephens, G. Johnson, and O. S. Taylor. "Open Metrics for Open Repositories. In: OR2012: the 7th International Conference on Open Repositories, 2012-07-09 - 2012-07-13, Edinburgh, Scotland, 2012.

Kruse, R. L. and D. R. Mehr. "Data management for prospective research studies using SAS® software." BMC Medical Research Methodology, Vol. 8, No. 61, 2008. http://www.biomedcentral.com/1471-2288/8/61. Last accessed August 16, 2012.

Lucas, H. *Information Systems Concepts for Management,* 5th ed. New York: McGraw-Hill, 1994.

Martin, J. *Strategic Data-Planning Methodologies.* Englewood Cliffs, NJ: Prentice Hall, 1982.

Sagheb-Tehrani, M. "Expert Systems Development and Some Ideas of Design Process." *ACM SIGSOFT Software Engineering Notes,* Vol. 30, No. 2, March 2005, pp. 1–5.

Subramaniam, V., and A. Hunt. *Practices of an Agile Developer.* Raleigh, NC: Pragmatic Bookshelf, 2006.

The CPU Case Episode and accompanying student files are available online at www.pearsonglobaleditions.com/kendall.

CHAPTER 9

Process Specifications and Structured Decisions

LEARNING OBJECTIVES

Once you have mastered the material in this chapter you will be able to:

1. Understand the purpose of process specifications.

2. Recognize the difference between structured and semistructured decisions.

3. Use structured English, decision tables, and decision trees to analyze, describe, and document structured decisions.

4. Choose an appropriate decision analysis method for analyzing structured decisions and creating process specifications.

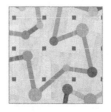

A systems analyst approaching process specifications and structured decisions has many options for documenting and analyzing them. In Chapters 7 and 8 you noted processes such as VERIFY AND COMPUTE FEES, but you did not explain the logic necessary to execute these tasks. The methods available for documenting and analyzing the logic of decisions include structured English, decision tables, and decision trees. It is important to be able to recognize logic and structured decisions that occur in a business and how they are distinguishable from semistructured decisions that tend to involve human judgment. Then it is critical to recognize that structured decisions lend themselves particularly well to analysis with systematic methods that promote completeness, accuracy, and communication.

CONSULTING OPPORTUNITY 9.1

Kit Chen Kaboodle, Inc.

"**I** don't want to get anyone stirred up, but I think we need to sift through our unfilled order policies," says Kit Chen. "I wouldn't want to put a strain on our customers. As you know already, Kit Chen Kaboodle is a Web and mail-order cookware business specializing in 'klassy kitsch for kitchens,' as our latest catalog says. I mean, we've got everything you need to do gourmet cooking and entertaining: nutmeg grinders, potato whisks, egg separators, turkey basters, placemats with cats on 'em, ice cube trays in shamrock shapes, and more.

"Here's how we've been handling unfilled orders. We search our unfilled orders file from the Internet as well as mail-order sales once a week. If the order was filled this week, we delete the record, and the rest is gravy. If we haven't written to the customer in four weeks, we send 'em this cute card with a chef peeking into the oven, saying, 'Not ready yet.' (It's a notification that their item is still on back order.)

"If the back order date changed to greater than 45 days from now, we send out a notice. If the merchandise is seasonal (as with Halloween treat bags, Christmas cookie cutters, or Valentine's Day cake molds) and the back order date is 30 days or more, though, we send out a notice with a chef glaring at his egg timer.

"If the back order date changed at all and we haven't sent out a card in the last two weeks, we send out a card with a chef checking his recipe. If the merchandise is no longer available, we send a notice (complete with chef crying in the corner) and delete the record. We haven't begun to use email in place of mailed cards, but I'd like to.

"Thanks for listening to all this. I think we've got the right ingredients for a good policy; we just need to blend them together and cook up something special."

Because you are the systems analyst whom Kit hired, go through the narrative of how Kit Chen Kaboodle, Inc., handles unfilled orders, drawing boxes around each action Kit mentions and circling each condition brought up. Make a list of any ambiguities you would like to clarify in a later interview, and then write five questions to address them.

Overview of Process Specifications

To determine the human information requirements of a decision analysis strategy, a systems analyst must first determine the users' objectives, along with the organization's objectives, using either a top-down approach or an object-oriented approach. The systems analyst must understand the principles of organizations and have a working knowledge of data-gathering techniques. The top-down approach is critical because all human decisions in the organization should be related, at least indirectly, to the broad objectives of the entire organization.

Process specifications—sometimes called *minispecs,* because they are a small portion of the total project specifications—are created for primitive processes on a data flow diagram as well as for some higher-level processes that explode to a child diagram. They also may be created for class methods in object-oriented design, and, in a more general sense, for the steps in a use case (as discussed in Chapters 2 and 10). These specifications explain the decision-making logic and formulas that will transform process input data into output. Each derived element must have process logic to show how it is produced from the base elements or other previously created derived elements that are input to the primitive process.

The three goals of producing process specifications are as follows:

1. To reduce the ambiguity of the process. This goal compels the analyst to learn details about how the process works. Any vague areas should be noted, written down, and consolidated for all process specifications. These observations form a basis and provide the questions for follow-up interviews with the user community.
2. To obtain a precise description of what is accomplished, which is usually included in a packet of specifications for the programmer.
3. To validate the system design. This goal includes ensuring that a process has all the input data flow necessary for producing the output. In addition, all input and output must be represented on the data flow diagram.

You will find many situations in which process specifications are not created. Sometimes the process is very simple or the computer code already exists. This eventuality would be noted in the process description, and no further design would be required. Categories of processes that generally *do not* require specifications are as follows:

FIGURE 9.1

How process specifications relate to a data flow diagram.

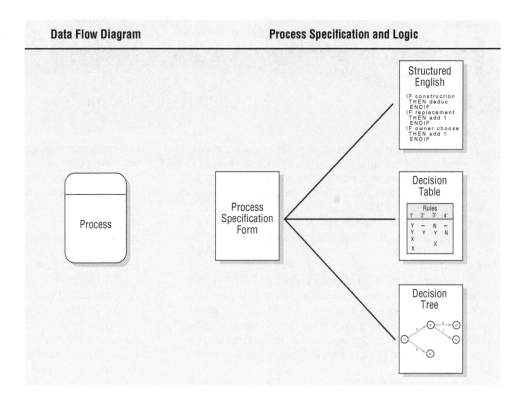

1. Processes that represent physical input or output, such as read and write. These processes usually require only simple logic.
2. Processes that represent simple data validation, which is usually fairly easy to accomplish. The edit criteria are included in the data dictionary and incorporated into the computer source code. Process specifications may be produced for complex editing.
3. Processes that use prewritten code. These processes are generally included in a system as procedures, methods, and functions or in class libraries (that are either purchased or available free on the Web).

These blocks are computer program code that is stored on the computer system. They usually perform a general system function, such as validating a date or a check digit. These general-purpose subprograms are written and documented only once but form a series of building blocks that may be used in many systems throughout the organization. Thus, these subprograms appear as processes on many data flow diagrams (or as class methods discussed in Chapter 10).

Process Specification Format

Process specifications link a process to a data flow diagram, and hence a data dictionary, as illustrated in Figure 9.1. Each process specification should be entered on a separate form or into a CASE tool screen such as the one used for Visible Analyst and shown in the CPU case at the end of this chapter. Enter the following information:

1. The process number, which must match the process ID on the data flow diagram. This specification allows an analyst to work on or review any process, and to locate the data flow diagram containing the process easily.
2. The process name, which again must be the same as the name displayed in the process symbol on the data flow diagram.
3. A brief description of what the process accomplishes.
4. A list of input data flows, using the names found on the data flow diagram. Data names used in the formula or logic should match those in the data dictionary to ensure consistency and good communication.
5. The output data flows, also using the data flow diagram and data dictionary names.

FIGURE 9.2

An example of a completed process specification form for determining whether an item is available.

Process Specification Form

Number __1.3__

Name __Determine Quantity Available__

Description __Determine if an item is available for sale. If it is not available, create a backordered item record. Determine the quantity available.__

Input Data Flow

Valid item from Process 1.2
Quantity on Hand from Item Record

Output Data Flow

Available Item (Item Number + Quantity Sold) to Processes 1.4 & 1.5
Backordered item to Inventory Control

Type of Process

☑ Online   ☐ Batch   ☐ Manual   Subprogram/Function Name

Process Logic:

IF the _Order Item Quantity_ is greater than _Quantity on Hand_
    Then Move _Order Item Quantity_ to _Available Item Quantity_
        Move _Order Item Number_ to _Available Item Number_
ELSE

    Subtract _Quantity on Hand_ from _Order Item Quantity_
        giving _Quantity Backordered_
    Move _Quantity Backordered_ to _Backordered Item Record_
    Move _Item Number_ to _Backordered Item Record_
    DO write _Backordered Record_
    Move _Quantity on Hand_ to _Available Item Quantity_
    Move _Order Item Number_ to _Available Item Number_
ENDIF

Refer to: Name: _____

☐ Structured English   ☐ Decision Table   ☐ Decision Tree

Unresolved Issues: Should the amount that is on order for this item be taken into account? Would this, combined with the expected arrival date of goods on order, change how the quantity available is calculated?

6. An indication of the type of process: batch, online, or manual. All online processes require screen designs, and all manual processes should have well-defined procedures for employees performing the process tasks.

7. If the process uses prewritten code, include the name of the subprogram or function containing that code.

8. A description of the process logic that states policy and business rules in everyday language, not computer language pseudo-code. Business rules are the procedures, or perhaps a set of conditions or formulas, that allow a corporation to run its business. The early problem definition (as explained in Chapter 3) that you completed initially may provide a starting place for this description. Common business rule formats include the following:

- Definitions of business terms
- Business conditions and actions

- Data integrity constraints
- Mathematical and functional derivations
- Logical inferences
- Processing sequences
- Relationships among facts about the business

9. If there is not enough room on the form for a complete structured English description, or if there is a decision table or tree depicting the logic, include the corresponding table or tree name.
10. A list of any unresolved issues, incomplete portions of logic, or other concerns. These issues form the basis of the questions used for follow-up interviews with users or business experts you have added to your project team.

These items should be entered to complete a process specification form, which includes a process number, process name, or both from the data flow diagram, as well as the eight other items shown in the World's Trend example (Figure 9.2). Notice that completing this form thoroughly facilitates linking the process to the data flow diagram and the data dictionary.

Structured English

When process logic involves formulas or iteration, or when structured decisions are not complex, an appropriate technique for analyzing the decision process is the use of structured English. As the name implies, structured English is based on (1) structured logic, or instructions organized into nested and grouped procedures, and (2) simple English statements such as add, multiply, and move. A word problem can be transformed into structured English by putting the decision rules into their proper sequence and using the convention of IF-THEN-ELSE statements throughout.

Writing Structured English

To write structured English, you may want to use the following conventions:

1. Express all logic in terms of one of these four types: sequential structures, decision structures, case structures, or iterations (see Figure 9.3 for examples).
2. Use and capitalize accepted keywords such as IF, THEN, ELSE, DO, DO WHILE, DO UNTIL, and PERFORM.

FIGURE 9.3

Examples of logic expressed in a sequential structure, a decision structure, a case structure, and an iteration.

Structured English Type	Example
Sequential Structure A block of instructions in which no branching occurs	Action #1 Action #2 Action #3

Decision Structure
    Only IF a condition is true, complete the following statements; otherwise, jump to the ELSE
	IF Condition A is True
	    THEN implement Action A
	ELSE implement Action B
	ENDIF

Case Structure
    A special type of decision structure in which the cases are mutually exclusive (if one occurs, the others cannot)
	IF Case #1 Implement Action #1
	ELSE IF Case #2
	          Implement Action #2
	ELSE IF Case #3
	          Implement Action #3
	ELSE IF Case #4
	          Implement Action #4
	ELSE print error
	ENDIF

Iteration Blocks of statements that are repeated until done	DO WHILE there are customers. Action #1 ENDDO

CONSULTING OPPORTUNITY 9.2

Kneading Structure

Kit Chen has risen to the occasion and answered your questions concerning the policy for handling unfilled orders at Kit Chen Kaboodle, Inc. Based on those answers and any assumptions you need to make, pour Kit's narrative (from Consulting Opportunity 9.1) into a new mold by rewriting the recipe for handling unfilled orders in structured English. In a paragraph, describe how this process might change if you used email for notification rather than regular mail.

3. Indent blocks of statements to show their hierarchy (nesting) clearly.
4. When words or phrases have been defined in a data dictionary (as in Chapter 8), underline those words or phrases to signify that they have a specialized, reserved meaning.
5. Be careful when using "and" and "or," and avoid confusion when distinguishing between "greater than" and "greater than or equal to" and like relationships. "A and B" means both A and B; "A or B" means either A or B, but not both. Clarify the logical statements now rather than waiting until the program coding stage.

FIGURE 9.4

Structured English for the medical-claim processing system. Underlining signifies that the terms have been defined in the data dictionary.

```
DO WHILE there are claims remaining
    IF claimant has not sent in a claim
        THEN set up new claimant record
    ELSE continue
    Add claim to YTD Claim
    IF claimant has policy-plan A
        THEN IF deductible of $100.00 has not been met
            THEN subtract deductible-not-met from claim
            Update deductible
        ELSE continue
        ENDIF
        Subtract copayment of 40% of claim from claim
    ELSE IF claimant has policy-plan B.
        THEN IF deductible of $50.00 has not been met
            THEN subtract deductible-not-met from claim
            Update deductible
        ELSE continue
        ENDIF
        Subtract copayment of 60% of claim from claim
    ELSE continue
    ELSE write plan-error-message
    ENDIF
    ENDIF
    IF claim is greater than zero
        THEN print check
    ENDIF
    Print summary for claimant
    Update accounts
ENDDO
```

FIGURE 9.5

Data structure for a shipping
statement for World's Trend.

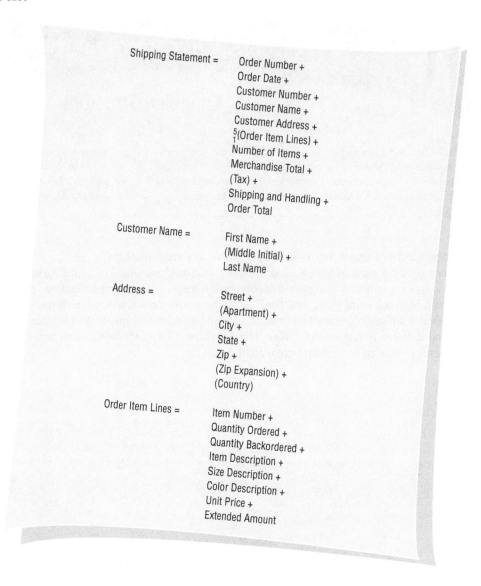

A STRUCTURED ENGLISH EXAMPLE. The following example demonstrates how a spoken procedure for processing medical claims is transformed into structured English:

> We process all our claims in this manner. First, we determine whether the claimant has ever sent in a claim before; if not, we set up a new record. The claim totals for the year are then updated. Next, we determine if a claimant has policy A or policy B, which differ in deductibles and copayments (the percentage of the claim claimants pay themselves). For both policies, we check to see if the deductible has been met ($100 for policy A and $50 for policy B). If the deductible has not been met, we apply the claim to the deductible. Another step adjusts for the copayment; we subtract the percentage the claimant pays (40 percent for policy A and 60 percent for policy B) from the claim. Then we issue a check if there is money coming to the claimant, print a summary of the transaction, and update our accounts. We do this until all claims for that day are processed.

In examining the foregoing statements, we notice some simple sequence structures, particularly at the beginning and end. There are a couple of decision structures, and it is most appropriate to nest them, first by determining which policy (A or B) to use and then by subtracting the correct deductibles and copayments. The last sentence points to an iteration: Either DO UNTIL all the claims are processed or DO WHILE there are claims remaining.

Realizing that it is possible to nest the decision structures according to policy plans, we can write the structured English for the foregoing example (see Figure 9.4). As we begin to work on the structured English, we find that some logic and relationships that seemed clear at one

FIGURE 9.6

Structured English for creating the shipping statement for World's Trend.

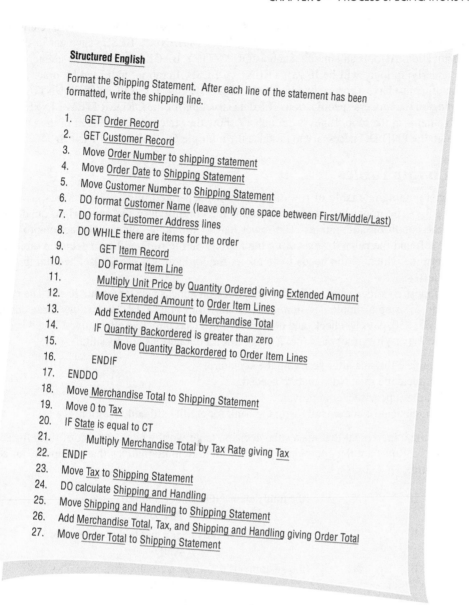

Structured English

Format the Shipping Statement. After each line of the statement has been formatted, write the shipping line.

1. GET Order Record
2. GET Customer Record
3. Move Order Number to shipping statement
4. Move Order Date to Shipping Statement
5. Move Customer Number to Shipping Statement
6. DO format Customer Name (leave only one space between First/Middle/Last)
7. DO format Customer Address lines
8. DO WHILE there are items for the order
9.   GET Item Record
10.   DO Format Item Line
11.   Multiply Unit Price by Quantity Ordered giving Extended Amount
12.   Move Extended Amount to Order Item Lines
13.   Add Extended Amount to Merchandise Total
14.   IF Quantity Backordered is greater than zero
15.   Move Quantity Backordered to Order Item Lines
16.   ENDIF
17. ENDDO
18. Move Merchandise Total to Shipping Statement
19. Move 0 to Tax
20. IF State is equal to CT
21.   Multiply Merchandise Total by Tax Rate giving Tax
22. ENDIF
23. Move Tax to Shipping Statement
24. DO calculate Shipping and Handling
25. Move Shipping and Handling to Shipping Statement
26. Add Merchandise Total, Tax, and Shipping and Handling giving Order Total
27. Move Order Total to Shipping Statement

time are actually ambiguous. For example, do we add the claim to the year-to-date (YTD) claim before or after updating the deductible? Is it possible that an error can occur if something other than policy A or B is stored in the claimant's record? We subtract 40 percent of what from the claim? These ambiguities need to be clarified at this point.

Besides the obvious advantage of clarifying the logic and relationships found in human languages, structured English has another important advantage: It is a communication tool. Structured English can be taught to and hence understood by users in the organization, so if communication is important, structured English is a viable alternative for decision analysis.

Data Dictionary and Process Specifications

All computer programs may be coded using the three basic constructs: sequence, selection (IF . . . THEN . . . ELSE and the case structure), and iteration or looping. The data dictionary indicates which of these constructs must be included in the process specifications.

If the data dictionary for the input and output data flow contains a series of fields without any iteration—{ }—or selection—[ ]—the process specification will contain a simple sequence of statements, such as MOVE, ADD, and SUBTRACT. Refer to the example of a data dictionary for the SHIPPING STATEMENT, illustrated in Figure 9.5. Notice that the data dictionary for the SHIPPING STATEMENT has the ORDER NUMBER, ORDER DATE, and CUSTOMER NUMBER as simple sequential fields. The corresponding logic, shown in lines 3 through 5 in the corresponding structured English in Figure 9.6, consists of simple MOVE statements.

A data structure with optional elements contained in parentheses or either/or elements contained in brackets will have a corresponding IF . . . THEN . . . ELSE statement in the process specification. Also, if an amount, such as QUANTITY BACKORDERED, is greater than zero, the underlying logic will be IF . . . THEN . . . ELSE. Iteration, indicated by braces on a data structure, must have a corresponding DO WHILE, DO UNTIL, or PERFORM UNTIL to control looping on the process specification. The data structure for the ORDER ITEM LINES allows up to five items in the loop. Lines 8 through 17 show the statements contained in the DO WHILE through the END DO necessary to produce the multiple ORDER ITEM LINES.

Decision Tables

A decision table is a table of rows and columns, separated into four quadrants, as shown in Figure 9.7. The upper-left quadrant contains the condition(s); the upper-right quadrant contains the condition alternatives. The lower half of the table contains the actions to be taken on the left and the rules for executing the actions on the right. When a decision table is used to determine which action needs to be taken, the logic moves clockwise, beginning from the upper left.

Suppose a store wanted to illustrate its policy on noncash customer purchases. The company could do so using a simple decision table, as shown in Figure 9.8. Each of the three conditions (sale under $50, pays by check, and uses credit cards) has only two alternatives. The two alternatives are Y (yes, it is true) or N (no, it is not true). Four actions are possible:

1. Complete the sale after verifying the signature.
2. Complete the sale. No signature needed.
3. Call the supervisor for approval.
4. Communicate electronically with the bank for credit card authorization.

The final ingredient that makes the decision table worthwhile is the set of rules for each of the actions. Rules are the combinations of the condition alternatives that precipitate an action. For example, Rule 3 says:

IF	N	(the total sale is NOT under $50.00)
		AND
IF	Y	(the customer paid by check and had two forms of ID)
		AND
IF	N	(the customer did not use a credit card)
		THEN
DO	X	(CALL THE SUPERVISOR FOR APPROVAL).

The foregoing example features a problem with four sets of rules and four possible actions, but that is only a coincidence. The next example demonstrates that decision tables often become large and involved.

Developing Decision Tables

To build decision tables, an analyst needs to determine the maximum size of the table; eliminate any impossible situations, inconsistencies, or redundancies; and simplify the table as much as

Conditions and Actions	Rules
Conditions	Condition Alternatives
Actions	Action Entries

Conditions and Actions	Rules 1	2	3	4
Under $50	Y	Y	N	N
Pays by check with two forms of ID	Y	N	Y	N
Uses credit card	N	Y	N	Y
Complete the sale after verifying signature.	X			
Complete the sale. No signature needed.		X		
Call supervisor for approval.			X	
Communicate electronically with bank for credit card authorization.				X

FIGURE 9.8

Using a decision table for illustrating a store's policy of customer checkout with four sets of rules and four possible actions.

possible. The following steps provide the analyst with a systematic method for developing decision tables:

1. Determine the number of conditions that may affect the decision. Combine rows that overlap, such as conditions that are mutually exclusive. The number of conditions becomes the number of rows in the top half of the decision table.
2. Determine the number of possible actions that can be taken. That number becomes the number of rows in the lower half of the decision table.
3. Determine the number of condition alternatives for each condition. In the simplest form of a decision table, there would be two alternatives (Y or N) for each condition. In an extended-entry table, there may be many alternatives for each condition. Make sure that all possible values for the condition are included. For example, if a problem statement calculating a customer discount mentions one range of values for an order total from $100 to $1,000 and another range of greater than $1,000, the analyst should realize that the range from 0 up to $100 should also be added as a condition. This is especially true when there are other conditions that may apply to the 0 up to $100 order total.
4. Calculate the maximum number of columns in the decision table by multiplying the number of alternatives for each condition. If there were four conditions and two alternatives (Y or N) for each of the conditions, there would be 16 possibilities, as follows:

$$\begin{array}{l} \text{Condition 1: } \times 2 \text{ alternatives} \\ \text{Condition 2: } \times 2 \text{ alternatives} \\ \text{Condition 3: } \times 2 \text{ alternatives} \\ \underline{\text{Condition 4: } \times 2 \text{ alternatives}} \\ 16 \text{ possibilities} \end{array}$$

5. Fill in the condition alternatives. Start with the first condition and divide the number of columns by the number of alternatives for that condition. In the foregoing example, there are 16 columns and two alternatives (Y or N), so 16 divided by 2 is 8. Then choose one of the alternatives, say Y, and write it in the first eight columns. Finish by writing N in the remaining eight columns, as follows:

Condition 1:   Y Y Y Y Y Y Y Y N N N N N N N N

Repeat this step for each condition, using a subset of the table:

Condition 1:   Y Y Y Y Y Y Y Y Y N N N N N N N N
Condition 2:   Y Y Y Y N N N N
Condition 3:   Y Y N N
Condition 4:   Y N

And continue the pattern for each condition:

Condition 1:   Y Y Y Y Y Y Y Y Y N N N N N N N N
Condition 2:   Y Y Y Y N N N N Y Y Y Y N N N N
Condition 3:   Y Y N N Y Y N N Y Y N N Y Y N N
Condition 4:   Y N Y N Y N Y N Y N Y N Y N Y N

CONSULTING OPPORTUNITY 9.3

Saving a Cent on Citron Car Rental

"We feel lucky to be this popular. I think customers feel we have so many options to offer that they ought to rent an auto from us," says Ricardo Limon, who manages several outlets for Citron Car Rental. "Our slogan is, 'You'll never feel squeezed at Citron.' We have five sizes of cars that we list as A through E:

A	Subcompact
B	Compact
C	Midsize
D	Full-size
E	Luxury

"Standard transmission is available only for A, B, and C. Automatic transmission is available for all cars."

"If a customer reserves a subcompact (A) and finds on arriving that we don't have one, that customer gets a free upgrade to the next-sized car, in this case a compact (B). Customers also get a free upgrade from their reserved car size if their company has an account with us. There's a discount for membership in any of the frequent-flyer clubs run by cooperating airlines, too. When customers step up to the counter, they tell us what size car they reserved, and then we check to see if we have it in the lot ready to go. They usually bring up any discounts, and we ask them if they want insurance and how long they will use the car. Then we calculate their rate and write out a slip for them to sign right there."

Ricardo has asked you to computerize the billing process for Citron so that customers can get their cars quickly and still be billed correctly. Draw a decision table that represents the conditions, condition alternatives, actions, and action rules you gained from Ricardo's narrative that will guide an automated billing process.

Ricardo wants to expand the ecommerce portion of his business by making it possible to reserve a car over the Web. Draw an updated decision table that shows a 10-percent discount for booking a car over the Web.

6. Complete the table by inserting an X where rules suggest certain actions.
7. Combine rules where it is apparent that an alternative does not make a difference in the outcome. For example,

| Condition 1: | YY |
Condition 2:	YN
Action 1:	XX

can be expressed as:

| Condition 1: | Y |
Condition 2:	—
Action 1:	X

The dash [—] signifies that Condition 2 can be either Y or N, and the action will still be taken.

8. Check the table for any impossible situations, contradictions, and redundancies. They are discussed in more detail later.
9. Rearrange the conditions and actions (or even rules) if it makes the decision table more understandable.

A DECISION TABLE EXAMPLE. Figure 9.9 is an illustration of a decision table developed using the steps previously outlined. In this example, a company is trying to maintain a meaningful mailing list of customers. The objective is to send out only the catalogs from which customers will buy merchandise.

The managers realize that certain loyal customers order from every catalog and that some people on the mailing list never order. These ordering patterns are easy to observe, but deciding which catalogs to send customers who order only from selected catalogs is more difficult. Once these decisions are made, a decision table is constructed for three conditions (C1: customer ordered from Fall catalog; C2: customer ordered from Christmas catalog; and C3: customer ordered from specialty catalog), each having two alternatives (Y or N). Three actions can be

Conditions and Actions	1	2	3	4	5	6	7	8
				Rules				
Customer ordered from Fall catalog.	Y	Y	Y	Y	N	N	N	N
Customer ordered from Christmas catalog.	Y	Y	N	N	Y	Y	N	N
Customer ordered from specialty catalog.	Y	N	Y	N	Y	N	Y	N
Send out this year's Christmas catalog.		X		X		X		X
Send out specialty catalog.			X				X	
Send out both catalogs.	X				X			

FIGURE 9.9

Constructing a decision table for deciding which catalog to send to customers who order only from selected catalogs.

taken (A1: send out this year's Christmas catalog; A2: send out the new specialty catalog; and A3: send out both catalogs). The resulting decision table has six rows (three conditions and three actions) and eight columns (two alternatives × two alternatives × two alternatives).

The decision table is now examined to see if it can be reduced. There are no mutually exclusive conditions, so it is not possible to get by with fewer than three condition rows. No rules allow the combination of actions. It is possible, however, to combine some of the rules, as shown in Figure 9.10. For instance, Rules 2, 4, 6, and 8 can be combined because they all have two things in common:

1. They instruct us to send out this year's Christmas catalog.
2. The alternative for Condition 3 is always N.

It doesn't matter what the alternatives are for the first two conditions, so it is possible to insert dashes [—] in place of the Y or N.

The remaining rules—Rules 1, 3, 5, and 7—cannot be reduced to a single rule because two different actions remain. Instead, Rules 1 and 5 can be combined; likewise, Rules 3 and 7 can be combined.

Checking for Completeness and Accuracy

Checking over your decision tables for completeness and accuracy is essential. Four main problems can occur in developing decision tables: incompleteness, impossible situations, contradictions, and redundancy.

Conditions and Actions	1	2	3	4	5	6	7	8
				Rules				
Customer ordered from Fall catalog.	Y	Y	Y	Y	N	N	N	N
Customer ordered from Christmas catalog.	Y	Y	N	N	Y	Y	N	N
Customer ordered from specialty catalog.	Y	N	Y	N	Y	N	Y	N
Send out this year's Christmas catalog.		X		X		X		X
Send out specialty catalog.			X				X	
Send out both catalogs.	X				X			

Conditions and Actions	1'	2'	3'
		Rules	
Customer ordered from Fall catalog.	—	—	—
Customer ordered from Christmas catalog.	Y	—	N
Customer ordered from specialty catalog.	Y	N	Y
Send out this year's Christmas catalog.		X	
Send out specialty catalog.			X
Send out both catalogs.	X		

FIGURE 9.10

Combining rules to simplify the customer-catalog decision table.

FIGURE 9.11

Adding a rule to the customer-catalog decision table changes the entire table.

Conditions and Actions	Rules			
	1'	2'	3'	4'
Customer ordered from Fall catalog.	—	—	—	—
Customer ordered from Christmas catalog.	Y	—	N	—
Customer ordered from specialty catalog.	Y	N	Y	—
Customer ordered $50 or more.	Y	Y	Y	N
Send out this year's Christmas catalog.		X		
Send out specialty catalog.			X	
Send out both catalogs.	X			
Do not send out any catalog.				X

Ensuring that all conditions, condition alternatives, actions, and action rules are complete is of utmost importance. Suppose an important condition—if a customer ordered less than $50—had been left out of the catalog store problem discussed earlier. The whole decision table would change because a new condition, new set of alternatives, new action, and one or more new action rules would have to be added. Suppose the rule is: IF the customer did not order more than $50, THEN do not send any catalogs. A new Rule 4 would be added to the decision table, as shown in Figure 9.11.

When building decision tables as outlined in the foregoing steps, it is sometimes possible to set up impossible situations. An example is shown in Figure 9.12. Rule 1 is not feasible, because a person cannot earn greater than $50,000 per year and less than $2,000 per month at the same time. The other three rules are valid. The problem went unnoticed because the first condition was measured in years and the second condition in months.

Contradictions occur when rules suggest different actions but satisfy the same conditions. The fault could lie with the way the analyst constructed the table or with the information the analyst received. Contradictions often occur if dashes [—] are incorrectly inserted into the table. Redundancy occurs when identical sets of alternatives require the exact same action. Figure 9.13 illustrates a contradiction and a redundancy. The analyst has to determine what is correct and then resolve the contradiction or redundancy.

Decision tables are an important tool in the analysis of structured decisions. One major advantage of using decision tables over other methods is that tables help the analyst ensure completeness. When using decision tables, it is also easy to check for possible errors, such as impossible situations, contradictions, and redundancy. Decision table processors, which take the table as input and provide computer program code as output, are also available.

Decision Trees

Decision trees are used when complex branching occurs in a structured decision process. Trees are also useful when it is essential to keep a string of decisions in a particular sequence. Although the decision tree derives its name from natural trees, decision trees are most often drawn on their

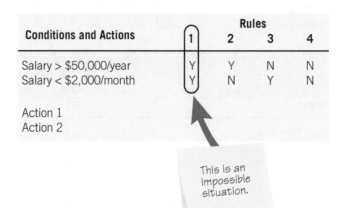

Conditions and Actions	Rules			
	1	2	3	4
Salary > $50,000/year	Y	Y	N	N
Salary < $2,000/month	Y	N	Y	N
Action 1				
Action 2				

This is an impossible situation.

FIGURE 9.13

Checking the decision table for inadvertent contradictions and redundancy is important.

side, with the root of the tree on the left side of the paper; from there, the tree branches out to the right. This orientation allows the analyst to write on the branches to describe conditions and actions.

Unlike the decision tree used in management science, the analyst's tree does not contain probabilities and outcomes. In systems analysis, trees are used mainly for identifying and organizing conditions and actions in a completely structured decision process.

Drawing Decision Trees

It is useful to distinguish between conditions and actions when drawing decision trees. This distinction is especially relevant when conditions and actions take place over a period of time and their sequence is important. For this purpose, use a square node to indicate an action and a circle to represent a condition. Using notation makes the decision tree more readable, as does numbering the circles and squares sequentially. Think of a circle as signifying IF, whereas the square means THEN.

When decision tables were discussed in an earlier section, a point-of-sale example was used to determine the purchase approval actions for a department store. Conditions included the amount of the sale (under $50) and whether the customer paid by check or credit card. The four

---

CONSULTING OPPORTUNITY 9.4

A Tree for Free

"I know you've got a plane to catch, but let me try to explain it to you once again, sir," pleads Glen Curtiss, a marketing manager for Premium Airlines. Curtiss has been attempting (unsuccessfully) to explain the airline's new policy for accumulating miles for awards (such as upgrades to first class and free flights) to a member of Premium's "Flying for Prizes" club.

Glen takes another pass at getting the policy off the ground, saying, "You see, sir, the traveler (that's you, Mr. Icarus) will be awarded the miles actually flown. If the actual mileage for the leg was less than 500 miles, the traveler will get 500 miles credit. If the trip was made on a Saturday, the actual mileage will be multiplied by two. If the trip was made on a Tuesday, the multiplication factor is 1.5. If this is the ninth leg traveled during the calendar month,

the mileage is doubled no matter what day, and if it is the 17th leg traveled, the mileage is tripled. If the traveler booked the flight on the Web or through a travel service such as Orbitz or Travelocity, 100 miles are added.

"I hope that clears it up for you, Mr. Icarus. Enjoy your flight, and thanks for flying Premium."

Mr. Icarus, whose desire to board the Premium plane has all but melted away during Glen's long explanation, fades into the sea of people wading through the security lanes, without so much as a peep in reply.

Develop a decision tree for Premium Airlines' new policy for accumulating award miles so that the policy becomes clearer, is easier to grasp visually, and hence is easier to explain.

FIGURE 9.14

Drawing a decision tree to show the noncash purchase approval actions for a department store.

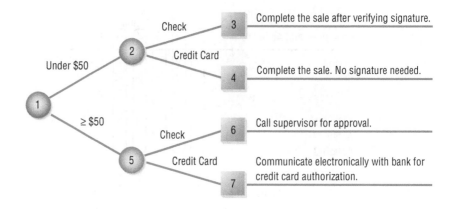

actions possible were to: complete the sale after verifying the signature; complete the sale with no signature needed; call the supervisor for approval; or communicate electronically with the bank for credit card authorization. Figure 9.14 illustrates how this example can be drawn as a decision tree. In drawing the tree:

1. Identify all conditions and actions and their order and timing (if they are critical).
2. Begin building the tree from left to right, making sure you list all possible alternatives before moving to the right.

This simple tree is symmetrical, and the four actions at the end are unique. A tree does not need to be symmetrical. Most decision trees have conditions that have a different number of branches. Also, identical actions may appear more than once.

A decision tree has three main advantages over a decision table. First, it takes advantage of the sequential structure of decision tree branches so that the order of checking conditions and executing actions is immediately noticeable. Second, conditions and actions of decision trees are found on some branches but not on others; in contrast, with decision tables, conditions and actions are all part of the same table. Conditions and actions that are critical are connected directly to other conditions and actions, whereas conditions that do not matter are absent. In other words, the tree does not have to be symmetrical. Third, compared with decision tables, decision trees are more readily understood by others in an organization. Consequently, they are more appropriate as a communication tool.

Choosing a Structured Decision Analysis Technique

We have examined the three techniques for analysis of structured decisions: structured English, decision tables, and decision trees. Although they need not be used exclusively, it is customary to choose one analysis technique for a decision rather than employing all three. The following guidelines provide you with a way to choose one of the three techniques for a particular case:

1. Use structured English when
   a. There are many repetitious actions,
      OR
   b. Communication to end users is important.
2. Use decision tables when
   a. Complex combinations of conditions, actions, and rules are found,
      OR
   b. You require a method that effectively avoids impossible situations, redundancies, and contradictions.
3. Use decision trees when
   a. The sequence of conditions and actions is critical,
      OR
   b. When not every condition is relevant to every action (the branches are different).

HYPERCASE® EXPERIENCE 9

"It's really great that you've been able to spend all of this time with us. One thing's for sure: We can use the help. And clearly, from your conversations with Snowden and others, you must realize we all believe that consultants have a role to play in helping companies change. Well, most of us believe it anyway.

"Sometimes structure is good for a person. Or even a company. As you know, Snowden is keen on any kind of structure. That's why some of the Training people can drive him wild sometimes. They're good at structuring things for their clients, but when it comes to organizing their own work, it's another story. Oh well, let me know if there's any way I can help you."

HYPERCASE Questions

1. Assume that you will create the specifications for an automated project tracking system for the Training employees. One of the system's functions will be to allow project members to update or add names, addresses, email addresses and phone numbers of new clients. Using structured English, write a procedure for carrying out the process of entering a new client name, address, email address and phone numbers. (*Hint:* The procedure should ask for a client name, check to see if the name is already in an existing client file, and let the user either validate and update the current client address, email address, and phone number [if necessary] or add a new client's address, email address, and phone number to the client file.)

Summary

Once an analyst has worked with users to identify data flows and begun constructing a data dictionary, it is time to turn to process specification and decision analysis. The three methods for decision analysis and describing process logic discussed in this chapter are structured English, decision tables, and decision trees. All three methods help an analyst develop the logic for structured decisions made in the organization. Making structured decisions need not involve human judgment.

Process specifications (or minispecs) are created for primitive processes on a data flow diagram as well as for some higher-level processes that explode to a child diagram. These specifications explain the decision-making logic and formulas that will transform process input data into output. The three goals of process specification are to reduce the ambiguity of the process, to obtain a precise description of what is accomplished, and to validate the system design.

One way to describe structured decisions is to use the method referred to as structured English, in which logic is expressed in sequential structures, decision structures, case structures, or iterations. Structured English uses accepted keywords such as IF, THEN, ELSE, DO, DO WHILE, and DO UNTIL to describe the logic used, and it indents to indicate the hierarchical structure of the decision process.

Decision tables provide another way to examine, describe, and document decisions. Four quadrants (viewed clockwise from the upper-left corner) are used to (1) describe the conditions, (2) identify possible condition (or decision) alternatives (such as Y or N), (3) indicate which actions should be performed, and (4) describe the actions. Decision tables are advantageous because the rules for developing the table itself, as well as the rules for eliminating redundancy, contradictions, and impossible situations, are straightforward and manageable. The use of decision tables promotes completeness and accuracy in analyzing structured decisions.

The third method for decision analysis is to use a decision tree, consisting of nodes (a square for actions and a circle for conditions) and branches. Decision trees are appropriate when actions must be accomplished in a certain sequence. There is no requirement that the tree be symmetrical, so only those conditions and actions that are critical to the decisions at hand are found on a particular branch.

Each of the decision analysis methods has advantages and should be used accordingly. Structured English is useful when many actions are repeated and when communicating with others is important. Decision tables provide a complete analysis of complex situations while limiting the need for change attributable to impossible situations, redundancies, or contradictions. Decision trees are important when proper sequencing of conditions and actions is critical and when each condition is not relevant to each action.

Keywords and Phrases

action	decision tree
action rule	minispecs
condition	process specifications
condition alternative	structured decision
decision table	structured English

Review Questions

1. List three reasons for producing process specifications.
2. Define what is meant by a structured decision.
3. What four elements must be known for a systems analyst to design systems for structured decisions?
4. What are the two building blocks of structured English?
5. List five conventions that should be followed when using structured English.
6. What is the advantage of using structured English to communicate with people in an organization?
7. Which quadrant of the decision table is used for conditions? Which is used for condition alternatives?
8. What is the first step to take in developing a decision table?
9. List the four main problems that can occur in developing decision tables.
10. What is one of the major advantages of decision tables over other methods of decision analysis?
11. What are the main uses of decision trees in systems analysis?
12. List the four major steps in building decision trees.
13. What three advantages do decision trees have over decision tables?
14. In which two situations should you use structured English?
15. In which two situations do decision tables work best?
16. In which two situations are decision trees preferable?

Problems

1. Clyde Clerk is reviewing his firm's expense reimbursement policies with the new salesperson, Trav Farr. "Our reimbursement policies depend on the situation. You see, first we determine if it is a local trip. If it is, we only pay mileage of 18.5 cents a mile. If the trip was a one-day trip, we pay mileage and then check the times of departure and return. To be reimbursed for breakfast, you must leave by 7:00 A.M., lunch by 11:00 A.M., and have dinner by 5:00 P.M. To receive reimbursement for breakfast, you must return later than 10:00 A.M., lunch later than 2:00 P.M., and have dinner by 7:00 P.M. On a trip lasting more than one day, we allow hotel, taxi, and airfare, as well as meal allowances. The same times apply for meal expenses." Write structured English for Clyde's narrative of the reimbursement policies.
2. Draw a decision tree depicting the reimbursement policy in Problem 1.
3. Draw a decision table for the reimbursement policy in Problem 1.
4. A computer supplies firm called True Disk has set up accounts for countless businesses in Dosville. True Disk sends out invoices monthly and will give discounts if payments are made within 10 days. The discounting policy is as follows: If the amount of the order for computer supplies is greater than $1,000, subtract 4 percent for the order; if the amount is between $500 and $1,000, subtract a 2 percent discount; if the amount is less than $500, do not apply any discount. All orders made via the Web automatically receive an extra 5 percent discount. Any special order (computer furniture, for example) is exempt from all discounting.

    Develop a decision table for True Disk discounting decisions, for which the condition alternatives are limited to Y and N.
5. Develop an extended-entry decision table for the True Disk company discount policy described in Problem 4.
6. Develop a decision tree for the True Disk company discount policy in Problem 4.
7. Write structured English to solve the True Disk company situation in Problem 4.
8. Premium Airlines has recently offered to settle claims for a class-action suit, which was originated for alleged price fixing of tickets. The proposed settlement is stated as follows:

    Initially, Premium Airlines will make available to the settlement class a main fund of $25 million in coupons. If the number of valid claims submitted is 1.25 million or fewer, the value of each claim will be the result obtained by dividing $25 million by the total number of valid claims submitted. For example, if there are 500,000 valid claims, each person submitting a valid claim will receive a coupon with a value of $50.

The denomination of each coupon distributed will be in a whole dollar amount not to exceed $50. Thus, if there are fewer than 500,000 valid claims, the value of each claim will be divided among two coupons or more. For example, if there are 250,000 valid claims, each person submitting a valid claim will receive two coupons, each having a face value of $50, for a total coupon value of $100.

If the number of valid claims submitted is between 1.25 million and 1.5 million, Premium Airlines will make available a supplemental fund of coupons, with a potential value of $5 million. The supplemental fund will be made available to the extent necessary to provide one $20 coupon for each valid claim.

If there are more than 1.5 million valid claims, the total amount of the main fund and the supplemental fund, $30 million, will be divided evenly to produce one coupon for each valid claim. The value of each such coupon will be $30 million divided by the total number of valid claims.

Draw a decision tree for the Premium Airlines settlement.

9. Write structured English for the Premium Airlines settlement in Problem 8.

10. "Well, it's sort of hard to describe," says Sharon, a counselor at Less Is More Nutrition Center. "I've never had to really tell anybody about the way we charge clients or anything, but here goes.

"When clients come into Less Is More, we check to see if they've ever used our service before. Unfortunately for them, I guess, we have a lot of repeat clients who keep bouncing back. Repeat clients get a reduced rate (pardon the pun) of $100 for the first visit if they return within a year of the end of their program.

"Everyone new pays an initial fee, which is $200 for a physical evaluation. The client may bring in a coupon at this time, and then we deduct $50 from the up-front fee. Half of our clients use our coupons and find out about us from them. We just give our repeaters their $100 off, though; they can't use a coupon, too! Clients who transfer in from one of our centers in another city get $75 off their first payment fee, but the coupon doesn't apply. Customers who pay cash get 10 percent off the $200, but they can't use a coupon with that."

Create a decision table with Y and N conditions for the client charge system at Less Is More Nutrition Center.

11. Reduce the decision table in Figure 9.EX1 to the minimum number of rules.

12. Azure Isle Resort has a pricing structure for vacationers in one of its three dwelling categories: the hotel, villas, and beach bungalows. The base price is for staying in the hotel. Beach bungalows have a 10 percent surcharge and villas have a 15 percent surcharge. The final price includes a discount of 4 percent for returning customers. Further conditions apply to how close to capacity the resort is and whether the requested date is within one month from the current date. If the resort is 50 percent full and the time is within one month, there is a 12 percent discount. If the resort is 70 percent full and the time is within one month, there is a 6 percent discount. If the resort is 85 percent full and it is within one month, there is a 4 percent discount.

Develop an optimized decision table for the Azure Isle Resort pricing structure.

13. Create a decision tree for Problem 12.

14. The base ticket price for Cloudliner Airlines is determined by the distance traveled and the day of the week a passenger is traveling. In addition, the airline adjusts its ticket prices based on a number of categories. If the seats remaining are greater than 50 percent of capacity and the number of days before the flight is less than 7, the price is deeply discounted with a special Web offer for the flight. If the seats remaining are greater than 50 percent and the flight date is from 7 to 21 days in the future,

FIGURE 9.EX1

A decision table for a warehouse.

Conditions and Actions	Rules															
	1	2	3	4	5	6	7	8	9	10	11	12	13	14	15	16
Sufficient quantity on hand	Y	Y	Y	Y	Y	Y	Y	Y	Y	N	N	N	N	N	N	N
Quantity large enough for discount	Y	Y	Y	Y	N	N	N	N	N	Y	Y	Y	Y	Y	N	N	N	N
Wholesale customer	Y	Y	N	N	Y	Y	N	N	N	Y	Y	N	N	Y	Y	N	N
Sales tax exemption filed	Y	N	Y	N	Y	N	Y	N	N	Y	N	Y	N	Y	N	Y	N
Ship items and prepare invoice	X	X	X	X	X	X	X	X	X							
Set up backorder										X	X	X	X	X	X	X	X
Deduct discount	X	X														
Add sales tax		X	X	X		X	X	X								

there is a medium price discount. If the seats remaining are greater than 50 percent and the number of days before travel are greater than 21, there is only a small discount.

If the seats remaining are from 20 to 50 percent and the days before the flight are fewer than 7, the ticket has a medium discount. If the seats remaining are from 20 to 50 percent and the flight date is from 7 to 21 days in the future, there is a low discount for prices. If the seats remaining are from 20 to 50 percent and the number of days before travel are greater than 21, there is no discount.

If the seats remaining are less than 20 percent and the number of days before the flight is less than 7, the ticket has the highest increase in price. If the seats remaining are less than 20 percent and the flight date is from 7 to 21 days in the future, there is a large increase in price. If the seats remaining are less than 20 percent and number of days before travel are greater than 21, there is a small increase in price.

Develop an optimized decision table for the Cloudliner Airlines ticket price adjustment policies.

15. Develop a decision tree for the situation in Problem 14.

Group Projects

1. Each group member (or each subgroup) should choose to become an "expert" and prepare to explain how and when to use one of the following structured decision techniques: structured English, decision tables, or decision trees. Each group member or subgroup should then make a case for the usefulness of its assigned decision analysis technique for studying the types of structured decisions made by Maverick Transport on dispatching particular trucks to particular destinations. Each group should make a presentation of its preferred technique.
2. After hearing each presentation, the group should reach a consensus on which technique is most appropriate for analyzing the dispatching decisions of Maverick Transport and why that technique is best in this instance.

Selected Bibliography

Anderson, D. R., D. J. Sweeney, T. A. Williams, and R. K. Martin. *An Introduction to Management Science,* 10th ed. Florence, KY: South-Western College Publishing (an imprint of Cengage), 2007.

Evans, J. R. *Applied Production and Operations Management,* 4th ed. St. Paul, MN: West, 1993.

Kress-Gazit, H., G.E. Fainekos, and G. J. Pappas. "Translating Structured English to Robot Controllers." Vol.22, No.12, 2008, pp. 1343–1359.

Wood, H. An Intro to Logic Trees and Structured Programming. Available at http://www.techopedia.com/2/28552/it-business/enterprise-applications/an-intro-to-logic-trees-and-structured-programming. Last accessed August 16, 2012.

www.wisegeek.com/what-is-structured-english.htm. Last accessed August 16, 2012.

CHAPTER 10

Object-Oriented Systems Analysis and Design Using UML[*]

LEARNING OBJECTIVES

Once you have mastered the material in this chapter you will be able to:

1. Understand what object-oriented systems analysis and design is and appreciate its usefulness.

2. Comprehend the concepts of Unified Modeling Language (UML), the standard approach for modeling a system in the object-oriented world.

3. Apply the steps used in UML to break down the system into a use case model and then a class model.

4. Diagram systems with the UML toolset so they can be described and properly designed.

5. Document and communicate the newly modeled object-oriented system to users and other analysts.

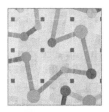

Object-oriented analysis and design can offer an approach that facilitates logical, rapid, and thorough methods for creating new systems that are responsive to a changing business landscape. Object-oriented techniques work well in situations in which complicated information systems are undergoing continuous maintenance, adaptation, and redesign.

In this chapter, we introduce Unified Modeling Language (UML), the industry standard for modeling object-oriented systems. The UML toolset includes diagrams that allow you to visualize the construction of an object-oriented system. Each design iteration takes a successively more detailed look at the design of the system, until the things and relationships in the system are clearly and precisely defined in UML documents. UML is a powerful tool that can greatly improve the quality of your systems analysis and design, and thereby help create higher-quality information systems.

When the object-oriented approach was first introduced, advocates cited reusability of the objects as the main benefit of the approach. It makes intuitive sense that the recycling of program parts should reduce the costs of development in computer-based systems. It has proven to be very effective in the development of GUIs and databases. Although reusability is the main goal, maintaining systems is also very important, and because the object-oriented approach creates objects that contain both data and program code, a change in one object has a minimal impact on other objects.

[*]By Julie E. Kendall, Kenneth E. Kendall, and Allen Schmidt.

Object-Oriented Concepts

Object-oriented programming differs from traditional procedural programming by examining the objects that are part of a system. Each object is a computer representation of some actual thing or event. General descriptions of the key object-oriented concepts of objects, classes, and inheritance are presented in this section, and further details on other UML concepts are introduced later in this chapter.

Objects

Objects are persons, places, or things that are relevant to the system we are analyzing. Object-oriented systems describe entities as objects. Typical objects may be customers, items, orders, and so on. Objects may also be GUI displays or text areas on the display.

Classes

Objects are typically part of a group of similar items called classes. The desire to place items into classes is not new. Describing the world as being made up of animals, vegetables, and minerals is an example of classification. The scientific approach includes classes of animals (such as mammals), and then divides the classes into subclasses (such as egg-laying animals and pouched mammals).

   The idea behind classes is to have a reference point and describe a specific object in terms of its similarities to or differences from members of its own class. In doing so, it is more efficient for someone to say "The koala bear is a marsupial (or pouched animal) with a large round head and furry ears" than it is to describe a koala bear by describing all its characteristics as a mammal. It is more efficient to describe characteristics, appearance, and even behavior in this way. When you hear the word *reusable* in the object-oriented world, it means you can be more efficient because you do not have to start at the beginning to describe an object every time it is needed for software development.

   Objects are represented by and grouped into classes that are optimal for reuse and maintainability. A class defines the set of shared attributes and behaviors found in each object in the class. For example, records for students in a course section have similar information stored for each student. The students could be said to make up a class (no pun intended). The values may be different for each student, but the type of information is the same. Programmers must define the various classes in the program they are writing. When the program runs, objects can be created from the established class. The term *instantiate* is used when an object is created from a class. For example, a program could instantiate a student named Peter Wellington as an object from the class labeled as **Student**.

   What makes object-oriented programming, and thus object-oriented analysis and design, different from classical programming is the technique of putting all of an object's attributes and methods within one self-contained structure, the class itself. This is a familiar occurrence in the physical world. For example, a packaged cake mix is analogous to a class since it has the ingredients as well as instructions on how to mix and bake the cake. A wool sweater is similar to a class because it has a label with care instructions sewn into it that caution you to wash it by hand and lay it flat to dry.

   Each class should have a name that differentiates it from all other classes. Class names are usually nouns or short phrases and begin with an uppercase letter. In Figure 10.1 the class is called **RentalCar.** In UML, a class is drawn as a rectangle. The rectangle contains two other important features: a list of attributes and a series of methods. These items describe a class, the unit of analysis that is a large part of what we call object-oriented analysis and design.

FIGURE 10.1

An example of a UML class. A class is depicted as a rectangle consisting of the class name, attributes, and methods.

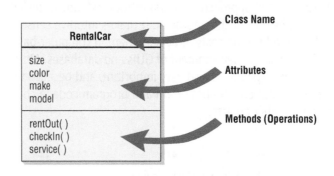

An attribute describes some property that is possessed by all objects of the class. Notice that the **RentalCar** class possesses the attributes size, color, make, and model. All cars possess these attributes, but each car will have different values for its attributes. For example, a car can be blue, white, or some other color. Later on we will demonstrate that you can be more specific about the range of values for these properties. When specifying attributes, the first letter is usually lowercase.

A method is an action that can be requested from any object of the class. Methods are the processes that a class knows to carry out. Methods are also called operations. For the class **RentalCar, rentOut(), checkIn(),** and **service()** are examples of methods. When specifying methods, the first letter is usually lowercase.

Inheritance

A key concept of object-oriented systems is inheritance. Classes can have children; that is, one class can be created out of another class. In UML, the original—or parent—class is known as a base class. The child class is called a derived class. A derived class can be created in such a way that it will inherit all the attributes and behaviors of the base class. A derived class, however, may have additional attributes and behaviors. For example, there might be a **Vehicle** class for a car rental company that contains attributes such as **size, color,** and **make.**

Inheritance reduces programming labor by using common objects easily. The programmer only needs to declare that the **Car** class inherits from the **Vehicle** class and then provide any additional details about new attributes or behaviors that are unique to a car. All the attributes and behaviors of the **Vehicle** class are automatically and implicitly part of the **Car** class and require no additional programming. An analyst can therefore define once but use many times, and this is similar to data that is in the third normal form, defined only once in one database table (as discussed in Chapter 13).

The derived classes shown in Figure 10.2 are **Car** or **Truck**. Here the attributes are preceded by minus signs and methods are preceded by plus signs. We will discuss this in more detail later in the chapter, but for now note that the minus signs mean that these attributes are private (not shared with other classes) and these methods are public (may be invoked by other classes).

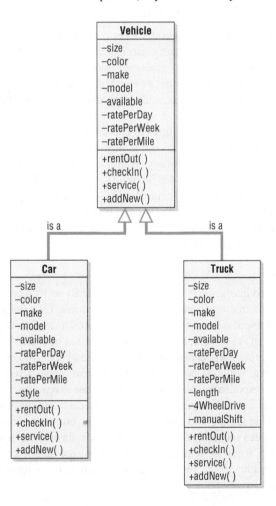

FIGURE 10.2

A class diagram showing inheritance. Car and Truck are specific examples of vehicles and inherit the characteristics of the more general class **Vehicle.**

CONSULTING OPPORTUNITY 10.1

Around the World in 80 Objects

Because you have described the advantages of using object-oriented (O-O) approaches, Jules and Vern, two top executives at World's Trend, would like you to analyze their business using this approach. You can find a summary of World's Trend business activities in Figure 7.15. Notice also the series of data flow diagrams in that chapter to help you conceptualize the problem and begin making the transition to Object Think.

Because you are such good friends with Jules and Vern and because you wouldn't mind getting a little practical experience using O-O thinking, you agree to apply what you know and give them a report. Once you have reread the business activities for World's Trend, provide a timely review by completing the following tasks:

- Use the CRC cards technique to list classes, responsibilities, and collaborators.
- Use the Object Think technique to list "knows" and corresponding attributes for the objects in those classes identified in the previous stage.

Write up both steps and fly over to World's Trend headquarters with your report in hand. Clearly, Jules and Vern are hoping for a fantastic voyage into the new world of object-oriented methods.

Program code reuse has been a part of structured systems development and programming languages (such as COBOL) for many years, and there have been subprograms that encapsulate data. Inheritance, however, is a feature that is found only in object-oriented systems.

CRC Cards and Object Think

Now that we have covered the fundamental concepts of object-oriented systems analysis and design, we need to examine ways to create classes and objects from the business problems and systems we are facing. One way to begin enacting the object-oriented approach is to start thinking and talking in this new way. One handy approach is to develop CRC cards.

CRC stands for class, responsibilities, and collaborators. An analyst can use these concepts when beginning to talk about or model a system from an object-oriented perspective. CRC cards are used to represent the responsibilities of classes and the interaction between the classes. Analysts create the cards based on scenarios that outline system requirements. These scenarios model the behavior of the system under study. If they are to be used in a group, CRC cards can be created manually on small note cards for flexibility, or they can be created using a computer.

We have added two columns to the original CRC card template: the Object Think column and the property column. The Object Think statements are written in plain English, and the property, or attribute, name is entered in its proper place. The purpose of these columns is to clarify thinking and help move toward creating UML diagrams.

Interacting During a CRC Session

CRC cards can be created interactively with a handful of analysts who can work together to identify the class in the problem domain presented by the business. One suggestion is to find all the nouns and verbs in a problem statement that has been created to capture the problem. Nouns usually indicate the classes in the system, and responsibilities can be found by identifying the verbs.

With your analyst group, brainstorm to identify all the classes you can. Follow the standard format for brainstorming, which is not to criticize any participant's response at this point but rather to elicit as many responses as possible. When all classes have been identified, the analysts can compile them, weed out the illogical ones, and write each one on its own card. Assign one class to each person in the group, who will "own" it for the duration of the CRC session.

Next, the group creates scenarios that are actually walk-throughs of system functions by taking desired functionality from the requirements document previously created. Typical systems methods should be considered first, with exceptions such as error recovery taken up after the routine ones have been covered.

As the group decides which class is responsible for a particular function, the analyst who owns the class for the session picks up that card and declares, "I need to fulfill my responsibility." When a card is held in the air, it is considered an object and can do things. The group then proceeds to refine the responsibility into smaller and smaller tasks, if possible. These tasks can be fulfilled by the object if it is appropriate, or the group can decide that it can be fulfilled by interacting with other things. If there are no other appropriate classes in existence, the group may need to create one.

The four CRC cards depicted in Figure 10.3 show four classes for course offerings. Notice that in a class called **Course,** the systems analyst is referred to four collaborators: the department, the textbook, the course assignment, and the course exam. These collaborators are then described as classes of their own on the other CRC cards.

The responsibilities listed will eventually evolve into what are called methods in UML. The Object Think statements seem elementary, but they are conversational so as to encourage a group of analysts during a CRC session to describe as many of these statements as possible. As

FIGURE 10.3

Four CRC cards for course offerings show how analysts fill in the details for classes, responsibilities, and collaborators, as well as for Object Think statements and property names.

Class Name: Department
Superclasses:
Subclasses:

Responsibilities	Collaborators	Object Think	Property
Add a new department	Course	I know my name	Department Name
Provide department information		I know my department chair	Chair Name

Class Name: Course
Superclasses:
Subclasses:

Responsibilities	Collaborators	Object Think	Property
Add a new course	Department	I know my course number	Course Number
Change course information	Textbook	I know my description	Course Description
Display course information	Assignment	I know my number of credits	Credits
	Exam		

Class Name: Textbook
Superclasses:
Subclasses:

Responsibilities	Collaborators	Object Think	Property
Add a new textbook	Course	I know my ISBN	ISBN
Change textbook information		I know my author	Author
Find textbook information		I know my title	Title
Remove obsolete textbooks		I know my edition	Edition
		I know my publisher	Publisher
		I know if I am required	Required

Class Name: Assignment
Superclasses:
Subclasses:

Responsibilities	Collaborators	Object Think	Property
Add a new assignment	Course	I know my assignment number	Task Number
Change an assignment		I know my description	Task Description
View an assignment		I know how many points I am worth	Points
		I know when I am due	Due Date

shown in the example, all dialog during a CRC session is carried out in the first person, so that even the **textbook** speaks: "I know my ISBN." "I know my author." These statements can then be used to describe attributes in UML. These attributes can be called by their variable names, such as **edition** and **publisher.**

Unified Modeling Language (UML) Concepts and Diagrams

The UML approach is well worth investigating and understanding, due to its wide acceptance and usage. UML provides a standardized set of tools to document the analysis and design of a software system. The UML toolset includes diagrams that allow people to visualize the construction of an object-oriented system, similar to the way a set of blueprints allows people to visualize the construction of a building. Whether you are working independently or with a large systems development team, the documentation that you create with UML provides an effective means of communication between the development team and the business team on a project.

UML consists of things, relationships, and diagrams, as illustrated in Figure 10.4. The first components, or primary elements, of UML are called things. You may prefer another word, such as object, but in UML they are called things. Structural things are most common. Structural things are classes, interfaces, use cases, and many other elements that provide a way to create models. Structural things allow the user to describe relationships. Behavioral things describe how things work. Examples of behavioral things are interactions and state machines. Group things are used to define boundaries. An example of a group thing is a package. Finally, annotational things are used to add notes to diagrams.

FIGURE 10.4

An overall view of UML and its components: things, relationships, and diagrams.

UML Category	UML Elements	Specific UML Details
Things	Structural Things	Classes Interfaces Collaborations Use Cases Active Classes Components Nodes
	Behavioral Things	Interactions State Machines
	Grouping Things	Packages
	Annotational Things	Notes
Relationships	Structural Relationships	Dependencies Aggregations Associations Generalizations
	Behavioral Relationships	Communicates Includes Extends Generalizes
Diagrams	Structural Diagrams	Class Diagrams Component Diagrams Deployment Diagrams
	Behavioral Diagrams	Use Case Diagrams Sequence Diagrams Communication Diagrams Statechart Diagrams Activity Diagrams

Relationships are the glue that holds the things together. It is useful to think of relationships in two ways. Structural relationships are used to tie things together in structural diagrams. Structural relationships include dependencies, aggregations, associations, and generalizations. Structural relationships show inheritance, for example. Behavioral relationships are used in the behavioral diagrams. The four basic types of behavioral relationships are communicates, includes, extends, and generalizes.

There are two main types of diagrams in UML: structural diagrams and behavioral diagrams. Structural diagrams are used, for example, to describe the relationships between classes. They include class diagrams, object diagrams, component diagrams, and deployment diagrams. Behavioral diagrams, on the other hand, can be used to describe the interaction between people (called actors in UML) and the thing we refer to as a use case, or how the actors use the system. Behavioral diagrams include use case diagrams, sequence diagrams, communication diagrams, statechart diagrams, and activity diagrams.

In the remainder of this chapter, we first discuss use case modeling, the basis for all UML techniques. Next, we look at how a use case is used to derive activities, sequences, and classes—the most commonly used UML diagrams. Because entire books are dedicated to the syntax and usage of UML (the actual UML specification document is over 800 pages long), we provide only a brief summary of the most valuable and commonly used aspects of UML.

The six most commonly used UML diagrams are:

1. A use case diagram, describing how a system is used. Analysts start with a use case diagram.
2. A use case scenario (although technically it is not a diagram). This scenario is a verbal articulation of exceptions to the main behavior described by the primary use case.
3. An activity diagram, illustrating the overall flow of activities. Each use case may create one activity diagram.
4. Sequence diagrams, showing the sequence of activities and class relationships. Each use case may create one or more sequence diagrams. An alternative to a sequence diagram is a communication diagram, which contains the same information but emphasizes communication instead of timing.
5. Class diagrams, showing the classes and relationships. Sequence diagrams are used (along with CRC cards) to determine classes. An offshoot of a class diagram is a gen/spec diagram (which stands for generalization/specialization).
6. Statechart diagrams, showing the state transitions. Each class may create a statechart diagram, which is useful for determining class methods.

How these diagrams relate to one another is illustrated in Figure 10.5. We will discuss each of these diagrams in the following sections.

Use Case Modeling

UML is fundamentally based on an object-oriented analysis technique known as use case modeling, which was introduced in Chapter 2. A use case model shows a view of the system from the user perspective, thus describing *what* a system does without describing *how* the system does it. UML can be used to analyze the use case model and to derive system objects and their interactions with each other and with the users of the system. Using UML techniques, you further analyze the objects and their interactions to derive object behavior, attributes, and relationships.

A use case provides developers with a view of what the users want. It is free of technical or implementation details. We can think of a use case as a sequence of transactions in a system. The use case model is based on the interactions and relationships of individual use cases.

A use case always describes three things: an actor that initiates an event, the event that triggers a use case, and the use case that performs the actions triggered by the event. In a use case, an actor using the system initiates an event that begins a related series of interactions in the system. Use cases are used to document a single transaction or event. An event is an input to the system that happens at a specific time and place and causes the system to do something. For more information about use case symbols and how to draw use case diagrams, see Chapter 2.

Figure 10.6 is a use case example of student enrollment at a university. Notice that only the most important functions are represented. The **Add Student** use case does not indicate how to add students,

FIGURE 10.5

An overall view of UML diagrams, showing how each diagram leads to the development of other UML diagrams.

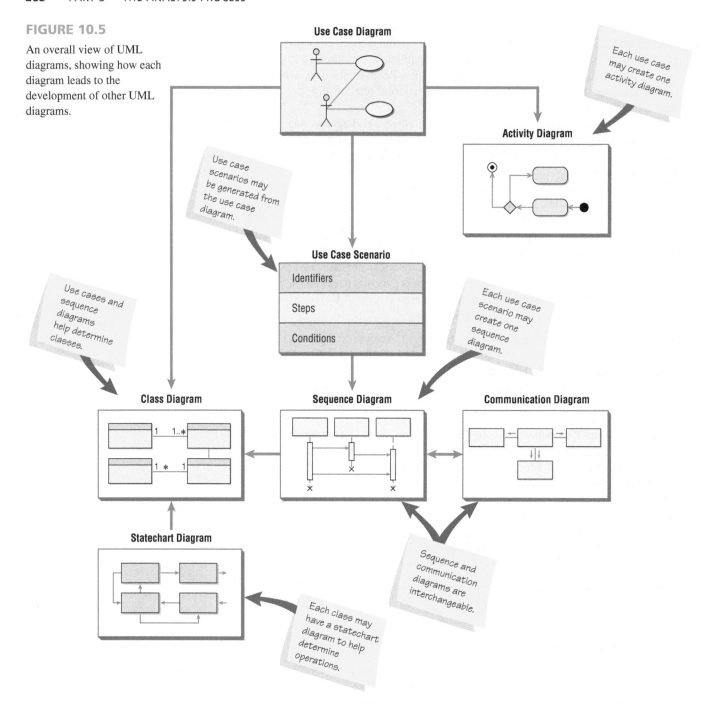

or the method of implementation. Students could be added in person, using the Web, using a touch-tone telephone, or any combination of these methods. The **Add Student** use case includes the **Verify Identity** use case to verify the identity of the student. The **Purchase Textbook** use case extends the **Enroll in Class** use case, and may be part of a system to enroll students in an online course.

It may seem as if the **Change Student Information** use case is a minor system feature and should not be included on the use case diagram, but because this information changes frequently, administration has a keen interest in allowing students to change their own personal information. The fact that the administrators deem this to be important not only justifies, but calls for, the use case to be written up.

Students would not be allowed to change grade point average, outstanding fees, and other information. This use case also includes the **Verify Identity** use case, and in this situation, it means having the student enter a user ID and password before gaining access to the system. **View Student Information** allows students to view their personal information, as well as courses and grades.

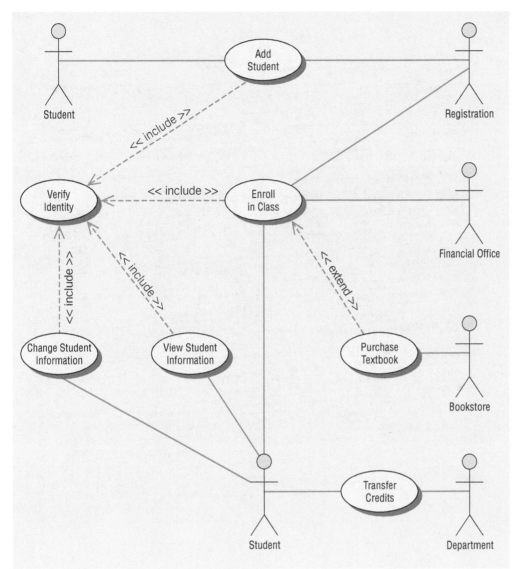

FIGURE 10.6

A use case example of student enrollment.

A use case scenario example is shown in Figure 10.7. Some of the areas included are optional and may not be used by all organizations. The three main areas are:

1. A header area containing case identifiers and initiators
2. Steps performed
3. A footer area that contains preconditions, assumptions, questions, and other information

In the first area, the use case is identified by its name, **Change Student Information;** the actor is identified as a **Student;** and the use case and triggering event are described. The second area contains a series of steps that are performed as long as no errors are encountered. Finally, in the third area, all the pre- and postconditions and assumptions are identified. Some of these are obvious, such as the precondition that the student is on the correct web page and the assumption that the student has a valid student ID and password. Others are not so obvious, such as the outstanding issue regarding how many times the student is allowed to log on to the system.

Use case diagrams provide the basis for creating other types of diagrams, such as class diagrams and activity diagrams. Use case scenarios are helpful in drawing sequence diagrams. Both use case diagrams and use case scenarios are powerful tools to help us understand how a system works in general.

FIGURE 10.7

A use case scenario is divided into three sections: identification and initiation; steps performed; and conditions, assumptions, and questions.

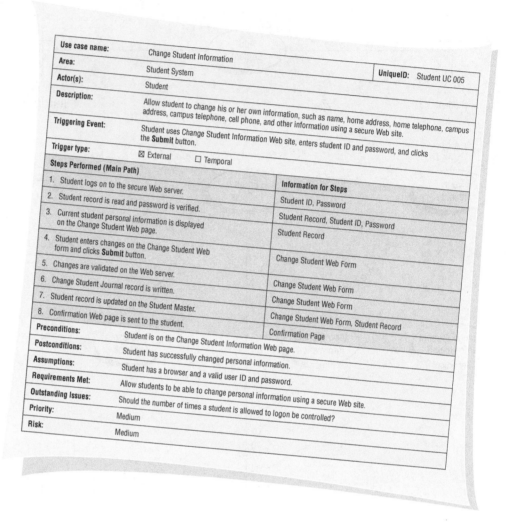

Use case name:	Change Student Information		
Area:	Student System		
Actor(s):	Student	**UniqueID:**	Student UC 005
Description:	Allow student to change his or her own information, such as name, home address, home telephone, campus address, campus telephone, cell phone, and other information using a secure Web site.		
Triggering Event:	Student uses Change Student Information Web site, enters student ID and password, and clicks the **Submit** button.		
Trigger type:	☒ External ☐ Temporal		

Steps Performed (Main Path)		Information for Steps
1. Student logs on to the secure Web server.		
2. Student record is read and password is verified.		Student ID, Password
3. Current student personal information is displayed on the Change Student Web page.		Student Record, Student ID, Password
4. Student enters changes on the Change Student Web form and clicks **Submit** button.		Student Record
5. Changes are validated on the Web server.		Change Student Web Form
6. Change Student Journal record is written.		Change Student Web Form
7. Student record is updated on the Student Master.		Change Student Web Form
8. Confirmation Web page is sent to the student.		Change Student Web Form, Student Record
		Confirmation Page

Preconditions:	Student is on the Change Student Information Web page.
Postconditions:	Student has successfully changed personal information.
Assumptions:	Student has a browser and a valid user ID and password.
Requirements Met:	Allow students to be able to change personal information using a secure Web site.
Outstanding Issues:	Should the number of times a student is allowed to logon be controlled?
Priority:	Medium
Risk:	Medium

Activity Diagrams

An activity diagram shows the sequence of activities in a process, including sequential and parallel activities, and decisions that are made. An activity diagram is usually created for one use case and may show the different possible scenarios.

The symbols on an activity diagram are illustrated in Figure 10.8. A rectangle with rounded ends represents an activity, either a manual one, such as signing a legal document, or an automated one, such as a method or program.

An arrow represents an event. Events represent things that happen at a certain time and place.

A diamond represents either a decision (also called a branch) or a merge. Decisions have one arrow going into the diamond and several going out. A guard condition, showing the condition values, may be included. Merges show several events combining to form one event.

A long, flat rectangle represents a synchronization bar. These are used to show parallel activities and may have one event going into the synchronization bar and several events going out of it, called a fork. A synchronization in which several events merge into one event is called a join.

There are two symbols that show the start and end of the diagram. The initial state is shown as a filled-in circle. The final state is shown as a black circle surrounded by a white circle.

Rectangles surrounding other symbols, called swimlanes, indicate partitioning and are used to show which activities are done on which platform, such as a browser, server, or mainframe computer, or to show activities done by different user groups. Swimlanes are zones that can depict logic as well as the responsibility of a class.

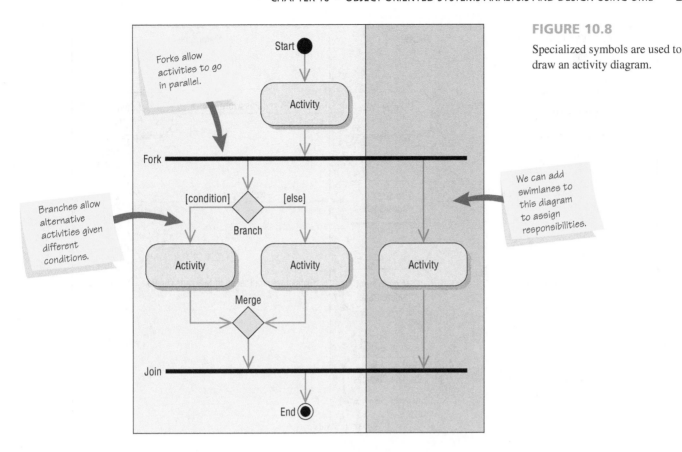

FIGURE 10.8

Specialized symbols are used to draw an activity diagram.

You can see an example of swimlanes in Figure 10.9, which illustrates an activity diagram for the **Change Student Information** use case. It starts with the student logging onto the system by filling out a Web form and clicking the **Submit** button. The form is transmitted to the web server, which then passes the data to the mainframe computer. The mainframe accesses the STUDENT database and passes either a "Not Found" message or selected student data to the web server.

The diamond below the **Get Student Record** state indicates this decision. If the student record has not been found, the web server displays an error message on the web page. If the student record has been found, the web server formats a new web page containing the current student data in a Web form. The student may cancel the change from either the **Logon System** or the **Enter Changes** states, and the activity halts.

If the student enters changes on the Web form and clicks the **Submit** button, the change data is transmitted to the server and a program starts running that validates the changes. If there are errors, an error message is sent to the web page. If the data are valid, the student record is updated and a Change Student Journal Record is written. After a valid update, a confirmation web page is sent to the browser, and the activity terminates.

Creating Activity Diagrams

Activity diagrams are created by asking what happens first, what happens second, and so on. You must determine whether activities are done in sequence or in parallel. If physical data flow diagrams (as described in Chapter 7) have been created, they may be examined to determine the sequence of activities. Look for places where decisions are made, and ask what happens for each of the decision outcomes. Activity diagrams may be created by examining all the scenarios for a use case.

Each path through the various decisions included on the use case is a different scenario. In the main path would be **Logon System, Receive Web Form, Get Student Record, Display Current Student Data, Enter Changes, Validate Changes, Update Student Record, Create Change Student Journal Record,** and **Display Confirmation.**

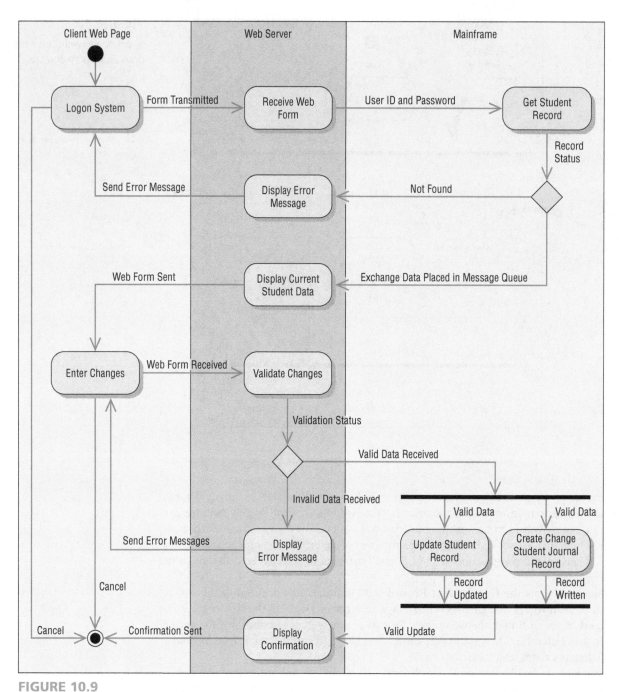

FIGURE 10.9

This activity diagram shows three swimlanes: Client Web Page, Web Server, and Mainframe.

This isn't the only scenario that comes from this use case. Other scenarios may occur. One possibility could be **Logon System, Receive Web Form, Get Student Record,** and **Display Error Message.** Another scenario could be **Logon System, Receive Web Form, Get Student Record, Display Current Student Data, Enter Changes, Validate Changes,** and **Display Error Message.**

The swimlanes are useful to show how the data must be transmitted or converted, such as from Web to server or from server to mainframe. For example, the **Change Student Record** activity diagram has three swimlanes.

The swimlane on the left shows activities that occur on the client browser. Web pages must be created for these activities. The middle swimlane shows activities that happen on the server. Events, such as **Form Transmitted,** represent data transmitted from the browser to the server, and there must be programs on the server to receive and process the client data.

The swimlane on the right represents the mainframe computer. In large organizations it is typical for many Web applications to work with a mainframe computer. Much of the data in

CONSULTING OPPORTUNITY 10.2

Recycling the Programming Environment

"I feel like I'm writing the same code over and over again," says Benito Pérez, a programmer working on a new automated warehouse design. "I have written so many programs lately that deal with robotic-type things that control themselves: automated mailroom trolleys, building surveillance robots, automatic pool cleaners, automatic lawn mowers, monorail trains, and now warehouse trolleys. They are all variations on a theme."

Lisa Bernoulli, the project manager, has heard this sort of complaint for years. She replies, "Oh, come on, Ben. These things aren't really that close. How can you compare a mailroom robot, an automated warehouse, and a monorail train? I'll bet less than 10 percent of the code is the same."

"Look," says Benito. "All three involve machines that have to find a starting point, follow a circuitous route, make stops for loading and unloading, and eventually go to a stopping point. All three have to make decisions at branches in their routes. All three have to avoid colliding with things. I'm tired of redesigning code that is largely familiar to me."

"Hmmm," Lisa muses as she looks over the basic requirements for the warehouse system and remembers the monorail system she and Benito had worked on last year. The requirements regarded a small-lot electronics manufacturing firm that was automating its warehouse and product movement system. The warehouse contains incoming parts, work in progress, and finished goods. The automated warehouse uses a flatbed robot trolley. This robot is a four-wheel electric cart, similar to a golf cart except that it has no seats. Flatbed robot trolleys have a flat 6′ × 4′ cargo surface about 3′ above ground level. These trolleys have a radio communications device that provides a real-time data link to a central warehouse computer. Flatbed trolleys have two sensors: a path sensor that detects a special type of paint and a motion sensor that detects movement. These trolleys follow painted paths around the factory floor. Special paint codes mark forks and branches in the paths, trolley start and stop points, and general location points.

The facility includes three loading dock stations and 10 workstations. Each station has a video terminal or computer connected to the central computer. When products are needed or are ready to be collected from a workstation, the central computer is informed by the worker at the station. The central computer then dispatches trolleys accordingly. Each station has a drop point and a pickup point. Flatbed trolleys move about the factory picking up work at pickup points and dropping off work at drop points. The program that will run the trolleys must interact heavily with the existing job-scheduling program that helps schedule workstation tasks.

How should Lisa go about reusing Benito Pérez's work on the monorail in their current task of creating a trolley object? Explain in two paragraphs.

large organizations exists on mainframe databases, and an enormous number of mainframe programs are in existence.

When an event crosses the swimlane from the server to the mainframe computer, there must be a mechanism for transmitting the event data between the two platforms. Servers use a different format to represent data (ASCII) than do mainframe computers (EBCDIC). Middleware must be present to take care of the conversion. IBM computers often use an mqueue (for message queue). The mqueue receives data from the server programs, places it in a holding area, and calls a mainframe program, usually written in a language called CICS. This program retrieves or updates the data, and sends the results back to the mqueue.

In the activity diagram shown in Figure 10.9, the decision below the **Get Student Record** state is made on the mainframe computer. This means that the message queue receives either a "Not Found" message or the database record for the student. If the mainframe simply placed the **Record Status Received** in the message queue and the decision was evaluated on the server, the server would have to call the mainframe again to obtain the valid data. This would slow down the response to the person waiting at the browser.

Swimlanes also help to divide up the tasks in a team. Web designers would be needed for the web pages displayed on the client browser. Other members would work with programming languages, such as Java, PHP, Ruby on Rails, Perl, or .NET, on the server. Mainframe CICS programmers would write programs that would work with the message queue. An analyst must ensure that the data that the various team members need are available and correctly defined. Sometimes the data in the message queue are contained in an XML document. If an outside organization is involved, the data also might be an XML document.

The activity diagram provides a map of a use case and allows an analyst to experiment with moving portions of the design to different platforms and ask "What if?" for a variety of decisions. The use of unique symbols and swimlanes makes this diagram one that people want to use to communicate with others.

Activity diagrams may be used to construct test plans. Each event must be tested to see whether the activity diagram goes to the next state. Each decision must be tested to see whether the correct path is taken when the decision conditions occur.

Activity diagrams are not used for all use cases. Use an activity diagram when:

1. It helps to understand the activities of a use case.
2. The flow of control is complex.
3. There is a need to model workflow.
4. All scenarios need to be shown.

An analyst would not need an activity diagram when the use case is simple or there is a need to model the change of state.

Activity diagrams may also be used to model a lower-level method, showing detailed logic.

Repository Entries for an Activity Diagram

Each state and event may be further defined using a text description in a repository, which is a collection of text descriptions for the project. We describe states with information about the state, such as the web page name, elements on the web page, and so on. We describe events with the information that is required to communicate with the next state, such as the data from the Web form, the data that is put into a message queue, or with a description of the event that caused the transition, such as a button click.

Sequence and Communication Diagrams

An interaction diagram is either a sequence diagram or a communication diagram, both of which show essentially the same information. These diagrams, along with class diagrams, are used in a use case realization, which is a way to achieve or accomplish a use case.

Sequence Diagrams

A sequence diagram can illustrate a succession of interactions between classes or object instances over time. Sequence diagrams are often used to illustrate the processing described in use case scenarios. In practice, sequence diagrams are derived from use case analysis and are used in systems design to derive the interactions, relationships, and methods of the objects in the system. Sequence diagrams are used to show the overall pattern of the activities or interactions in a use case. Each use case scenario may create one sequence diagram, although sequence diagrams are not always created for minor scenarios.

The symbols used in sequence diagrams are shown in Figure 10.10. Actors and classes or object instances are shown in boxes along the top of the diagram. The leftmost object is the starting object and may be a person (for which a use case actor symbol is used), window, dialog box, or other user interface. Some of the interactions are physical only, such as signing a contract. The top rectangles use indicators in the name to indicate whether the rectangle represents an object, a class, or a class and object.

objectName:	A name with a colon after it represents an object.
:class	A colon with a name after it represents a class.
objectName:class	A name, followed by a colon and another name, represents an object in a class.

A vertical line represents the lifeline for the class or object, which corresponds to the time from when it is created through when it is destroyed. An X on the bottom of the lifeline represents when the object is destroyed. A lateral bar or vertical rectangle on the lifeline shows the focus of control when the object is busy doing things.

Horizontal arrows show messages or signals that are sent between the classes. Messages belong to the receiving class. There are some variations in the message arrows. Solid arrowheads

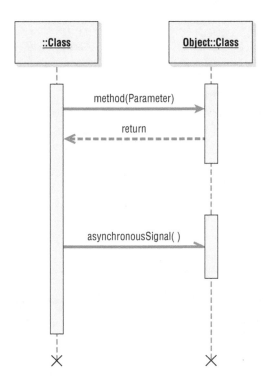

FIGURE 10.10

Specialized symbols used to draw a sequence diagram.

represent synchronous calls, which are the most common. These are used when the sending class waits for a response from the receiving class, and control is returned to the sending class when the class receiving the message finishes executing. Half (or open) arrowheads represent asynchronous calls, or those that are sent without an expectation of returning to the sending class. An example would be using a menu to run a program. A return is shown as an arrow, sometimes with a dashed line. Messages are labeled using one of the following formats:

- The name of the message followed by empty parentheses: **messageName( )**.
- The name of the message followed by parameters in parentheses:
  messageName(parameter1, parameter2 . . .).
- The message name followed by the parameter type, parameter name, and any default value for the parameter in parentheses:
  messageName(parameterType:parameterName(defaultValue). Parameter types indicate the type of data, such as string, number, or date.
- The message may be a stereotype, such as **«Create»,** indicating that a new object is created as a result of the message.

Timing in the sequence diagram is displayed from top to bottom; the first interaction is drawn at the top of the diagram, and the interaction that occurs last is drawn at the bottom of the diagram. The interaction arrows begin at the bar of the actor or object that initiates the interaction, and they end pointing at the bar of the actor or object that receives the interaction request. The starting actor, class, or object is shown on the left. This may be the actor that initiates the activity or it may be a class representing the user interface.

Figure 10.11 is a simplified example of a sequence diagram for a use case that admits a student to a university. On the left is the **newStudentUserInterface** class that is used to obtain student information. The **initialize( )** message is sent to the **Student** class, which creates a new student record and returns the student number. To simplify the diagram, the parameters that are sent to the **Student** class have been omitted, but would include the student name, address, and so on. The next activity is to send a **selectDorm** message to the **Dorm** class. This message would include dorm selection information, such as a health dorm or other student requirements. The **Dorm** class returns the dorm name and room number. The third activity is to send a **selectProgram** message to the **Program** class, including the program name and other course of study information. The program advisor name is returned to the **newStudentUserInterface** class. A **studentComplete** message is sent to the **Student** class with the dorm, advisor name, and other information.

FIGURE 10.11

A sequence diagram for student admission. Sequence diagrams emphasize the time ordering of messages.

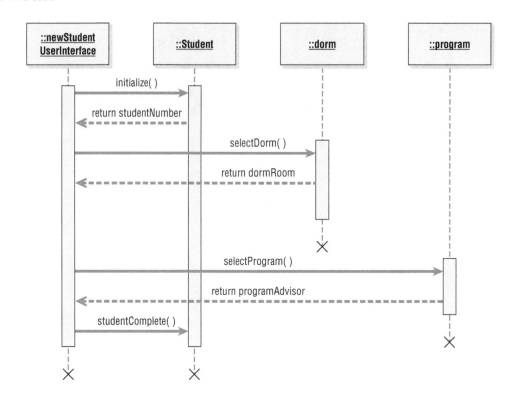

A sequence diagram can be used to translate a use case scenario into a visual tool for systems analysis. The initial sequence diagram used in systems analysis shows the actors and classes in the system and the interactions between them for a specific process. You can use this version of the sequence diagram to verify processes with the business area experts who have assisted you in developing the system requirements. A sequence diagram emphasizes the time ordering (sequence) of messages.

During the systems design phase, sequence diagrams are refined to derive the methods and interactions between classes. Messages from one class are used to identify class relationships. The actors in the earlier sequence diagrams are translated to interfaces, and class interactions are translated to class methods. Class methods used to create instances of other classes and to perform other internal system functions become apparent in the system design using sequence diagrams.

Communication Diagrams

Communication diagrams were introduced in UML 2.0. Their original name in UML 1.x was collaboration diagrams. A communication diagram describes the interactions of two or more things in the system that perform a behavior that is more than any one of the things can do alone. For instance, a car can be broken down into several thousand individual parts. The parts are put together to form the major subsystems of the vehicle: the engine, the transmission, the brake system, and so forth. The individual parts of the car can be thought of as classes, because they have distinct attributes and functions. The individual parts of the engine form a collaboration, because they "communicate" with each other to make the engine run when the driver steps on the accelerator.

A communication diagram is made up of three parts: objects (also called participants), communication links, and messages that can be passed along those links. Communication diagrams show the same information as a sequence diagram but may be more difficult to read. In order to show time ordering, you must indicate a sequence number and describe the message.

A communication diagram emphasizes the organization of objects, whereas a sequence diagram emphasizes the time ordering of messages. A communication diagram will show a path to indicate how one object is linked to another.

Some UML modeling software, such as IBM's Rational Rose, will automatically convert a sequence diagram to a communication diagram or a communication diagram to a sequence diagram

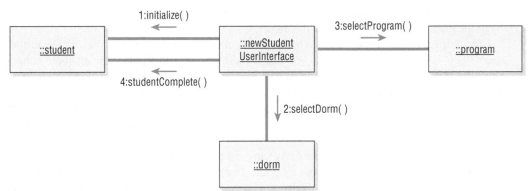

FIGURE 10.12

A communication diagram for student admission. Communication diagrams show the same information that is depicted in a sequence diagram but emphasize the organization of objects rather than the time ordering.

with the click of a button. A communication diagram for the student admission example is illustrated in Figure 10.12. Each rectangle represents an object or a class. Connecting lines show the classes that need to collaborate or work with each other. The messages sent from one class to another are shown along connecting lines. Messages are numbered to show the time sequence. Return values may also be included and numbered to indicate when they are returned within the time sequence.

Class Diagrams

Object-oriented methodologies work to discover classes, attributes, methods, and relationships between classes. Because programming occurs at the class level, defining classes is one of the most important object-oriented analysis tasks. Class diagrams show the static features of the system and do not represent any particular processing. A class diagram also shows the nature of the relationships between classes.

A class is represented by a rectangle on a class diagram. In the simplest format, the rectangle may include only the class name, but it may also include the attributes and methods. Attributes are what the class knows about characteristics of the objects, and methods (also called operations) are what the class knows about how to do things. Methods are small sections of code that work with the attributes.

Figure 10.13 illustrates a class diagram for course offerings. Notice that the name is centered at the top of the class, usually in boldface type. The area directly below the name shows the attributes, and the bottom portion lists the methods. The class diagram shows data storage

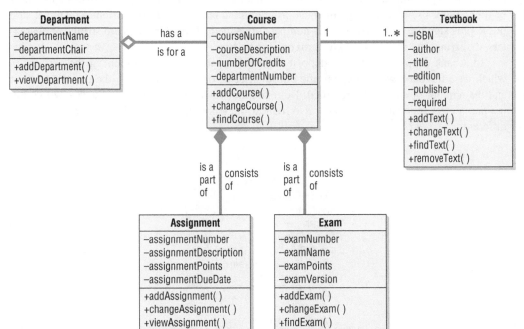

FIGURE 10.13

A class diagram for course offerings. The filled-in diamonds show aggregation, and the empty diamond shows a whole/part relationship.

requirements as well as processing requirements. Later in the chapter we will discuss the meaning of the diamond symbols shown in this figure.

The attributes (or properties) are usually designated as private, or only available in the object. This is represented on a class diagram by a minus sign in front of the attribute name. Attributes may also be protected, indicated with a pound symbol (#). These attributes are hidden from all classes except immediate subclasses. Under rare circumstances, an attribute is public, meaning that it is visible to other objects outside its class. Making attributes private means that the attributes are only available to outside objects through the class methods, a technique called encapsulation, or information hiding.

A class diagram may show just the class name; or the class name and attributes; or the class name, attributes, and methods. Showing only the class name is useful when the diagram is very complex and includes many classes. If the diagram is simpler, attributes and methods may be included. When attributes are included, there are three ways to show the attribute information. The simplest is to include only the attribute name, which takes the least amount of space.

The type of data (such as string, double, integer, or date) may be included on a class diagram. The most complete descriptions would include an equal sign (=) after the type of data, followed by the initial value for the attribute. Figure 10.14 illustrates class attributes.

If the attribute must take on one of a finite number of values, such as a student type with values of F for full-time, P for part-time, and N for nonmatriculating, these may be included in curly brackets separated by commas as shown here — **studentType:char{F,P,N}**.

Information hiding means that objects' methods must be available to other classes, so methods are often public, meaning that they may be invoked from other classes. On a class diagram, public messages (and any public attributes) are shown with a plus sign (+) in front of them. Methods also have parentheses after them, indicating that data may be passed as parameters along with the message. The message parameters, as well as the type of data, may be included on the class diagram.

There are two types of methods: standard and custom. Standard methods are basic things that all classes of objects know how to do, such as create a new object instance. Custom methods are designed for a specific class.

Method Overloading

Method overloading refers to including the same method (or operation) several times in a class. The method signature includes the method name and the parameters included with the method. The same method may be defined more than once in a given class, as long as the parameters sent as part of the message are different; that is, there must be a different message signature. There may be a different number of parameters, or the parameters might be a different type, such as a number in one method and a string in another method. An example of method overloading may be found in the use of a plus sign in many programming languages. If the attributes on either side of the plus sign are numbers, the two numbers are added. If the attributes are strings of characters, the strings are concatenated to form one long string.

In a bank deposit example, a deposit slip could contain just the amount of the deposit, in which case the bank would deposit the entire amount, or it could contain the deposit amount and the amount of cash to be returned. Both situations would use a deposit check method, but the parameters (one situation would also request the amount of cash to be returned) would be different.

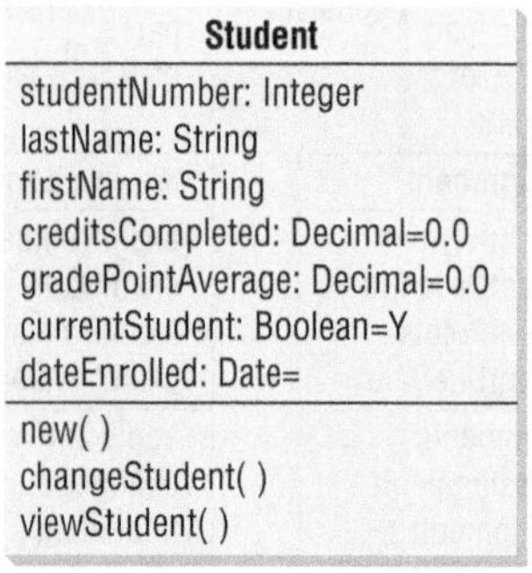

FIGURE 10.14

An extended **Student** class that shows the type of data and, in some cases, its initial value or default value.

Student
studentNumber: Integer
lastName: String
firstName: String
creditsCompleted: Decimal=0.0
gradePointAverage: Decimal=0.0
currentStudent: Boolean=Y
dateEnrolled: Date=
new( )
changeStudent( )
viewStudent( )

Types of Classes

Classes fall into four categories: entity, interface, abstract, and control. These categories are explained below.

ENTITY CLASSES. Entity classes represent real-world items, such as people, things, and so on. Entity classes are the entities represented on an entity-relationship diagram. CASE tools such as Visible Analyst will allow you to create a UML entity class from an entity on an E-R diagram.

An analyst needs to determine which attributes to include in the classes. Each object has many attributes, but a class should include only those that are used by the organization. For example, when creating an entity class for a student at a college, you would need to know attributes that identify the student, such as home and campus address, as well as grade point average, total credits, and so on. If you were keeping track of the same student for an online clothing store, you would have to know basic identifying information, as well as other descriptive attributes such as measurements or color preferences.

BOUNDARY, OR INTERFACE, CLASSES. Boundary, or interface, classes provide a means for users to work with the system. There are two broad categories of interface classes: human and system.

A human interface may be a display, window, Web form, dialog box, menu, list box, or other display control. It may also be a touch-tone telephone, bar code, or other ways for users to interact with the system. Human interfaces should be prototyped (as described in Chapter 6), and often a storyboard is used to model the sequence of interactions.

System interfaces involve sending data to or receiving data from other systems. This may include databases in the organization. If data are sent to an external organization, they are often in the form of XML files or other well-published interfaces with clearly defined messages and protocols. External interfaces are the least stable, because there is often little or no control over an external partner who may alter the format of the message or data.

XML helps to provide standardization because an external partner may add new elements to the XML document, but a corporation transforming the data to a format that may be used to append to an internal database may simply choose to ignore the additional elements without any problems.

The attributes of these classes are those found on the display or report. The methods are those required to work with the display, or to produce the report.

ABSTRACT CLASSES. Abstract classes are classes that cannot be directly instantiated. Abstract classes are those that are linked to concrete classes in a generalization/specialization (gen/spec) relationship. The name of an abstract class is usually denoted in italics.

CONTROL CLASSES. Control, or active, classes are used to control the flow of activities, and they act as a coordinator when implementing classes. To create reusable classes, a class diagram may include many small control classes. Control classes are often derived during system design.

Often a new control class will be created just to make another class reusable. An example would be the logon process. There might be one control class that handles the logon user interface, containing the logic to check the user ID and password. The problem that arises is that the logon control class is designed for a specific logon display. By creating a logon control class that handles just the unique logon display, the data may be passed to a more general validation control class, which performs a check on user IDs and passwords received from many other control classes receiving messages from specific user interfaces. This increases reusability and isolates the logon verification methods from the user interface handling methods.

The rules for creating sequence diagrams are that all interface classes must be connected to a control class. Similarly, all entity classes must be connected to a control class. Interface classes, unlike the other two, are never connected directly to entity classes.

Defining Messages and Methods

Each message may be defined using a notation similar to that described for the data dictionary (as shown in Chapter 8). The definition would include a list of the parameters passed with the message as well as the elements contained in the return message. The methods may have logic defined using structured English, a decision table, or a decision tree, as depicted in Chapter 9.

The analyst can use the techniques of horizontal balancing with any class method. All the data returned from an entity class must be obtained either from the attributes stored in the entity class, from the parameters passed on the message sent to the class, or as a result of a calculation performed by the method of the class. The method logic and parameters must be examined to ensure that the method logic has all the information required to complete its work. Horizontal balancing is further described in Chapter 7.

Enhancing Sequence Diagrams

Once a class diagram is drawn, it may be desirable to go back to the sequence diagram and include special symbols for each of the different types of classes introduced in the last section. Sequence diagrams in particular can be overbearing if an analyst doesn't have a systematic approach to drawing them. The following steps are a useful approach to enhancing a sequence diagram:

1. Include the *actor* from the use case diagram in the enhanced sequence diagram. This will be a stick figure from the use case diagram. There may be an additional actor on the right side of the diagram, such as a credit card company or bank.
2. Define one or more *interface classes* for each actor. Each actor should have his or her own interface class.
3. Create prototype web pages for all human interfaces.
4. Ensure that each use case has one *control class,* although more may be created during the detailed design. Look for that control class and include it in the sequence diagram.
5. Examine the use case to see what *entity classes* are present. Include these on the diagram.
6. Realize that the sequence diagram may be modified again when doing detailed design, such as creating additional web pages or control classes (one for each Web form submitted).
7. To obtain a greater degree of reuse, consider moving methods from a control class to an entity class.

A Class Example for the Web

Classes may be represented using special symbols for entity, boundary (or interface), and control classes. These are called stereotypes, an extension to UML, which are special symbols that may be used during analysis, but are often used when performing object-oriented design. They give an analyst freedom to play with the design to optimize reusability.

The different types of classes are often used when working in the systems design phase. Figure 10.15 is an example of a sequence diagram representing a student viewing his or her personal and course information. In the diagram, **:View Student User Interface** is an example of an interface class; **:Student**, **:Section**, and **:Course** are examples of entity classes; and **:View Student Interface Controller** and **:Calculate Grade Point Average** are control classes.

The student is shown on the left as an actor, and he or she provides a **userLogon** to the **:View Student User Interface** class. This is a Web form that obtains the student's user ID and password. When the student clicks the **Submit** button, the Web form is passed to the **:View Student Interface Controller**. This class is responsible for the coordination of sending messages and receiving returned information from all the other classes.

The **:View Student Interface Controller** sends a **getStudent( )** message to the **:Student** class, which reads a database table and proceeds to return the **studentData**.

The **studentWebPage** is returned to the **:View Student User Interface**. which displays the information in the Web browser. At the bottom of the page is a **nextButton** that the student clicks to view courses. When the user clicks this button, it sends a Web form to the **:View Student Interface Controller**. This form contains the **studentNumber()**, sent along with the **studentWebPage**, and is used to send a message to the **:Section** class to obtain the section grade. If the **studentNumber()** was not automatically sent, it would mean that the student would have to enter his or her **studentNumber()** again, which would not be a satisfactory user interface because it involves redundant keying. Notice that the **:Student** class is not involved and that the focus of control (the vertical bar that is connected to the **:Student** class) ends before the second set of activities (the horizontal arrows pointing to the right) begins.

The **:View Student Interface Controller** class sends a **getSection()** message to the **:Section** class, which returns a **sectionGrade**. The **:Section** class also sends a **calculateGPA()** message to the **:Calculate Grade Point Average** class, which sends a message back to the **:Course** class.

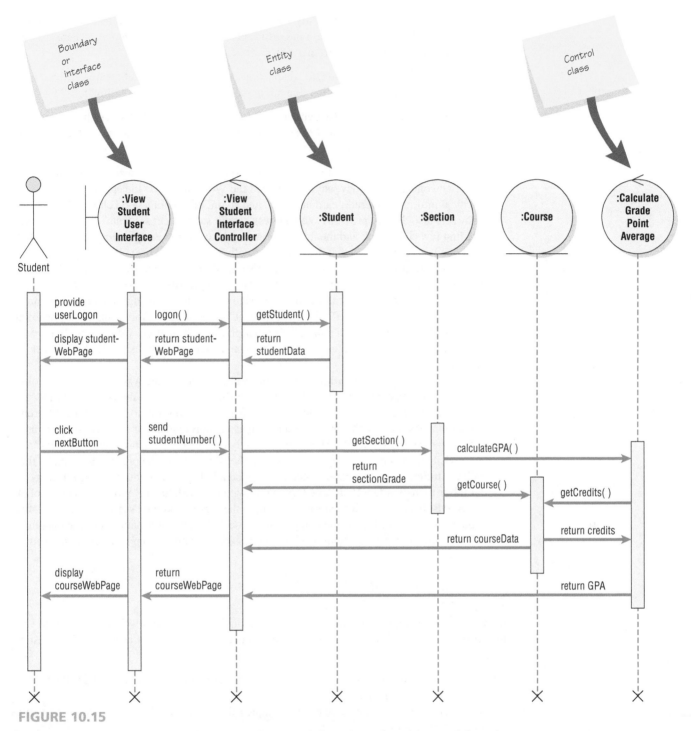

FIGURE 10.15

A sequence diagram for using two web pages: one for student information and one for course information.

The **:Course** class returns the **credits**, which enables the **:Calculate Grade Point Average** class to determine the GPA and return it to the **:View Student Interface Controller**.

The **:View Student Interface Controller** would repeat sending messages to the **:Section** class until all sections for the student have been included. At this time, the **:View Student Interface Controller** would send the **courseWebPage** to the **:View Student User Interface** class, which would display the information in the browser.

Using the user interface, control, and entity classes also allows the analyst to explore and play with the design. The design mentioned previously would display all the student personal information on one page and the course information on a second page. The analyst may modify the design so that the student personal information and the course information appear on one web page. These two possible scenarios would be reviewed with users to determine the best option.

One of the difficulties for the analyst is to determine how to include the **studentNumber** after clicking the **Next** button, because the **:Student** class is no longer available. There are three ways to store and retransmit data from a web page:

1. Include the information in the URL displaying in the address or location area of the browser. In this case, the location line might read something like the following:

http://www.cpu.edu/student/studentinq.html?studentNumber=12345 Everything after the question mark is data that may be used by the class methods. This means of storing data is easy to implement and is often used in search engines.

There are several drawbacks to using this method, and an analyst must use due caution. The first concern is privacy; anyone can read the Web address. If the application involves medical information, credit card numbers, and so on, this is not a good choice. Most browsers also display previous Web address data in subsequent sessions if the user enters the first few characters, and the information may be compromised, leading to identity theft. A second disadvantage is that the data are usually lost after the user closes the browser.

2. Store the information in a cookie, a small file stored on the client (browser) computer. Using cookies is the only way to store data that have persistence, existing beyond the current browser session. This enables the web page to display a message such as "Welcome back, Robin. If you are not Robin, click here." Cookies usually store primary key account numbers, but not credit card numbers or other private information. Cookies are limited to 20 per domain (such as www.cpu.edu), and each cookie must be 4,000 or fewer characters.

   An analyst must work with other business units to determine who needs to use cookies, and there must be some central control over the names used in the cookies. If the organization needs to have more than 20 cookies, a common solution is to create different domain names used by the organization, such as support.cpu.edu or instruction.cpu.edu.

3. Use hidden Web form fields. These fields usually contain data that are sent by the server, are invisible, and do not occupy any space on the web page. In the preceding example, the **:View Student Interface Controller** class added a hidden field containing the **student-Number** to the **studentWebPage** form along with the **nextButton**. When the student clicks the **nextButton**, the **studentNumber** is sent to the server, and the **:View Student Interface Controller** knows which student to obtain course and grade information for. The data in hidden forms is not saved from one browser session to another, so privacy is maintained.

Presentation, Business, and Persistence Layers in Sequence Diagrams

In the previous example, we showed all the classes in the same diagram. When it comes to writing code for systems, it has been useful to look at sequence diagrams as having three distinct layers:

1. The presentation layer, which represents what the user sees. This layer contains the interface or boundary classes.
2. The business layer, which contains the unique rules for this application. This layer contains the control classes.
3. The persistence, or data access, layer, which describes obtaining and storing data. This layer contains the entity classes.

Ideally, program code would be written separately for each of these layers.

With the introduction of Ajax, the lines became blurred. Ajax, an acronym for Asynchronous JavaScript and XML, is a collection of techniques that allows Web applications to retrieve information from the server without altering the display of the current page. This turns out to be an advantage because the entire web page does not need to be reloaded when it gets additional data from the server.

Before Ajax was created, a user visiting a website would answer some questions by entering data on a Web-based form and then wait until a new page loaded. This was necessary because the code to validate, get the data, and then answer the user resided on the server. With the advent of Ajax, the web page is updated rapidly because much of the validation and other control logic is now included in the browser JavaScript code or on the client side. This means that business rules are included in both the boundary and the control class, so it might not be possible to have three distinct layers.

Enhancing Class Diagrams

The class symbols also may be used on class and communication diagrams. Figure 10.16 illustrates a class diagram for a student viewing personal and course information on web pages. Each class has attributes and methods (which are not shown on diagrams that use this notation).

If the class is a user interface type of class, the attributes are the controls (or fields) on the screen or form. The methods would be those that work with the screen, such as submit or reset. They might also be JavaScript for a web page because the code works directly with the web page.

If the class is a control class, the attributes would be those needed to implement the class, such as variables used just in the control class. The methods would be those used to perform calculations, make decisions, and send messages to other classes.

If the class is an entity class, the attributes represent those stored for the entity and the methods working directly with the entity, such as creating a new instance, modifying, deleting, obtaining, or printing.

A website may use a combination of many different classes to accomplish user objectives. For example, a website may use JavaScript to prevalidate data, then pass data to the server control classes, which perform thorough validation, including obtaining data. The server control classes may in turn send JavaScript back to the web page to do some formatting. It is not uncommon to have a Web application involve many classes, some of them containing only one line of code in a method, in order to achieve the goal of reusability.

Relationships

Another way to enhance class diagrams is to show relationships. Relationships are connections between classes, similar to those found on an entity-relationship diagram. These are shown as lines connecting classes on a class diagram. There are two categories of relationships: associations and whole/part relationships.

ASSOCIATIONS. The simplest type of relationship is an association, or a structural connection between classes or objects. Associations are shown as a simple line on a class diagram. The end points of the line are labeled with a symbol indicating multiplicity, which is the same as

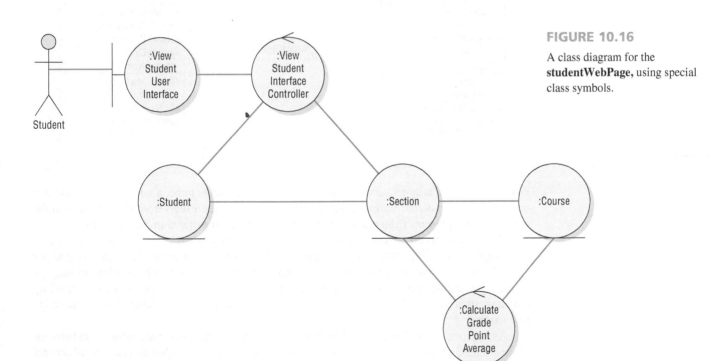

FIGURE 10.16

A class diagram for the **studentWebPage,** using special class symbols.

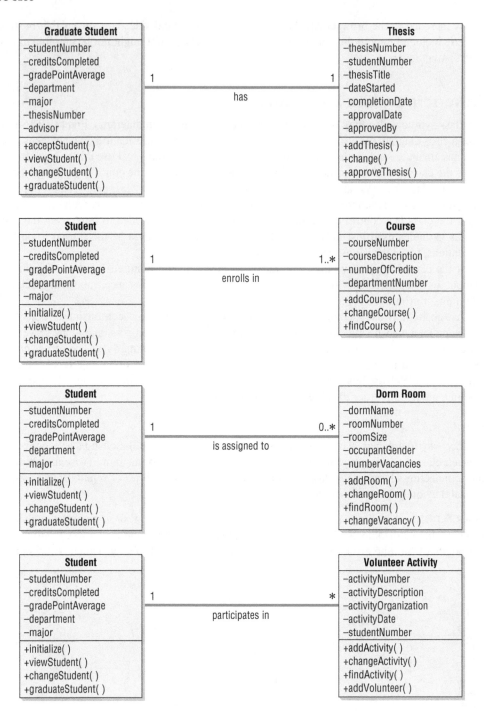

cardinality on an entity-relationship diagram. A zero represents none, a one represents one and only one, and an asterisk represents many. The notation 0..1 represents from zero to one, and the notation 1..* represents from one to many. Associations are illustrated in Figure 10.17.

Class diagrams do not restrict the lower limit for an association. For example, an association might be 5..*, indicating that a minimum of five must be present. The same is true for upper limits. For example, the number of courses a student is currently enrolled in may be 1..10, representing from 1 to 10 courses. It can also include a range of values separated by commas, such as 2, 3, 4. In the UML model, associations are usually labeled with a descriptive name.

Association classes are classes that are used to break up a many-to-many association between classes. These are similar to associative entities on an entity-relationship diagram. **Student** and **Course** have a many-to-many relationship, which is resolved by adding an association class

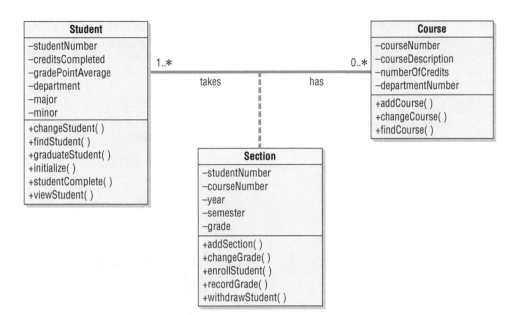

FIGURE 10.18

An example of an associative class in which a particular section defines the relationship between a student and a course.

called **Section** between the classes of **Student** and **Course**. Figure 10.18 illustrates an association class called **Section**, shown with a dotted line connected to the many-to-many relationship line.

An object in a class may have a relationship to other objects in the same class; this is called a reflexive association. An example would be a task having a precedent task or an employee supervising another employee. This is shown as an association line connecting the class to itself, with labels indicating the role names, such as task and precedent task.

WHOLE/PART RELATIONSHIPS. A whole/part relationship occurs when one class represents a whole object and other classes represent parts of that object. The whole acts as a container for the parts. These relationships are shown on a class diagram by a line with a diamond on one end. The diamond is connected to the object that is the whole. Whole/part relationships (as well as aggregation, discussed later) are shown in Figure 10.19.

A whole/part relationship may be an entity object that has distinct parts, such as a computer system that includes the computer, printer, display, and so on, or an automobile that has an engine, a brake system, a transmission, and so on. Whole/part relationships may also be used to describe a user interface, in which one GUI screen contains a series of objects such as lists, boxes, or radio buttons, or perhaps a header, body, and footer area. Whole/part relationships have three categories: aggregation, collection, and composition.

Aggregation An aggregation is often described as a "has a" relationship. Aggregation provides a means of showing that the whole object is composed of the sum of its parts (other objects). In the student enrollment example, the department *has a* course, and the course *is for a* department. This is a weaker relationship because a department may be changed or removed and the course may still exist. A computer package may not be available any longer, but the printers and other components still exist. The diamond at the end of the relationship line is not filled in.

Collection A collection consists of a whole and its members. This may be a voting district with voters or a library with books. The voters or books may change, but the whole retains its identity. This is a weak association.

Composition Composition, a whole/part relationship in which the whole has a responsibility for the part, is a stronger relationship, and is usually shown with a filled-in diamond. Keywords for composition are one class "always contains" another class. If the whole is deleted, all parts are deleted. An example would be an insurance policy with riders. If the policy is canceled, the insurance riders are also canceled. In a database, the referential integrity would be set to delete cascading child records. In a university there is a composition relationship between a course and an assignment as well as between a course and an exam. If the course is deleted, assignments and exams are deleted as well.

FIGURE 10.19

An example of whole/part and
aggregation relationships.

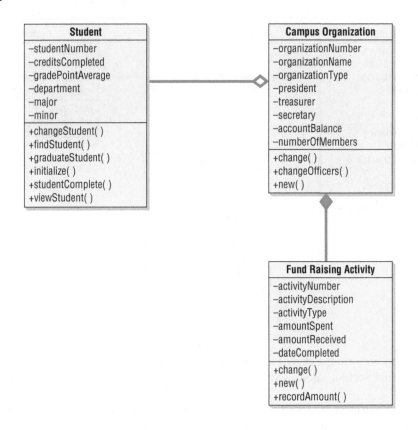

Generalization/Specialization (Gen/Spec) Diagrams

A generalization/specialization (gen/spec) diagram may be considered to be an enhanced class diagram. Sometimes it is necessary to separate the generalizations from the specific instances. As we mentioned at the beginning of this chapter, a koala bear is part of a class of marsupials, which is part of a class of animals. Sometimes we need to distinguish whether a koala bear is an animal or a koala bear is a type of animal. Furthermore, a koala bear can be a stuffed toy animal. We often need to clarify these subtleties.

GENERALIZATION. A generalization describes a relationship between a general kind of thing and a more specific kind of thing. This type of relationship is often described as an "is a" relationship. For example, a car *is a* vehicle and a truck *is a* vehicle. In this case, vehicle is the general thing, whereas car and truck are the more specific things. Generalization relationships are used for modeling class inheritance and specialization. A general class is sometimes called a superclass, base class, or parent class; a specialized class is called a subclass, derived class, or child class.

INHERITANCE. Several classes may have the same attributes and/or methods. When this occurs, a general class is created, containing the common attributes and methods. The specialized class inherits or receives the attributes and methods of the general class. In addition, the specialized class has attributes and methods that are unique and only defined in the specialized class. Creating generalized classes and allowing the specialized class to inherit the attributes and methods helps to foster reuse because the code is used many times. It also helps to maintain existing program code. This allows the analyst to define attributes and methods once but use them many times, in each inherited class.

One of the special features of the object-oriented approach is the creation and maintenance of large class libraries that are available in multiple languages. So, for instance, a programmer using Java, .NET, or C# will have access to a huge number of classes that have already been developed.

POLYMORPHISM. Polymorphism (meaning "many forms"), or method overriding (not the same as method overloading), is the capability of an object-oriented program to have several versions of the same method with the same name within a superclass/subclass relationship. The subclass inherits a parent method but may add to it or modify it. The subclass may change the type of data, or change how the method works. For example, there might be a customer who receives an additional volume discount, and the method for calculating an order total is modified. The subclass method is said to override the superclass method.

When attributes or methods are defined more than once, the most specific one (the lowest in the class hierarchy) is used. The compiled program walks up the chain of classes, looking for methods.

ABSTRACT CLASSES. Abstract classes are general classes and are used when gen/spec is included in the design. The general class becomes the abstract class. The abstract class has no direct objects or class instances, and is only used in conjunction with specialized classes. Abstract classes usually have attributes and may have a few methods.

Figure 10.20 is an example of a gen/spec class diagram. The arrow points to the general class, or superclass. Often the lines connecting two or more subclasses to a superclass are joined using one arrow pointing to the superclass, but these could be shown as separate arrows as well. Notice that the top level is Person, representing any person. The attributes describe qualities that all people at a university have. The methods allow the class to change the name and the address (including telephone and email address). This is an abstract class, with no instances.

Student and Employee are subclasses because they have different attributes and methods. An employee does not have a grade point average and a student does not have a salary. This is a

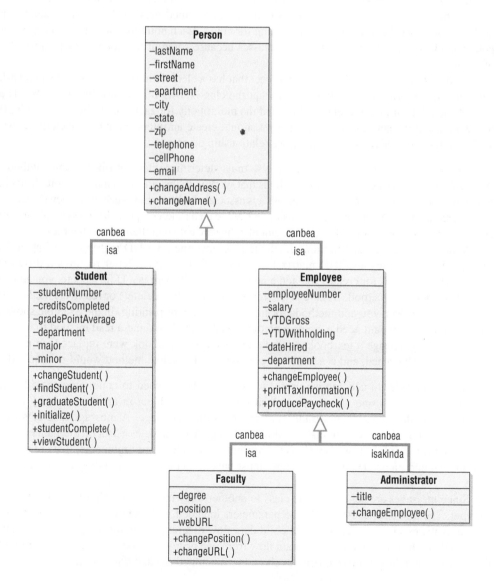

FIGURE 10.20

A gen/spec diagram is a refined form of a class diagram.

simple version and does not include employees who are students and students who work for the university. If these were added, they would be subclasses of the **Employee** and **Student** classes. **Employee** has two subclasses, **Faculty** and **Administrator**, because there are different attributes and methods for each of these specialized classes.

Subclasses have special verbs to define them. These are often run-on words, such as *isa* for "is a," *isakinda* for "is a kind of," and *canbea* for "can be a." There is no distinction between "is a" and "is an"; they both use *isa*.

isa	Faculty *isa* Employee
isakinda	Administrator *isakinda* Employee
canbea	Employee *canbea* Faculty

IDENTIFYING ABSTRACT CLASSES. You may be able to identify abstract classes by looking to see if a number of classes or database tables have the same elements or if a number of classes have the same methods. You can create a general class by pulling out the common attributes and methods, or you might create a specialized class for the unique attributes and methods. Using a banking example, such as a withdrawal, a payment on a loan, or a check written, will all have the same method—subtracting money from the customer balance.

FINDING CLASSES. There are a number of ways to determine classes. They may be discovered during interviewing or JAD sessions (described in Chapter 4), during facilitated team sessions, or from brainstorming sessions. Analyzing documents and memos may also reveal classes. One of the easiest ways is to use the CRC method described previously in this chapter. An analyst should also examine use cases, looking for nouns. Each noun may lead to a candidate, or potential, class. They are called candidate classes because some of the nouns may be attributes of a class.

Each class should exist for a distinct object that has a clear definition. An analyst should ask what the class knows (the attributes) and what the class knows how to do (the methods). The analyst should identify class relationships and the multiplicity for each end of the relationship. If the relationship is many-to-many, the analyst should create an intersection or associative class, similar to the associative entity in an entity-relationship diagram.

DETERMINING CLASS METHODS. An analyst must determine class attributes and methods. Attributes are easy to identify, but the methods that work with the attributes may be more difficult. Some of the methods are standard and are always associated with a class, such as the new() method or the «create» method, which is an extension to UML called a stereotype. (The « » symbols are not simply pairs of greater than and less than symbols; they are called guillemots or chevrons.)

Another useful way to determine methods is to examine a CRUD matrix (see Chapter 7). Figure 10.21 illustrates a CRUD matrix for course offerings. Each letter requires a different method. If there is a C for create, you add a new() method. If there is a U for update, you add an update() or change() method. If there is a D for delete, you add a delete() or remove() method. If there is an R for read, you add methods for finding, viewing, or printing. In the example shown, the **textbook** class would need a create a method to add a textbook and a read method to initiate a course inquiry, change a textbook, or find a textbook. If a textbook were replaced, an update method would be needed, and if a textbook were removed, a delete method would be required.

MESSAGES. In order to accomplish useful work, most classes need to communicate with one another. An object in one class can send information to an object in another class by using a message, similar to a call in a traditional programming language. A message also acts as a command, telling the receiving class to do something. A message consists of the name of the method in the receiving class, as well as the attributes (parameters or arguments) that are passed with the method name. The receiving class must have a method corresponding to the message name.

Since messages are sent from one class to another, they may be thought of as an output or an input. The first class must supply the parameters included with the message, and the second class uses the parameters. If a physical child data flow diagram exists for the problem domain, it may help to discover methods. The data flow from one primitive process to another represents the message, and the primitive processes should be examined as candidate methods.

Activity	Department	Course	Textbook	Assignment	Exam
Add Department	C				
View Department	R				
Add Course	R	C			
Change Course	R	U			
Course Inquiry	R	R	R	R	R
Add Textbook	R	R	C		
Change Textbook		R	RU		
Find Textbook		R	R		
Remove Textbook		R	D		
Add Assignment		R		C	
Change Assignment		R		RU	
View Assignment		R		R	
Add Exam		R			R
Change Exam		R			RU
View Exam		R			R

FIGURE 10.21

A CRUD matrix can be used to help determine what methods are needed. This CRUD matrix is used to determine the methods and operations for course offerings.

Statechart Diagrams

Using a statechart, or state transition, diagram is another way to determine class methods. It is used to examine the different states that an object may have.

A statechart diagram is created for a single class. Typically objects are created, go through changes, and are deleted or removed.

Objects exist in these various states, which are the conditions of an object at a specific time. An object's attribute values define the state that the object is in, and sometimes there is an attribute, such as Order Status (pending, picking, packaged, shipped, received, and so on) that indicates the state. A state has a name in which each word is capitalized. The name should be unique and meaningful to the users. A state also has entry and exit actions, the things the object must do every time it enters or leaves a given state.

An event is something that happens at a specific time and place. Events cause a change of the object state, and it is said that a transition "fires." States separate events, such as an order that is waiting to be filled, and events separate states, such as an Order Received event or an Order Complete event.

An event causes the transition and happens when a guard condition has been met. A guard condition is something that evaluates to either true or false and may be as simple as "Click to confirm order." It also may be a condition that occurs in a method, such as an item that is out of stock. Guard conditions are shown in square brackets next to the event label.

There are also deferred events, or events that are held until an object changes to a state that can accept them. A user keying something in when a word processor is performing a timed backup is an example of a deferred event. After the timed backup has completed, the text appears in the document.

Events fall into three different categories:

1. Signals or asynchronous messages, which occur when the calling program does not wait for a returning message, such as a feature run from a menu.

2. Synchronous messages, which are calls to functions or subroutines. The calling object stops and waits for control to be returned to it, along with an optional message.
3. Temporal events, which occur at a predetermined time. These usually do not involve an actor or any external event.

Material objects have persistence; that is, they exist for a long period of time. Airplane flights, concerts, and sporting events have shorter persistence (they may have states that transition in a shorter time). Some objects, called transient objects, do not survive the end of a session. These include main memory, Web URL (or location) data, web pages, CICS displays, and so on. The only way to save transient objects is to store information about them, such as storing Web data in a cookie.

Each time an object changes state, some of the attributes change their values. Furthermore, each time an object's attributes change, there must be a method to change the attributes. Each of the methods would need a display or Web form to add or change the attributes. These become the interface objects. The display or Web form would often have more controls (or fields) on them than just the attributes that change. They would usually have primary keys, identifying information (such as a name or address), and other attributes that are needed for a good user interface. The exception is a temporal event, which may use database tables or a queue containing the information.

A State Transition Example

Consider a student enrolling at a university and the various states that he or she would go through. Three of the states are listed below in detail:

State:	Potential Student
Event:	Application Submitted
Method:	new( )
Attributes changed:	Number
	Name
	Address
User interface:	Student Application Web Form
State:	Accepted Student
Event:	Requirements Met
Method:	acceptStudent( )
Attributes changed:	Admission Date
	Student Status
	Return Acceptance Letter
User interface:	Accept Student Display
State:	Dorm Assigned Student
Event:	Dorm Selected
Method:	assignDorm( )
Attributes changed:	Dorm Name
	Dorm Room
	Meal Plan
User interface:	Assign Student Dorm Display

The other states are **Program Student, Current Student, Continuing Student,** and **Graduated Student**. Each state would have an event, methods, attributes changed, and a user interface associated with it. This series of states can be used to determine the attributes and methods that make up part of the class.

The states and events that trigger the changes may be represented on a statechart diagram (or a state transition diagram). The statechart diagram for **Student** is illustrated in Figure 10.22. States are represented by rectangles, and events or activities are the arrows that link the states and cause one state to change to another state. Transition events are named in the past tense because they have already occurred to create the transition.

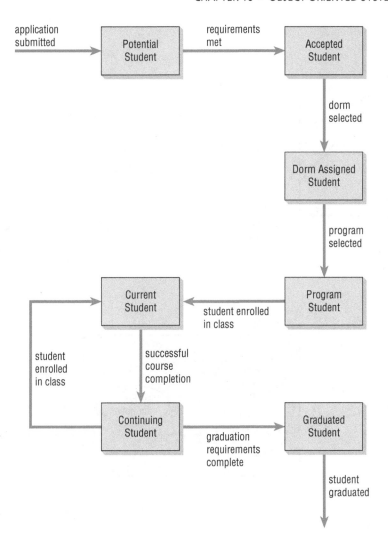

FIGURE 10.22

A statechart diagram that shows how a student progresses from being a potential student to a graduated student.

Statechart diagrams are not created for all classes. They are created when:

1. A class has a complex life cycle.
2. An instance of a class may update its attributes in a number of ways through the life cycle.
3. A class has an operational life cycle.
4. Two classes depend on each other.
5. The object's current behavior depends on what happened previously.

When you examine a statechart diagram, use the opportunity to look for errors and exceptions. Inspect the diagram to see whether events are happening at the wrong time. Also check that all events and states have been represented. Statechart diagrams have only two problems to avoid. Check to see that a state does not have all transitions going into the state or all transitions coming out of the state.

Each state should have at least one transition in and out of it. Some statechart diagrams use the same start and terminator symbols that an activity diagram uses: a filled-in circle to represent the start and concentric circles with the center filled in to signify the end of the diagram.

Packages and Other UML Artifacts

Packages are containers for other UML things, such as use cases or classes. Packages can show system partitioning, indicating which classes or use cases are grouped into a subsystem, called logical packages. They may also be component packages, which contain physical system components, or use case packages, containing a group of use cases. Packages use a folder symbol with the package name either in the folder tab or centered in the folder. Packaging can occur during systems analysis, or later when the system is being designed. Packages may also have relationships, similar to class diagrams, which may include associations and inheritance.

FIGURE 10.23

Use cases can be grouped into packages.

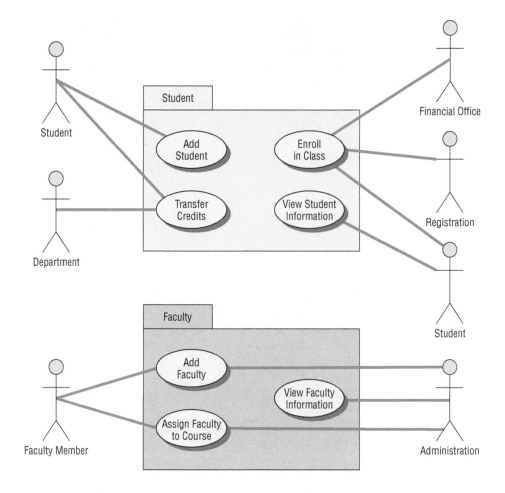

Figure 10.23 is an example of a use case package diagram. It shows that four use cases, **Add Student**, **Enroll in Class**, **Transfer Credits**, and **View Student Information**, are part of the **Student** package. Three use cases, **Add Faculty**, **View Faculty Information**, and **Assign Faculty to Course**, are part of the **Faculty** package.

As you continue constructing diagrams, you will want to make use of component diagrams, deployment diagrams, and annotational things. These permit different perspectives on the work being accomplished.

A component diagram is similar to a class diagram, but it is more of a bird's-eye view of the system architecture. The component diagram shows the components of the system, such as a class file, a package, shared libraries, a database, and so on, and how they are related to each other. The individual components in a component diagram are considered in more detail within other UML diagrams, such as class diagrams and use case diagrams.

A deployment diagram illustrates the physical implementation of the system, including the hardware, the relationships between the hardware, and the system on which it is deployed. The deployment diagram may show servers, workstations, printers, and so on.

Annotational things give developers more information about the system. These consist of notes that can be attached to anything in UML: objects, behaviors, relationships, diagrams, or anything that requires detailed descriptions, assumptions, or any information relevant to the design and functionality of the system. The success of UML relies on the complete and accurate documentation of your system model to provide as much information as possible to the development team. Notes provide a source of common knowledge and understanding about your system to help put your developers on the same page. Notes are shown as a paper symbol with a bent corner and a line connecting them to the area that needs elaboration.

CONSULTING OPPORTUNITY 10.3

Developing a Fine System That Was Long Overdue: Using Object-Oriented Analysis for the Ruminski Public Library System*

As Dewey Dezmal enters the high-ceilinged, wood-paneled reading room of the Ruminski Public Library, a young woman, seated at a long, oak table, pokes her head out from behind a monitor, sees him, and stands, saying, "Welcome. I'm Peri Otticle, the director of the library. I understand you are here to help us develop our new information system."

Still in awe of the beauty of the old library building and the juxtaposition of so much technology amid so much history, Dewey introduces himself as a systems analyst with a small IT consulting firm, People and Objects, Inc.

"It's the first time I've been assigned to this type of project, although it's actually interesting for me, because my degree is from the Information Studies School at Upstate University. You can major in library science or IT there, so lots of my classmates went on to work in public libraries. I opted for the IT degree."

"We should work well together, then," Peri says. "Let's go to my office so we don't disturb any patrons, and I can talk you through a report I wrote."

As they pass the beautiful, winding staircase seemingly sculpted in wood, Peri notices Dewey looking at the surroundings and says, "You may wonder about the grandeur of the building, because we are a public institution. We are fortunate. Our benefactor is Valerian Ruminski. In fact, he has donated so much money to so many libraries that the staff affectionately calls him 'Valerian the Librarian.'"

As they pass several patrons, Peri continues, "As you can see, it's a very busy place. And, regardless of our old surroundings, we don't dwell in the past."

Dewey reads the report Peri has handed him. One large section is titled "Summary of Patrons' Main Requirements," and the bulleted list states:

- A library patron who is registered in the system can borrow books and magazines from the system.
- The library system should periodically check (at least once per week) whether a copy of a book or journal borrowed by a patron has become overdue. If so, a notice will be sent to the patron.
- A patron can reserve a book or journal that has been lent out or is in the process of purchase. The reservation should be canceled when the patron checks out the book or journal or through a formal canceling service.

*Based on a problem written by Dr. Wayne Huang.

As Dewey looks up from the report, he says to Peri, "I'm beginning to understand the patron (or user) requirements. I see lots of similarities between my old university library and yours. One item I didn't see covered, though, was how you decide what the library should collect and what it should get rid of."

Peri chuckles and replies, "That's an insightful question. The library staff handles the purchase of new books and journals for the library. If something is popular, more than two copies are purchased. We can create, update, and delete information about titles and copies of books and journals, patrons, loan of materials, and reservations in the system."

Dewey looks up from his note pad and says, "I'm still a little confused. What's the difference between the terms *title* and *copy*?"

Peri responds, "The library can have several copies of a title. Title normally refers to the name of a book or journal. Copies of a title are actually lent out from the library."

Based on Dewey's interview with Peri and the requirements description in her report, as well as your own experience using library services, use UML to answer the following questions. (*Note:* It is important to make sure your solutions are logical and workable. State your assumptions clearly whenever necessary.)

1. Draw a use case diagram to represent actors and use cases in the system.
2. For each use case, describe the steps (as we did to organize the use cases).
3. Describe scenarios for the steps. In other words, create a patron and write up an example of the patron as he or she goes through each step.
4. Develop a list of things.
5. Create sequence diagrams for use cases based on steps and scenarios.
6. Complete the class diagram by determining relationships between classes and defining the attributes and methods of each class. Use the grouping thing called package to simplify the class diagram.

Putting UML to Work

UML provides a useful toolset for systems analysis and design. As with any other product created with the help of tools, the value of UML deliverables in a project depends on the expertise with which the systems analyst wields the tools. The analyst will initially use the UML toolset to break down the system requirements into a use case model and an object model. The use case model describes the use cases and actors. The object model describes the objects and object associations, and the responsibilities, collaborators, and attributes of the objects.

1. Define the use case model:

   - Find the actors in the problem domain by reviewing the system requirements and interviewing some business experts.
   - Identify the major events initiated by the actors and develop a set of primary use cases at a very high level that describe the events from the perspective of each actor.
   - Develop the use case diagrams to provide understanding of how the actors relate to the use cases that will define the system.
   - Refine the primary use cases to develop a detailed description of system functionality for each primary use case. Provide additional details by developing the use case scenarios that document the alternate flows of the primary use cases.
   - Review the use case scenarios with the business area experts to verify processes and interactions. Make modifications as necessary until the business area experts agree that the use case scenarios are complete and accurate.

2. Continue UML diagramming to model the system during the systems analysis phase:

   - Derive activity diagrams from use case diagrams.
   - Develop sequence and communication diagrams from use case scenarios.
   - Review the sequence diagrams with the business area experts to verify processes and interactions. Make modifications as necessary until the business area experts agree that the sequence diagrams are complete and accurate. This additional review of the graphical sequence diagrams often provides the business area experts an opportunity to rethink and refine processes in more atomic detail than the review of the use case scenarios.

3. Develop the class diagrams:

   - Look for nouns in use cases and list them. They are potential objects. Once you identify the objects, look for similarities and differences in the objects due to the objects' states or behavior, and then create classes.
   - Define the major relationships between the classes. Look for "has a" and "is a" relationships between classes.
   - Examine use case and sequence diagrams in order to determine classes.
   - Beginning with the use cases that are the most important to the system design, create class diagrams that show the classes and relationships that exist in the use cases. One class diagram may represent the classes and relationships described in several related use cases.

4. Draw statechart diagrams:

   - Develop statechart diagrams for certain class diagrams to provide further analysis of the system at this point. Use statechart diagrams to aid in understanding complex processes that cannot be fully derived by the sequence diagrams.
   - Determine methods by examining statechart diagrams. Derive state (data) class attributes from use cases, business area experts, and class methods. Indicate whether the methods and attributes of the class are public (accessible externally) or private (internal to the class). The statechart diagrams are extremely useful in modifying class diagrams.

5. Begin systems design by refining UML diagrams and using them to derive classes and their attributes and methods:

   - Review all existing UML diagrams for the system. Write class specifications for each class that include the class attributes, methods, and their descriptions. Review sequence diagrams to identify other class methods.
   - Develop methods specifications that detail the input and output requirements for the method, along with a detailed description of the internal processing of the method.
   - Create another set of sequence diagrams (if necessary) to reflect the actual class methods and interactions with each other and the system interfaces.
   - Create class diagrams, using the specialized class symbols for boundary or interface class, entity class, and control class.

CONSULTING OPPORTUNITY 10.4

C-Shore++

"They want the core of the customer service representative's user interface to be radically reprogrammed again!" says Bradley Vargo, the information systems development director at C-Shore Mutual Funds, based in Burton, Michigan. "Only eight months ago, we completed a two-year development project of the Customer Service Representative (CSR) system. During that entire project, we endured a parade of moving requirements. Every month, Chelsea and Jonathan from Marketing would invent some competitive new customer service feature, and within a week, the CSR group would be down here with vast changes to the CSR system specifications. I thought we'd never finish that project! Now it looks as if we will have to start a new reprogramming project on a system less than a year old. We had forecast this system for a seven-year life span! Now I think it may be going into eternal reconstruction."

Bradley is talking with Rachael Ciupek, the senior application systems analyst responsible for the CSR system, and Bridget Ciupek, her sister and the programmer who wrote most of the user interface. "Calm down, Bradley," says Rachael. "It is not the fault of the people in Marketing or CSR. The nature of our business has been affected by fast-paced competition. Marketing doesn't invent these changes out of boredom. They are often responding to new, Web-based customer services offered by our competition. We have to stay ahead or at least keep up, or we'll all be looking for new jobs."

"Bradley, Rachael, I think you better know that the situation may be worse than you think," Bridget chips in. "The programmers have actually been making small changes in the CSR user interface for the past eight months anyway. The CSR users have been calling us directly and begging for help. They usually want just a small change to one isolated part of the system, but that has created a high labor drain because we have to recertify the entire system. You know how the effects of a small change can ripple through a large program. We've billed the time to program maintenance on the grounds that we thought we were just fine-tuning the completed system. Although the changes have been gradual, in eight months we've pretty much rewritten about a quarter of the CSR user interface code already. The work has not been falling off. It's still pretty steady."

"So what you're telling me," says Bradley, "is that we have system needs in this area that have been changing constantly while we tried to write specifications, tried to write program code, and tried to make a fixed solution work against a fluid problem. How can we afford to write programs if they will only last a few months?

How can Bradley manage a systems development process that no longer has fixed or constant business processes as part of its goal set? Is there a way for Rachael to manage specification and control maintenance costs when programmers are constantly asked to tinker with isolated parts of a large program? Keep in mind that an important goal is to provide good support for the users' needs and the organization's business strategies.

- Analyze the class diagrams to derive the system components; that is, functionally and logically related classes that will be compiled and deployed together as a .DLL, a .COM object, a Java Bean, a package, and so forth.
- Develop deployment diagrams to indicate how your system components will be deployed in the production environment.

6. Document your system design in detail. This step is critical. The more complete the information you provide the development team through documentation and UML diagrams, the faster the development and the more solid the final production system.

The Importance of Using UML for Modeling

UML is a powerful tool that can greatly improve the quality of your systems analysis and design, and it is hoped that the improved practices will translate into higher-quality systems.

By using UML iteratively in analysis and design, you can achieve a greater understanding between the business team and the IT team regarding the system requirements and the processes that need to occur in the system to meet those requirements.

The first iteration of analysis should be at a very high level, to identify the overall system objectives and validate the requirements through use case analysis. Identifying the actors and defining the initial use case model are part of this first iteration. Subsequent iterations of analysis further refine the system requirements through the development of use case scenarios, class diagrams, sequence diagrams, statechart diagrams, and so on. Each iteration takes a successively

HYPERCASE® EXPERIENCE 10

"I hope you still feel as if you're learning new things about MRE every day. I understand you've been talking to some of the systems people: Melissa, Todd, Roger (and even Lewis, our new intern) about using some different diagramming methods to understand us better. I hope you see us as a family, not just a collection of people, though. We all certainly feel as if we've 'inherited' some great wisdom from Jimmy Hyatt and Warren's father. I'm all for using your new approach, if it helps us improve our project reporting. Of course, Snowden is eager to see your object-oriented work. Can you have something on his desk in a couple of weeks?"

HYPERCASE Questions

1. Create an activity diagram for the Report Project Progress use case. Refer to the use case specifications in Melissa Smith's office for details and a prototype.
2. Create an activity diagram for the Add Client use case. Refer to the use case specifications in Melissa Smith's office for details and a prototype that can be found in Todd Taylor's office.
3. Create a sequence diagram for the main path of the Report Project Progress use case. Refer to the use case specifications in Melissa Smith's office for details and a prototype.
4. Create a sequence diagram for the main path of the Add Client use case. Refer to the use case specifications in Melissa Smith's office for details and a prototype that can be found in Todd Taylor's office.
5. Create a statechart diagram for the Assignment class. Assignments are created for tasks, resources are selected, hours are updated, and assignments are finished.
6. Create a statechart diagram for the Task class. Tasks are created, but not started; planned; sometimes put on hold; currently being worked on; and are completed.

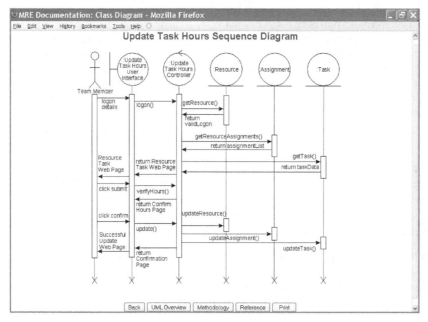

FIGURE 10.HC1

Sequence diagrams can be found in HyperCase.

more detailed look at the design of the system until the things and relationships in the system are clearly and precisely defined in UML documents.

When your analysis and design are complete, you should have an accurate and detailed set of specifications for the classes, scenarios, activities, and sequencing in the system. In general, you can relate the thoroughness of the analysis and design of a system to the amount of time required to develop the system and the resultant quality of the delivered product.

Often overlooked in the development of a new system is that the further a project progresses, the costlier the changes are to the business requirements of a system. Changing the design of a system using a CASE tool, or even on paper, during the analysis and design phases of a project is easier, faster, and much less expensive than doing so during the development phase of the project.

Unfortunately, some employers are shortsighted, believing that only when a programmer or analyst is coding is that employee actually working. Some employers erroneously assume that programmer productivity can be judged solely by the amount of code produced, without recognizing that diagramming ultimately saves time and money that might otherwise be wasted if a project is prototyped without proper planning.

An analogy to building a house is very apt in this situation. You do not want to live in a structure built without planning, one in which rooms and features are randomly added without regard to function or cost. You want a builder to build your agreed-upon design from blueprints containing specifications that have been carefully reviewed by everyone concerned. As a member of an analyst team so accurately observed, "Putting a project on paper before coding will wind up costing less in the long run. It's much cheaper to erase a diagram than it is to change coding."

When business requirements change during the analysis phase, you may have to redraw some UML diagrams. If the business requirements change during the development phase, however, a substantial amount of time and expense may be required to redesign, recode, and retest the system. By confirming your analysis and design on paper (especially through the use of UML diagrams) with users who are business area experts, you help to ensure that correct business requirements will be met when the system is completed.

Summary

Object-oriented systems describe entities as objects. Objects are part of a general concept called classes, the main unit of analysis in object-oriented analysis and design. When the object-oriented approach was first introduced, advocates cited reusability of the objects as the main benefit of the approach. Although reusability is the main goal, maintaining systems is also very important.

Analysts can use CRC cards to begin the process of object modeling in an informal way. Object Think can be added to the CRC cards to assist the analyst in refining responsibilities into smaller and smaller tasks. CRC sessions can be held with a group of analysts to determine classes and responsibilities interactively.

Unified Modeling Language (UML) provides a standardized set of tools to document the analysis and design of a software system. UML is fundamentally based on an object-oriented technique known as use case modeling. A use case model describes *what* a system does without describing *how* the system does it. A use case model partitions system functionality into behaviors (called use cases) that are significant to the users of the system (called actors). Different scenarios are created for each different set of conditions of a use case.

The main components of UML are things, relationships, and diagrams. Diagrams are related to one another. Structural things are most common; they include classes, interfaces, use cases, and many other elements that provide a way to create models. Structural things allow the user to describe relationships. Behavioral things describe how things work. Group things are used to define boundaries. Annotational things permit the analyst to add notes to the diagrams.

Relationships are the glue that holds the things together. Structural relationships are used to tie the things together in structural diagrams. Structural relationships include dependencies, aggregations, associations, and generalizations. Behavioral diagrams use the four basic types of behavioral relationships: communicates, includes, extends, and generalizes.

The toolset of UML is composed of UML diagrams. They include use case diagrams, activity diagrams, sequence diagrams, communication diagrams, class diagrams, and statechart diagrams. In addition to using the diagrams, analysts can describe a use case by writing a use case scenario.

Using UML iteratively in analysis and design allows for greater understanding between a business team and an IT team regarding the system requirements and the processes that need to occur in a system to meet those requirements.

Keywords and Phrases

abstract class	Ajax
activity diagram	annotational thing
actor	association
aggregation	asynchronous message

boundary class   method overriding
branch   object
class   object-oriented
class diagram   package
collaboration   polymorphism
communication diagram   primary use case
control class   relationship
CRC cards   sequence diagram
dependencies   state
deployment diagram   statechart diagram
entity class   swimlane
event   synchronization bar
fork   synchronous message
generalization/specialization (gen/spec)   temporal event
inheritance   Unified Modeling Language (UML)
join   unified process
main path   use case diagram
merge   use case scenario
message   whole/part relationship
method overloading

Review Questions

1. List two reasons for taking an object-oriented approach to systems development.
2. Describe the difference between a class and an object.
3. Explain the concept of inheritance in object-oriented systems.
4. What does CRC stand for?
5. Describe what Object Think adds to the CRC card.
6. What is UML?
7. What are the three major elements of UML?
8. List what the concept of structural things includes.
9. List what the concept of behavioral things includes.
10. What are the two main types of diagrams in UML?
11. List the diagrams included in a structural diagram.
12. List the diagrams included in a behavioral diagram.
13. What does a use case model describe?
14. Would you describe a use case model as a logical or physical model of a system? Defend your answer in a paragraph.
15. Define what an actor is in a use case diagram.
16. What three things must a use case always describe?
17. What does an activity diagram depict?
18. Write a paragraph that describes the use of swimlanes on activity diagrams.
19. What can be depicted on a sequence or communication diagram?
20. Why is defining classes such an important object-oriented analysis task?
21. What can be shown on a class diagram?
22. Define *method overloading*.
23. List the four categories into which classes fall.
24. What are the steps in creating a sequence diagram?
25. What are the two categories of relationships between classes?
26. What are gen/spec diagrams used for?
27. What is another term for polymorphism?
28. What does a statechart diagram depict?
29. What is a package in the UML approach?
30. Why is using UML for modeling important?

Problems

1. Create a series of CRC cards for World's Trend Catalog Division. Once an order is placed, the order fulfillment crew takes over and checks for availability, fills the order, and calculates the total amount of the order. Use five CRC cards, one for each of the following classes: order, order fulfillment, inventory, product, and customer. Complete the section on classes, responsibilities, and collaborators.
2. Finish the CRC cards in Problem 1 by creating Object Think statements and property names for each of the five classes.

3. Draw a use case diagram for World's Trend Catalog Division.

4. Draw four pictures, showing examples of four types of behavioral relationships for Joel Porter's BMW automobile dealership. What type of relationship is involved when a customer must arrange financing? Are there common activities involved when a person either leases or buys an automobile? What type of relationship is between an employee who is a manager or one who is a salesperson?

5. Draw a communication diagram for a student taking a course from a teacher, who is part of the faculty.

6. Coleman County has a phone exchange that handles calls between callers and those receiving the call. Given these three actors, draw a simple sequence diagram for making a simple phone call.

7. You are ready to begin UML modeling for the Kirt Clinic. Draw a class diagram that includes a physician, a patient, an appointment, and a patient's bill. Do not get the insurance company involved.

8. Use UML to draw examples of the four structural relationships for the Kirt Clinic.

9. Write a sample use case scenario for a patient who sees a physician in the Kirt Clinic.

10. Woody's Supermarket, a small chain of grocery stores, is building a website to allow customers to place orders for groceries and other items. The customer places a Web order, the customer master is updated, and an order record is created. The order prints at a local store, and store employees pick the goods from the shelves. The customer is then sent an email notification that his or her order is ready. When the customer picks up the order, frozen goods, chilled products, and other items are assembled. Draw an activity diagram showing the customer using the website to place an order, verification of the order, order confirmation, order details sent to the local store, and a customer email sent to the customer.

11. Sludge's Auto (refer to Chapter 12) is as an auto parts recycling center that is using Ajax on websites for customers to browse for parts. Ajax allows the website to obtain data from the server while the user stays on the original web page. The customer needs to know the make, model, and year of a car as well as the part. If the part is in stock, the description, condition of the part, price, and shipping cost are displayed, with the quantity available for each condition of the part, along with a picture of the part. Draw a sequence diagram using boundary, control, and entity classes for the Auto Part Query for Sludge's Auto.

12. Musixscore.com is an online service providing sheet music to customers. On the "browse music" web page, customers select a genre of music from a drop-down list. The web page uses Ajax to obtain a list of performers, musicians, or groups that match the genre, which is formatted as a drop-down list. When a selection is made from the performer's drop-down list, the web page uses Ajax to display a third drop-down list displaying all the CDs or other works of the performer. When a CD is selected, the web page uses Ajax to obtain all the songs on the CD in a fourth drop-down list. The viewer may make multiple selections. When the **Add to Shopping Cart** image is clicked, the songs are added to the shopping cart. The viewer may change any of the drop-down lists to select additional sheet music, and the process is repeated.

   a. Write a use case description for the Browse Music Score use case, representing this activity.

   b. Draw a sequence diagram using boundary, control, and entity classes for the Musixscore web page.

   c. Write a list of the messages, names, and the parameters, along with the data types, that would be passed to the classes and the values (with data types) that are included with the return message. Make any assumptions you need to make about the data.

   d. Create a class diagram for the entity classes used in the sequence diagram.

Selected Bibliography

Beck, K., and W. Cunningham. "A Laboratory for Teaching Object-Oriented Thinking," OOPSLA '89 http://c2.com/doc/oopsla89/paper.html. Last accessed November 12, 2012.

Bellin, D., and S. Suchman Simone. *The CRC Card Book.* Indianapolis: Addison-Wesley Professional, 1997.

Booch, G., I. Jacobson, and J. Rumbaugh. *The Unified Modeling Language User Guide,* 2nd ed. Indianapolis: Addison-Wesley Professional, 2005.

Cockburn, A. *Writing Effective Use Cases.* Boston: Addison-Wesley Publishing Co., 2001.

Dobing, B., and J. Parsons. "How UML Is Used." *Communications of the ACM,* Vol. 49, No. 5, May 2006, pp. 109–113.

Fowler, M. *UML Distilled: A Brief Guide to the Standard Object Modeling Language,* 3rd ed. Indianapolis: Addison-Wesley Professional, 2003.

Kulak, D., and E. Guiney. *Use Cases: Requirements in Context,* 2nd ed. Indianapolis: Addison-Wesley Professional, 2004.

Miles, R., and K. Hamilton. *Learning UML 2.0.* Indianapolis: O'Reilly Media, Inc., 2006.

Sahraoudi, A. E. K., and T. Blum. "Using Object-Oriented Methods in a System Lifecycle Process Model." *ACM SIGSOFT Software Engineering Notes,* Vol. 28, No. 2, March 2003.

The CPU Case Episode and accompanying student files are available online at www.pearsonglobaleditions.com/kendall.

Designing Effective Output

LEARNING OBJECTIVES

Once you have mastered the material in this chapter you will be able to:

1. Understand the objectives for effective output design.
2. Relate output content to output methods inside and outside the organization.
3. Realize how output bias affects users.
4. Design display output.
5. Design dashboards, widgets, and gadgets.
6. Design websites for ecommerce and corporate uses that include Web 2.0 technologies.
7. Understand the development process for apps used on smartphones and tablets.

Output is information delivered to users through an information system by way of intranets, extranets, or the Web. Some data require extensive processing before they become suitable output; other data are stored, and when they are retrieved, they are considered output, with little or no processing. Output can take many forms, including soft copy such as display screens, smartphones, tablets, microforms, and video and audio output, as well as the traditional hard copy of printed reports. Users rely on output to accomplish their tasks, and they often judge the merit of a system solely by its output. To create the most useful output possible, a systems analyst must work closely with users through an interactive process until the result is considered to be satisfactory.

Output Design Objectives

Because useful output is essential to ensuring the use and acceptance of the information system, there are six objectives that a systems analyst tries to attain when designing output:

1. Designing output to serve the intended purpose
2. Designing output to fit the user
3. Delivering the appropriate quantity of output
4. Making sure the output is where it is needed
5. Providing the output on time
6. Choosing the right output method

Designing Output to Serve the Intended Purpose

All output should have a purpose. During the information requirements determination phase of analysis, a systems analyst finds out what user and organizational purposes exist. Output is then designed based on those purposes.

You will have numerous opportunities to supply output simply because an application permits you to do so. Remember the rule of purposiveness, however: If the output is not functional, it should not be created because there are costs of time and materials associated with all output from a system.

Designing Output to Fit the User

With a large information system serving many users for many different purposes, it is often difficult to personalize output. On the basis of interviews, observations, cost considerations, and perhaps prototypes, it will be possible to design output that addresses what many, if not all, users need and prefer.

Generally speaking, it is most practical to create user-specific or user-customizable output when designing for a decision support system or other highly interactive applications such as those using the Web as a platform. It is still possible, however, to design output to fit a user's tasks and function in the organization, which leads us to the next objective.

Delivering the Appropriate Quantity of Output

Part of the task of designing output is deciding what quantity of output is correct for users. A useful heuristic is that a system must provide what each person needs in order to complete his or her work. This answer is still far from a total solution because it may be appropriate to display a subset of that information at first and then provide a way for the user to access additional information easily.

The problem of information overload is so prevalent that it is a cliché, but it remains a valid concern. No one is served if excess information is given only to flaunt the capabilities of the system. Always keep the decision makers in mind. Often they will not need great amounts of output, especially if there is an easy way to access more via a hyperlink or drill-down capability.

Making Sure the Output Is Where It Is Needed

Output is often produced at one location and then distributed to the user. The increase in online, screen-displayed output that is personally accessible has cut down somewhat on the problem of distribution, but appropriate distribution is still an important objective for a systems analyst. To be used and useful, output must be presented to the right user. No matter how well designed reports are, if the pertinent decision makers do not see them, they have no value.

Providing Output on Time

One of the most common complaints of users is that they do not receive information in time to make necessary decisions. Although timing isn't everything, it does play a large part in how useful output will be. Many reports are required on a daily basis, some only monthly, others annually, and others only by exception. Using well-publicized, Web-based output can alleviate some problems with the timing of output distribution as well. Accurate timing of output can be critical to business operations.

Choosing the Right Output Method

Choosing the right output method for each user is another objective in designing output. Much output now appears on display screens, and users have the option of printing it out with their

FIGURE 11.1

A turnaround document for
Minigasco's data processing.

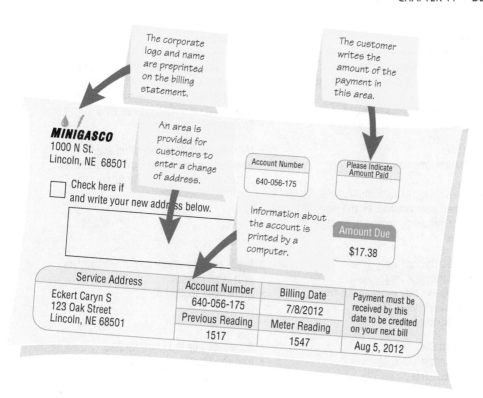

own printer. An analyst needs to recognize the trade-offs involved in choosing an output method. Costs differ; for the user, there are also differences in the accessibility, flexibility, durability, distribution, storage and retrieval possibilities, transportability, and overall impact of the data. The choice of output methods is not trivial, nor is it usually a foregone conclusion.

Relating Output Content to Output Method

The content of output from information systems must be considered as interrelated to the output method. Whenever you design output, you need to think of how function influences form and how the intended purpose will influence the output method that you choose.

Output should be thought of in a general way so that any information generated by the computer system that is useful to people in some way can be considered output. It is possible to conceptualize output as either external (going outside the business), such as information that appears to the public on the Web, or internal (staying within the business), such as material available on an intranet.

External output is familiar to you through utility bills, advertisements, paychecks, annual reports, and myriad other communications that organizations have with their customers, vendors, suppliers, industry, and competitors. Systems analysts sometimes design external output, such as utility bills, to serve double duty as turnaround documents. Figure 11.1 is a gas bill that is a turnaround document for a gas company's data processing. The output for one stage of processing becomes the input for the next. When the customer returns the designated portion of the document, it is optically scanned and used as computer input. With the advent of green IT (also called green computing, or ICT sustainability), many organizations are trying to encourage customers to think about environmental savings, ease of use, and 24-hour access to accounts so that they switch from paper statements to online bills. Electronic transactions would then lessen the environmental impact of paper output, and would further computerize the business to consumer relationship.

External output differs from internal output in its distribution, design, and appearance. Many external documents must include instructions to the recipient if they are to be used correctly. Many external outputs are placed on preprinted forms or websites bearing the company logo and corporate colors.

Output Method	Advantages	Disadvantages
Printer	• Affordable for most organizations • Flexible in types of output, location, and capabilities • Handles large volumes of output • Highly reliable with little down time	• Still requires some operator intervention • Compatibility problems with computer software • May require special, expensive supplies • Depending on model, may be slow • Environmentally unfriendly
Display screen	• Interactive • Online, real-time transmission • Quiet • Takes advantage of computer capabilities for movement within databases and files • Good for frequently accessed, ephemeral messages	• May require cabling and setup space • Requires system for taking "snapshots" of screen and storing them for future use
Audio output and podcasts	• Good for individual user • Good for transient messages • Good where worker needs hands free • Good if output needs to be widely distributed	• Needs earbuds where output will interfere with other tasks • Has limited application
Mobile devices	• Highly portable • Very interactive using gestures • Zoom is possible	• Screen may be too small for text • Icons and buttons may be confusing • May be lost more easily
Electronic output (email, Web sites, blogs, and RSS feeds)	• Reduces paper • Can be updated very easily • Can be "broadcast" • Can be made interactive	• Is not conducive to formatting (email) • Is difficult to convey context of messages (email) • Web sites need diligent maintenance

FIGURE 11.2

A comparison of output methods.

Internal outputs include various reports to decision makers. They range from short summary reports to lengthy, detailed reports. An example of a summary report is a report summarizing monthly sales totals. A detailed report might give weekly sales by salesperson.

Other kinds of internal reports include historical reports and exception reports that are output only at the time an exception occurs. Examples of exception reports are a listing of all employees with no absences for the year, a listing of all salespeople who did *not* meet their monthly sales quota, or a report on consumer complaints made in the past six months.

Output Technologies

Producing different types of output requires different technologies. For printed output, the options include a variety of printers, including standard document printers, specialized label printers, and 3-D printers. For screen output, the options include attached or stand-alone displays as well as touch-sensitive screens found on smartphones and tablets. Audio output can be amplified over a loudspeaker or designed for individual listening on mobile devices. Electronic output includes web pages, email, RSS (Real Simple Syndication) feeds, and blogs. Figure 11.2 is a comparison of output methods.

Factors to Consider When Choosing Output Technology

There are several factors to consider when choosing output technology. Although the technology changes rapidly, certain usage factors remain fairly constant in relation to technological breakthroughs. These factors, some of which present trade-offs, must be considered. They include the following:

1. Who will use (see) the output (requisite quality)?
2. How many people need the output?
3. Where is the output needed (distribution, logistics)?

4. What is the purpose of the output? What user and organizational tasks are supported?
5. What is the speed with which output is needed?
6. How frequently will the output be accessed?
7. How long will (or must) the output be stored?
8. Under what special regulations is the output produced, stored, and distributed?
9. What are the initial and ongoing costs of maintenance and supplies?
10. What are the human and environmental requirements (accessibility, noise absorption, controlled temperature, space for equipment, cabling, and proximity to Wi-Fi transmitters or access points—that is, hot spots) for output technologies?

Increasingly, organizations are taking up green IT initiatives as part of sustainability efforts. These initiatives may limit the quantity of paper reports that are printed or may discourage employees from printing out copies of email messages by adding a green IT notification to the bottom of each corporate email message. Examining each output factor separately will allow you to see the interrelationships and how they may be traded off for one another in a particular system.

WHO WILL USE (SEE) THE OUTPUT? Discovering who will use the output is important because job requirements help dictate what output method is appropriate. For example, when district managers must be away from their desks for extended periods, they need printed output that can travel with them or technology that can access appropriate websites and databases as they visit the managers in their region. Screen output or interactive Web documents are excellent for people such as truck dispatchers who are deskbound for long periods.

External recipients of output (clients and customers, vendors and suppliers, shareholders, and regulatory agencies) and users within the business will require different output. Clients, vendors, and suppliers can be part of several extranets, which are networks of computers built by the organization, providing applications, processing, and information to users on the network.

Examine the website shown in Figure 11.3 for an ecommerce company called Merchants Bay. The Web designer is attuned to the intended users of the wholesale gift site. The ecommerce company's website is powered by a patented negotiating algorithm in which users submit bids (for 1 item or 400) on an array of merchandise. The company's strategy is based on the president's personal experience with flea markets and the observation that people are powerfully attracted to bargaining for a deal.

The website intentionally invokes a cluttered feel, similar to what one gets walking through a flea market. The site is intended for customers who would frequent flea markets in person: They are known to be collectors, gregarious and curious by nature. The website is a profusion

FIGURE 11.3

When designing a website, it is important to choose a metaphor that can be used throughout the site. This example from Merchants Bay (www.merchantsbay.com) employs a nautical theme. (Screenshot of Merchant Bay website. Copyright © 2000 by Merchants Bay LLC. Reprinted with permission of Troy Pappas.)

CONSULTING OPPORTUNITY 11.1

Your Cage or Mine?

"Why can't they get this right? It's driving me to distraction. The zoo in Colombia is writing to me about a tiger that has been on loan from our place since 2008. They should be writing to Tulsa," trumpets Ella Fant, waving a letter in the air. Ella is general curator in charge of the animal breeding program at the nonprofit Gotham Zoo.

She is talking with members of the zoo's five-person committee about the proposals before them. The committee meets every month to decide which animals to loan to other zoos and which animals to get on loan so as to breed them. The committee is composed of Ella Fant, the general curator; Ty Garr, the zoo's director; two zoo employees, Annie Malle and Mona Key; and a layperson, Rex Lyon, who is in business in the community.

Ty paces in front of the group and continues the meeting, saying, "We have the possibility of loaning out two of our golden tamarins, and we have the opportunity to play matchmaker for two lesser pandas. Because three of you are new to the committee, I'll briefly discuss your responsibilities. As you know, Ella and I would pounce on any chance to lure animals in for the breeding program. Your duties are to assess the zoo's financial resources and to look at our zoo's immediate demands. You also must consider the season and our shipping capability, as well as that of the zoos we're considering. The other zoos charge us nothing for the loan of their animals for the breeding program. We pay the shipping for any animal being loaned to us and then maintain them, and that gets expensive."

"There is a special database on the Web that we subscribe to that includes selected species with 164 other zoos," says Ella as she picks up the story from Ty. "I have the password for the database stored on my computer. I can access the records of all captive animals in the system, including those from the two zoos we are negotiating with right now."

As the committee members work, they begin asking questions. "I need to read some information, get some meat to sink my teeth into, before I'm ready to decide whether the loan of the lesser pandas is a good idea. Where are the data on the animals we're considering?" growls Rex.

Annie replies, "We have to go to Ella's office to access the database. Mostly, the other employees who need to know just use her computer."

Mona gets into the swing of the discussion and says, "Some information on the current state of the budget would be divine, too. I'll go bananas with new expenditures until we at least have a summary of what we're spending. I bet it's a bunch."

Ty answers, "We don't mean to monkey around, but frankly we feel trapped. Costs of reproducing all the financial data seem high to us. We'd rather put our money into reproducing rare and endangered species! Paperwork multiplies on its own."

The group laughs nervously together, but there is an air of expectancy in the room. The consensus is that the committee members need more internal information about the zoo's financial status and the prospective loan animals.

Ella, aware that the group cannot be tamed in the way the previous one was, says, "The old committee preferred to get their information informally, through chatting with us. Let's spend this first meeting discovering what kinds of information you think you need to do your work as a committee. Financial data are stored on a laptop that our financial director uses. It's his baby, of course."

What are some of the problems related to output that the committee is experiencing? What suggestions do you have for improving output to the committee? How can the budget constraints of the zoo be met while still allowing the committee to receive the output it needs to function? Comment on the adequacy of the output technology that is currently in use at the zoo. Suggest alternatives or modifications to output and output technology that would enhance what is being done. (*Hint:* Consider ways in which the committee can leverage its use of the Web—to get the output that it needs and that it needs to share.) Analyze both internal and external output requirements.

of colors, includes a variety of sale signs in a mixture of lettering, and even incorporates a video that provides new layers of color and action. Colloquial language is used throughout the site.

Notice that the company's catchphrase is "purveyor of good stuff." The Web designer has carried out a nautical metaphor throughout the site. The user is invited to "search the Bay" for merchandise. In addition, the company's logo includes a wave and a sun on the horizon, and an icon of a ship's steering wheel is placed above a column that invites the user to "navigate" for products, services, and customer service.

To complete a transaction on the site, a customer has an opportunity to accept the "Captain's price" as posted or to submit a bid. If the bid submitted is too low according to the stored negotiation algorithm, a natural language response is returned in a pop-up window stating: "Thanks for your offer, mate. You don't like to part with your money if you don't have to, heh? Yet hey, I like ya, mate. Please try again by offering a better price or by ordering a larger quantity." In this way, the bid is rejected in a friendly, humorous way, and bidders are even given two hints on how to improve the chances that their next bids will be successful. The Web designer clearly had a solid profile of the intended customer in mind when designing the site.

HOW MANY PEOPLE NEED THE OUTPUT? Choice of output technology is also influenced by how many users need the output. If many people need output, Web-based documents with a print option or printed copies are probably justified. Some external customers may want a printed copy of specific documents, such as a stockholder report or a monthly billing statement, but others may prefer Web-based documents with an email notification. If only one user needs the output, a screen or audio may be more suitable.

WHERE IS THE OUTPUT NEEDED (DISTRIBUTION, LOGISTICS)? The choice of output technology is influenced by the physical destination of the output. Information that will remain close to its point of origin, that will be used by only a few users in the business, and that may be stored or referred to frequently can safely be printed or mounted on an intranet. An abundance of information that must be transmitted to users at great distances in branch operations may be better distributed electronically, via the Web or extranets, with the recipient customizing it.

Sometimes federal or state regulations dictate that a printed form remain on file at a particular location for a specified period of time. In those instances, it is the responsibility of the systems analyst to see that the regulation is observed for any output that is designed.

WHAT IS THE PURPOSE OF THE OUTPUT? What user and organizational tasks are supported? Consider the purpose of the output when choosing output technology. If it is intended to be a report created to attract shareholders to the business by allowing them to peruse corporate finances at their leisure, well-designed, printed output such as an annual report is desirable. A variety of media may also be used so that the annual report is available on the Web as well as in printed form. The web page shown in Figure 11.4 provides a series of six brief video clips to document an actual event, the Decision Sciences Institute (DSI) Knowledge Bowl. Video output is useful here because the event was held to commemorate an important anniversary in the organization's history. It serves a historical purpose.

If the purpose of output may be to provide instant updates on stock market quotes, and if the material is highly encoded and changeable, screen crawls, web pages, or audio presentations are preferable. Output must support user tasks, such as performing analysis, or determining ratios, so software tools, including calculators and embedded formulas, could be part of output. Output must also support organizational tasks such as tracking, scheduling, and monitoring.

WHAT IS THE SPEED WITH WHICH OUTPUT IS NEEDED? As we go through the three levels of strategic, middle, and operations management in an organization, we find that decision makers

FIGURE 11.4

Streaming video can be used effectively for telling a story or sharing an event. This web page chronicles an event called the DSI Knowledge Bowl.

at the lowest level of operations management need output rapidly so that they can quickly adjust to events such as a stopped assembly line, raw materials not arriving on time, or a worker being absent unexpectedly. Online screen output may be useful here.

As we ascend the management levels, we observe that strategic managers are more in need of output for a specific time period, which helps in forecasting business cycles and trends.

HOW FREQUENTLY WILL THE OUTPUT BE ACCESSED? The more frequently output is accessed, the more important is the capability to view it on a display connected to local area networks or the Web. Infrequently accessed output that is needed by only a few users is well suited to a CD-ROM archive.

Output that is accessed frequently is a good candidate for incorporation into Web-based or other online systems or networks with displays. Adopting this type of technology allows users easy access and alleviates physical wear and tear that cause frequently handled printed output to deteriorate.

HOW LONG WILL (OR MUST) THE OUTPUT BE STORED? Output printed on paper deteriorates rapidly with age. Output preserved on microforms or digitized in archives is not as prone to succumb to environmental disturbances such as light, humidity, and human handling. However, if hardware to access the archived material becomes hard to acquire or obsolete, this output method can become problematic.

A business may be subject to governmental regulations on local, state, or federal levels that dictate how long output must be kept. As long as the corporation is willing to maintain it and it is nonproprietary, archival information, it can be maintained in Web documents as part of the organization's website. Organizations can enact their own internal policies about how long output must be retained.

UNDER WHAT SPECIAL REGULATIONS IS THE OUTPUT PRODUCED, STORED, AND DISTRIBUTED? The appropriate format for some output is actually regulated by the government. For example, in the United States, the statement of an employee's wages and tax withholding, called a W-2 form, must be printable, even if it exists in an ERP-supported payroll system such as Oracle. Each business in each country exists within a different complex of regulations under which it produces output. To that extent, law may dictate appropriate technology for some functions.

Much of this regulation, however, is industry dependent. For example, in the United States a regional blood system is required by federal law to keep a medical history of a blood donor—as well as his or her name—on file. The exact output form is not specified, but the content is strictly spelled out.

WHAT ARE THE INITIAL AND ONGOING COSTS OF MAINTENANCE AND SUPPLIES? The initial costs of purchasing or leasing equipment must be considered as yet another factor that enters into the choice of output technology. Most vendors will help you estimate the initial purchase or lease costs of computer hardware, including the cost of printers and displays, the cost of access to online service providers (Web access), or the costs of building intranets and extranets. Many vendors, however, do not provide information about how much it costs to keep a printer or other technologies working. Therefore, it falls to the analyst to research the costs of operating different output technologies or of maintaining a corporate website over time.

WHAT ARE THE HUMAN ENVIRONMENTAL REQUIREMENTS FOR OUTPUT TECHNOLOGIES? Analysts need to factor into their output decisions accessibility, noise absorption, controlled temperature, space for equipment, cabling, and proximity to Wi-Fi transmitters or access points, called "hot spots." When humans interact with technologies, specific environments help systems run more effectively and efficiently. Users need accessibility and support in accessing web pages as well as other output.

Printers require a dry, cool environment to operate properly. Displays require space for setup and viewing. Audio and video output require a quiet environment if they are to be heard, and they should be audible only to employees (or customers) who are using them. Thus, the analyst should not specify audio output for a work situation in which many employees or customers are engaged in a variety of tasks unrelated to the output.

In order to set up wireless local area networks so users can access the Web wirelessly, Wi-Fi access points need to be made available. These work when computers are within a few hundred feet of transmitters but can be subjected to interference by other devices.

CONSULTING OPPORTUNITY 11.2

A Right Way, a Wrong Way, and a Subway

"So far so good. Sure, there have been some complaints, but any new subway will have those. The 'free ride' gimmick has helped attract some people who never would have ridden otherwise. I think there are more people than ever before interested in riding the subway," says Bart Rayl. "What we need is an accurate fix on what ridership has been so far so we can make some adjustments on our fare decisions and scheduling of trains."

Rayl is an operations manager for S.W.I.F.T., the newly built subway for Western Ipswich and Fremont Transport that serves a major northeastern city in the United States. He is speaking with Benton Turnstile, who reports to him as operations supervisor of S.W.I.F.T. The subway system is in its first month of operation, offering limited lines. Marketing people have been giving away free rides on the subway to increase public awareness of S.W.I.F.T.

"I think that's a good idea," says Turnstile. "It's not just a token effort. We'll show them we're really on the right track. I'll get back to you with ridership information soon," he says.

A month later, Rayl and Turnstile meet to compare the projected ridership with the new data. Turnstile proudly presents a 2-inch-high stack of computer printouts to Rayl. Rayl looks a little surprised but proceeds to go through it with Turnstile. "What all is in here?" Rayl asks, fingering the top page of the stack hesitantly.

"Well," says Turnstile, training his eyes on the printout, "it's a list of all the tickets that were sold from the computerized vending machines. It tells us how many tickets were bought and what kinds of tickets were bought. The guys from Systems That Think, Inc., told me this report would be the most helpful for us, just like it was for the operations people in Buffalo and Pittsburgh," says Turnstile, turning quickly to the next page.

"Maybe, but remember those subway systems began with really limited service. We're bigger. And what about the sales from the three manned ticket booths in the Main Street Terminal?" asks Rayl.

"The clerks in the booth can get information summarizing ticket sales onscreen any time they want it, but it's not included

here. Remember that we projected that only 10 percent of our sales would be from the booths anyway. Let's go with our original idea and add that to the printout," suggests Turnstile.

Rayl replies, "But I've been observing riders. Half of them seem to be afraid of the automated vending machines. Others start using them, get frustrated reading the directions, or don't know what to do with the ticket that comes out, and they wind up at the ticket booth blowing off steam. Furthermore, they can't understand the routine information posted on the kiosks, which is all in graphics. They wind up asking clerks what train goes where." Rayl pushes the printout holding the ticket sales to one side of the conference table and says, "I don't have much confidence in this report. I feel as if we're sitting here trying to operate the most sophisticated subway system in the United States by peering down a tunnel instead of at the information, like we should be. I think we need to think seriously about capturing journey information on magnetically stripped cards like the New York Transit Authority is doing. Every time you insert the card to take a ride, the information is stored."

What are some of the specific problems with the output that the systems consultants and Benton Turnstile gave to Bart Rayl? Evaluate the media that are being used for output as well as the timing of its distribution. Comment on the external output that users of the automated ticket machines are apparently receiving. Suggest some changes in output to help Rayl get the information he needs to make decisions on fares and scheduling of trains, and to help users of the subway system get the information they need. What are some decisions facing organizations like the New York Transit Authority if they collect and store input concerning an individual's destinations each time a trip is taken? What changes would S.W.I.F.T. have to make to its output and its tickets if it adopted this technology?

Some output technologies are prized for their unobtrusiveness. Libraries and hospitals, which emphasize silence in the workplace, make extensive use of displays for Web documents and other networked database information, but printers might be scarce. Analysts working for a company that has sustainability as part of its mission may also want to include the idea of green IT in their output decisions. This might translate into putting more transactions, reports, and documents online, and discouraging printing of email messages.

Realizing How Output Bias Affects Users

Output is not just a neutral product that is subsequently analyzed and acted on by decision makers. Output affects users in many different ways. Systems analysts must put great thought and care into designing output so as to avoid biasing it.

Recognizing Bias in the Way Output Is Used

It is a common error to assume that once a systems analyst has signed off on a system project, his or her impact is ended. Actually, the analyst's influence is long-lasting. Much of the information on which organizational members base their decisions is determined by what analysts perceive is important to the business.

Bias is present in everything that humans create. This statement is not to judge bias as bad but to make the point that it is inseparable from what we (and consequently our systems) produce. The concerns of systems analysts are to avoid unnecessarily biasing output and to make users aware of the possible biases in the output they receive.

Presentations of output are unintentionally biased in three main ways:

1. How information is sorted
2. Setting of acceptable limits
3. Choice of graphics

INTRODUCING BIAS WHEN INFORMATION IS SORTED. Bias is introduced to output when analysts and users make choices about how information is sorted for a report. Common sorts include alphabetical, chronological, and cost.

Information presented alphabetically may overemphasize the items that begin with the letters A and B because users tend to pay more attention to information presented first. For example, if past suppliers are listed alphabetically, companies such as Aardvark Printers, Advent Supplies, and Barkley Office Equipment are shown to the purchasing manager first. When certain airlines created the SABRE and APOLLO reservations systems, they listed their own flights first, until the other airlines complained that this type of sorting was biased.

INTRODUCING BIAS BY SETTING LIMITS. A second major source of bias in output is the predefinition of limits for particular values being reported. Many reports are generated on an exception basis only, which means that when limits on values are set beforehand, only exceptions to those values will be output. Exception reports make the decision maker aware of deviations from satisfactory values.

For example, limits that are set too low for exception reports can bias the user's perception. An insurance company that generates exception reports on all accounts one week overdue has set too low a limit on overdue payments. The decision maker receiving the output will be overwhelmed with "exceptions" that are not really cause for concern. The one-week overdue exception report leads to the user's misperception that there are a great many overdue accounts. A more appropriate limit for generating an exception report would be accounts 30 days or more overdue.

INTRODUCING BIAS THROUGH GRAPHICS. Output is subject to a third type of presentation bias, which is brought about by an analyst's (or users') choice of graphics for output display. Bias can occur in the selection of the graph size, its color, the scale used, and even the type of graphic.

Graph size must be proportional so that the user is not biased as to the importance of the variables that are presented. For example, Figure 11.5 shows a column chart comparing the number of no-shows for hotel bookings in 2011 with no-shows for hotel bookings in 2012. Notice that the vertical axis is broken, and it appears that the number of no-shows for 2012 is twice as much as the number of no-shows in 2011, although the number of no-shows has actually gone up only slightly.

FIGURE 11.5

A misleading graph will most likely bias the user.

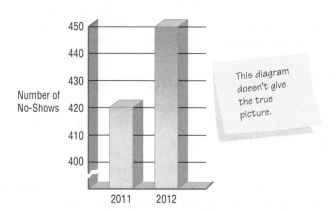

CONSULTING OPPORTUNITY 11.3

Should This Chart Be Barred?

"Gee, I'm glad they hired you guys. I know the Redwings will be better next season because of you. My job'll be a lot easier, too," says Andy Skors, ticket manager for the Kitchener, Ontario, hockey team, the Kitchener Redwings. Andy has been working with your systems analysis team on analyzing the systems requirements for computerizing ticket sales.

Recall that when we last heard from the systems analysis team, consisting of Hy Sticking (your leader), Rip Shinpadd, Fiona Wrink, and you, you were wrestling with whether to expedite the project and set team productivity goals (in Consulting Opportunity 3.5, "Goal Tending").

Andy is talking with the team about what to include in the systems proposal to make it as persuasive as possible to the Redwings' management. "I know they're going to like this chart," Andy continues. "It's a little something I drew up after you asked me all those questions on past ticket sales, Rip."

Andy hands the bar chart to Rip, who looks at it and suppresses a slight smile. "As long as we have you here, Andy, why don't you explain it to us?"

Like a player fresh out of the penalty box, Andy skates smoothly into his narrative of the graph. "Well, our ticket sales reached an all-time high in 2010. We were real crowd pleasers that

year. Could've sold seats on the scoreboard if they let me. Unfortunately, ticket sales were at an all-time low in 2012. I mean, we're talking about a disaster. Tickets moved slower than a glacier. I had to convince the players to give tickets away when they made appearances at the shopping mall. Why, just look at this table, it's terrible.

"I think computerizing the ticket sales will help us pick out who our season supporters are. We've got to figure out who they are and get them back. Get them to stick with us. That would be a good goal to shoot for," Andy concludes.

As Andy's presentation finally winds down, Hy looks as if he thought the 20-minute period would never end. Picking up on his signal, Fiona says, "Thanks for the data, Andy. We'll work on getting them into the report somehow."

As Fiona and Rip head out of the room with Andy, Hy realizes the bench has emptied, so he asks you, the fourth team member, to coach Andy on his bar chart by making a list of the problems you see in it. Hy would also like you to sketch some alternative ways to graph the data on ticket sales so that a correct and persuasive graph of ticket sales can be included in the systems proposal. Draw two alternatives.

FIGURE 11.C1

An incorrectly drawn graph.

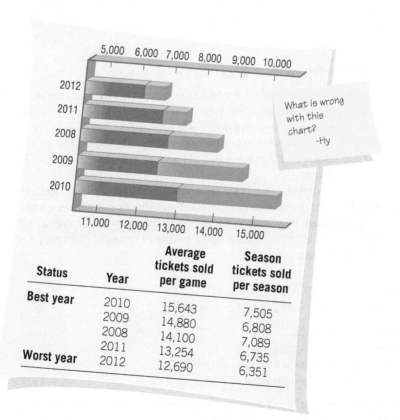

Avoiding Bias in the Design of Output

Systems analysts can use specific strategies to avoid biasing the output they and others design:

1. Be aware of the sources of bias.
2. Create an interactive design of output during prototyping that includes users and a variety of differently configured systems when testing the appearance of Web documents.
3. Work with users so that they are informed of the output's biases and can recognize the implications of customizing their displays.
4. Create output that is flexible and that allows users to modify limits and ranges.
5. Train users to rely on multiple outputs for conducting "reality tests" on system output.

All these strategies except the first focus on the relationship between a systems analyst and users as it involves output. Systems analysts need to recognize the potential impact of output and be aware of the possible ways in which output is unintentionally biased. They then need to be proactive in helping users design output with minimal, but identifiable, biases.

Designing Printed Output

The source of information to be included in reports is the data dictionary, the compilation of which was covered in Chapter 8. Recall that the data dictionary includes names of data elements as well as the required field length of each entry.

Reports fall into three categories: detailed, exception, and summary. Detailed reports print a report line for every record on the master file. They are used for mailings to customers, sending student grade reports, printing catalogs, and so on. Inquiry screens have replaced many detailed reports.

Exception reports print a line for all records that match a set of conditions, such as which holiday decorations will be discounted the day after the holiday or which students are on the dean's list. They are usually used to help operations managers and clerical staff run a business. Summary reports print one line for a group of records and are used to make decisions, such as which items are not selling and which are hot selling.

A systems designer must understand that some website visitors might prefer to print out content. Consider inserting PDF files that can be downloaded and try printing each page in different browsers to ensure they will have a professional look if a user prints them.

Designing Output for Displays

Chapter 12 covers designing displays for human or computer input, and the same guidelines also apply for designing output, although the contents will change. Notice that output for displays differs from printed output in a number of ways. It is ephemeral (that is, a display is not permanent in the same way that printouts are), it can be more specifically targeted to the user, it is available on a more flexible schedule, it is *not* portable in the same way, and sometimes it can be changed through direct interaction.

In addition, users must be instructed on which keys to press, which links to click, and how to scroll when they want to continue reading additional displays, when they want to know how to end the display, and when they want to know how to interact with the display (if possible). User access to displays may be controlled through a password, whereas distribution of printed output is controlled by other means.

Guidelines for Display Design

Four guidelines facilitate the design of displays:

1. Keep the display simple.
2. Keep the presentation consistent.
3. Facilitate user movement among displayed output.
4. Create an attractive and pleasing display.

Just as with printed output, good displays are not created in isolation. Systems analysts need the feedback of users to design worthwhile displays. Once approved by users after successive prototypes and refinements, the display layout can be finalized.

```
                        New Zoo Order Status
        Retailer           Order #    Order Date        Order Status
Animals Unlimited          933401     09/05/2012      Shipped On 09/29
                           934567     09/11/2012      Shipped On 09/21
                           934613     09/13/2012      Shipped On 09/21
                           934691     09/14/2012      Shipped On 09/21
Bear Bizarre               933603     09/02/2012      Partially Shipped
                           933668     09/08/2012      Scheduled For 10/03
                           934552     09/18/2012      Scheduled For 10/03
                           934683     09/18/2012      Shipped On 09/28
Cuddles Co.                933414     09/12/2012      Shipped On 09/18
                           933422     09/14/2012      Shipped On 09/21
                           934339     09/16/2012      Shipped On 09/26
                           934387     09/18/2012      Shipped On 09/21
                           934476     09/25/2012      Backordered
Stuffed Stuff              934341     09/14/2012      Shipped On 09/26
                           934591     09/18/2012      Partially Shipped
                           934633     09/26/2012      Backordered
                           934664     09/29/2012      Partially Shipped

Press any key to see the rest of the list; ESC to end; ? for help.
For more detail place cursor over the order number and hit the Enter key.
```

FIGURE 11.6

The New Zoo output display screen is uncluttered and orients users well. (Three screenshots from App Cooker application. Copyright © 2011 by Hot Apps Factory SARL. Reprinted with permission.)

The output produced from the design display is pictured in Figure 11.6. Notice that it is uncluttered, but it still gives a basic summary of the shipping status. The display orients users as to what they are looking at with the use of a heading. Instructions at the bottom of the display provide users with several options, including continuing the present display, ending the display, getting help, or getting more detail. This display provides context for users attempting to complete a task such as checking on the status of an order.

Output displays in an application should show information consistently from page to page. Figure 11.7 shows the display that results when the user positions the cursor over the order number for a particular retailer. The new display presents more details on Bear Bizarre. In the body of the display, the user can see the retailer's order number, complete address, the order date, and the status. In addition, a detailed breakdown of the shipment and a detailed status of each part of the shipment are given. A contact name and phone number are supplied, along with the account balance, credit rating, and shipment history. Notice that the bottom portion of the display advises the user of options, including more details, ending the display, or getting help. Users are provided control over what they might do next while viewing the display.

```
Order #              Retailer           Order Date           Order Status
933603          Bear Bizarre           09/02/2011         Partially Shipped
                1001 Karhu Lane
                Bern, Virginia 22024

Units   Pkg     Description          Price   Amount    Detailed Status
 12     Each    Floppy Bears         20.00   240.00    Backordered Due 10/15
  6     Each    Growlers             25.00   150.00    Backordered Due 10/15
  2     Each    Special Edition      70.00   140.00    Shipped 09/02
  1     Box     Celebrity Mix       150.00   150.00    Shipped 09/02
 12     Each    Santa Bears          10.00   120.00    Backordered Due 10/30
                                             800.00

Contact        Account Balance     Credit Rating    Last Order    Shipped
Ms. Ursula Major      0.00           Excellent       08/21/2011    On Time
703-484-2327


Press any key to see the rest of the list; ESC to end; ? for help.
```

FIGURE 11.7

If users want more details regarding the shipping status, they can call up a separate screen.

FIGURE 11.8

A bar chart display for onscreen inspection of troop time response.

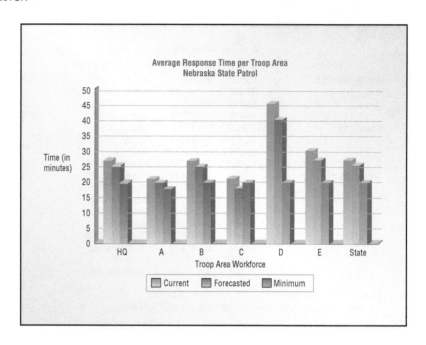

Rather than crowding all retailer information onto one page, the analyst has made it possible for the user to bring up a particular retailer if a problem or question arises. If, for example, the summary indicates that an order was only partially shipped, the user can check further on the order by calling up a detailed retailer display and then following up with appropriate action.

Using Graphical Output in Screen Design

Graphical output can be powerful. It is much easier to identify a trend or notice a pattern when the right graph is displayed. Most people notice differences in graphs more easily than they notice differences in tables. It is important to collaborate with users in choosing the correct style of graph to communicate your meaning.

As with the presentation of tabular output, graphical output needs to be accurate and easy to understand and use if it is to be effective in communicating information to users. Decision makers using the graphs need to know the assumptions (biases) under which the graphs are being constructed so that they can adjust to or compensate for them.

In designing graphical output, a systems analyst and any users involved in design prototyping must determine (1) the purpose of the graph, (2) the kind of data that need to be displayed, (3) its audience, and (4) the effects on the audience of different kinds of graphical output. In the instance of a decision support system, the purposes of graphical displays are to support any of the three phases of problem solving a user experiences: intelligence, design, or choice. An example from the Nebraska State Patrol workforce planning DSS is shown in Figure 11.8. Here, current response times, forecasted response times, and minimum requirements are graphed as differently shaded bars.

Dashboards

Decision makers need output that helps them make decisions effectively and quickly. It helps executives and other decision makers if all the information they need to make decisions is displayed in front of them. When given a written report, a decision maker would prefer all the information to be contained in that one report rather than searching for information in other places. The same principle applies to screen design.

A dashboard, similar to the dashboard in a car, has many different gauges. Each gauge can display a graph (similar to the speed in miles or kilometers per hour), a problem light (similar to a light showing that the automatic braking system is not functioning), or even text (like an odometer that simply counts the miles traveled).

An executive can find a dashboard to be extremely useful in making decisions, but only if the dashboard is designed properly. The dashboard in Figure 11.9 shows that a considerable amount of information can be included on a single screen.

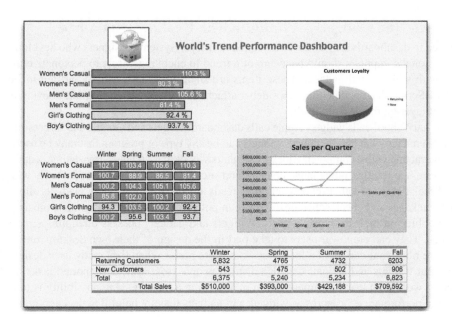

FIGURE 11.9

This dashboard has a variety of displays depicting performance measurements to help make decisions.

Dashboards are all about communicating measurements to the user. An executive uses a dashboard to review performance measures and to take action if the information on the screen calls for it. Here are some rules of thumb that can help you make the dashboards you design more attractive and more effective:

1. Make sure the data have context. If you design a screen stating that sales last month were $851,235, what does that mean? Are sales above or below average?
2. Display the proper amount of summarization and precision. It will clutter the screen if you display last month's sales as $851,235.32 instead of $851,235 or even $851K.
3. Choose appropriate performance measures for display. For example, plotting the difference in actual versus expected sales in a deviation chart is much more meaningful than using a line chart to plot actual and expected sales.
4. Present data fairly. If you introduce bias into the dashboard, it will hinder rather than support good decisions.
5. Choose the correct style of graph or chart for display. Using the correct chart is important. While a pie chart may be an excellent graph to persuade someone, it may not be a good way for an executive to monitor the performance of regional offices, for example.
6. Use well-designed display media. Even if you choose the very best type of graph, you still need to draw, size, color, and label the graph in a meaningful and pleasing way.
7. Limit the variety of item types. Keep the number of graph, chart, and table styles to a minimum so that the information can be communicated quickly and accurately.
8. Highlight important data. Use bright colors and bold fonts only for important data. You can highlight key performance measures or important exceptions that are occurring but not both. Choose what to emphasize.
9. Arrange the data in meaningful groups. Performance measures are almost always associated with other performance measures because of the data displayed or the type of graph. Learn how to group associated items together.
10. Keep the screen uncluttered. Avoid photographs, ornate logos, or themes that can distract users from the data.
11. Keep the entire dashboard on a single screen. All the performance measures are meant to be on the same screen. If forced to switch screens, a user will not see two relevant measures at the same time.
12. Allow flexibility. If an executive wants a different graph or chart, consider replacing it. Prototyping the dashboard and refining it based on the user's feedback makes sense. Decision makers often know best when it comes to getting the right information in the most appropriate form for their decision style.

Widgets and Gadgets—Changing the Desktop Metaphor

Related to dashboards are new, user-designed desktops. Systems designers who develop software for personal computers should be aware of a trend to encourage users to personalize their desktops with widgets and gadgets. These items are small programs, usually written in JavaScript and VBScript, that reside either in a sidebar attached to a browser or program or even reside in a special layer on the desktop itself.

What Yahoo! calls widgets Apple calls dashboard widgets and Google and Microsoft call gadgets. No matter what they're called, widgets can be any type of program that may be useful to anyone interacting with a computer. Clocks, calculators, bookmark helpers, translators, search engines, easy access to utilities, quick launch panels, and sticky notes are popular productivity widgets.

Stock tickers, weather reports, and RSS feeds are also useful widgets. Gadgets allow users to track packages and check schedules. Users can put amusements such as games, music podcasts, and hobbies on their desktops as well. Widgets and gadgets possess dual, almost paradoxical natures. They can empower users to take part in the design of their own desktop, and designers who are observant can learn a lot about what users prefer when they study user-designed desktops. But widgets and gadgets can also distract people from system-supported tasks. Designers need to work with users to support them in achieving a balance. One possibility is to add user-specific performance measures as widgets and gadgets that are helpful to decision makers.

Designing a Website

You can borrow some design principles from other displays when you design a website. Remember, though, that the key word here is *site*. The first documents displayed on the Internet using the HTTP protocol were called home pages, but it became apparent very quickly that companies, universities, governments, and people were not going to be displaying just one page. The term *website* replaced *home page*, indicating that the array of pages would have to be organized, coordinated, designed, developed, and maintained in an orderly process.

Printing is a highly controlled medium, and the analyst has a very good idea of what the output will look like. GUI and CHUI (character-based user interface) screens are also highly controlled. The Web, however, is a very uncontrolled environment for output.

Different browsers display images differently, and screen resolution has a large impact on the look and feel of a website. There is no "standard" resolution anymore. For many years displays had a 4:3 aspect ratio and common resolutions were 1024 × 768 pixels or 1600 × 1200 pixels. Now displays are made for HD and have a 16:9 aspect ratio. The issue is further complicated by the use of handheld devices, such as smartphones and tablets, that are used to browse the Web. The complexity deepens when you realize that each person may set a browser to use different fonts and may disable the use of JavaScript, cookies, and other Web programming elements. Analysts and users face many decisions when designing a website.

General Guidelines for Designing Websites

In addition to the general design elements discussed earlier in this chapter, there are specific guidelines appropriate for the design of professional-quality websites.

USE PROFESSIONAL TOOLS. Consider using Web editor software such as Adobe Dreamweaver (for Windows) or Freeway Pro (for Mac OS). These tools are definitely worth the price. You will be more creative and you'll get the website finished much faster than if you work directly with HTML (Hypertext Markup Language).

STUDY OTHER WEBSITES. Look at websites you and other users think are engaging. Analyze what design elements are being used and see how they are functioning and then try to emulate what you see by creating prototype pages. (It is not ethical or legal to cut and paste pictures or code, but you still can learn from the other sites.)

Firefox, which is part of the open source software movement, is a useful browser for studying other websites. It has a number of extensions created by third-party developers that are available as free downloads. Run Firefox and click "Tools/Extensions" and "Get More Extensions." There are pages of extensions, but one called Web Developer is very useful to designers and Webmasters. It allows you to outline tables and styles and to view JavaScript and cookies; it provides form information as well as a wealth of other useful items from which to choose. Palette Grabber is

CONSULTING OPPORTUNITY 11.4

Is Your Work a Grind?

"**I** want everything I can get my hands on, and the tighter the information is packed, the better. Forget that stuff you hear about information overload. It's not in my vocabulary. I want it all, and not in a bunch of pretty-looking, half-page reports either. I want it all together, packed on one sheet that I can take into a meeting in case I need to look something up. And I need it every week," proclaims Stephen Links, vice president of a large, family-owned sausage company.

During an interview, Links has been grilling Paul Plishka, who is part of the systems analysis team that is busy designing an information system for Links Meats. Although Paul is hesitant about what Links has told him, he proceeds to design a printed report that includes all the important items the team has settled on during the analysis phase.

When a prototype of the new report, designed to his specifications, is handed to Stephen, however, there appears to be a change of heart. Links says in no uncertain terms that he can't find what he needs.

"This stuff looks terrible. It looks like scraps. My kindergartner makes better reports in crayon. Look at it. It's all ground up together. I can't find anything. Where's the summary of the number of beef items sold in each outlet? Where is the total volume of items sold for *all* outlets? How about the information on our own shop downtown?" says Links, slicing at the report.

The report clearly needs to be redesigned. Design a report (or reports) that better suits Stephen Links. What approach can the analyst take in suggesting more reports with a less-crowded format? Comment on the difficulty of implementing user suggestions that go against your design training. What are the trade-offs involved (as far as information overload goes) in generating numerous reports as opposed to generating one large report containing all the information Stephen wants? Devise a heuristic concerning the display of report information on one report in contrast to the generation of numerous reports. Consider advocating a Web-based or dashboard solution that would permit hyperlinks to all the information Stephen desires. How feasible is that?

---

another extension that allows Web developers to see a display of color codes just by picking any color on a website. There are also tools for working with XML. Figure 11.10 is an example of the Web Developer toolbar used to highlight table cells. Notice the red border around each individual cell.

EXAMINE THE WEBSITES OF PROFESSIONAL DESIGNERS. As you look at professionally designed pages, ask yourself, "What works? What doesn't work? In what ways can users interact with the site?" For example, does the site have hot links to email addresses, interactive forms to fill in, consumer surveys, games, quizzes, chat rooms, blogs, and so on? What about color schemes and pervasive metaphors?

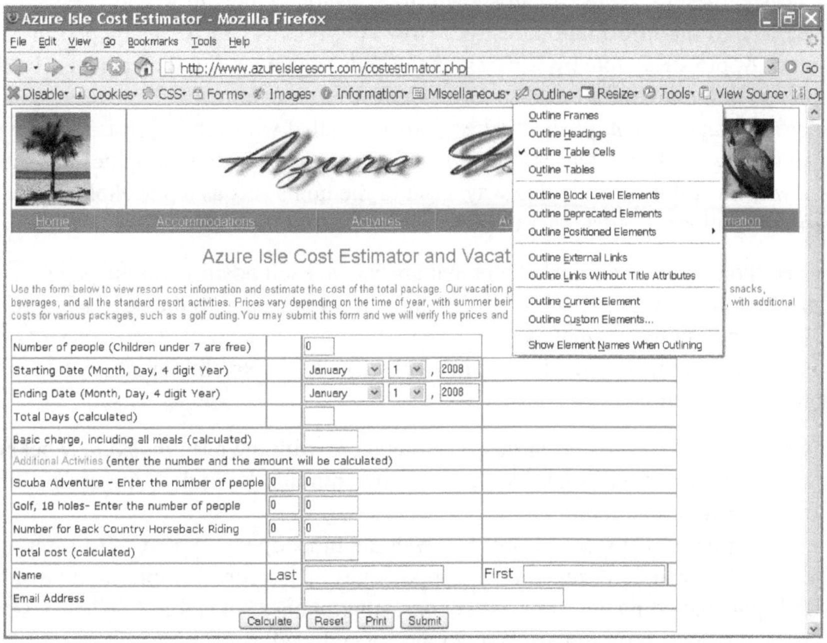

FIGURE 11.10

A Web developer can outline table cells when designing a web page, as shown in this example.

FIGURE 11.11

A website evaluation form.

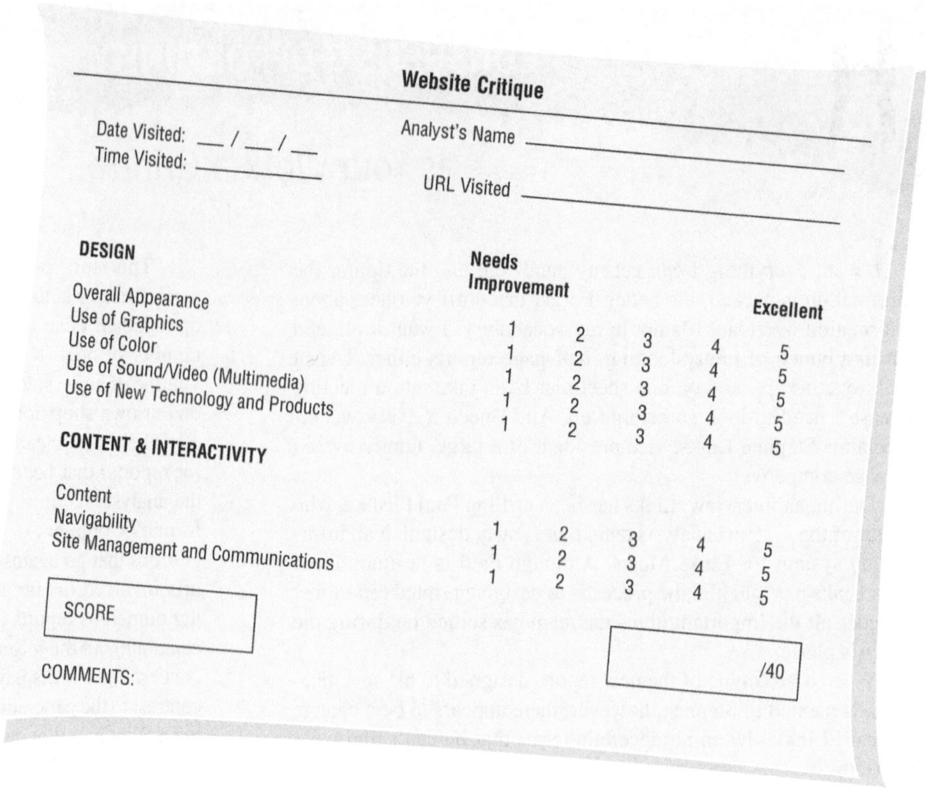

Website Critique

Date Visited: __ / __ / __
Time Visited: _____

Analyst's Name _____

URL Visited _____

DESIGN	Needs Improvement				Excellent
Overall Appearance	1	2	3	4	5
Use of Graphics	1	2	3	4	5
Use of Color	1	2	3	4	5
Use of Sound/Video (Multimedia)	1	2	3	4	5
Use of New Technology and Products	1	2	3	4	5

CONTENT & INTERACTIVITY					
Content	1	2	3	4	5
Navigability	1	2	3	4	5
Site Management and Communications	1	2	3	4	5

SCORE

COMMENTS: /40

Also look at websites that give hints on design. One such site is useit.com.

USE THE TOOLS YOU'VE LEARNED. Figure 11.11 provides a form that Web designers have used successfully to evaluate web pages systematically. You might want to use copies of the form to help compare and contrast the many websites you will visit as you go about learning web page design.

USE STORYBOARDING, WIREFRAMING, AND MOCKUPS. Designing for the Web brought back a practice known in the film industry as storyboarding. In the 1930s at Disney studios, each scene was drawn on a piece of paper and then tacked to a bulletin board in order of the episodes.

In developing a website (or any app, for that matter), a storyboard could be used to show the differences between screens. It can show how a visitor to the site would navigate the website in order to find the information desired. Now storyboards can be developed in Microsoft PowerPoint or Apple's Keynote by clicking on the slide organizer, but they can also be drawn in Microsoft Visio or OmniGraffle.

Page design can be accomplished using a process called wireframing. It is called wireframing because it shows only the basics. There is no color, no type style. Graphics are shown as a simple box with an X drawn in. In this way, each of the items acts as a placeholder.

Wireframing allows the designer to plan:

1. The overall design, showing what element appears at each position on the page
2. The navigational design, showing how to move from one page to the next using buttons, tabs, links, and pull-down menus
3. The interface design, showing how to interact with the website by inputting data or responding to questions

Today the term *wireframe* has largely been replaced with the term *mockup*. Mockups show what the output and input will look like before programming the underlying code to make the program or application functional.

Software that helps a systems designer develop a mockup is abundant. Built into the software are objects that can be dragged and dropped onto the screen. Templates are available for any type of display including desktops, notebooks, smartphones, and tablets. When designing for smartphones and tablets, both screen orientations are included.

CONSULT THE BOOKS. Something that can add to your expertise in this new field is to read about Web design. The following are some books on website design:

Eckerson, W. W. *Performance Dashboards: Measuring, Monitoring, and Managing Your Business,* 2nd ed. New York: John Wiley & Sons, Inc., 2010.

Few, S. *Information Dashboard Design: The Effective Visual Communication of Data.* Sebastopol, CA: O'Reilly Media, Incorporated, 2006.

Flanders, V., and D. Peters. *Son of Web Pages That Suck: Learn Good Design by Looking at Bad Design.* Alameda, CA: Sybex, 2002.

McNeil, P. *The Web Designer's Idea Book, Vol. 2: More of the Best Themes, Trends & Styles in Website Design.* New York: F+W Media, 2010.

LOOK AT SOME POOR EXAMPLES OF WEB PAGES, TOO. You can learn from others' mistakes by critiquing poor web pages and remembering to avoid those mistakes. Examine the website found at www.webpagesthatsuck.com. Despite its "counterculture" name, this is a wonderful site that provides links to many poorly designed sites and points out the errors that designers have made on them.

CREATE TEMPLATES OF YOUR OWN. If you adopt a standard-looking page for most of the pages you create, you'll get the website up and running quickly, and it will consistently look good. Websites may be made using Cascading Style Sheets (CSS), a language that allows a designer to specify the color, font size, font type, and many other attributes only once. These attributes are stored in a style sheet file and then are applied to many web pages. If a designer changes a specification in the style sheet file, all the web pages using that style sheet will be updated to reflect the new style.

USE PLUG-INS, AUDIO, AND VIDEO VERY SPARINGLY. It is wonderful to have features that the professional pages have, but remember that everyone looking at your site doesn't have every new plug-in. Don't discourage visitors to your page.

Specific Guidelines for Website Design

Good websites are well thought out. Pay attention to the following:

1. Structure
2. Content
3. Text
4. Graphics
5. Presentation style
6. Navigation
7. Promotion

STRUCTURE. Planning the structure of a website is one of the most important steps in developing a professional website. Think about your goals and objectives. Each page in the overall Web structure should have a distinct message or other related information. Sometimes it is useful to examine professional sites to analyze them for content and features. Figure 11.12 is a screen capture from the DinoTech website. The purpose for the site and the Web medium work well together. Notice that there is great attention to supporting users on the site. There are words, graphics, JPEG images, and icons. In addition, there are many kinds of links: to RSS feeds, video, sub-Webs, chat rooms, a search engine, and many other features.

To help plan and maintain a solid structure, a Webmaster can benefit from using one of the many website diagramming and mapping tools available. Many software packages, including Microsoft Visio, have Web charting options built into the software. Although helpful for development, these tools become even more important when maintaining a website. Given the dynamic nature of the Web, sites that are linked to your site may move at any time, requiring you or your Webmaster to update the links.

In Figure 11.13, a map of a section of the authors' website is shown in the Microsoft Visio window. In this example, we explore the website down to all the existing levels. Notice the links to HTML pages, documents, images (JPEG, GIF, or PNG files), and mail-tos (a way to send email to a designated person). The links can be either internal or external. If a link is broken, a red X appears, and the analyst can investigate further. This Visio file can be printed out in sections and posted on the wall to get an overall picture of the website.

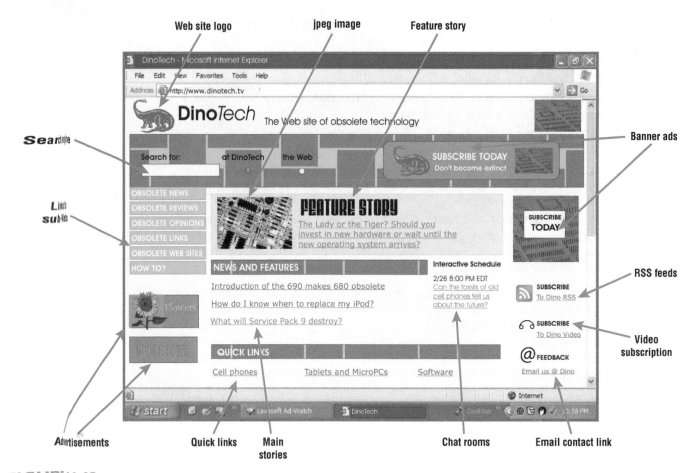

FIGURE 11.12

The DinoTech website makes the most of links, RSS feeds, video subscriptions, and banner ads.

CONTENT. Provide something important to website users. Exciting animation, movies, and sounds are fun, but you have to include appropriate content to keep the user interested. Supply some timely advice, important information, a free offer, a top ten list of tips on the page's topic, or any activity that you can provide that is interactive and moves users away from a browsing mode and into an interactive one.

"Stickiness" is a quality a website can possess. If a user stays at your site for a long period of time, your site has a high degree of stickiness. That is why a merchant includes many items of interest on a site. A wine merchant, for example, may post lessons on how to uncork a bottle, taste the wine, or choose a proper glass.

Use a metaphor or images that provide a metaphor for your site. You can use a theme, such as a storefront, with additional pages having various metaphors related to the storefront, such as a deli. Avoid the overuse of cartoons, and don't be repetitive.

Every website should include an FAQ (frequently asked questions) page. Often these are created based on the experiences of users and technical support people who identify the topics of continuing concern. Eighty percent of the questions will fall into the FAQ category. By having answers readily available, 24 hours a day, you will save valuable employee time and also save user time. FAQ pages also demonstrate to users of your site that you are in concert with them and have a good idea of what they would like to know.

On the Web, COTS software takes on another meaning. A website may take advantage of prewritten software. Examples include search engines (such as Google, Bing, or Yahoo! search), mapping software (such as Google Maps or MapQuest), weather information, and news and stock tickers. Website designers value these packages because they can increase the functionality of the site, and the additional features encourage users to bookmark their clients' websites because they provide valuable bonus content.

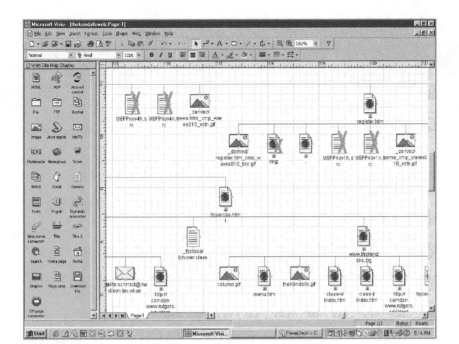

FIGURE 11.13

A website can be evaluated for broken links by using a package such as Microsoft Visio.

TEXT. Remember that text is important, too. Each web page should have a title. Place meaningful words in the first sentence appearing on your web page. Let people know that they have indeed navigated to the right website. Clear writing is especially important.

Content on ecommerce sites needs to be constantly updated. Content management systems (CMSs) are powerful software tools that can enable the analyst to develop and maintain websites and other online applications. An increasingly popular CMS is Joomla!, which can be found at www.joomla.org/about-joomla.html. It is based on PHP and MySQL. Unlike proprietary CMSs, which are expensive and not widely available, Joomla! is an open source solution that is made freely available to any developer under a General Public License (GPL).

GRAPHICS. The following list provides details about creating effective graphics for websites:

1. Use one of the most commonly used image formats: JPEG, GIF, or PNG. JPEGs are best for photographs, and GIFs are best for artwork images. GIFs are limited to 256 colors but may include a transparent background, pixels that allow the background to show through the GIF image. GIF images may also be interlaced, meaning that the Web browser will show the image in successive stages, presenting a clearer image with each stage. PNG (portable network graphics) files have greater color depth than GIF files and are able to have a transparent background.
2. Keep the background simple and make sure users can read the text clearly. When using a background pattern, make sure that you can see the text clearly on top of it.
3. Create a few professional-looking graphics for use on your pages.
4. Keep graphic images small, and reuse bullet or navigational buttons such as BACK, TOP, EMAIL, and NEXT. These images are stored in a cache, an area on the browsing computer's hard drive. Once an image has been received, it will be taken from the cache whenever it is used again. Using cached images improves the speed with which a browser can load a web page.
5. Include text in what is called a title attribute for images and image hot spots. The text displays when the user moves the mouse over the image. An alt attribute provides text for screen readers and is essential to support Web accessibility for visually impaired site visitors.
6. Examine your website on a variety of displays and screen resolutions as well as smartphones and tablets. Scenes and text that look great on a high-end video display may not look good to others with poorer-quality equipment.

PRESENTATION STYLE. The following list gives added details about how to design engaging entry displays for websites:

MAC APPEAL

There are many different approaches to creating websites. Coders want as much control over the HTML code as possible, but many designers aren't very interested in tweaking the code. Good designers want to be able to include many different items in both graphics and text, rotate and enhance images, format them in carefully thought-out designs, and make them appear just right in any browser and in any resolution. There are many WYSIWYG (what you see is what you get) packages available that allow designers to do this, both on a Mac and a PC. Some of these packages work well, but others don't.

Softpress Systems, the developers of Freeway Pro, have created Web design software that has a different approach. Unlike other software, Freeway Pro does not create code while a designer works. Once a designer is pleased with the design, Freeway Pro generates the code. The code is therefore extremely efficient. This is also a highly useful tool for prototyping. Freeway Pro assumes that when standards change, updates to the software will allow Web designers to simply republish the entire site using the updated standard.

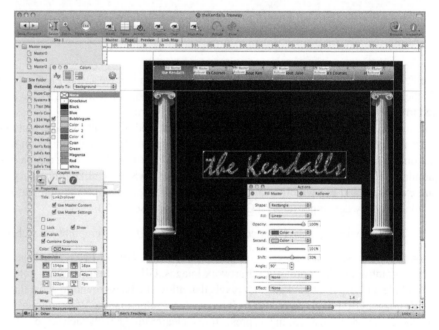

FIGURE 11.MAC

Freeway Pro, from Softpress Systems, offers a unique approach for website designers. (Screenshot from FREEWAY PRO. Copyright © by Softpress Systems, Inc. Reprinted by permission; Screenshot from www.thekendalls.org. Copyright © by Kenneth and Julie Kendall. Reprinted with permission.)

1. Provide a home page that introduces the visitor to the website. The page must be designed to load quickly. A useful rule of thumb is to design a page that will load in 14 seconds. (Although you may be designing the page on a workstation at the university, a visitor to your website may be accessing it from home, using a slower Internet connection.) This entry display should be 100 kilobytes or less, including all graphics.

   The home page should contain a number of choices, much like a menu. An easy way to accomplish that is to design a set of links or buttons and position them on the left side or the top of the screen. These links can be linked to other pages on the same website or linked to different websites. An example of this is shown in Figure 11.14, an entry page that contains images and content but that directs the visitor to journey elsewhere in the site. This page was constructed using the template-based Web app called Weebly. Design is mainly accomplished by dragging and dropping items to create a web page, but Weebly

FIGURE 11.14

Using a visual HTML editor (in this example, Visual Page), a website designer can see what a page looks like in a browser and the HTML code (see bottom of screen) at the same time. (Screenshot from Weebly.com. Copyright © 2012 by Weebly, Inc. Reprinted with permission. Top photo: Oleg Gekman / Shutterstock.com; Bottom photo: Chin Kit Sen / Shutterstock.com)

allows the designer more direct control by allowing a designer to see HTML code (at the top of the screen) at the same time they see what the page would look like in a browser.

2. Keep the number of graphics to a reasonable minimum. It takes additional download time to transfer a graphics-intensive site.

3. Use large and colorful fonts for headings.

4. Use interesting images and buttons for links. A group of images combined into a single image is called an image map, which contains various hot spots that act as links to other pages.

5. Use Cascading Style Sheets (CSS) to control the formatting and layout of the web page. CSS separates the content (the text and images) from how they look (the presentation). Cascading Style Sheets information is commonly stored in a file external to the web page, and one style sheet may control the formatting of many pages. An advantage of using external style sheets is making a change in the style sheet; for example, changing the color of bold text will change the formatting of all the web pages that use the style sheet. Cascading Style Sheets may also be used in a single web page, and any duplicate styles will override an external style sheet if one is used. This allows a designer to vary from the standard look and feel of a website, perhaps for a "special sale" web page or some other exception. Styles may be added to individual items on a web page, overriding any other style sheets.

6. Use divisions and cascading styles or tables to enhance a layout. Tables are easy to use and provide adequate layout. However, tables are not well suited for visually impaired visitors. Screen reading software reads across the page, not necessarily in a table column. Divisions control the layout by providing blocks of text on the web page. Each block may be defined with a position from the top and left of the screen or a larger block, and it may have a width and height, as well as border style and background color. Divisions eliminate the need for tables within tables and simplify design; screen reading software will read all the text in the division, making the site accessible for visually impaired viewers. Designers may use a division to control how the page looks if the viewer has a wide-screen display, by setting a smaller width with a border area on either side.

7. Use the same graphics images on several web pages. Consistency will be improved, and the pages will load more quickly because the computer stores the image in a cache and doesn't have to load it again.

8. Use JavaScript to enhance the web page layout by having images that change when a mouse is moved over them, having menus expand, and so on. JavaScript may be used to reformat a web page based on the height and width of the screen. If the website is multinational, JavaScript can detect the language being used (a browser setting) and redirect the viewer to a different web page in a different language.

9. Avoid overusing animation, sound, and other elements.

NAVIGATION. Is it fun for you to follow links on the Web? The answer is most likely that it depends. When you discover a website that loads easily, has meaningful links, and allows you to easily return to the places you want to go back to, chances are you think it is fun. Fun is not just play; it can be an important part of work, too. Recent research shows that fun can have a powerful effect on making computer training effective.

If, on the other hand, you can't decide which button or hot spot to click on, and you are afraid to choose the wrong one because you might get into the wrong page that takes a long time to load, navigation is more painful than fun. An example is visiting a software company's page to find information about the features of the latest version of a product. You have choices such as products, download, FAQ, and tech support. Which button will lead to the answers you're looking for?

Make sure you include a navigation bar and links to the home page on every page on the website. A visitor to the site may have found a particular page using a search engine, and it is important that the visitor can easily find their way to the home page.

Most importantly, observe the three-clicks rule: Users should be able to move from the page they are currently on to the page containing the information they want in three clicks of the mouse button.

PROMOTION. Promote your site. Don't assume that search engines will find you right away. Submit your site every few months to various search engines. Include keywords, called metatags, that search engines will use to link search requests to your site. You can find general information about metatags at searchenginewatch.com/showPage.html?page=2167931. In addition, you can download free metatag-generating software from www.siteup.com/meta.html and a metatag builder from vancouver-webpages.com/META/mk-metas.html. You can also purchase software to make this process easier. If you try to use email to promote your site, others will consider it junk email or spam.

Encourage your readers to bookmark your website. If you link to and suggest that they go to affiliated websites that feature the "best movie review page in the world" or to the "get music for free" website, don't assume that they'll be coming back to your site in the near future. You will encourage them to revisit if they bookmark your site (bookmarks are called "favorites" in Microsoft Internet Explorer). You may add a **Click here to bookmark this page** link to your web page to automate the process. You may also want to design a "favicon," or favorite icon, so that users can identify your site in their lists of favorites.

Web 2.0 Technologies and Social Media Design

Of increasing importance for an analyst working on organizational websites is the strategic inclusion of Web 2.0 technologies that focus on enabling and facilitating user-generated content and collaboration via the Web. Familiar types of technologies you should think about including on both public-facing and internal websites you design include blogs, wikis, links to social networks on which the company has a presence, and tagging (also called social bookmarking) that provides useful pointers to online resources such as websites, content on corporate intranets, corporate documents, or photos that are relevant to the organization and to users. Notice that Web 2.0 technologies can be externally faced to clients, customers, vendors, or suppliers, or internally faced to corporate intranets.

For outward-facing Web technologies, the reasons for adding these important collaborative tools to organizational websites are clear. Companies use collaborative tools to communicate an integrated branding and messaging strategy across multiple platforms, to gauge consumer opinion, to gather feedback, to create a community of users, and so on. Inward-facing Web technologies can be useful in building employee relationships, maintaining trust, sharing knowledge, innovating among employees and groups of employees, locating corporate resources more readily, and nurturing corporate culture and subcultures inside the organization.

Your requirements determination phase informed you about user preferences. As you are prototyping, you might also want to add Web 2.0 technologies to internal and external web pages if there is a well-thought-out, strategic reason for doing so. Part of your recommended design solutions may include adding collaborative tools. From our research and practice with for-profit and nonprofit organizations attempting to use Web 2.0 technologies, we have found that there are five aspects an analyst should consider:

CONSULTING OPPORTUNITY 11.5

A Field Day

"The thing of it is, I get impatient," says Seymour Fields, owner of a chain of 15 highly successful florist shops/indoor floral markets called Fields that are located in three midwestern cities. "See this thing here?" He taps his computer display irritatedly. "We do all the payroll and all the accounting with these things, but I don't use it like I should. I actually feel a little guilty about it. See?" he says, as he makes a streak on the display with his finger. "It's even got dust on it. I'm a practical person, though. If it's sitting here, taking up space, I want to use it. Or smell it, or at least enjoy looking at it, like flowers, right? Or weed it out, that's what I say. The one time I tried something with it, it was a real disaster. Well, look, I can show you if I still remember how." Seymour proceeds to try to open a program but can't seem to get it working.

Clay Potts, a systems analyst, has been working on a systems project for the entire Fields chain. Part of the original proposal was to provide Seymour and his vice presidents with a decision support system that would help them devise a strategy to determine which European markets to visit to set up purchase agreements for fresh flowers, which outlets to ship particular kinds of flowers to, and how much general merchandise, such as planters, vases, note cards, and knickknacks, to stock in each outlet.

Seymour continues, "I can tell you what we disliked about the reports I worked with. There were too many darn layers, too much foliage, or whatever you call it, to go through. Even with a screen in front of me, it was like paging through a thick report. What do you call that?"

"Menus?" Potts suggests helpfully. "The main point is that you didn't like having to go through lots of information to get to the display you needed."

Seymour Fields looks happily at Potts and says, "You've got it. I want to see more fields on each screen."

How should Potts design screen output so that Fields and his group can get what they want on each screen while observing the guidelines for good display design? Remember that the group members are busy and that while they wouldn't be caught without their smartphones, they don't use their desktop systems very often. Design a hyperlinked page that would work well in a DSS for the vice presidents. What should be included in the first display, and what should be stored in hyperlinks? List elements for each and explain in a paragraph why you have decided on this strategy.

1. Realize differences between corporate objectives and objectives of key stakeholders. There will be differences in why each group values collaborative technologies that you discover in the outlooks of both groups.
2. Serve as the voice of the customer to your client organization. In recommending the inclusion of Web 2.0 technologies for externally facing websites, you need to be able to articulate the needs of the customer to the organization.
3. Recognize the importance of visual page design for effectively displaying collaborative tools. Whenever users have an expectation of placement on the web page (for Facebook links, Twitter icons, tagging capability, and so forth), you must observe the convention, or at least reinforce the emerging convention. So users will expect to be able to "Make a Comment," "Like us on Facebook," or hit an "in share" button to share content on LinkedIn, or touch a "g+ 1" icon to share comments or content on Google +, click a birdy-labeled button to tweet, or click a button showing an icon of an envelope to email the story or content to a friend, or subscribe to RSS content with a colored hyperlink simply labeled RSS. Often these options appear at the top of a web page or in tabbed format at the top of the page and then are repeated at the bottom of the page. Another popular display convention for providing collaborative tools is along the left side of the page. Users are getting more accustomed to seeing this array of interactive options displayed in a predictable pattern on corporate websites. See Figure 11.15 for an example.
4. Revise and update the Web 2.0 technologies offered frequently. Develop a plan (and a set of tools) for revising and updating collaborative tools offered on both internal and external websites when styles, customs, and the tools change.
5. Work to integrate Web 2.0 technologies with the existing branding. Ensure that messages are consistent throughout outwardly facing websites and in all public-facing communications.

FIGURE 11.15

A page containing many different Web 2.0 features.

Name of Icon Set	Designer or Supplier	Facebook	LinkedIn	RSS Feed	Twitter	YouTube
Social Networks Pro Icons	Artbees					
Circular Social Media	BlogPerfume.com					
Free Social Media icon Set	Elegant Themes					
Amazing 3D social icons	pinkmoustache.net					
Vintage Icons for Bloggers	Nikola Lazarevic					

Eventually, one goal you will have for developing a strategy for addressing evolving user standards and conventions is to use text analytics (TA) software to interpret the qualitative data captured in blogs, wikis, and through other social media. Indeed, this is an approach you can recommend to your client organization as you close the feedback loop created by establishing Web 2.0 technologies for consumers and employees.

Designing Apps for Smartphones and Tablets

As smartphones and tablets become more powerful and ubiquitous in organizations, systems analysts will need to conceptualize their software as apps. In the early days of computing, software was called "programs." Apple preferred the term "application." When software was designed for the iPhone and iPod, software was simply called an "app." The word *app* became mainstream with the introduction of iPhones and iPads that can run these small programs. Apple sells these apps on iTunes much the same way it sells music. You can download an app and install it on your iPhone or iPad.

Creating an app for a mobile phone or larger device like the iPad involves brainstorming, imagining, preliminary screen design, user interface decisions, and detailed screen design. Many designers like to work in Adobe Illustrator, but others prefer Adobe Photoshop. But before one even attempts to use one of these packages, using a large whiteboard with a marker can be the very best way to begin designing an app. Later in this chapter we will examine apps that are used to create mockups, the preferred name for smartphone and tablet prototypes.

It is interesting to recognize that an analyst who develops apps will be trying to fill or create a user need in a unique way. The motivation for creating the app often arises from the developer, and usually not from formal requirements analysis performed for an organization. However, there are numerous approaches and tools you have learned for display and Web design that apply to app development as well as larger systems projects.

In this section, we'll talk about designing for the small screens of smartphones and tablets. It is interesting to note, however, that Apple requires developers to send in a 512 × 512-pixel icon for the App Store. When you think about the fact that requirements for the original Macintosh computer only necessitated a 512 × 342-pixel display, you take output design for smartphones and tablets seriously.

Throughout this section, we discuss concepts useful in any operating system, whether for Apple, Android, or Microsoft. We use examples from Apple's iPad, which is the dominant tablet, and iPhone, the dominant smartphone, at the time of writing this section.

1. Set up a developer account.
2. Choose a development process.

MAC APPEAL

Creating Blogs

Blogs are being written by corporate users for both internal and external communication. Blogs are informal and personal, and they often invite comments and feedback. They are easy to create and update and are designed to change daily. Companies are using blogs for advertising and to build social networks for consumers, clients, and vendors around their products, building trust and customer relationships.

Corporate blogs are monitored out of a sense of responsibility for the participants. Guidelines, policies, and laws that shape monitoring practices include shared cultural, ethical, and legal values such as respecting other employees and customers; not publishing any sensitive or secret corporate information or anything protected by copyright (without permission); and excluding anything that is hateful or profane or that violates anyone's privacy.

Even with all of the preceding guidelines, you still need to ensure that blog posts are written in a human voice, not immersed in legal language. The latest entry should be at the start of the blog. It should contain the following elements:

1. The permalink, or permanent link, specific for the blog post, which should never change
2. The headline or title of the post
3. The primary link, which connects the reader to the subject under discussion
4. An optional summary, often appearing after the link
5. The blog text or commentary
6. An optional image
7. A block quote containing quotations or other material from other sources that contributes to the discussion (often indented or in a different font to set it apart from the main text)
8. Links for comments by other people
9. Other blog software features, such as a calendar, search form, and other universal features.

One highly acclaimed desktop blogging editor for the Mac is MarsEdit by Red Sweater. It is compatible with WordPress, Blogger, Tumblr, and many other blog publishing platforms.

---

3. Be an original.
4. Determine how you will price the app.
5. Follow the rules for output design.
6. Design your icon.
7. Choose an appropriate name for the app.
8. Design for a variety of devices.
9. Design the output for the app.
10. Design the output a second time for different orientation.
11. Design the logic.
12. Create the user interface using gestures.
13. Protect your property.
14. Market your app.

We'll concentrate on the steps that are design oriented in this chapter.

Set Up a Developer Account

Apple charges a software developer a $99 fee to develop and list an app in the App Store and then takes 30 percent commission on everything it sells. The author receives 70 percent. Some think this is too small a cut for the author, but it depends on your perspective.

Apple insists on approving everything sold in its App Store. This means there are some restrictions that frustrate software developers and, sometimes, consumers. Here's where the walled garden (in this case the term refers to an array of pre-approved apps, that are closed to outside developers) comes in: You won't find any app with objectionable content in the App Store. And you won't find a great app that lets you reprogram the mute button on your iPhone to function as a camera shutter either.

Choose a Development Process

Unless you are developing a specialized app that will cost more than $19.95, prototyping is most likely the best way to develop your app. Quick releases are important, and there is a strong advantage to being the first to introduce a certain type of app. You can draw on many of the principles and values of agile development in designing an app.

Quality should not be sacrificed, but it is possible to introduce an app and then add features in subsequent releases. Numerous app developers have adopted this strategy. It has a number of advantages. It allows you to gain an advantage by being the first to introduce a specific app. It also allows you to revise the app adding new features and users approve wholeheartedly when improvements are made. And it increases visibility because the app appears on a list of apps that have been updated.

Be an Original

Never copy other developers' software. Doing so is illegal and unethical. Make sure your app has different functionality and looks very different from other apps on the market. Observe all licensing restrictions.

This may seem obvious, but some developers don't realize that you can't use Apple trademarks in your app. You should avoid using words like iPod, iPad, and iPhone in your app and even avoid the name Apple, for that matter.

Determine How You Will Price the App

The pricing decision you make will affect the design of the app. If your app will contain advertising, the ads will take up valuable space on your screen and may distract the users. If you adopt a pricing strategy that allows upgrading of features, you need to consider the interface that asks the user to upgrade.

There are six basic options for pricing:

1. Choose a low-cost strategy.
2. Introduce an app as a "premium" app.
3. Adopt a "freemium" model.
4. Offer an app for free.
5. Promote an app by reducing its price.
6. Accept advertising.

CHOOSING A LOW-COST STRATEGY. A very large number of iPhone apps are $0.99. For the iPad, $1.99 rather than $0.99 is the standard.

This strategy is often used when offering a version of software "ported," or adapted from a previous version on another platform. Apple's Productivity Suite costing $99.00 was unbundled in the Mac store at $19.95 each for Pages, Keynote, and Numbers. Versions of these apps are available for iPhones and iPads for $9.95.

INTRODUCING AN APP AS A "PREMIUM" APP. Premium apps cost $19.95 and up. These apps are intended for professionals and are often iPad versions of desktop software. Omni group offers a suite of apps that accomplish much of the functionality of Mac apps. If you price your app this way, you will have fewer users adopting the app, but they will be serious users. They will demand that you add features and functionally and provide adequate support.

The Hot Apps Factory team introduced an app for developing apps as version .983 with an introductory price of $19.95. They are planning to add six additional features and incrementally increase the price by $5.00 every time a feature is added. Version 1.0.0 will cost $49.99 when it is introduced.

ADOPTING A "FREEMIUM" MODEL. Apps that are offered at no cost for basic features but allow in-app purchases for premium features and functionality are call freemium apps. In the early days of app development, authors would introduce a "lite" version and a regular version of the same app, but in-app purchases allow users to add features at will. There is somewhat of a backlash to this approach, as many users feel they are forced to pay for each feature separately. So consider all options before choosing this business model.

OFFERING AN APP FOR FREE. Sometimes it is advisable to offer an app at no charge to the user. If you are the *n*th person to write an app for this situation, that may be the only way to get users to adopt your app. It may be useful if you have other apps for sale. Adopters may appreciate your work and buy something else you have developed.

PROMOTING AN APP BY REDUCING ITS PRICE. A mixed strategy is sometimes the best. Occasionally you can reduce the price of an app for a short period of time or even offer it for free. Apple iTunes now features a free app of the week, and other sites feature daily free offerings. Ideally you want to be on one of Apple's top lists, and offering an app for free might accomplish this.

ACCEPTING ADVERTISING. If you have a very useful app that is open much of the time (a weather app, a stock market app, or even a popular game, for example), it might be advantageous to have a free app with advertising. If you read the comments of users about a certain weather app on the App Store pages, you will notice that customers keep asking for an ad-free option. Use advertising cautiously.

Follow the Rules

Follow the design rules set forth by the developer of the operating system. Apple rejects many apps designed for Apple's iOS because they violate the Human Interface Guidelines (HIG). Interfaces are discussed in detail in Chapter 14.

Design Your Icon

On a smartphone or tablet, the icon is just as import as the app name. Try to design a simple, easily identifiable, and easy-to-remember icon. Figure 11.16 shows an example from App Cooker, developed by Hot Apps Factory. It allows you to see how your icon will look in various different formats.

Choose an Appropriate Name for the App

Keep your app name to 12 or fewer characters. It will appear under the icon on the home screen, and if it is longer than 12 letters, some letters will be dropped when the icon is displayed on a screen.

Design for a Variety of Devices

Even staying within Apple's set of iPads, iPhone, and iPods may mean that you need to design apps that are flexible. Some of the differences among displays are shown in Figure 11.17. You need to supply Apple with a number of different icon sizes, so take that into consideration when you are developing an icon. Icons will look different from one device to another if they are too complex. Simplicity is a virtue when designing icons.

FIGURE 11.16

App Cooker (by Hot Apps Factory) allows a designer to see how an app icon will appear in its many forms.

FIGURE 11.17

Apple smartphones and tablets come in different sizes.

Icon Required	Size for Older iPhones in (pixels)	Size for High-Resolution iPhone (pixels)	Size for Older iPad (pixels)	Size for High-Resolution iPad (pixels)
Display size	320 × 480	640 × 1136	1024 × 768	2048 × 1536
Application icon (Home screen icon)	57 × 57	114 × 114	72 × 72	144 × 144
App icon for the app store	512 × 512	1024 × 1024 recommended	512 × 512	1024 × 1024 recommended
Launch image	320 × 480	640 × 960	768 × 1004 and 1024 × 748	1536 × 2008 and 2048 × 1496
Spotlight search icon	29 × 29	58 × 58	50 × 50	100 × 100
Settings icon	29 × 29	58 × 58	29 × 29	58 × 58
Documents icon	22 × 29	44 × 58	64 × 64 and 320 × 320	128 × 128 and 640 × 640

You should create icons in a vector-based graphic program like Adobe Illustrator. If you use vector-based graphics, you can scale your icons rather than create bitmap images in many different sizes. Start with a 1024 × 1024 design, and you can shrink the graphic to smaller sizes. If you create a 1024 × 1024 design, it will print as a 3.5 × 3.5-inch item at the 300 dots per inch resolution needed for magazines.

Design the Output for the App

While it is possible to use Adobe Illustrator, Visio (on the PC), or OmniGraffle (on a Mac), you might consider using one of the many apps designed specifically for creating apps on handheld devices. Some apps especially useful for designing for Apple iPhones and iPads are AppCraftHD, iMockups, and App Cooker.

Figure 11.18 shows how the App Cooker app allows a designer to pick from a series of widgets. This app, as well as many others, provides templates for iPhones and iPads so that a designer can plan precisely what will appear on the screen when the app is running. The principles for designing smartphone and tablet screens are similar to the ones mentioned earlier in the chapter.

Mockup programs are useful in workflow as well as understanding which pages are connected to other pages. An example of this flow is shown in Figure 1.19. Here, the App Cooker app facilitates design by highlighting the relationship between pages.

Design the Output a Second Time for a Different Orientation

Once you have designed all the screens for an app, you should do it again, but this time turn the devices 90 degrees. If you designed the app in landscape mode, turn the device and design it in portrait mode.

Let's face it. Some apps look better in one mode than another. If you read a full page of text on a smartphone in a browser like Safari, you may be familiar with looking at it in portrait mode. On a tablet, however, the text may appear to look better in landscape mode, even if the number of words are exactly the same.

That is because when text is justified, the gaps between words seem to be larger in portrait mode, making it more difficult to read. On a device like Kindle, changing the orientation changes the number of words on a page and also changes the column structure. Users like to choose their screen orientation.

Developers can use different orientation to their advantage. In portrait mode, a calculator app can present a very simple calculator, but putting it in landscape mode may transform it into a powerful scientific calculator. Giving this easy option to a customer is usually welcomed.

Design the App's Logic

Tablets and smartphones fit in well with the prototyping method of development, but sometimes the best way to approach an app is to sketch out the logic using some of the principles discussed in Chapter 9 on creating process specifications and diagramming structured decisions.

Create the User Interface Using Gestures

Smartphones and tablets have innovative user interfaces (technically called touchscreen capacitive sensing), and you need to design your apps assuming that users will demand touch-sensitive interfaces that use gestures such as swipes, pinches, tugs, and shakes. They will expect different types of feedback and will demand the ability to opt out of some of the features.

Gestures and feedback for tablets and smartphones are discussed in greater detail in Chapter 14, in the context of HCI (human–computer interaction), providing user feedback, and usability.

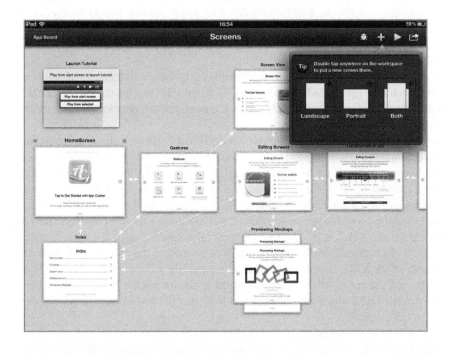

Protect Your Intellectual Property

If you put the effort into design, you need to protect it. Trademark your icons and logos. Recently there have been competitors who introduced apps with one-off logos or icons that were very similar to already popular ones. You might also want to copyright your app. Doing so is not very expensive.

You are also advised to create your own end user license agreement (EULA). Your end user license agreement gives others the right to use your app and limits what a user can do with the app.

When you write a EULA, consider including a warranty disclaimer. Some users will not like an app and will want a refund. This is often unreasonable, especially when a user may have paid only 99 cents to buy the app. Still, users' expectations are not always aligned with yours, so protect yourself. Apple has a default EULA, but our advice is to consider a custom EULA.

The smartphone market and tablet markets are constantly evolving. New operating systems will change the way apps are designed in the future.

Market Your App

If you are fortunate enough to get a potential user to visit your app's product page, you need to convince the person to pay for and download your app. Pay attention to the content on Apple's App Store, which is divided into many sections. To market your app, you will need a large icon, a description, a section explaining what is new in the current version, and a sample set of screen shots. The screen shots selected are very important. All this is mentioned here because good app design is essential to putting your best face forward.

There are many other ways to market your app, but that discussion is beyond the scope of this book. There are many books and websites devoted to marketing apps.

Output Production and XML

Output production varies depending on the platform used. There are many different ways to create output, ranging from database software such as Microsoft Access and FileMaker Pro, to statistical packages such as SAS, to document creators such as Adobe Acrobat.

We discussed XML in Chapter 8. One of the advantages of using XML is that an XML document can be transformed into different output media types. This is done using Cascading Style Sheets (CSSs) or Extensible Stylesheet Language Transformations (XSLT). These methods reinforce the idea that data can be defined once and used many times in different formats.

Using Cascading Style Sheets is an easy way to transform an XML document. A style sheet provides a series of styles, such as font family, size, color, border, and so on, that are linked to the elements of the XML document. These styles may vary for different media, such as a screen, printed output, or a handheld device. The transforming software detects the type of device and applies the correct styles to control the output.

For example, a style used for a flat-panel display might use a rich palette of colors and a sans serif font, which is easier to read on a screen. A different style using a serif font and black or gray color may be used to define a printed report for the same data. A smaller font size might be used for a handheld device or mobile phone.

The drawback of using Cascading Style Sheets is that it does not allow an analyst to manipulate the data, such as rearranging the order of the elements or sorting, and only a limited amount of identifying text, such as captions, may be added. CSS is basically used for formatting.

Using Extensible Stylesheet Language Transformations (XSLT) is a more powerful means of transforming an XML document. XSLT allows an analyst to select the elements and insert them into a web page or another output medium. Figure 11.20 illustrates the transformation process. XSLT is not a programming language but uses a series of statements to define which elements should be output, the sort sequence, the selection of data, and so on. An example of an XML transformation is illustrated in Figure 11.21. The XML is shown on the left, and the result of the transformation is shown on the right. Notice that only the data between the tags (which are contained in less than [<] and greater than [>] symbols) are included in the output.

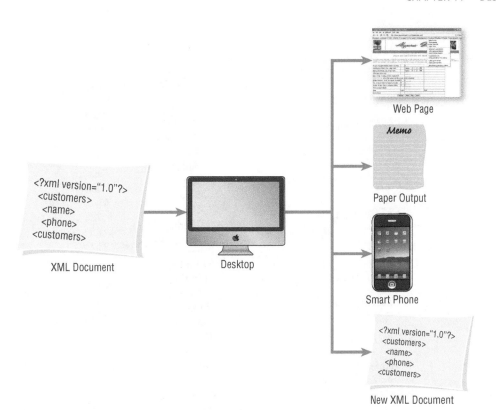

FIGURE 11.20

Extensible Stylesheet Language Transformations (XSLT) software can be used to make XML documents and transform them into many different formats for a variety of platforms.

Ajax

Ajax is a technique that uses both JavaScript and XML to obtain small amounts of data, either plain text or XML, from a server without leaving the web page. This is a big advantage because it means that the entire web page does not need to be reloaded. It works by allowing the web page to reformat itself based on choices that a user inputs. Since Ajax is also related to user input, see additional details in Chapter 12.

Ajax is discussed here because the output implications are important as well. It is up to analysts and designers to determine when data should be added or changed on a web page and to identify the conditions that cause the change. The order in which the questions are asked plays into this design as well.

An example of a web page using Ajax is shown in Figure 11.22, which demonstrates that Ajax makes it possible to display much less data on a page, thereby making the output less cluttered and less confusing. In this example, the user entered one of four ways to narrow down the search to view a list of current customers. The options the user had available were (1) enter the first three digits of a zip code (postal code), (2) enter a telephone area code, (3) select the state, or (4) select a country. The user may not know the postal code or area code and may therefore need to search by state or country, so the options are very useful.

After entering one of the location choices—in this case, the first three digits of the postal code—the user clicked on the **Get Customers** button. The value of the postal code is sent to the server, along with data indicating that it was a postal code. The server then finds all customer records for the selected location, creates an XML document, and sends it to the same web page.

When designing output, a systems analyst has many different options regarding how to display this data on the web page. In this case, the analyst specified that the XML document would be used to create a drop-down list containing all current customers for the desired location. Once a user selects a customer from the drop-down list, more information about the particular customer is displayed, as shown in the example.

The advantage of using Ajax for displaying data is that the user does not have to wait for a new web page to display after making a selection. The Ajax philosophy is to display limited questions for the user to answer on an incremental basis. This eliminates screen clutter. Once the user responds to a question by making a choice, a new question may be generated.

FIGURE 11.21

An XML transformation, with XML on the left and the result of the transformation on the right. Only data between the tags are included in the output on the right.

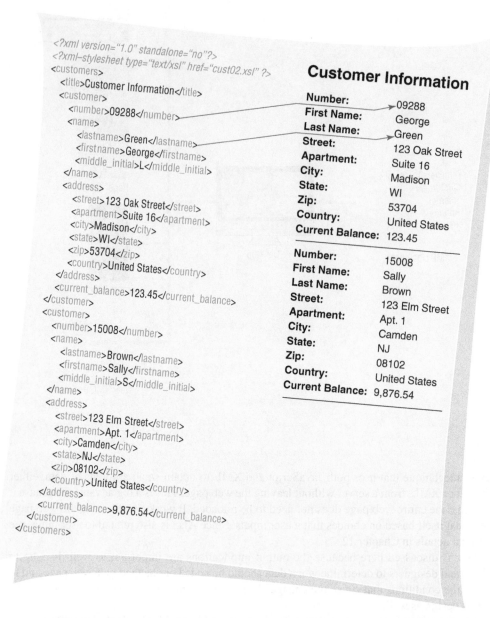

FIGURE 11.22

A web page using Ajax makes it possible to display much less data on a page, allowing an uncluttered display.

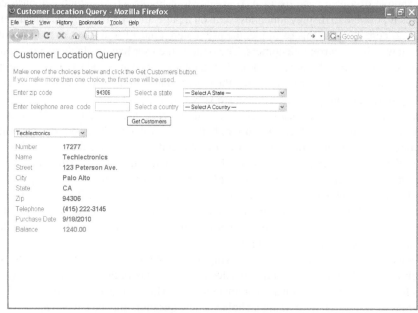

HYPERCASE® EXPERIENCE 11

"I'd say the reception you received, or should I say your team received, for your proposal presentation was quite warm. How did you like meeting Mr. Hyatt? What? He didn't come? Oh [*laughing*], he's his own man. Anyway, don't worry about that too much. The reports I got from Snowden were encouraging. In fact, now he wants to see some preliminary designs from all of you. Can you have something on his desk or send it as an attachment to his email in two weeks? He'll be in Singapore on business next week, but then when he recovers from the jet lag, he'll be looking for those designs. Thanks."

HYPERCASE Questions

1. Consider the reports from the Training Unit. What are Snowden's complaints about these reports? Explain in a paragraph.
2. Using either a layout paper form, Microsoft Visio, or a CASE tool, design a prototype output display based on the Training

Unit's reports that will summarize the following information for Snowden:

Number of accepted projects in the Training Unit.
Number of projects currently being reevaluated.
Training subject areas for which a consultant is being requested.

3. Design an additional output display that you think will support Snowden in the kind of decision making he does frequently.
4. Show your designs to three classmates. Get written feedback from them about how to improve the output displays you have designed.
5. Redesign the displays to capture the improvements suggested by your classmates. In a paragraph, explain how you have addressed each of their concerns.

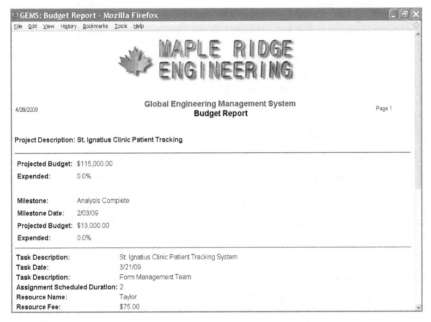

FIGURE 11.HC1

You have the ability to view and critique output screens in HyperCase.

Summary

Output is any useful information or data delivered by the information system, decision support system, the Web, or smartphones or tablets to the user. Output can take virtually any form, including print, computer, smartphone, and tablet displays, audio, microforms, and Web-based documents.

A systems analyst has six main objectives in designing output. They are to design output to serve the intended human and organizational purpose, to fit the user, to deliver the right quantity of output, to deliver it to the right place, to provide output on time, and to choose the right output method.

Analysts should recognize that output content is related to output method. Output of different technologies affects users in different ways. Output technologies also differ in their speed, cost, portability, flexibility, accessibility, storage, and retrieval possibilities. All these factors must be considered when deciding among output options.

The presentation of output can bias users in their interpretation of it. Analysts and users must be aware of the sources of bias. Analysts should interact with users to design and customize output, inform users of the possibilities of bias in output, create flexible and modifiable output, and train users to use multiple outputs to help verify the accuracy of any particular report.

Designing output for user displays is important, especially for DSS, the Web, and smartphones and tablets. Web 2.0 technologies (including social media) should be incorporated into both public-facing and internal web pages to permit user communication and collaboration.

Aesthetics and usefulness are critical when creating well-designed output for displays. It is important to produce prototypes of screens and Web documents that encourage users to interact with them and make changes where desired.

Developing applications, or "apps," that are fun or useful programs for smartphones and tablets has its own distinct development process, but many aspects of agile development, such as quick releases or limited features, as well as prototyping, can be extended to app development. App designers who design apps for market-dominant Apple smartphones and tablets should consider pricing options for apps, developing apps for specific user gestures, designing for different screen orientations on different devices, and legally protecting the intellectual property they create.

Keywords and Phrases

app	green IT
Ajax	Java
audio output	mockups
blog	orientation
bookmark	output bias
browser	output design
Cascading Style Sheets (CSS)	plug-in
collaborative technologies	podcasting
content management system (CMS)	RSS feeds
dashboard	social media
display screen	stickiness
electronic output	storyboarding
email	walled garden
end user license agreement (EULA)	Web 2.0 technologies
Extensible Stylesheet Language Transformations (XSLT)	Webmaster
	web page
frequently asked questions (FAQ)	wiki
General Public License (GPL)	wireframing
gestures	

Review Questions

1. List six objectives an analyst pursues in designing system output.
2. Contrast external outputs with internal outputs produced by a system. Remember to consider differences in external and internal users.
3. List potential electronic output methods for users.
4. What are the drawbacks of electronic and Web-based output?
5. List 10 factors that must be considered when choosing output technology.
6. What output type is best if frequent updates are necessary?
7. What kind of output is desirable if many readers will be reading, storing, and reviewing output over a period of years?
8. What are two of the drawbacks of audio output?
9. List three main ways in which presentations of output are unintentionally biased.
10. What are five ways an analyst can avoid biasing output?
11. Why is it important to show users a prototype output report or display?
12. List three categories of printed reports.
13. Give one difference between exception and summary reports.

14. In what ways do displays, printed output, and Web-based documents differ?
15. List four guidelines to facilitate the design of good display output.
16. What differentiates output for a DSS from that of a more traditional MIS?
17. What are the four primary considerations of an analyst in designing graphical output for decision support systems?
18. Define *stickiness*.
19. List seven guidelines for creating good websites.
20. List five guidelines for using graphics in designing websites.
21. List seven ideas for improving the presentation of corporate websites that you design.
22. What is the "three-clicks" rule?
23. In what ways can you encourage companies to promote their websites that you have developed?
24. What are Web 2.0 technologies? List four of the many in existence.
25. List the five factors analysts should consider when including Web 2.0 technologies in organizational web pages.
26. What is another word for *app*?
27. List 7 of the 14 app development steps that deal with design.
28. List two programs that can help you design output for a smartphone or tablet app.
29. What principles of agile development apply to developing an app?
30. Why is it important to design an app for different orientations (portrait and landscape)?
31. List the six basic options for pricing an app you develop.
32. What are *gestures* in designing smartphones and tablets?
33. What are three ways to protect your app as intellectual property?
34. How do Cascading Style Sheets allow an analyst to produce output?
35. What are the advantages of using XSLT instead of Cascading Style Sheets?
36. What are RSS feeds?
37. What are dashboards mainly used for?
38. What are widgets (or gadgets)?
39. Why should a systems designer be aware of the popularity of widgets (or gadgets)?
40. How do Cascading Style Sheets allow an analyst to produce output?
41. What are the advantages of using Extensible Stylesheet Language Transformations instead of Cascading Style Sheets?
42. How does Ajax help to build effective web pages?

Problems

1. "I'm sure they won't mind if we start sending them the report on these oversized computer sheets. All this time we've been condensing it, retyping it, and sending it to our biggest accounts, but we just can't now. We're so understaffed, we don't have the time," says Otto Breth. "I'll just write a comment here telling them how to respond to this report, and then we can send it out."
   a. What potential problems do you see in casually changing external output? List them.
   b. Discuss in a paragraph how internal and external output can differ in appearance and function.
2. "I don't need to see it very often, but when I do, I have to be able to get at it quickly. I think we lost the last contract because the information I needed was buried in a stack of paper on someone's desk somewhere," says Luke Alover, an architect describing the company's problems to one of the analysts assigned to the new systems project. "What I need is instant information about how much a building of that square footage cost the last time we bid it; what the basic materials such as steel, glass, and concrete now cost from our three top suppliers; who our likely competition on this type of building might be; and who comprises the committee that will be making the final decision on who gets the bid. Right now, though, it's in a hundred reports somewhere. I have to look all over for it."
   a. Given the limited details you have here, write a paragraph to suggest an output method for Luke's use that will solve some of his current problems. In a second paragraph, explain your reasons for choosing the output method you did. (*Hint:* Be sure to relate output method to output content in your answer.)
   b. Luke's current thinking is that no paper record of the output discussed need be kept. In a paragraph, discuss what factors should be weighed before displayed output is used to the exclusion of printed reports.
   c. Make a list of five to seven questions concerning the output's function in the organization that you would ask Luke and others before deciding to do away with any printed reports currently being used.
3. Here are several situations calling for decisions about output content, output methodology, distribution, and so on. For each situation, note the appropriate output decision.

a. A large, well-regarded supplier of key raw materials to your company's production process requires a year-end summary report of totals purchased from it.

b. Internal brainstorming memos are circulated through the staff regarding plans for a company picnic and fund-raiser.

c. A summary report of the company's financial situation is needed by a key decision maker, who will use it when presenting a proposal to potential external backers.

d. A listing of the current night's hotel room reservations is needed for front desk personnel.

e. A listing of the current night's hotel room reservations is needed by the local police.

f. A real-time count of people passing through the gates of Wallaby World (an Australian theme park) will be used by parking lot patrols.

g. An inventory system must register an item each time it has been scanned by a wand.

h. A summary report of merit pay increases allotted to each of 120 employees will be used by 22 supervisors during a joint supervisors' meeting, and subsequently when explaining merit pay increases to the supervisors' own departmental employees.

i. Competitive information is needed by three strategic planners in the organization, but it is industrially sensitive if widely distributed.

j. A casual style of conversation is needed to inform customers about powerful but seldom-used features of a product.

k. A historic district of a city wants to let visitors know about historical buildings and events.

l. Storm warnings must be delivered to subscribers in a large geographical area.

4. "I think I see now where that guy was coming from, but he had me going for a minute there," says Miss deLimit. She is discussing a prototype of display output, one designed by the systems analyst, that she has just seen. "I mean, I never considered it a problem before if even as much as 20 percent of the total class size couldn't be fit into a class," she says. "We know our classes are in demand, and because we can't hire more faculty to cover the areas we need, the adjustment has to come in the student demand. He's got it highlighted as a problem if only 5 percent of the students who want a class can't get in, but that's okay. Now that I know what he means, I'll just ignore it when the computer beeps."

a. In a sentence or two, describe the problem Miss deLimit is experiencing with the display output.

b. Is her solution to "ignore the beeps" a reasonable one given that output is in the prototype stage?

c. In a paragraph, explain how the display output for this particular problem can be changed so that it better reflects the rules of the system Miss deLimit is using.

5. Following is a log sheet for a patient information system used by nurses at a convalescent home to record patient visitors and activities during their shifts. Design a printed report using form design software that provides a summary for the charge nurse of each shift and a report for the activities coordinator at the end of a week. Be sure to use proper conventions to indicate constant data, variable data, and so on. These reports will be used to determine staffing patterns and future activities offerings.

Date	Patient	Visitors	Relationship	Activities
2/14	Clarke	2	Mother, father	Walked about halls, attended chapel, meals in cafeteria
	Coffey	6	Coworkers	Played games, party in room
	Martine	0	—	Meals in room
	Laury	4	Husband and friends	Games in sunroom, watched TV
	Finney	2	Parents	Conversation, meals in cafeteria
	Cartwright	1	Sister	Conversation, crafts room
	Goldstein	2	Sister, brother	Conversation, games out of room, whirlpool

6. Design display output for Problem 5 using form design software. Make any assumptions about system capability necessary and follow display design conventions for onscreen instructions. (*Hint:* You can use more than one display screen if you wish.)

a. In a paragraph, discuss why you designed each report as you did in this problem and Problem 5. What are the major differences in your approach to each one? Can the printed reports be successfully transplanted to displays without changes? Why or why not?

b. Some of the nurses are interested in a Web-based system that patients' families can access from home with a password. Design an output screen for the Web. In a paragraph, describe how your report had to be altered so that it could be viewed by one patient's family.

7. Clancy Corporation manufactures uniforms for police departments worldwide. Its uniforms are chosen by many groups because of their low cost and simple but dignified design. You are helping to design a DSS for Clancy Corporation, and it has asked for tabular output that will help it in making various decisions about what designers to use, where to market its uniforms, and what changes to make to uniforms to keep them looking up-to-date. The following table lists some of the data the company would like to see in tables, including uniform style names, an example of a buyer group for each style, and which designers design which uniform styles. Prepare an example of tabular output for display that incorporates these data about Clancy's. Follow proper conventions for tabular output displays. Use codes and a key, where appropriate.

Style Name	Example of Buyer	Designers
Full military	NYPD	Claudio, Rialtto, Melvin Mine
Half military	LAPD	Rialtto, Calvetti, Duran, Melvin Mine
Formal dress	Australian Armed Forces	Claudio, Dundee, Melvin Mine
Casual dress	"Miami Vice"	Johnson, Melvin Mine

8. Clancy's is interested in graphical output for its DSS. It wants to see a graphical comparison of how many of each style of uniform are being sold each year.
   a. Choose an appropriate graph style and design a graph for display that incorporates the following data:

	Full Military (percent of total)	Half Military	Formal Dress	Casual Dress
2008	50	20	20	10
2009	55	15	20	10
2010	60	15	15	10
2011	62	15	15	8
2012	65	10	15	10

   Be sure to follow proper design conventions for displays. Use codes and a key, if necessary.
   b. Choose a second method of graphing that might allow the decision makers at Clancy's to see a trend in the purchase of particular uniform styles over time. Draw a graph for display as part of the output for Clancy's DSS. Be sure to follow proper design conventions for displays. Use codes and a key if necessary.
   c. In a paragraph, discuss the differences in the two onscreen graphs you have chosen. Defend your choices.
9. Michael Cerveris owns a number of cars used for racing. What performance measures does he need to develop to keep track of the performance of his driver, pit crews, and support staff (not to mention any bald tires his cars experience)?
10. Design a DSS dashboard for Michael from Problem 9. Use appropriate types of charts and graphs to illustrate performance.
11. Design a dashboard for keeping track of a person's stock and portfolio. Think about how the dashboard could be used to make decisions about buying and selling stock. Remember that a client can have more than one stockbroker.
12. Gabriel Shanks runs a nonprofit theatre that produces seven plays per year in three theatres. Each play lasts eight weeks but can be extended four weeks if the show is a success. Design a dashboard for Gabriel, taking into consideration the different phases of putting on a performance as well as the need to sell as many tickets as possible. Don't forget that Gabriel is involved in theatre and is very visual. He doesn't like tables, however.
13. While Gabriel from Problem 12 is taking care of various details during an ordinary day, he would like to keep up on theatre news in Manhattan, at the same time having some simple tools around to help him with his computer-related activities. What sort of widgets and gadgets would Gabriel need to do his job while having some simple computer-based tools always available?
14. Browse the Web to view well-designed and poorly designed websites. Choose three examples of each. Comment on what makes the sites excellent or poor, using the critique form presented earlier in the chapter to compare and contrast them.
15. Propose a website for Clancy's, the uniform company described in Problems 7 and 8. Sketch by hand or use form design software to create a prototype of Clancy's home page. Indicate hyperlinks, and

include a sketch of one hyperlink document. Remember to include graphics, icons, and even sound or other media if appropriate. In a paragraph, describe who the intended users of the website are and state why it makes sense for Clancy's to have a Web presence.

16. Elonzo's Department Stores is a chain of about 50 retail stores, specializing in kitchen, bath, and other household items, including many decorative and fashionable items. Recently Elonzo's decided to automate its gift registry to allow wedding and other event guests to be able to browse for items that were selected by the wedding couple or others.

    a. Design a web page that would allow customers to enter a zip code (postal code) and find the nearest store.

    b. Design a web page for customers to browse gifts and order them online. Do not include the actual ordering forms, simply the products. What sort of options should be available for customers? Include buttons or links to change the sort sequence in your design.

    c. Design a printed list that customers could request when they go to one of the stores. What sequences would be optimal for a customer trying to find items? Would all items requested by the wedding couple be included on the list? (*Hint:* Some may have been purchased already.)

17. Design an outline of a podcast for someone touring your university, college, or business. What sequence would you place the topics in? How much time would you allow for each campus or building location? Assume that the party will arrive in the morning and sequence lunch into the podcast.

18. Draw a diagram that compares the flow of app development through the 14 steps with the seven phases of the SDLC.

19. Write two paragraphs that compare and contrast the process of app development and the SDLC as you diagrammed it in Problem 18.

20. Design an airline flight reminder screen for a smartphone using portrait orientation.

21. Design an airline flight reminder screen for tablet using landscape orientation.

22. Design an Ajax-style web page that would allow a dean at a community college to select part-time instructors. The dean should be able to select a discipline or a course and have the server send an XML document containing all the potential part-time instructors for the selection. The XML document should be used to populate a drop-down list of the instructor names. Clicking an instructor's name would display information about the potential instructor. Decide what information to include that would help the dean make a decision on whom to hire. (*Hint:* Part-time instructors may be able to teach only on certain days or only in the morning, afternoon, or evening.)

Group Projects

1. Brainstorm with your team members about what types of output are most appropriate for a variety of executives and high-level managers of Dizzyland, a large theme park in Florida. Include a list of environments or decision-making situations and types of output. In a paragraph, discuss why the group suggested particular options for output.

2. Have each group member design an output display or form for the output situations you listed in Group Project 1. (Use either Microsoft Visio, a CASE tool, or a paper layout form to complete each display or form.)

3. Create a dashboard for Dizzyland managers in Group Project 1.

4. Design a website, either on paper or using software with which you are familiar, for Dizzyland in Group Project 1. Although you may sketch documents or graphics for three levels of pages and required hyperlinks on paper, create a prototype home page for Dizzyland, indicating hyperlinks where appropriate. Obtain feedback from other groups in your class and modify your design accordingly. In a paragraph, discuss how designing a website is different from designing displays for other online systems.

5. Without looking at your phone or tablet, design three sample pages for the following apps. You do not need to use a mock-up app for this assignment. Just draw them on paper. (Note that each group may be assigned a different app to do).

    a. A grocery list helper

    b. A hotel and room finder

    c. An app to help you get to sleep

    d. A task management (to-do) list

    e. An apartment finder

    f. A weather app

6. Explore the Joomla! website, at www.joomla.org. How could this open source application be helpful in implementing your designs from Group Project 4? Summarize your findings in a paragraph. Find another CMS on the Web and write a paragraph comparing it to Joomla!. Be sure to address cost, ease of use, support, and availability in your comparison.

7. Use brainstorming to develop a new set of widgets (gadgets) to be more productive. Come up with a list of your top five bright ideas for new widgets.

Selected Bibliography

Davenport, T. H. "Saving IT's Soul: Human-Centered Information Management." *Harvard Business Review,* March–April 1994, pp. 119–131.

Davis, G. B., and H. M. Olson. *Management Information Systems, Conceptual Foundations, Structure, and Development,* 2nd ed. New York: McGraw-Hill, 1985.

Kendall, K. E., and J. E. Kendall. "DSS Systems Analysis and Design: The Role of the Analyst as Change Agent from Early DSS to Mashups," in *Handbook of Decision Support Systems 2*, edited by F. Burstein and C. W. Holsapple, pp. 293–312. Berlin: Springer, 2008.

Kendall, J. E., K. E. Kendall, and H. Mirakula. "Systems Design Considerations for Web 2.0 Technologies," research presentation, Decision Sciences Institute Annual Meeting, Boston, November 2011.

Nevo, D., I. Benbasat, and Y. Wand. "Understanding Technology Support for Organizational Transactive Memory: Requirements, Application, and Customization." *Journal of Management Information Systems,* Vol. 28, No. 4, 2012, pp. 69–98.

Souders, S. "High-Performance Websites." *Communications of the ACM,* Vol. 51, No. 12, December 2008, pp. 36–41.

The CPU Case Episode and accompanying student files are available online at www.pearsonglobaleditions.com/kendall.

CHAPTER 12

Designing Effective Input

LEARNING OBJECTIVES

Once you have mastered the material in this chapter you will be able to:

1. Design functional input forms for users of business systems.
2. Design engaging input displays for users of information systems.
3. Design useful input forms for people interacting on the Web.
4. Design useful input pages for users of intranets, the Web, smartphones, and tablets.

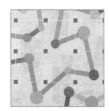

Users deserve quality output. The quality of system input determines the quality of system output. It is vital that input forms, displays, and interactive Web documents be designed with this critical relationship in mind.

Well-designed input forms, displays, and interactive Web fill-in forms should meet the objectives of effectiveness, accuracy, ease of use, consistency, simplicity, and attractiveness. All these objectives are attainable through the use of basic design principles, knowledge of what is needed as input for the system, and an understanding of how users respond to different elements of forms and displays.

Effectiveness means that input forms, input displays, and fill-in forms on the Web all serve specific purposes for users of the information system, whereas accuracy refers to design that ensures proper completion. Ease of use means that forms and displays are straightforward and require no extra time for users to decipher. Consistency means that all input forms, whether they are input displays or fill-in forms on the Web, group data similarly from one application to the next, whereas simplicity refers to keeping those same designs uncluttered in a manner that focuses the user's attention. Attractiveness implies that users will enjoy using input forms because of their appealing design.

Good Form Design

A systems analyst should be capable of designing a complete and useful form. Unnecessary forms that waste an organization's resources should be eliminated.

Forms are important instruments for steering the course of work. They are preprinted papers that require people to fill in responses in a standardized way. Forms elicit and capture information required by organizational members that will often be input to the computer. Through this process, forms often serve as source documents for users or for input to ecommerce applications that humans must enter.

To design forms that people find useful, four guidelines for form design should be observed:

1. Make forms easy to fill in.
2. Ensure that forms meet the purpose for which they are designed.
3. Design forms to ensure accurate completion.
4. Keep forms attractive.

Each of the four guidelines is considered separately in the following sections.

Making Forms Easy to Fill In

To reduce error, speed completion, and facilitate the entry of data, it is essential that forms be easy to fill in. The cost of forms is minimal compared with the cost of the time employees spend filling them in and then entering data into the information system. It is often possible to eliminate the process of transcribing data that are entered on a form into the system by using electronic submission. That method often features data keyed in by users themselves, who visit websites set up for informational or ecommerce transactions.

FORM FLOW. Designing a form with proper flow can minimize the time and effort employees expend in form completion. Forms should flow from left to right and top to bottom. Illogical flow takes extra time and is frustrating. A form that requires people to go directly to the bottom of the form and then skip back up to the top for completion has poor flow.

SEVEN SECTIONS OF A FORM. A second method that makes it easy for people to fill out forms correctly is logical grouping of information. The seven main sections of a form are the following:

1. Heading
2. Identification and access
3. Instructions
4. Body
5. Signature and verification
6. Totals
7. Comments

Ideally, these sections should appear on a page grouped as they are on the Bakerloo Brothers Employee Expense Voucher in Figure 12.1. Notice that the seven sections cover the basic information required on most forms. The top quarter of the form is devoted to three sections: the heading, the identification and access section, and the instructions section.

The heading section usually includes the name and address of the business originating the form. The identification and access section includes codes that may be used to file the report and gain access to it at a later date. (In Chapter 13, we discuss in detail how to access specially keyed information in a database.) This information is very important when an organization is required to keep the document for a specified number of years. The instructions section tells how the form should be filled out and where it should be routed when complete.

The middle of the form is its body, which composes approximately half of the form. This part of the form requires the most detail and development from the person completing it. The body is the part of the form most likely to contain explicit, variable data.

The bottom quarter of the form is composed of three sections: signature and verification, totals, and comments. Requiring ending totals and a summary of comments is a logical way to provide closure for the person filling out the form.

There is one more feature to notice about the Bakerloo Brothers form. The form design provides an internal double-check, with column totals and row totals expected to add up to the same number. If the row and column totals don't add up to the same number, the employee filling out the form

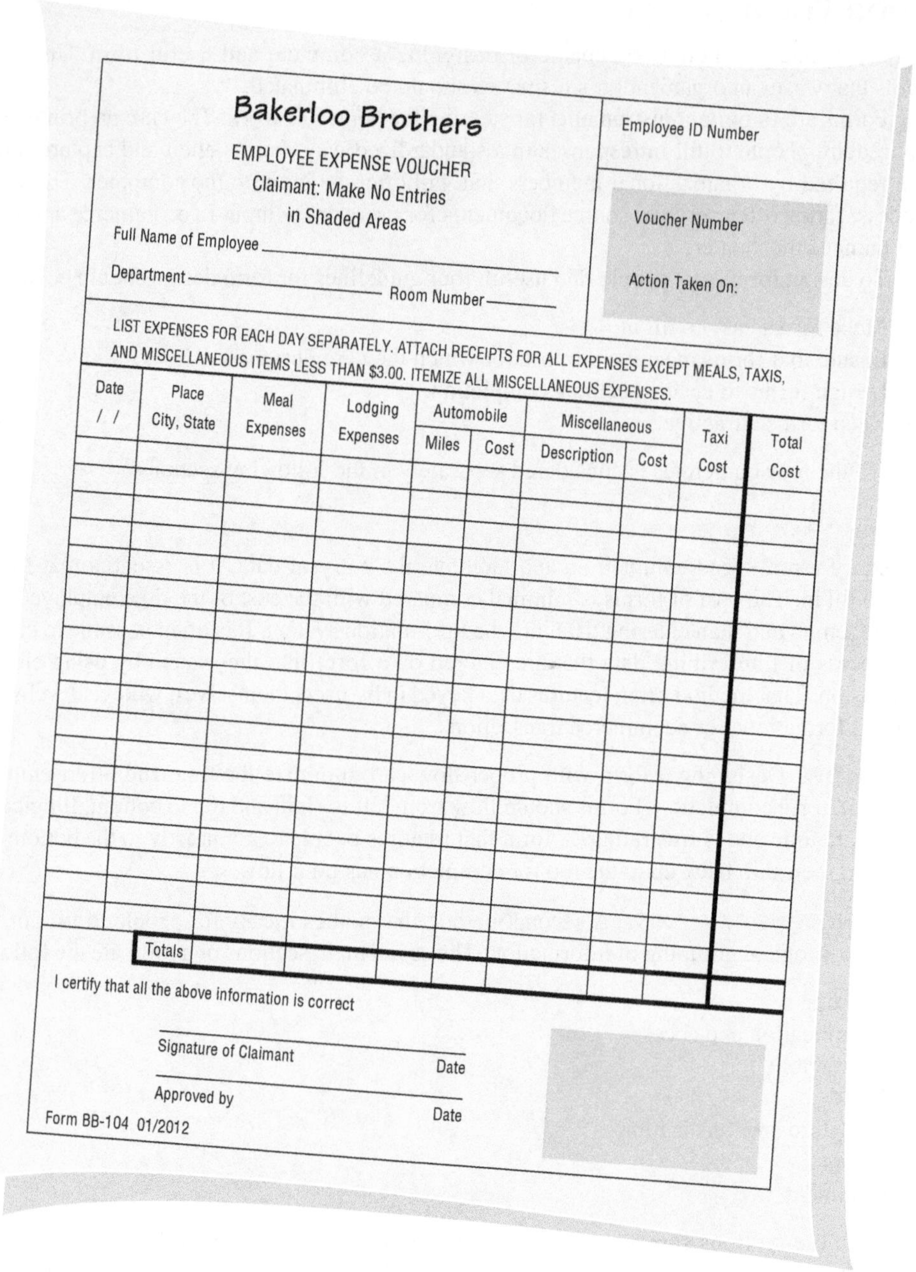

Bakerloo Brothers

EMPLOYEE EXPENSE VOUCHER

Claimant: Make No Entries in Shaded Areas

Employee ID Number

Voucher Number

Action Taken On:

Full Name of Employee ____________

Department ____________ Room Number ______

LIST EXPENSES FOR EACH DAY SEPARATELY. ATTACH RECEIPTS FOR ALL EXPENSES EXCEPT MEALS, TAXIS, AND MISCELLANEOUS ITEMS LESS THAN $3.00. ITEMIZE ALL MISCELLANEOUS EXPENSES.

Date / /	Place City, State	Meal Expenses	Lodging Expenses	Automobile		Miscellaneous		Taxi Cost	Total Cost
				Miles	Cost	Description	Cost		
Totals									

I certify that all the above information is correct

Signature of Claimant ____________ Date

Approved by ____________ Date

Form BB-104 01/2012

FIGURE 12.1

Seven sections found in a well-designed form help encourage completion.

knows there is a problem and can correct it on the spot. An error is prevented, and the employee can be reimbursed the amount due; both outcomes are attributable to the suitable form design.

CAPTIONING. Clear captioning is another technique that can make easy work of filling out a form. Captions tell the person completing the form what to put in a blank line, space, or box. Several options for captioning are shown in Figure 12.2. Two types of line captions, two types of check-off captions, and examples of a boxed caption and table caption are shown.

The advantage of putting the caption below the line is that there is more room on the line itself for data. The disadvantage is that it is sometimes unclear which line is associated with the caption—the line above or below the caption.

Line captions can be to the left of blanks and on the same line, or they can be printed below the line on which data will be entered.

Another way to caption is to provide a box for data instead of a line. Captions can be placed inside, above, or below the box. Boxes on forms help people enter data in the correct place, and they also make reading the form easier for the form's recipient. The caption should use a small type size so that it does not dominate the entry area. Small vertical tick marks may be included

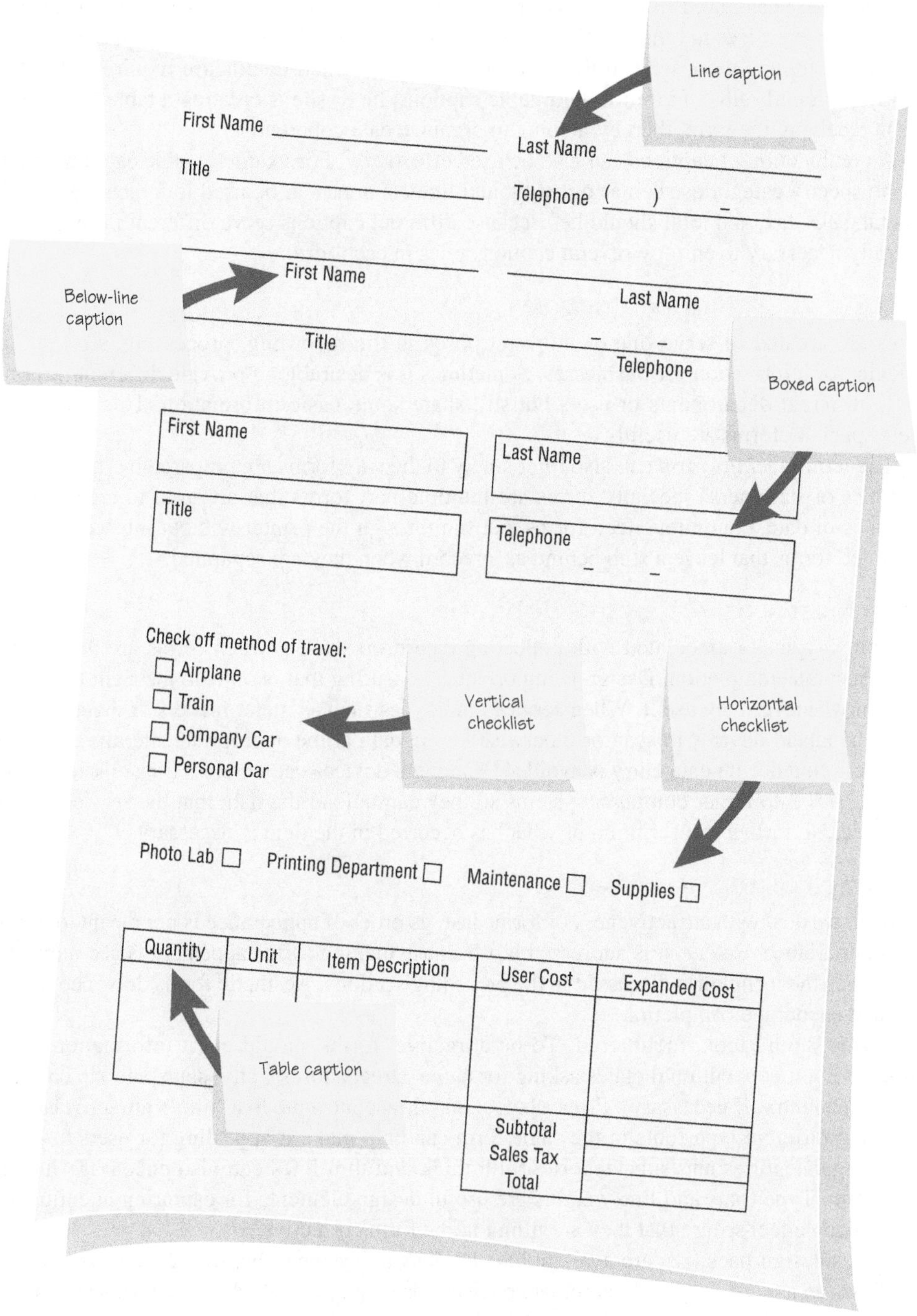

FIGURE 12.2

Major captioning alternatives.

in the box if the data is intended for entry into a computer system. If there is not enough room on a record for the data, the person filling out the form, rather than the data entry operator, has the freedom to determine how the data should be abbreviated. Captions may also include small clarification notes to help the user correctly enter the information, such as Date (MM/DD/YYYY) or Name (Last, First, Middle Initial).

Whatever styles of line caption are chosen, it is important to employ them consistently. For instance, it is confusing to fill out a form that has both above- and below-line captions.

Check-off captions are superior when response options are necessarily restricted. Notice the list of travel methods shown for the vertical check-off example in Figure 12.2. If employee expenses for business travel are reimbursed only for the travel methods listed, a check-off system is more expedient than a blank line. This method has the added advantage of reminding the person who is verifying the data to look for an airline ticket stub or other receipt.

A horizontal check-off caption is also superior to a line caption when information required is routine and constant. An example is a form that would request services from one of the following departments: Photo Lab, Printing Department, Maintenance, or Supplies. The departments routinely provide services to others in the organization and are not likely to change quickly.

Table captions work well in the body of a form on which details are required. When an employee properly fills out a form with table captions, he or she is creating a table for the next person receiving the form, thereby helping to organize data coherently.

A combination of captions can also be used effectively. For example, table captions can be used to specify categories such as quantity, and line captions can be used to indicate where the subtotal, sales tax, and total should be. Because different captions serve different purposes, it is generally necessary to employ several caption styles in each form.

Meeting the Intended Purpose

Forms are created to serve one or more purposes in the recording, processing, storing, and retrieving of information for businesses. Sometimes it is desirable to provide different information to different departments or users but still share some basic information. This situation is where specialty forms are useful.

The term *specialty form* can also refer solely to the way forms are prepared by the stationer. Examples of stationers' specialty forms are multiple-part forms that are used to create instant triplicates of data, continuous-feed forms that run through the printer without intervention, and perforated forms that leave a stub behind as a record when they are separated.

Ensuring Accurate Completion

Error rates typically associated with collecting data drop sharply when forms are designed to ensure accurate completion. Design is important for ensuring that people do the right thing with the form whenever they use it. When service employees such as meter readers or inventory takers use handheld devices to scan or otherwise key in data at the appropriate site, the extra step of transcription during data entry is avoided. Handheld devices use wireless transmission, or are plugged back into larger computer systems so they can upload the data that the service worker has stored. No further transcription of what has occurred in the field is necessary.

Keeping Forms Attractive

Although we deal with attractiveness of forms last, its order of appearance is not meant to diminish its importance. Rather, it is addressed last because making forms appealing is accomplished by applying the techniques discussed in the preceding sections. Aesthetic forms draw people into them and encourage completion.

Forms should look uncluttered. To be attractive, forms should elicit information in the expected order; convention dictates asking for name, street address, city, state, and zip or postal code (and country, if necessary). Proper layout and flow contribute to a form's attractiveness.

Using different type fonts in the same form can help make it appealing for users to fill in. Separating categories and subcategories with thick and thin lines can also encourage interest in the form. Type fonts and line weights are useful design elements for capturing attention and making people feel secure that they are filling in the form correctly.

Forms design packages are available for all desktop systems. Figure 12.3 shows software that allows an analyst to quickly automate business processes for which paper forms are already in existence. The analyst can use a set of tools to set up fields, check boxes, lines, boxes, and many other features. Paper forms can also be scanned in and then published to the Web.

Controlling Business Forms

Controlling business forms is an important task. Businesses often have a forms specialist who controls forms, but sometimes this job falls to a systems analyst, who sets up and implements forms control.

The basic duties for controlling forms include making sure that each form in use fulfills its specific purpose in helping workers accomplish their tasks and that the specified purpose is integral to organizational functioning, preventing duplication of the information that is collected and of the forms that collect it, designing effective forms, deciding on how to reproduce forms in the most economical way, and establishing procedures that make forms available (when needed) at the lowest possible cost. Often this entails making forms available on the Web for printing out. A unique form number and revision date (month/year) should be included on each form, regardless of whether it is completed and submitted manually or electronically. This helps users be organized and efficient.

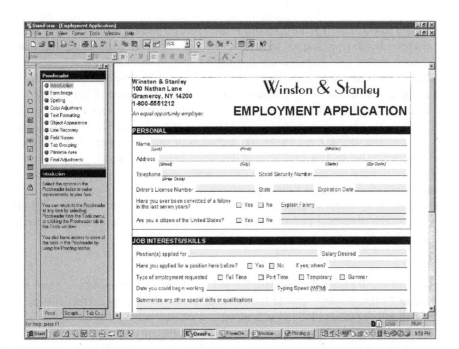

FIGURE 12.3

Software allows a user to scan an existing form into the computer and define fields so that the form can be easily filled out on a PC. (Screenshots courtesy of www.tableausoftware.com. Reprinted with permission.)

Good Display and Web Forms Design

Much of what we have already said about good form design is transferable to display design and the design of websites, web pages, smartphones, and tablets. Once again, the users must remain foremost in an analyst's thoughts during the design of displays.

There are differences, however, and systems analysts should strive to realize the unique qualities of displays rather than to adopt blindly the conventions of paper forms. One big difference is the constant presence of a cursor on the display, which orients the user to the current data entry position. As data are entered onscreen, the cursor moves one character ahead, pointing the way.

Another major difference among electronic, Web, and static forms is that designers can include context-sensitive user help in any electronic fill-in form. This practice can reduce the need for instructions being shown for each line, thus reducing the clutter of the form and cutting down on calls to technical support. Using a Web-based approach also permits the designer to take advantage of hyperlinks, thus ensuring that the forms are filled out correctly by providing users with hyperlinked examples of correctly completed forms.

In this section, we present guidelines for effective display design. They are presented in order to aid the attainment of the overall input design goals of effectiveness, accuracy, ease of use, simplicity, consistency, and attractiveness. The following four guidelines for display design are important but not exhaustive (refer to Chapter 11):

1. Keep the display simple.
2. Keep the display presentation consistent.
3. Facilitate user movement among display screens and pages.
4. Create an attractive and pleasing display.

In the next subsections, we develop each of these guidelines and present many design techniques for observing the four guidelines.

Keeping the Display Simple

The first guideline for good display design is to keep the display simple. The display should show only that which is necessary for the particular action being undertaken. For the occasional user, 50 percent of the display area should contain useful information.

THREE SCREEN SECTIONS. Display output should be divided into three sections. The top of the screen features a heading section. The heading contains titles of software and open files, pull-down menus, and icons that do certain tasks.

The middle section is called the body of the display. The body can be used for data entry and is organized from left to right and top to bottom, because people in Western cultures move their eyes over a page in this way. Captions and instructions should be supplied in this section to help the user enter the pertinent data in the right place. Context-sensitive help can also be made available by having the user click the right mouse button in the body section of the display.

The third section of the display is the comments and instructions section. This section may display a short menu of commands that remind the user of basics such as how to change pages or functions, save the file, or terminate entry. Inclusion of such basics can make inexperienced users feel infinitely more secure about their ability to complete their task.

Other ways to keep the display simple are to use context-sensitive help, roll-over buttons that reveal more information, and other pop-up windows. Also, keep in mind that users can minimize or maximize windows as needed. In this way, users start with a simple, well-designed display that they can customize and control through the use of multiple windows. Hyperlinks on a Web-based fill-in form serve a similar purpose.

Keeping the Display Consistent

The second guideline for good display design is to keep the display consistent. If users are working from paper forms, displays should follow what is shown on paper. Displays can be kept consistent by locating information in the same area each time a new display is accessed. Also, information that logically belongs together should be consistently grouped together: Name and address go together, not name and zip code. Although the display should have a natural movement from one region to another, information should not overlap from one group to another. You would not want name and address in one area and zip code in another.

Facilitating Movement

The third guideline for good display design is to make it easy to move from one page to another. The "three-clicks" rule says that users should be able to get to the pages they need within three mouse clicks or keyboard strokes. Web-based forms facilitate movement with the use of hyperlinks to other relevant web pages. Another common method for movement is to have users feel as if they are physically moving to a new page. This illusion of physical movement among screens can be obtained by using scrolling arrows, context-sensitive pop-up windows, or onscreen dialog.

Designing an Attractive and Pleasing Display

The fourth guideline for good display design is to create an attractive display. If users find displays appealing, they are likely to be more productive, need less supervision, and make fewer errors. Displays should draw users into them and hold their attention. This goal is accomplished with the use of plenty of open area surrounding data entry fields so that the display achieves an uncluttered appearance. You should never crowd a form; similarly, you should never crowd a display. You are far better off using multiple windows or hyperlinks than jamming everything onto one page.

Use logical flows in the plan to your display pages. Organize material to take advantage of the way people conceptualize their work so that they can easily find their way around. With the advent of GUIs, it is possible to make input displays very attractive. By using color or shaded boxes and creating three-dimensional boxes and arrows, you can make forms user friendly and fun to use.

When contemplating the use of different font styles and sizes, ask yourself if they truly assist the user in understanding and approving of the display. If they draw undue attention to the art of display design or if they serve as a distraction, leave them out. Be aware that not all web pages are viewed identically by different browsers. Test your prototype forms with a variety of combinations to see if users declare preferences for combinations or whether they are distressing to the majority of users. For Web fonts, use Verdana or Arial.

Using Icons in Display Design

Icons are pictorial, onscreen representations symbolizing computer actions that users may select using a mouse, keyboard, lightpen, touch screen, or joystick. Icons serve functions similar to those of words and may replace them in many menus, because their meaning is more quickly grasped than words. Mobile devices such as Apple's iPhone and iPod have popularized the use of icons on touch screens and made this a familiar interface for many businesses and other users.

There are some guidelines for the design of effective icons. Shapes should be readily recognizable so that the user is not required to master a new vocabulary. Numerous icons are already known to most users. Use of standard icons can quickly tap into this reservoir of common meaning. A user

CONSULTING OPPORTUNITY 12.1

This Form May Be Hazardous to Your Health

Figure 12.C1 is a printed medical history form that Dr. Mike Robe, a family practitioner, has his receptionist give to all new patients. All patients must fill it out before they see the doctor.

The receptionist is getting back many incomplete or confusing responses, which makes it difficult for Dr. Robe to review the forms and understand why the new patient is there. In addition, the poor responses make it time consuming for the receptionist to enter new patients into the files.

Redesign the form on 8½" × 11" paper so that pertinent new patient data can be collected in a logical and inoffensive way. Make sure the form is self-explanatory to new patients. It should also be easy for Dr. Robe to read and easy for the receptionist to enter into the patient database, which is sorted by patient name and unique seven-digit account ID number. Recently all of the doctors associated with Mike Robe's practice decided to experiment with patients filling out this form using an iPad issued by the office. Redesign the form for both landscape and portrait orientation for an iPad. Which office procedures would you have to change to accommodate patient data entry with an iPad?

Medical History Form

Name _____ Employer _____ Age _____

Address _____ Zip _____ Phone _____ Office _____

Insurer _____ Is this [ ] your policy [ ] your spouse's policy

Blue Cross [ ]   State Physician's Service [ ]   Other [ ]   (state) _____

Have you ever had surgery? Yes____ No____ If so, when? _____

Describe the surgery _____

Have you ever been hospitalized? Yes____ No____ If so, when? _____

Why? _____

Complete the following.

	I have had	Family history
Diabetes	☐	☐
Heart trouble	☐	☐
Cancer	☐	☐
Seizure	☐	☐
Fainting	☐	☐

What have you been immunized for?

Family: _____
            Spouse or next of kin   Relationship   Address

Date of last exam ___/___ Who referred you? _____

Why are you seeing the doctor today? _____

Are you currently having pain? _____ Constant _____ Sporadic _____

How long does it last? _____ Please give us your soc. sec. # _____

IMPORTANT! We need your correct insurance carrier number _____

FIGURE 12.C1

Your help in improving this form is greatly appreciated.

CONSULTING OPPORTUNITY 12.2

Squeezin' Isn't Pleasin'

The Audiology Department in a large veteran's hospital is using a PC and monitor so that audiology technicians can enter data directly into the patient records system. After talking with Earl Lobes, one of the technicians, you determine that the screen design is a major problem.

"We used a form at one time, and that was decent," said Mr. Lobes. "The display didn't make sense, though. I guess they had to squeeze everything on there, and that ruined it."

You have been asked to redesign the display (see Figure 12.C2) to capture the same information but simplify it, and by doing so reduce the errors that have been plaguing the technicians. You realize that squeezing isn't the only problem with the display.

Explain your reasons for changing the display as you did. You may use more than one display page if you think it is necessary.

```
                    AUDIOLOGICAL EXAMINATION REPORT

    Patient Last Name              First              Middle Initial
    Examining Station              Date of Exam
    Patient Number                 Social Security Number
    First Exam          Claim number

                            AIR CONDUCTION
           Right ear                         Left ear
    500  1000  2000  4000  6000      500  1000  2000  4000  6000
    □    □     □     □     □          □    □     □     □     □

                           BONE CONDUCTION
           Right ear                         Left ear
    500  1000  2000  4000  6000      500  1000  2000  4000  6000
    □    □     □     □     □          □    □     □     □     □

    SPEECH AUDIOMETRY SECT.         Comments [
    SPEECH RECEP.  THRESHOLD
    Right Ear [  ]
    Left Ear [  ]                   Referred by [              ]
    RIGHT EAR DISCR.               Reason for referral
    % [  ] Masking [  ]            Examining Audiologist
    LEFT EAR DISCRIM. Exam. Audiologist's No.
    % [  ] Masking [  ]            Next Appt.
```

FIGURE 12.C2

This screen can be designed to be more user friendly.

may point to a file cabinet, "pull out" a file folder icon, "grab" a piece of paper icon, and "throw" it in the wastebasket icon. By employing standard icons, designers and users all save time.

Icons for a particular application should be limited to approximately 20 recognizable shapes so that icon vocabulary is not overwhelming and so that a worthwhile coding scheme can still be realized. Use icons consistently throughout applications where they will appear together to ensure continuity and understandability. Generally, icons are worthwhile for users if they are meaningful.

Graphical User Interface Design

Users interface with an operating system (OS), such as Windows or the Mac OS, through a graphical user interface (GUI, pronounced "GOO-ey"), also referred to as a point-and-click interface. Users can use a mouse to click on an object and drag it into position. Graphical user interfaces take advantage of additional features in display design such as text boxes, check boxes, option buttons, list and drop-down list boxes, sliders and spin buttons, tab control dialog boxes, and image maps. Figure 12.4 is a Microsoft Access input display showing a variety of GUI controls.

TEXT BOXES. A rectangle represents a text box and is used to outline data entry and display fields. Care must be taken to ensure that a text box is large enough to accommodate all the

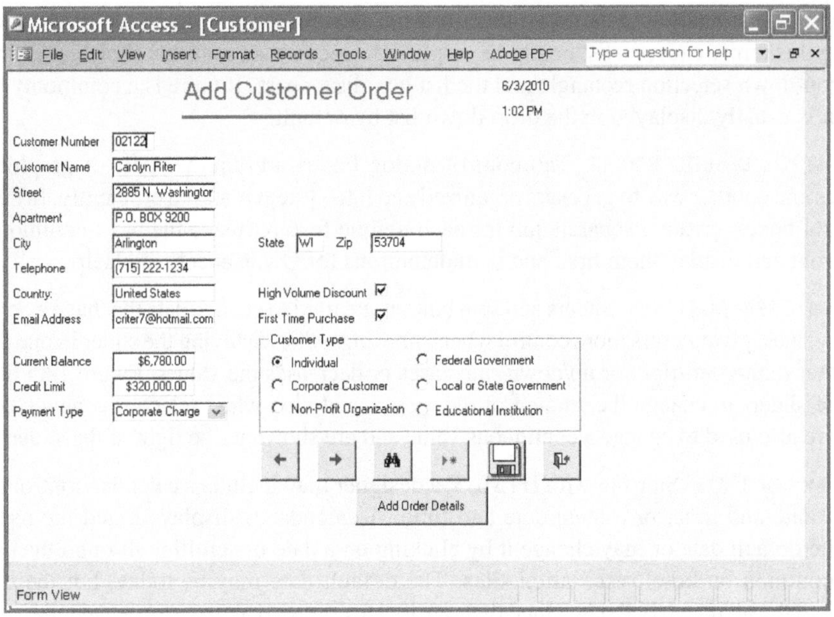

FIGURE 12.4

The many available GUI components allow flexibility in designing input screens for the Web or other software packages. This example is from Microsoft Access.

characters that must be entered. Each text box should have a caption to the left, identifying what is to be entered or what is displayed in the box. In Microsoft Access, character data are left-aligned, and numeric data are right-aligned.

Starting with HTML5 browsers, a feature called a *placeholder* allows a designer to put a small amount of text into a text box, which will display in a lighter color, informing the viewer what to key into the text box. When the cursor is placed in the field (clicking or tabbing into it), the text vanishes.

Several new types of text boxes are available with HTML5. These are email, telephone, and URL (a Web address). These appear as normal text boxes on a computer, but when used on a tablet or smartphone, they can be used to customize the pop-up keyboard. For example, if the text box is used for a telephone number, the virtual keyboard layout changes to a number pad. If the text box is used to enter a URL, the virtual keyboard includes a *.com* button. Finally, if the text box is used to enter an email address, the virtual keyboard includes an @ symbol. This customization helps the user enter data quickly and accurately.

A datalist is a new feature included in HTML5 that displays a drop-down list of predefined suggestions to make entry easier for the user. As the user begins to type the first few letters, the datalist is displayed. The user can choose one of the items in the list to make a selection. This is used in the autocomplete function.

CHECK BOXES. In the GUI controls example, a check box is used to indicate a new customer. Check boxes contain an X or are empty, depending on whether the user selected the option; they are used for nonexclusive choices in which one or more of the options may be checked. An alternative notation is to use a square button with a check mark (✓) to indicate that the option has been selected. Note that check box text, or a label, is usually placed to the right of the box. If there is more than one check box, the labels should have some order to them, either alphabetic or with the most commonly checked item appearing first in a list. If there are more than 10 check boxes, you can group them together in a bordered box.

OPTION BUTTONS. An option button, also called a radio button, is used to select exclusive choices. Only one of several options can be chosen. By using option buttons, you can make it clear to users that they must decide among options. Choices are again listed to the right of the button, usually in some sequence. If there is a commonly selected option, it is usually selected as a default when the page first displays. Often there is a rectangle, called an option group, surrounding the radio buttons. If there are more than six option buttons, you should consider using a list box or a drop-down list box.

LIST AND DROP-DOWN LIST BOXES. A list box displays several options that may be selected with the mouse. A drop-down list box is used when there is little room available on the page. A single rectangle with an arrow points down toward a line located on the right side of the rectangle. Selecting this arrow causes a list box to be displayed. Once a user makes a choice, it is displayed in the drop-down selection rectangle and the list box disappears. If there is a commonly selected choice, it is usually displayed in the drop-down list by default.

TAB CONTROL DIALOG BOXES. Tab control dialog boxes are another part of graphical user interfaces and another way to get users organized and into system material efficiently. In designing tab control boxes, create a separate tab for each unique feature, place the most commonly used tabs in front and display them first, and include buttons for OK, Cancel, and Help.

SLIDERS AND SPIN BUTTONS. Sliders and spin buttons are used to change data that have a continuous range of values, giving users more control when choosing values. Moving the slider in one direction or the other (either left/right or up/down) increases or decreases the values. Figure 12.5 illustrates the use of sliders to change the amount of red, green, and blue when selecting a new color. Spin buttons are also used to change a continuous value and are shown to the right of the sliders.

CALENDAR CONTROL. Starting with HTML5, a designer may include a calendar control to select a date, a date and time, or a local date and time. A calendar is displayed, and the user either selects the default date or may change it by clicking on a date or scrolling through the calendar. This is common on hotel reservation sites. The default date may be today, but the user will change it to the actual arrival day. Once the user locates the arrival date on the calendar, a second calendar pops up with the next day as the default departure date. The user again changes this to the correct departure date, but the process of selecting dates from a pop-up calendar is still easier and less error-prone than entering text.

IMAGE MAPS. Image map fields are used to select values within an image. The user clicks on a point within an image, and the corresponding *x*- and *y*-coordinates are sent to the program. Image maps are used when creating web pages containing maps with instructions to click in a certain area in order to view a detailed map of the region.

TEXT AREAS. A text area is used for entering a larger amount of text. These areas include a number of rows, columns, and scroll bars that allow the user to enter and view text greater than the size of the box area. There are two ways to handle this text. One is to avoid the use of word wrap, forcing the user to press the Enter key to move to the next line; the text will scroll to the right if it exceeds the width of the text area. The other option is to allow word wrap.

MESSAGE BOXES. Message boxes are used to warn users and provide other feedback messages in a dialog box, often overlapping the display. Message boxes have different formats. Each should appear in a rectangular window and should clearly spell out the message so that the user knows precisely what is happening and what actions are possible.

COMMAND BUTTONS. A command button performs an action when the user selects it with the mouse. Calculate Total, Add Order, and OK are all examples. The text is centered inside the button, which has a rectangular shape. If there is a default action, the text is surrounded with a dashed line. The button may also be shaded to indicate that it is the default. Users press the Enter key to select the default button.

FIGURE 12.5

Sliders and spin buttons are two GUI components an analyst can use to design input screens.

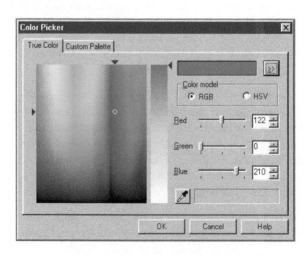

Form Controls and Values

Each of the controls included in a GUI interface must have some way of storing the data associated with the control. On a web page this is done using a name and value pair that is transmitted to the server or in an email sent along with the form (such as the name city and the value Paris). The name is defined on the web page form, and the server software must recognize the name to understand what to do with the value or data sent with the Web form.

How the value is obtained differs for each Web form control. In text boxes or text areas, the value consists of the characters keyed into the boxes. In radio buttons and check boxes, the text that displays to the right of each radio button or check box is for human use only. The value is defined in the Web form and is transmitted when the form is sent. If these data are used to update a database, the values are often codes that are sent and then stored at the server, and the analyst must decide what the appropriate values should be when each radio button or check box is clicked. Drop-down lists are somewhat different from radio buttons or check boxes in that there are many options for a given drop-down list. Values must be decided for each of the drop-down list options and, when an option is chosen, the selected value is sent with the form. Form values may also be used in calculations that are performed using JavaScript on the browser. These may be used to multiply, add, and make decisions.

Figure 12.6 is an example of a form used to obtain prices and to register for a cruise. The text in the Name, Address, City, State, Zip, Phone, and Email areas is sent to the server when the form is submitted. Only one of the radio buttons for the 4-day, 7-day, or 14-day cruise may be selected. The values sent are S for short if the 4-day cruise is selected, A for average length if 7 days has been selected, and L for a long cruise if the 14-day cruise is selected. In addition, when one of these cruises is selected, the dollar amount is inserted into one of the text boxes on the left side of the Web form, and any previously selected radio buttons and amounts are cleared. If the ocean side room check box is checked, a value of Y for yes is transmitted to the server, the amount is inserted into the left-side text box, and the total is updated. If the customer tries to change the amounts in the calculated text boxes, they are reset. When the submit button is clicked, the amounts are sent to the server along with all the other data.

Hidden Fields

Another type of control found on Web forms is a hidden field. These are not visible to the viewer, do not take up any space on the web page, and can contain only a name and a value. Often hidden fields are used to store values sent from one Web form to the server. These typically need to be included on a second form when multiple forms are required to capture all the transaction data. Sometimes they are used to retain information about the type of browser being used, the viewer's

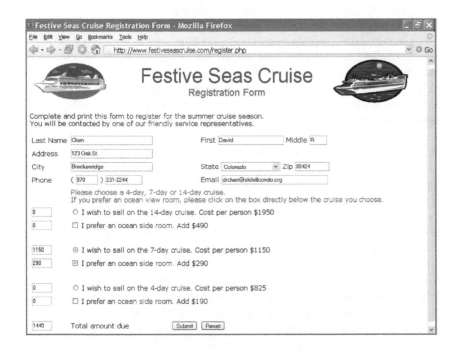

FIGURE 12.6

A Web-based input form for users to register for a cruise.

FIGURE 12.7

A website that permits users to estimate the cost of staying at Azure Islé Resort. The cost depends on the number of people included, length of stay, and extra activities added.

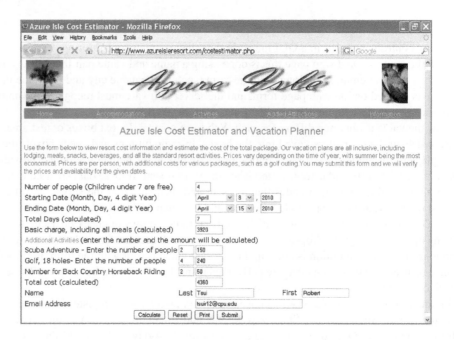

operating system, and so on. Sometimes a hidden field contains a key field used to locate a record for the customer or the browsing session.

Event-Response Charts

When there are complicated interactions on a Web form (or any other GUI form), an event-response chart may be used to list the variety of events that can occur. Event-response charts may be used at a high level to model business events and responses (covered in Chapter 7), but the events that occur on a Web form or other display are usually limited to user actions. These events may be clicking a button, changing a value, focusing the field (moving the cursor inside the field or to a radio button, check box, or other control), blurring a field (the user moves the cursor out of the field), loading the web page, detecting keystrokes, and many other events. The response lists how the web page should react when the event occurs. Events are for a particular object, such as a button, a text field, the whole web page, and so on.

Figure 12.7 is a website used to estimate the cost of staying at Azure Islé Resort. The user can enter the number of people, change the starting and ending dates, and enter the number of people for a variety of extra activities, such as scuba diving or golf.

The event-response chart is shown in Figure 12.8. Notice that there may be a number of events for each Web form control. Since the user may do any number of actions in any order, the event-response chart is useful to show what should happen. For example, the user may click the Calculate button first, change the starting and ending dates, or change the number of people. The event-response chart is also useful for building a Web form that requires minimal action from the user. An example of this is when the user changes the starting month or day; the ending month or day is then changed to match the starting month or day. The year changes when the month is earlier than the current month, since people cannot stay at the resort prior to the current day in the same year.

Sometimes the event-response chart may be used to explore improvements to the web page. Suppose that Azure Islé Resort determined that most of its customers stayed for 7 days. When the starting month or day changes, the ending date could be set for 7 days in the future as a default. It might also be a good idea to have radio buttons that allow the customer to select a stay of 4, 7, or 14 days and calculate the ending date. Other improvements to a web page might be detecting when a number of characters have been entered, for example the three digits that comprise a U.S. telephone area code, and then moving the cursor into the next field.

Events are not limited to working within a single web page. They may also be used to control navigation among web pages. This can happen when changing a selection in a drop-down list or clicking a radio button. Events may also be used to change the contents of drop-down lists. For example, on a job search page, by selecting one category of job, detailed positions for that job appear in a second drop-down list.

Form Control	Event	Response
Web Page	Page loads	Place the current year in the **Starting Year** and **Ending Year** fields. Place the cursor in the **Number of People** field.
Number of People	Value changes	Verify that **Number of People** contains a number greater than zero.
Starting Month	Selection changes	Set the **Ending Month** in the drop-down list to the **Starting Month**. If the month is less than the current month, change the **Starting Year** and **Ending Year** values to the next year.
Starting Day	Selection changes	Set the **Ending Day** in the drop-down list to the **Starting Day**. Use the **Starting Month** value to set the **Starting Year** and **Ending Year** values.
Starting Year	Receives focus	Use the **Starting Month** value to calculate the **Starting Year** and **Ending Year** values.
Number of Days	Receives focus; value changes	Calculate the **Number of Days** that the customer is staying. If the number is less than or equal to zero, display an error message.
Basic Charge	Receives focus	Calculate the **Basic Charge** and put the cursor in the **Scuba Adventure Number of People** field.
Scuba Adventure Number of People	Receives focus	Select the current amount displayed (zero) so the customer may replace it.
Scuba Adventure Number of People	Value changes	Calculate the **Scuba Cost** based on the value of **Scuba Adventure Number of People** and put the cursor in the **Golf Number of People** field.
Scuba Cost	Value changes	Recalculate the **Scuba Cost** and put the cursor in the **Golf Number of People** field.
Golf Number of People	Receives focus	Select the current amount (zero) so the customer may replace it.
Golf Number of People	Value changes	Calculate the **Golf Cost** and put the cursor in the **Horseback Riding Number of People** field.
Golf Cost	Value changes	Recalculate the **Golf Cost** and put the cursor in the **Horseback Riding Number of People** field.
Horseback Riding Number of People	Receives focus	Select the current amount displayed (zero) so the customer may replace it.
Horseback Riding Number of People	Value changes	Calculate the **Horseback Riding Cost** based on the value of **Horseback Riding Number of People** and put the cursor in the **Last Name** field.
Horseback Riding Cost	Value changes	Recalculate the **Horseback Riding Cost** and put the cursor in the **Last Name** field.
Total Cost	Value changes	Calculate the total cost and put the cursor in the **Last Name** field.
Calculate Button	Button clicked	Validate the form data and display an error message if any errors occur. Calculate the **Total Cost** if there are no errors.
Reset Button	Button clicked	Clear the form and place the current year in the **Starting Year** and **Ending Year** fields. Place the cursor in the **Number of People** field.
Print Button	Button clicked	Validate the form data and display an error message if any errors occur. Calculate **Total Cost** if there are no errors. Use a Web cookie to pass the data to a confirmation page that does not allow the users to change any data.
Submit Button	Button clicked	Validate the form data and display an error message if any errors occur. Calculate **Total Cost** if there are no errors. Send the form to the server and send confirmation to the user.

FIGURE 12.8

An event-response chart that lists the form control, event, and response for a number of events that can occur as a user interacts with the Azure Islé Resort cost estimator screen.

Dynamic Web Pages

Dynamic web pages change themselves as the result of user action. They often use JavaScript to modify some part of the web page or a style. Changing an image when the mouse moves over it or rotating random images at a given time interval are common examples of dynamic web pages. The web page may detect the width of the browser window and modify the page accordingly. Menus that expand when the user clicks a small plus sign to the left of the menu or when the mouse moves over a menu are other examples of dynamic web pages.

The power of dynamic web pages has been greatly expanded in recent Web browsers. By using JavaScript, a Web form may morph or change itself to add new fields or remove old fields, or change field attributes, such as the length of a field or a radio button changing into a check box. This makes the web page more responsive to user actions and often will eliminate the need to load new web pages based on user choices.

An analyst should think about the information that would make sense to the website viewer. For example, placing the country selection list on a web page before other address elements would allow the user to change the country list and then to change captions to reflect the country. If the person selected United States from the drop-down list, the captions would say State and Zip Code. If the country were Canada, the captions would say Province and Postal Code. If the country were Japan, the captions would say Prefecture and Mail Code.

Three-Dimensional Web Pages

Dynamic web pages can be used to temporarily display information, such as a block of help information, a calendar with clickable dates used to assist date entry fields, airport codes, and other information. This information may be stored by using a series of stacked layers (using the cascading styles z-index property) in the web page design, each on top of another. The main web page is the base plane, the standard layer of the web page that displays or obtains information, while others below the page are not visible.

When help is requested or the user clicks in the date field, the layer is either moved to the top and becomes visible or is generated by JavaScript code and appears. The position of the layer is determined by the designer or analyst, such as a calendar appearing on the right side of a date field. When a date is selected, a close link is clicked, or the user clicks outside of the calendar, the layer then moves below the surface of the web page or is removed. An analyst must determine when it makes sense to include a layer, often examining each field on a web page to determine whether additional information would help to ensure accurate information and good communication with the user.

The analyst should determine the following:

1. *How is the layer built?* Is it created using JavaScript code, such as a calendar, or is additional information required to build the layer? If additional information is required, where is the data located and how should it be obtained? Ideally the information is obtained from only one database table on the server using Ajax techniques.
2. *What events cause the layer to be created?* These include a user clicking or tabbing into a field, clicking a link, or counting the number of keystrokes entered into a field. An example would be a hotel chain with many locations. To include all the locations in a drop-down list would make the list too long. When the user enters three characters, a block surfaces listing hotels starting with those three letters, including the country, city, state or province, and other information. If the event was detecting only two letters, the list may be too large for the display block.
3. *What events remove the layer, such as a close button, clicking a date, clicking outside the region, or selecting a hotel from the list?*
4. *Where should the surfaced block be placed?* Typically next to the field that created the block, in x- and y-coordinates.
5. *How large should the block be, measured in pixels?* If the information is too large for the block, such as a list of hotels, the analyst should decide how to handle the additional information. Options include adding scroll bars or positioning a link at the bottom to the next page of information.
6. *What should the region formatting attributes, such as color and border, be?* If there is a series of links, review their appearance with the users. Ask the users if they would like the color to change as the mouse moves over each row.
7. *What should happen when an option is selected?* In the case of a reservation calendar, when a date is clicked, the date is placed in the starting date. If a hotel is selected, the city,

state or province, and country information, along with a link to the hotel, should populate the Web form fields.

Using layers is an effective way to build websites since it does not require any pop-up pages (which may be blocked by Internet security software). Additionally, a new web page does not have to load, and, because the information is contained in a layer, it does not take up any space on the main web page.

Figure 12.9 is an example of a Web form used by an insurance company to change client information; to add a new location for a client, such as a new store or restaurant for an existing client; or to remove a store for the client. If the **Corporate** check box is checked, the **Last Name, First Name,** and **Middle** are changed to a **Company** name field, with the caption text changing as well. If the **Add New Property** button is clicked, a new set of fields for the third property is added. Care must be taken to generate unique names that the server will recognize for the additional fields. When the form is submitted, the server updates the database tables for the additional fields.

The analyst must decide when the use of dynamic web pages is appropriate. If the data change when other parts of the web page change (such as clicking a radio button or selecting an item from a drop-down list), it may be good policy to design the web pages as a dynamic form. If, however, some parts of the Web form are unsecured and other parts require encryption, it is probably best not to use dynamic forms.

A good example of a form that modifies itself may be found at Expedia.com (www.expedia.com). Clicking radio buttons for a flight, hotel, car, or cruise causes the form to change to gather the data appropriate to reserving a flight, hotel, and so on.

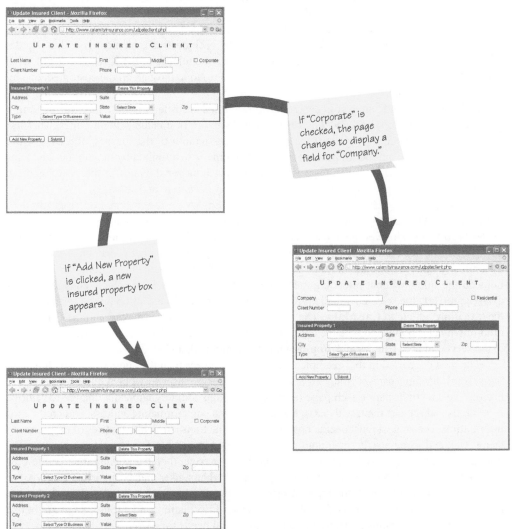

FIGURE 12.9

An example of a dynamic web page from an insurance company. If a user clicks on Add New Property, a new insured property box appears.

Dynamic web pages have the advantage of modifying themselves quickly, with fewer interruptions to send and receive data from the server. However, there are several disadvantages when creating dynamic web pages. One is that they will not work if JavaScript is turned off. An analyst must decide what to do in this situation.

If a person must use a website (as in a corporate intranet environment, in a site used to obtain student loans, or in the case of processing government or other transactions), the web page can state clearly that it will not function if JavaScript is turned off and then direct the user on how to turn it on. Most commerce websites do not require JavaScript to be turned on and have an alternate website for customers.

A second disadvantage when using dynamic web pages is that they may not be compliant with the Americans with Disabilities Act. (For more on Web accessibility for all users, please see Chapter 14 on designing for human–computer interaction.)

Ajax (Asynchronous JavaScript and XML)

Ajax is a technique that works in recent Web browsers. It involves the use of JavaScript and Extensible Markup Language (XML). Traditionally, each time a web page needed data from a different database table, a request was sent to the server, and a whole new page was loaded. This is effective but slow because an entire page must be loaded just to provide additional data for a drop-down list or some other Web form control based on the one selected previously.

Ajax allows Web developers to build a web page that works more like a traditional desktop program. As new data are needed, the browser sends a request to the server, and the server sends a small amount of data back to the browser, which updates the current page. This means that the viewer does not experience an interruption of work and the web page does not reload. The page is dynamically updated with the new data.

The data may be either a small text file or an XML document that contains many customers or other repeating data. If the data is an XML file, each customer element is called a node, and each node is numbered (starting with zero) from the beginning of the XML document. This allows the web page to go to the first or last customer or to loop through all the customers one by one with a button click.

Let's say a systems analyst is designing a traditional website, without Ajax, for making a reservation for a European ferry. The resultant website might contain several pages. The first page would ask the customer about the origin and destination of the journey, the date of the planned trip, and the number of passengers. Since pricing is determined by the number and ages of the passengers, a second web page would display asking for the ages of the passengers. A third would ask the type of vehicle desired for land transportation, and so on.

The same information may be obtained by using Ajax techniques, as illustrated in Figure 12.10. The same starting and ending destinations, as well as dates, are entered on the top of the Web form. The web page uses the destinations and dates to determine whether there is any available space on the ferry. After the customer changes the number of passengers, the form dynamically changes to add the three drop-down lists for each passenger, along with instructions on the side—without reloading the entire page. When the type of vehicle changes, in this example to **Car,** the selected vehicle type is sent to the server. The caption on the form changes the text from vehicle to **Car Make.** The server sends the possible car makes, and the **Car Make** drop-down list is populated with the data. When the car make is selected, the chosen value is sent to the server and the **Car Model** drop-down list is populated, and so on.

The web page used in this example responds much faster than the alternative, which is having several different pages displayed, and it is easier for the user to work with. There is still a need to have a confirmation web page (although the page could dynamically change to remove form fields and replace them with text), and another web page for name, address, and credit card information. The analyst must decide how to partition the transaction into a series of pages, some using Ajax and some not. If small amounts of data need to be obtained to continue the transaction, and the data logically fits on a single Web form or page, then using Ajax may be the best approach. At times it is better to use several pages, as when a user makes an airline reservation. One Web form can obtain all the flight information, and another page can be used to display the flights. A third page might be used to obtain passenger information, and a fourth page might use Ajax to select seat locations, meals, and other individual needs for each passenger.

Ajax has the advantage of making the Web work faster and providing a smoother viewing experience for users. The disadvantages are that JavaScript must be enabled and that the web

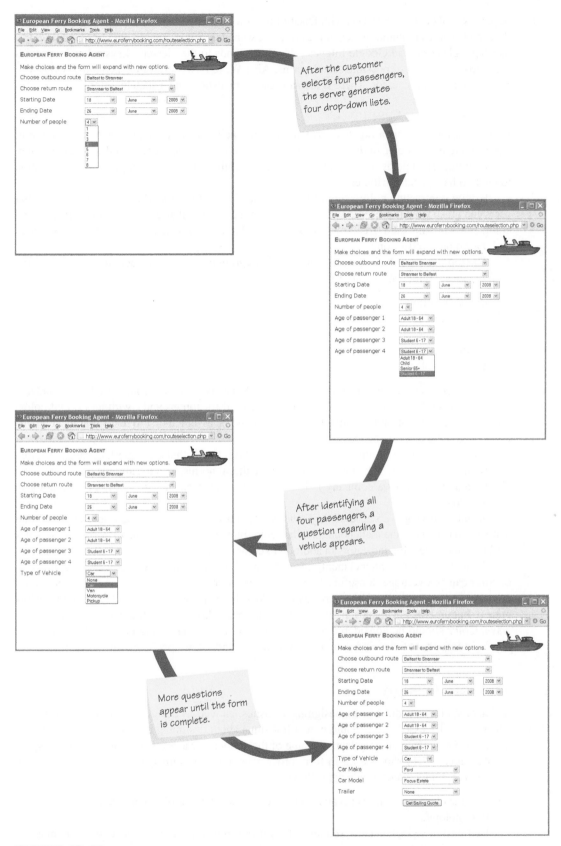

FIGURE 12.10

When analysts use Ajax techniques, a dynamic web page responds more rapidly to short user input than it would if several different pages were required for display.

page may violate the Americans with Disabilities Act. Security must be taken into consideration, if needed. There are numerous examples of Ajax websites. Some notable ones are Google Earth (earth.google.com) and Google Suggest, which responds to the viewer's keystrokes by providing a drop-down list of possible search terms. Ajax Write (www.ajaxlaunch.com/ajaxwrite) is a Web-based word processor. There is also an Ajax spreadsheet and a sketching tool.

Using Color in Display Design

Using color is an appealing and proven way to facilitate tasks requiring computer input. Appropriate use of color in display screens allows you to contrast foreground and background, highlight important fields on forms, feature errors, highlight special code input, and call attention to many other special attributes.

Highly contrasting colors should be used for display foreground and background so that users can grasp what is presented quickly. Background color will affect perception of foreground color. For example, dark green may look like a different color if taken off a white background and placed on a yellow one.

The top five most legible combinations of foreground lettering on background are (starting with the most legible combination):

1. Black on yellow
2. Green on white
3. Blue on white
4. White on blue
5. Yellow on black

The least legible are red on green and blue on red. As can be gathered from these foreground and background combinations, bright colors should be used for foregrounds, with less bright colors for the background. Strongly contrasting colors should be assigned first to fields that must be differentiated; then other colors can be assigned.

Use color to highlight important fields on displays. Fields that are important can be colored differently than the rest. Take into consideration cultural norms. Red usually means danger, and "in the red" also means a company is losing money. Green means "go" and is a safe color in Western countries.

When observing Web accessibility guidelines, you will also want to take into consideration that 8 to 10 percent of the male population has color blindness, but fewer than 1 percent of females suffer from it. Use other indicators in addition to color to support users in completing their tasks.

As with any other enhancement, designers need to question the value of using color. Use of color can be overdone; a useful heuristic is no more than four colors for new users and only up to seven for experienced ones. Irrelevant colors distract users and detract from their performance. In numerous instances, however, color has been shown to facilitate use in very specific ways. Color should be considered an important way to contrast foreground and background, highlight important fields and data, point out errors, and allow special coding of input.

Website Design

In Chapter 11, the rudiments of designing websites, as well as designing for smartphones and tablets were discussed. You should note a few hints about designing a good Internet or intranet fill-in form now that you have learned some of the elementary aspects of input form and display design. Figure 12.11 shows a fill-in form order page that shows many elements of good design for the Web. Guidelines include the following:

1. Provide clear instructions because Web users may not be familiar with technical terminology.
2. Demonstrate a logical entry sequence for fill-in forms, especially because the users may have to scroll down to a region of the page that is not visible at first.
3. Use a variety of text boxes, push buttons, drop-down menus, check boxes, and radio buttons to serve specific functions and to create interest in the form.
4. Provide a scrolling text box if you are uncertain about how much space users will need to respond to a question, or about what language, structure, or form users will use to enter data.
5. Prepare two basic buttons on every Web fill-in form: Submit and Clear Form.

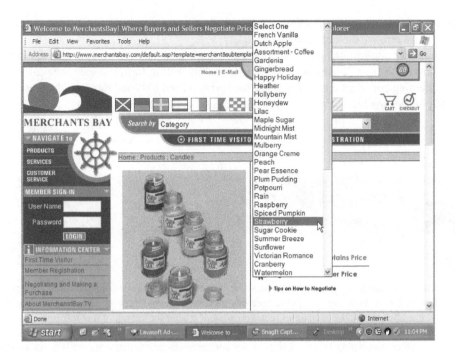

FIGURE 12.11

The order screen from the Merchants Bay website (www. merchantsbay.com) is a good example of how to design an input form that is clear, easy to use, and functional. (Screenshot of Merchant Bay website. Copyright © 2000 by Merchants Bay LLC. Reprinted with permission of Troy Pappas.)

6. If the form is lengthy and the users must scroll excessively, divide the form into several simpler forms on separate pages.
7. Create a feedback screen that refuses submission of a form unless mandatory fields are filled in correctly. The returned form screen can provide detailed feedback to the user in a different color. Red is appropriate here. For example, a user may be required to fill in a country in the country field, or indicate a credit card number if that type of payment has been checked off. Often a required field is denoted on an initial input screen with a red asterisk.

Ecommerce applications involve more than just good design of websites. Customers need to feel confident that they are buying the correct quantity, that they are getting the right price, and that the total cost of an Internet purchase, including shipping charges, is what they expect. The most common way to establish this confidence is to use the metaphor of a shopping cart or shopping bag. Figure 12.12 shows the contents of a shopping cart for a customer making a purchase.

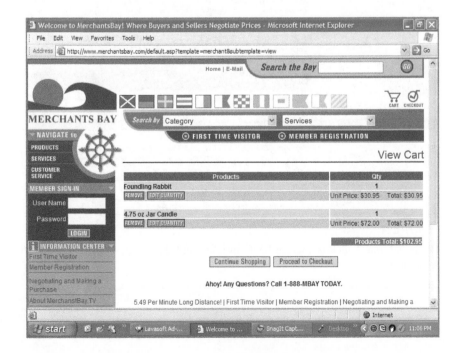

FIGURE 12.12

The Merchants Bay website (www.merchantsbay.com) is a good example of a shopping cart. (Screenshot of Merchant Bay website. Copyright © 2000 by Merchants Bay LLC. Reprinted with permission of Troy Pappas.)

An important feature of the shopping cart is that the customer can edit the quantity of the item ordered or can remove the item entirely.

Ecommerce applications place additional demands on an analyst who must design websites to meet several user and business objectives, including setting forth the corporate mission and values regarding confidentiality, preserving user privacy, and easy and rapid product returns; the efficient processing of transactions; and building good customer relationships.

MAC APPEAL

Ecommerce has changed the world by asking users to enter their own information directly to Web-based input forms; in doing so, ecommerce has increased the accuracy of data entry. While this approach is efficient for the companies receiving data, it pushes the keying of that information to the user. Fortunately, software is available to automate the process so users can merely make a couple of clicks rather than typing in long strings of alphanumeric characters that make up IDs, passwords, and credit card numbers. On a PC, RoboForm by SiberSystems is a good alternative. On the Mac, 1Password by AgileBits appears to be the current leader.

1Password allows a user to automate logins, complete credit card information, fill in an identity complete with street and email addresses, and key secure notes. Like every other good password program, 1Password includes important features such as strong password generators, antiphishing technology, and built-in protection from keyloggers. 1Password is also an app for the iPhone and Android devices, so users can take their passwords with them.

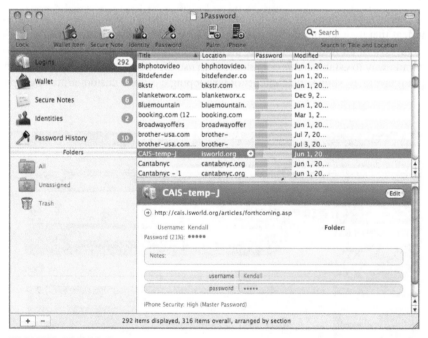

FIGURE 12.MAC

1Password from AgileBits. (Screenshot from 1Password application. Copyright © 2012 by AgileBits Inc. Reprinted with permission.)

HYPERCASE® EXPERIENCE 12

"**I**sn't spring the most beautiful season here? The architect really captured the essence of the landscape, didn't he? I mean, you can't go anywhere in the building without seeing another beautiful vista through those huge windows. When Snowden came back, he looked at your output displays. The good news is that he thinks they'll work. The project is blossoming, just like the flowers and trees. When Snowden returns from Finland, would you have some input display screens ready to demonstrate? He doesn't want things to slow down just because he's out of the country. By the way, the Singapore trip was very successful. Maybe MRE will be worldwide someday."

HYPERCASE Questions

1. Using either a paper layout form, Microsoft Visio, or form design software, design a prototype paper form that captures client information for the Training Unit.

2. Test your form on three classmates by having each of them fill it out. Ask them for a written critique of the form.
3. Redesign your input form to reflect your classmates' comments.
4. Using either a paper layout form, Microsoft Visio, or form design software, design a prototype display form that captures client information for the Training Unit.
5. Test your input display on three classmates by having each of them try it out. Ask them for a written critique of the display's design.
6. Redesign the input display based on the comments you receive. In a paragraph, explain how you have addressed each comment.

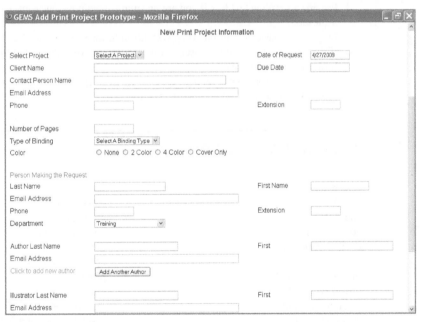

FIGURE 12.HC1

Take a look at some of the input screens in HyperCase. You may want to redesign some of the electronic forms.

Summary

This chapter has covered elements of input design for forms, displays, and Web fill-in forms. Well-designed input should meet the goals of effectiveness, accuracy, ease of use, simplicity, consistency, and attractiveness. Knowledge of many different design elements will allow the systems analyst to reach these goals.

The four guidelines for well-designed input forms are the following: (1) Make forms easy to fill in, (2) ensure that forms meet the purpose for which they are designed, (3) design forms to ensure accurate completion, and (4) keep forms attractive.

Design of useful forms, displays, and Web fill-in forms overlaps in many important ways, but there are some distinctions. Displays may show a cursor that continually orients the user. Displays often provide assistance with input, whereas with the exception of preprinted instructions, it may be difficult to get additional assistance with a form. Web-based documents have additional capabilities, such as embedded hyperlinks, context-sensitive help functions, and feedback forms, to correct input before final submission.

The four guidelines for well-designed displays are as follows: (1) Keep the display simple, (2) keep the display presentation consistent, (3) facilitate user movement among display screens and pages, and (4) create an attractive and pleasing display. Many different design elements allow the systems analyst to meet these guidelines.

The proper flow of paper forms, display screens, and fill-in forms on the Web is important. Forms should group information logically into seven categories, and displays should be divided into three main sections. Captions on forms and displays can be varied, as can font types and the weights of lines dividing subcategories of information. Multiple-part forms are another way to ensure that forms meet their intended purposes. Designers can use windows, pop-ups, dialog boxes, and defaults onscreen to ensure the effectiveness of design.

Event-response charts help the analyst to document what should happen when events occur. Dynamic web pages modify the web page in response to events. These can be constructed as three-dimensional web pages. Ajax techniques request and receive a small amount of data from the server and use the data to modify the web page on the fly.

Web fill-in forms should be constructed with the following seven guidelines in mind, as well as those in Chapter 11:

1. Provide clear instructions.
2. Demonstrate a logical entry sequence for fill-in forms.
3. Use a variety of text boxes, push buttons, drop-down menus, check boxes, and radio buttons.
4. Provide a scrolling text box if you are uncertain about how much space users will need to respond to a question.
5. Prepare two basic buttons on every Web fill-in form: Submit and Clear Form.
6. If the form is lengthy and the users must scroll extensively, divide the form into several simpler forms on separate pages.
7. Create a feedback screen that highlights errors in an appropriate color and refuses submission of the form until mandatory fields are correctly filled in.

Keywords and Phrases

Ajax
base plane
box caption
check box
command button
control of business forms
cursor
display color combinations
drop-down list box
dynamic web pages
event-response chart
facilitating movement on pages
form flow
form values
hidden field
horizontal check-off caption
image map
Internet/intranet fill-in form
layer

line caption
list box
message box
onscreen color
onscreen dialog
onscreen icon
option button
prompt
radio button
response time
seven sections of a form
slider
specialty form
spin button
table caption
text box
three sections of a display
vertical check-off caption

Review Questions

1. What are the design objectives for paper input forms, input screens, and Web-based fill-in forms?
2. List the four guidelines for good form design.
3. What is proper form flow?
4. What are the seven sections of a good form?

5. List four types of captioning for use on forms.
6. What is a specialty form? What are some disadvantages of using specialty forms?
7. List the four guidelines for good display design.
8. What are the three sections useful for simplifying a display?
9. What are the advantages of using onscreen windows?
10. What are the disadvantages of using onscreen windows?
11. List two ways display screens can be kept consistent.
12. List three ways to facilitate movement between display pages.
13. List four graphical interface design elements. Alongside each one, describe when it would be appropriate to incorporate each of them in a display design or on a Web-based fill-in form.
14. When should check boxes be used?
15. When should option buttons be used?
16. What are two different ways that form values are used?
17. What are hidden fields used for on a Web form?
18. List four different types of events.
19. What are dynamic web pages?
20. What are three-dimensional web pages?
21. How does Ajax improve a web page that changes based on user actions?
22 List the five most legible foreground and background color combinations for display use.
23. What are four situations in which color may be useful for display and Web-based fill-in form design?
24. List seven design guidelines for a Web-based fill-in form.

Problems

1. Here are captions used for a U.S. state census form:
   Name

   _____

   Occupation

   _____

   Address

   _____

   Zip code

   _____

   Number of people in household

   _____

   Age of head of household

   _____

   a. Redo the captions so that the state census bureau can capture the same information requested on the old form without confusing respondents.
   b. Redesign the form so that it exhibits proper flow. (*Hint:* Make sure to provide an access and identification section so that the information can be stored in the state's computer system.)
   c. Redesign the form so it can be filled in by citizens who visit the state's website. What changes were necessary in moving from a paper form to one that will be submitted electronically?
2. Elkhorn College needs to keep better track of students and others who use the many computers available in the Buck Memorial Library.
   a. Design and draw a representation of a display screen for students to use in signing into the computers in the library. Label the three sections of a display that you included.
   b. Design a paper form to be left alongside each computer that users who are part of the community (but are not students) are required to fill out daily. The form should ask for name, date and time of visit, general purpose for computer use (i.e., word processing, Web surfing, examining real estate documents online), and the time they have logged off. Label the seven sections of a form that you included.
3. Speedy Spuds is a fast-food restaurant offering all kinds of potatoes. The manager has a 30-second rule for serving customers. Servers at the counter say they could achieve that rule if the form they must fill out and give to the kitchen crew were simplified. The information from the completed form is entered into the computer system at the end of the day, when the data entry person needs to enter the kind of potato purchased, additional toppings purchased, the quantity, and the price charged. The current form is difficult for servers to scan and fill out quickly.
   a. Design and draw a form (you choose the size, but be sensible) that lists possible potatoes and toppings in a manner that is easy for counter servers and kitchen crew to scan, and can also be used

as input for the inventory/reorder system that is on the extranet connecting Speedy Spuds and Idaho potato growers. (*Hint:* Remember to observe *all* the guidelines for good form design.)

b. Design and draw a representation of a display screen that can be used by the servers and clerks to fill in the information captured on the form.

c. Design a display screen based on the display you designed in part b. This time, it should function as a display that shows a kitchen crew member what to prepare for each Spuds order. List three changes to the existing display that you made to adapt it to function as an output display.

4. Sherry's Meats, a regional meat wholesaler and retailer, needs to collect up-to-date information on how much of each meat product it has in each store. It will then use that information to schedule deliveries from its central warehouse. Currently, customers entering the store fill out a detailed form specifying their individual orders. The form lists over 150 items; it includes meat and meat products available in different amounts. At the end of the day, between 250 and 400 customer orders are tabulated and deducted from the store's inventory. Then the office worker in each store phones in an order for the next day. Store employees have a difficult time tabulating sales because of the mistakes customers make in filling out their forms.

a. It is not possible to have the solitary office worker in each store fill out the numerous customer order forms. Change the form ($3^{1}/_{2}$" × 6" *either* horizontal *or* vertical) and draw it so that it is easier for customers to fill out correctly and for office workers to tabulate.

b. Design and draw a specialty form of the same size that will meet the needs of Sherry's customers, office workers, and warehouse workers.

c. Design and draw two different forms of the same size to meet the purposes in part b because Sherry's carries both poultry and beef products. (*Hint:* Think about ways to make forms easy to distinguish visually.)

d. Design a fill-in form for onscreen display. When a customer submits an order, it is entered into Sherry's inventory system by any person who is serving customers at the counter. This information will be captured and sent to the central warehouse computer to help control inventory.

e. In a paragraph, describe the drawbacks of having lots of different people at different locations enter data. In a paragraph, list steps you can take as the designer so that the fill-in form is designed to ensure accuracy of entry.

f. Design a web page used by a customer to enter an order directly to Sherry's.

g. Design a web page to obtain credit card information for a Web order. Partition the data onto two web pages for additional security.

h. Design a three-dimensional dynamic web page that allows Sherry's to customize certain products, such as requesting specific ingredients in a meatloaf or a salad. When the customer selects a product from a drop-down list, the ingredients must be displayed with a means of selecting which ones should be included in the product.

5. R. George's, a fashionable clothing store that also has a catalog business, would like to keep track of the customers coming into the store so as to expand its mailing list.

a. Design and draw a simple form that can be printed on 3" × 5" cards and given to in-store customers to fill out. (*Hint:* The form must be aesthetically appealing to encourage R. George's upscale clientele to complete it.)

b. Design and draw a representation of a display screen that captures in-store customer information from the cards in part a.

c. Design and draw a second onscreen tab control dialog box that compares in-store customers with catalog customers.

d. The owner is having you help enhance his catalog business by setting up an ecommerce site. Design a Web-based form to capture information from those who visit the website. In a paragraph, explain how it will differ from the printed form.

6. Recently, an up-and-coming discount brokerage house expressed an interest in developing its own Web-based portfolio management software that clients could use at home on their PCs to make trades, get real-time stock quotes, and so on.

a. Design two input displays that make data entry easy for the client. The first display should allow users to enter stock symbols for the stocks they want to track on a daily basis. The second display should allow the client to use an icon-based system to design a customized report showing stock price trends in a variety of graphs or text.

b. Suggest two other input displays that should be included in this new portfolio management software.

7. My Belle Cosmetics is a large business that has sales well ahead of any other regional cosmetics firm. As an organization, it is very sensitive to color, because it introduces new color lines in its products every fall and spring. The company has recently begun using technology to electronically show in-store customers how they appear in different shades of cosmetics without requiring them to actually apply the cosmetics.

a. Design and draw a representation of a display screen that can be used by sales clerks at a counter to try many shades of lipstick and makeup on an individual customer very quickly and with a high degree of accuracy. Input from customers should be their hair color, the color of their favorite clothing, and their typical environmental lighting (fluorescent, incandescent, outdoor, and so on).

b. Design and draw a representation of a display screen that is equivalent to the one in part a but that vividly demonstrates to decision makers in My Belle how color improves the understandability of the screen.

c. One of the affiliates My Belle has on the Web is a large department store chain. In a paragraph, describe how the display screen in part a can be altered so that an individual can use it and My Belle can put it on the department store's ecommerce site to attract customers.

8. The Home Finders Realty Corporation specializes in locating homes for prospective buyers. Home information is stored in a database and is to be shown on an inquiry display screen. Design a GUI interface, Web-based display to enter the following data fields, which are used to select and display homes matching the criteria. Keep in mind the features available for a GUI display. The design elements (which are not in any particular sequence) are as follows:

a. Minimum size (in square feet)

b. Maximum size (in square feet, optional)

c. Minimum number of bedrooms

d. Minimum number of bathrooms

e. Garage size (number of cars, optional)

f. School district (a limited number of school districts are available for each area)

g. Swimming pool (yes/no, optional)

h. Setting (either city, suburban, or rural)

i. Fireplace (yes/no, optional)

j. Energy efficient (yes/no)

In addition, describe the hyperlinks necessary to achieve this type of interaction.

9. Design a Web entry page for the Home Finders Realty Corporation display screen created in Problem 8.

10. Sludge's Auto is an auto parts recycling center, including classic and antique cars. Rhode Wheeler, the owner, would like to get his bearings on a website for customers to browse for parts. Design an Ajax web page used to find parts. The customer needs to know the make, model, and year of a car as well as the part. If the part is in stock, the description, condition of the part, price, and shipping cost are displayed, with the quantity available for each part, along with a picture of the part. Provide a button for each part that may be clicked to purchase the part.

11. Design the Add Customer web page for Sludge's auto (see Problem 10). Include a profile that would allow Sludge's to send the customer an email if a certain part becomes available.

12. Design the Purchase web page for Sludge's Auto (see Problem 10). Assume that the customer has been added and has been logged on. Display some information about the customer. Split credit card information (type of credit card, credit card number, expiration date, and the security code found on the back of the card) between two web pages.

13. Design a web page using Ajax for registering an electronic product, either hardware or software. The form should have the purchaser's name and address, telephone number, email address, and a drop-down list of product categories. When the category is changed, send the category value to the server, which returns an XML document containing the products for the category, used to create a drop-down list of products. When the customer selects a product, the product value is sent to the server, which returns an XML document used to create a model or version of the product.

Group Projects

1. Maverick Transport is considering updating its input display screens. With your team, brainstorm about what should appear on input screens of computer operators who are entering delivery load data as loads are approved. Fields will include date of delivery, contents, weight, special requirements (for example, whether contents are perishable), and so on.

2. Each team member should design an appropriate input display using either a CASE tool, a drawing tool such as Microsoft Visio, or paper and pencil. Share your results with your team members.

3. Make a list of other input displays that Maverick Transport should develop. Remember to include dispatcher screens as well as screens to be accessed by customers and drivers. Indicate which should be PC screens or displays on wireless handheld devices.

4. Design a Web-based screen that will allow Maverick Transport customers to track the progress of a shipment. Brainstorm with team members for a list of elements, or perform an interview with a local trucking company to find out its requirements. List what hyperlinks will be essential. How will you control access so that customers can track only their own shipments?

Selected Bibliography

Direct Ferries. "Web Site Uses Ajax in Their Application." www.directferries.co.uk/poirishsea.htm. Last accessed August 21, 2012.

Garrett, J. J. "Ajax: A New Approach to Web Applications." February 18, 2005. http://skm.zoomquiet.org/data/20050225112101/. Last accessed August 21, 2012.

Gube, J. "20 Excellent Websites for Learning Ajax." December 11, 2008. http://sixrevisions.com/ajax/20-excellent-websites-for-learning-ajax/. Last accessed August 21, 2012.

Ives, B. "Graphical User Interfaces for Business Information Systems." *MIS Quarterly* (Special Issue), December 1982, pp. 15–48.

Kyng, M., and L. Mathiassen. *Computers and Design in Context.* Cambridge, MA: MIT Press, 1997.

Nielsen, J., and K. Pernice, K. Eyetracking Web Usability. Indianapolis: New Riders Press, 2010.

Nielsen, J., R. Molich, C. Snyder, and S. Farrell. *E-Commerce User Experience.* Fremont, CA: Nielsen Norman Group, 2001.

Reisner, P. "Human Factors Studies of Data Base Query Languages: A Survey and Assessment." *Computing Surveys,* Vol. 4, No. 1, 1981.

Schmidt, A., and K. E. Kendall. "Using Ajax to Clean Up a Web Site: A New Programming Technique for Web Site Development." *Decision Line,* October 2006, pp. 11–13.

CHAPTER 13

Designing Databases

LEARNING OBJECTIVES

Once you have mastered the material in this chapter you will be able to:

1. Understand database concepts.
2. Use normalization to efficiently store data in a database.
3. Use databases for presenting data.
4. Understand the concept of data warehouses.
5. Comprehend the usefulness of publishing databases to the Web.
6. Understand the relationship of business intelligence to data warehouses, big data, business analytics and text analytics in helping systems and people make decisions.

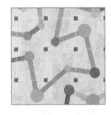

Some consider data storage to be the heart of an information system. The data have to be available when the user wants to use them. In addition, the data must be accurate and consistent (they must possess integrity). The objectives of database design include efficient storage of data as well as efficient updating and retrieval. Finally, it is necessary that information retrieval be purposeful. The information obtained from stored data must be in a form that is useful for managing, planning, controlling, or making decisions.

There are two approaches to the storage of data in a computer-based system. The first is to store data in individual files, each unique to a particular application. The second approach involves building a database. A database is a formally defined and centrally controlled store of data intended for use in many different applications.

Individual files are often designed with only immediate needs in mind, so it becomes important to query the system for a combination of some of the attributes; these attributes may be contained in separate files or may not even exist. Databases need to be planned so that data are organized for efficient storage and effective retrieval. Data warehouses are very large databases that store summarized data relating to a specific subject so that queries are answered very efficiently. Business intelligence is built around the concept of processing large volumes of data, sometimes called big data. Business analytics uses quantitative tools to analyze big data and inform decisions of managers and computer systems. Text analytics is software that analyzes unstructured or soft data collected through Web 2.0 technologies like wikis, blogs, chat rooms, social networking sites and so on in order to afford decision makers insights about qualitative aspects of customer and vendor interactions with the organization.

CONSULTING OPPORTUNITY 13.1

Hitch Your Cleaning Cart to a Star

The Marc Schnieder Janitorial Supply Company has asked for your assistance in cleaning up its data storage. As soon as you begin asking Marc Schnieder detailed questions about his database, his face gets flushed. "We don't really have a database as you describe it," he says with some embarrassment. "I've always wanted to clean up our records, but I couldn't find a capable person to head the effort."

After talking with Mr. Schnieder, you walk down the hall to the closet-sized office of Stan Lessink, the chief programmer. Stan fills you in on the historical development of the current information system. "The Marc Schnieder Janitorial Supply Company is a rags-to-riches story," Stan remarks. "Mr. Schnieder's first job was as a janitor in a bowling alley. He saved enough money to buy some products and started selling them to other alleys. Soon he decided to expand the janitorial supply business. He found out that as his business grew, he had more product lines and types of customers. Salespeople in the company are assigned to different major product

lines (stores, offices, and so on); some are in-house sales, and some specialize in heavy equipment, such as floor strippers and waxers. Records were kept in separate files."

You recall Mr. Schnieder saying, "The problem is that we have no way to compare the profits of each division. We would like to set up incentive programs for salespeople and provide better balance in allocating salespeople to each product line."

When you talk with Stan, however, he adds, "Each division has its own incentive system. Commissions vary. I don't see how we can have a common system. Besides, I can get our reports out quickly because our files are set up the way we want them. We have never issued a paycheck late."

Describe how you would go about analyzing the data storage needs of the Marc Schnieder Janitorial Supply Company. Would you trash the old system or just polish it up a bit? Discuss the implications of your decision in two paragraphs.

Databases

A database is not merely a collection of files. Rather, a database is a central source of data meant to be shared by many users for a variety of applications. The heart of a database is the database management system (DBMS), which allows the creation, modification, and updating of the database; the retrieval of data; and the generation of reports and displays. A person who ensures that a database meets its objectives is called a database administrator.

The effectiveness objectives of a database include the following:

1. Ensuring that data can be shared among users for a variety of applications.
2. Maintaining data that are both accurate and consistent.
3. Ensuring that all data required for current and future applications will be readily available.
4. Allowing the database to evolve as the needs of the users grow.
5. Allowing users to construct their personal view of the data without concern for the way the data are physically stored.

The foregoing list of objectives provides us with a reminder of the advantages and disadvantages of the database approach. First, the sharing of the data means that data need to be stored only once. That in turn helps achieve data integrity because changes to data are accomplished more easily and reliably if the data appear once rather than in many different files.

When a user needs particular data, a well-designed database anticipates the need for such data (or perhaps the data has already been used for another application). Consequently, the data have a higher probability of being available in a database than in a conventional file system. A well-designed database can also be more flexible than separate files; that is, a database can evolve as the needs of users and applications change.

Finally, the database approach has the advantage of allowing users to have their own view of the data. Users need not be concerned with the actual structure of the database or its physical storage.

Many users are extracting parts of the central database from mainframes and downloading them onto PCs or handheld devices. These smaller databases are then used to generate reports or answer queries specific to the end user.

Relational databases for PCs have improved dramatically over the last few years. One major technological change has been the design of database software that takes advantage of the GUI.

With the advent of programs such as Microsoft Access, users can drag and drop fields between two or more tables. Developing relational databases with these tools has been made relatively easy.

Data Concepts

It is important to understand how data are represented before considering the use of files or the database approach. In this section, critical definitions are covered, including the abstraction of data from the real world to the storage of data in tables and database relations.

Reality, Data, and Metadata

Data collected about people, places, or events in reality will eventually be stored in a file or database. (In this section, we refer to the real world as *reality*.) To understand the form and structure of the data, information about the data themselves is required. The information that describes data is referred to as *metadata*.

The relationship between reality, data, and metadata is pictured in Figure 13.1. Within the realm of reality are entities and attributes; within the realm of actual data are record occurrences and data item occurrences; and within the realm of metadata are record definitions and data item definitions. The meanings of these terms are discussed in the following subsections.

ENTITIES. Any object or event about which someone chooses to collect data is an entity. An entity may be a person, place, or thing (for example, a salesperson, a city, or a product). An entity can also be an event or a unit of time, such as a machine breakdown, a sale, or a month or year. In addition to the entities discussed in Chapter 2 is an additional minor entity called an entity subtype. Its symbol is a smaller rectangle within the entity rectangle.

An entity subtype is a special one-to-one relationship used to represent additional attributes (fields) of another entity that may not be present on every record of the first entity. Entity subtypes eliminate the situation in which an entity may have null fields stored on database tables.

An example is the primary entity of a customer. Preferred customers may have special fields containing discount information, and this information would be in an entity subtype. Another example is students who have internships. The STUDENT MASTER should not have to contain information about internships for each student, because perhaps only a small number of students have internships.

RELATIONSHIPS. Relationships are associations between entities (sometimes referred to as *data associations*). Figure 13.2 shows a number of entity-relationship (E-R) diagrams that portray a variety of relationships.

The first type of relationship is a one-to-one relationship (designated as 1:1). The diagram shows that there is only one PRODUCT PACKAGE for each PRODUCT. The second one-to-one relationship shows that each EMPLOYEE has a unique OFFICE. Notice that all these entities can be described further (for example, a product price would not be an entity, nor would a phone extension).

Another type of relationship is a one-to-many (1:M) or many-to-one association. As shown in the figure, a PHYSICIAN in a health maintenance organization is assigned many PATIENTS, but a PATIENT is assigned only one PHYSICIAN. Another example shows that an EMPLOYEE is a member of only one DEPARTMENT, but each DEPARTMENT has many EMPLOYEES.

Finally, a many-to-many relationship (designated as M:N) describes the possibility that entities may have many associations in either direction. For example, a STUDENT can have many

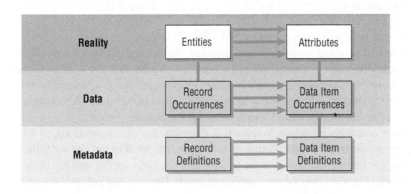

FIGURE 13.1

Reality, data, and metadata.

FIGURE 13.2

Entity-relationship (E-R) diagrams can show one-to-one, one-to-many, many-to-one, or many-to-many associations.

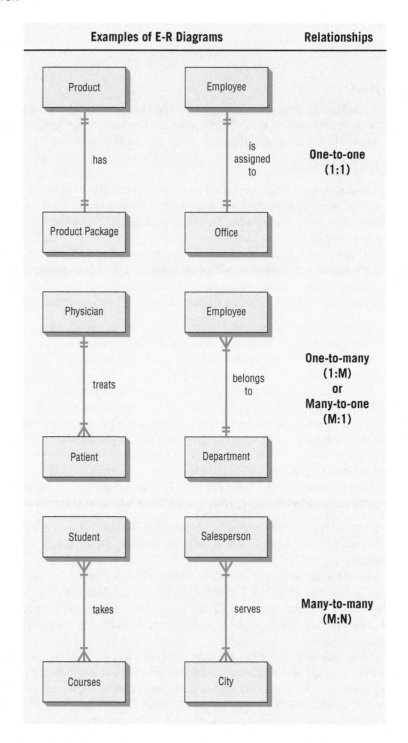

COURSE(s), and at the same time a COURSE may have many STUDENT(s) enrolled in it. The second M:N example in Figure 13.2 shows that a SALESPERSON can call on many CITY(s) and a CITY can be a sales area for many SALESPERSON(s).

The standard symbols for crow's foot notation, the official explanation of the symbols, and what they actually mean, are all given in Figure 13.3. Notice that the symbol for an entity is a rectangle. An entity is defined as a class of a person, place, or thing. A rectangle with a diamond inside stands for an associative entity, which is used to join two entities. A rectangle with an oval in it stands for an attributive entity, which is used for repeating groups.

The other notations necessary to draw E-R diagrams are the connections, of which there are five different types. In the lower portion of the figure, the meaning of the notation is explained. When a straight line connects two plain entities and the ends of the line are both marked with two short marks (‖), a one-to-one relationship exists. Following that you will notice a crow's foot

Symbol	Official Explanation	What It Really Means
	Entity	A class of persons, places, or things
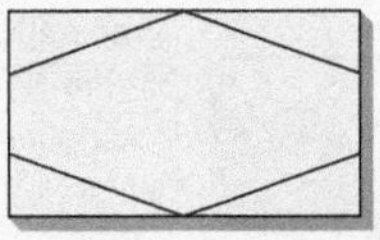	Associative entity	Used to join two entities
	Attributive entity	Used for repeating groups
———⊦⊦	To 1 relationship	Exactly one
———⟨	To many relationship	One or more
———O⊦	To 0 or 1 relationship	Only zero or one
 ———O⟨	To 0 or more relationship	Can be zero, one, or more
 ———⟨	To more than 1 relationship	Greater than one

FIGURE 13.3

Entity-relationship diagram symbols and their meanings.

with a short mark (|); when this notation links entities, it indicates a relationship of one-to-one or one-to-many (to one or more).

Entities linked with a straight line plus a short mark (|) and a zero (which looks more like a circle, O) are depicting a relationship of one-to-zero or one-to-one (only zero or one). A fourth type of link for relating entities is drawn with a straight line marked on the end with a zero (O) followed by a crow's foot. This type shows a zero-to-zero, zero-to-one, or zero-to-many relationship. Finally, a straight line linking entities with a crow's foot at the end depicts a relationship to more than one.

An entity may have a relationship connecting it to itself. This type of relationship is called a *self-join relationship*; the implication is that there must be a way to link one record in a file to another record in the same file. An example of a self-join relationship can be found in the HyperCase simulations found throughout the text. A task may have a precedent task (that is, one that must be completed before starting the current task). In this situation, one record (the current task) points to another record (the precedent task) in the same file.

The relationships in words can be written along the top or the side of each connecting line. In practice, you see the relationship in one direction, although you can write relationships on both sides of the line, each representing the point of view of one of the two entities. (See Chapter 2 for more details about drawing E-R diagrams.)

AN ENTITY-RELATIONSHIP EXAMPLE. An entity-relationship diagram containing many entities, many different types of relations, and numerous attributes is featured in Figure 13.4. In this E-R diagram, we are concerned about a billing system, and in particular with the prescription part of the system. (For simplicity, we assume that office visits are handled differently and are outside the scope of this system.)

The entities are PRESCRIPTION, PHYSICIAN, PATIENT, and INSURANCE CARRIER. The entity TREATMENT is not important for the billing system, but it is part of the E-R diagram

FIGURE 13.4

An entity-relationship diagram for patient treatment. Attributes can be listed alongside the entities. In each case, the key is underlined.

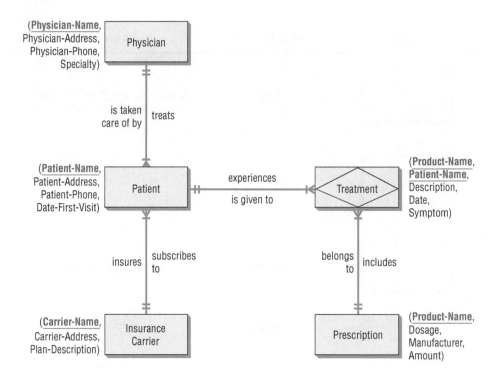

because it is used to bridge the gap between PRESCRIPTION and PATIENT. We therefore drew it as an associative entity in the figure.

Here, a PHYSICIAN treats many PATIENT(s) (1:M), who each subscribe to an individual INSURANCE CARRIER. Of course, the PATIENT is only one of many patients that subscribe to that particular INSURANCE CARRIER (M:1).

To complete the PHYSICIAN's records, the physician needs to keep information about the treatments a PATIENT has. Many PATIENT(s) experience many TREATMENT(s), making it a many-to-many (M:N) relationship. TREATMENT is represented as an associative entity because it is not important in our billing system by itself. TREATMENT(s) can include the taking of PRESCRIPTION(s), and thus is also an M:N relationship, because many treatments may call for combinations of pharmaceuticals and many drugs may work for many treatments.

Some detail is then filled in for the attributes. The attributes are listed next to each of the entities, and the key is underlined. For example, the entity PRESCRIPTION has a PRODUCT-NAME, DOSAGE, MANUFACTURER, and AMOUNT. Ideally, it would be beneficial to design a database in this fashion, using entity-relationship diagrams and then filling in the details concerning attributes. This top-down approach is desirable, but it is sometimes very difficult to achieve.

ATTRIBUTES. An attribute is some characteristic of an entity. There can be many attributes for each entity. For example, a patient (entity) can have many attributes, such as last name, first name, street address, city, state, and so on. The date of the patient's last visit as well as the prescription details are also attributes. When the data dictionary was constructed in Chapter 8, the smallest particular described was called a data element. When files and databases are discussed, these data elements are generally referred to as data items. Data items are in fact the smallest units in a file or database. The term *data item* is also used interchangeably with the word *attribute*.

Data items can have values. These values can be of fixed or variable length; they can be alphabetic, numeric, special characters, or alphanumeric. Examples of data items and their values can be found in Figure 13.5.

Sometimes a data item is also referred to as a *field*. A field, however, represents something physical, not logical. Therefore, many data items can be packed into a field; the field can be read and converted to a number of data items. A common example of this is to store the date in a single field as MM/DD/YYYY. To sort the file in order by date, three separate data items are extracted from the field and sorted first by YYYY, then by MM, and finally by DD.

Entity	Data Item	Value
Salesperson	Salesperson Number	87254
	Salesperson Name	Kaytell
	Company Name	Music Unlimited
	Address	45 Arpeum Circle
	Sales	$20,765
Package	Width	2
	Height	6
	Length	6
	Weight	1
	Mailing Address	765 Dulcinea Drive
	Return Address	P.O. Box 341, Spring Valley, MN
Order	Product(s)	B521
	Description(s)	"My Fair Lady" compact disc
	Quantity Ordered	1
	Last Name of Customer	Kiley
	First Initial	R.
	Street Address	765 Dulcinea Drive
	City	La Mancha
	State	CA
	Zip Code	93407
	Credit Card Number	65-8798-87
	Date Order Was Placed	01/03/2013
	Amount	$6.99
	Status	Backordered

FIGURE 13.5

Typical values assigned to data items may be numbers, alphabetic characters, special characters, and combinations of all three.

RECORDS. A record is a collection of data items that have something in common with the entity described. Figure 13.6 is an illustration of a record with many related data items. The record shown is for an order placed with a mail-order company. ORDER-#, LAST NAME, INITIAL, STREET ADDRESS, CITY, STATE, and CREDIT CARD are all attributes. Most records are of fixed length, so there is no need to determine the length of the record each time.

Under certain circumstances (for instance, when space is at a premium), variable-length records are used. A variable-length record is used as an alternative to reserving a large amount of space for the longest possible record, such as the maximum number of visits a patient has made to a physician. Each visit would contain many data items that would be part of the patient's full record (or file folder in a manual system). Later in this chapter, normalization of a relation is discussed. Normalization is a process that eliminates repeating groups found in variable-length records.

KEYS. A key is one of the data items in a record that is used to identify a record. When a key uniquely identifies a record, it is called a *primary key*. For example, ORDER-# can be a primary key because only one number is assigned to each customer order. In this way, the primary key identifies the real-world entity (customer order).

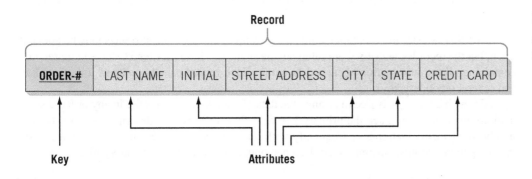

FIGURE 13.6

A record has a primary key and may have many attributes.

Special care must be taken when designing the primary key. Often it is a sequential number or a sequential number with a self-checking number (called a check digit) at the end of the digits. At times there is some meaning built into the primary key, but defining a primary key based on an attribute is considered a risk. If the attribute changes, the primary key will also change, creating a dependency between the primary key and the data.

An example of a primary key based on data is a state abbreviation for the state name or an airline luggage code for an airport name. An attribute or a collection of attributes that can serve as a primary key is called a candidate key. A primary key should also be minimal and contain no extra attributes than are necessary to identify a record.

A key is called a secondary key if it cannot uniquely identify a record. Secondary keys either may be unique or may identify multiple records in a database. Secondary keys can be used to select a group of records that belong to a set (for example, orders from the state of New Jersey).

When it is not possible to identify a record uniquely by using one of the data items found in a record, a key can be constructed by choosing two or more data items and combining them. This key is called a concatenated, or composite, key. When a data item is used as a key in a record, the description is underlined. Therefore, in the ORDER record (<u>ORDER-#</u>, LAST NAME, INITIAL, STREET ADDRESS, CITY, STATE, CREDIT CARD), the key is <u>ORDER-#</u>. If an attribute is a key in another file, it should be underlined with a <u>dashed line</u>.

Some databases allow a developer to use an object identifier (OID), which is a unique key for each record in the database, not just in a table. Given an object identifier, one record will be obtained regardless of the table on which it exists. This may be included with an order or a payment confirmation, along with a message like, "This is your confirmation number."

METADATA. Metadata are data about the data in a file or database. Metadata describe the name given and the length assigned to each data item. Metadata also describe the length and composition of each of the records.

Figure 13.7 is an example of metadata for a database for some generic software. The length of each data item is indicated according to a convention, where 7.2 means that seven spaces are reserved for the number, two of which are to the right of the decimal point. The letter N signifies "numeric," and the A stands for "alphanumeric." The D stands for "date" and is automatically in the form MM/DD/YYYY. Some programs, such as Microsoft Access, use plain English for metadata, so words such as *text, currency,* and *number* are used. Microsoft Access provides a default of 50 characters as the field length for names, which is fine when working with small systems. If, however, you are working with a large database for a bank or a utility company, for example, you do not want to devote that much space to that field. Otherwise, the database would become quite large and filled with wasted space. That is when you can use metadata to plan ahead and design a more efficient database.

Files

A file contains groups of records used to provide information for operations, planning, management, and decision making. The types of files used are discussed next, followed by a description of the many ways conventional files can be organized.

FILE TYPES. Files can be used for storing data for an indefinite period of time, or they can be used to store data temporarily for a specific purpose. Master files and table files are used to store data for a long period. The temporary files are usually called transaction files, work files, or report files.

Master Files Master files contain records for a group of entities. The attributes may be updated often, but the records themselves are relatively permanent. These files tend to have large records containing all the information about a data entity. Each record usually contains a primary key and several secondary keys.

Although an analyst is free to arrange the data elements in a master file in any order, a standard arrangement is to place the primary key field first, followed by descriptive elements, and finally by elements that change frequently with business activities. Examples of a master file include patient records, customer records, a personnel file, and a parts inventory file.

Data Item	Value	
Salesperson Number	N	5
Salesperson Name	A	20
Company Name	A	26
Address	A	36
Sales	N	9.2
Width	N	2
Height	N	2
Length	N	2
Weight	N	2
Mailing Address	A	36
Return Address	A	36
Product(s)	A	4
Description(s)	A	30
Quantity Ordered	N	2
Last Name of Customer	A	24
First Initial	A	1
Street Address	A	28
City	A	12
State	A	2
Zip Code	N	9
Credit Card Number	N	10
Date Order Was Placed	D	8	MM/DD/YYYY
Amount	$	7.2
Status	A	22

Fields

N	Numeric
A	Alphanumeric or text
D	Date MM/DD/YYYY
$	Currency
M	Memo

7.2 means that the field takes up 7 digits, two of which are right of the decimal.

Special formats for fields may be specified.

FIGURE 13.7

Metadata include a description of what the value of each data item looks like.

Table Files A table file contains data used to calculate more data or performance measures. One example is a table of postage rates used to determine the shipping costs of a package. Another example is a tax table. Table files usually are read only by a program.

Transaction Files A transaction file is used to enter changes that update the master file and produce reports. Suppose a newspaper subscriber master file needs to be updated; the transaction file would contain the subscriber number, and a transaction code such as E for extending the subscription, C for canceling the subscription, or A for address change. Then only information relevant to the updating needs to be entered; that is, the length of renewal if E, and the address if A. No additional information would be needed if the subscription were canceled. The rest of the information already exists in the master file. Transaction files may contain several different types of records, such as the three used for updating the newspaper subscription master, with a code on the transaction file indicating the type of transaction.

Report Files When it is necessary to print a report when no printer is available (for example, when the printer is busy printing other jobs), a report file is used. Sending the output to a file rather than a printer is called spooling. Later, when the device is ready, the document can be printed. Report files are very useful, because users can take files to other computer systems and output to specialty devices.

Relational Databases

Databases can be organized in several ways. The most common type of database is a relational database. A relational database is organized in meaningful tables, which minimizes the repetition of data, which in turn minimizes errors and storage space.

FIGURE 13.8

Database design includes synthesizing user reports, user views, and logical and physical designs.

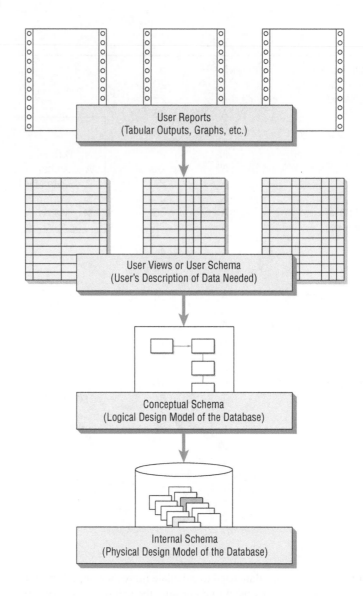

LOGICAL AND PHYSICAL VIEWS OF DATA. A database, unlike a file, is intended to be shared by many users. It is clear that the users all see the data in different ways. We refer to the way a user pictures and describes the data as a user view. The problem, however, is that different users have different user views. These views need to be examined by the systems analyst, and an overall logical model of the database developed. Finally, the logical model of the database must be transformed into a corresponding physical database design. Physical design is involved with how data are stored and related, as well as how they are accessed.

In database literature, views are referred to as *schema*. Figure 13.8 shows how the user reports and user views (user schema) are related to the logical model (conceptual schema) and physical design (internal schema).

There are three main types of logically structured databases: hierarchical, network, and relational. The first two types may be found in legacy (older) systems. An analyst today would typically design a relational database.

RELATIONAL DATA STRUCTURES. A relational data structure consists of one or more two-dimensional tables, which are referred to as *relations*. The rows of the table represent the records, and the columns contain attributes.

Figure 13.9 shows the relational structure for a music CD ordering database. Here, three tables are needed to (1) describe the items and keep track of the current price of CDs (ITEM PRICE), (2) describe the details of the order (ORDER), and (3) identify the status of the order (ITEM STATUS).

FIGURE 13.9

In a relational data structure, data
are stored in many tables.

ITEM PRICE

ITEM-#	TITLE	PRICE
B235	Guys and Dolls	8.99
B521	My Fair Lady	6.99
B894	42nd Street	10.99
B992	A Chorus Line	10.99

ORDER

ORDER-#	LAST NAME	I	STREET ADDRESS	CITY	ST	CHARGE ACCT
10784	MacRae	G	2314 Curly Circle	Lincoln	NE	45-4654-76
10796	Jones	S	34 Dream Lane	Oklahoma City	OK	44-9876-74
11821	Preston	R	1008 Madison Ave.	River City	IA	34-7642-64
11845	Channing	C	454 Harmonia St.	New York	NY	34-0876-87
11872	Kiley	R	765 Dulcinea Drive	La Mancha	CA	65-8798-87

ITEM STATUS

ITEM-#	ORDER-#	STATUS
B235	10784	Shipped 5/12
B235	19796	Shipped 5/14
B235	11872	In Process
B521	11821	In Process
B894	11845	Backordered
B894	11872	Shipped 5/12
B992	10784	Shipped 5/12

To determine the price of an item, we need to know the item number to be able to find it in the relation ITEM PRICE. To update G. MacRae's credit card number, we can search the ORDER relation for MacRae and correct it only once, even though he ordered many CDs. To find out the status of part of an order, however, we must know the ITEM-# and ORDER-#, and then we must locate that information in the relation ITEM STATUS.

Maintaining the tables in a relational structure is usually quite simple compared to maintaining a hierarchical or network structure. One of the primary advantages of relational structures is that ad hoc queries are handled efficiently.

When relational structures are discussed in database literature, different terminology is often used. A file is called either a table or relation, a record is usually referred to as a tuple, and the attribute value set is called a domain.

For relational structures to be useful and manageable, the relational tables must first be normalized. Normalization is detailed in the following section.

Normalization

Normalization is the transformation of complex user views and data stores to a set of smaller, stable data structures. In addition to being simpler and more stable, normalized data structures are more easily maintained than other data structures.

The Three Steps of Normalization

Beginning with either a user view or a data store developed for a data dictionary (see Chapter 8), an analyst normalizes a data structure in three steps, as shown in Figure 13.10. Each step involves an important procedure that simplifies the data structure.

FIGURE 13.10

Normalization of a relation is accomplished in three major steps.

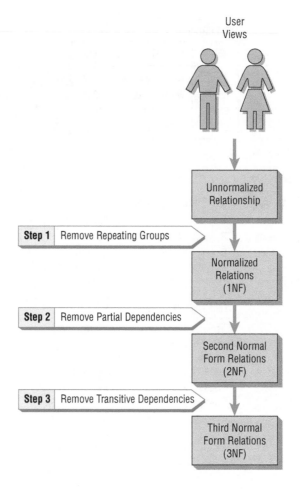

The relation derived from the user view or data store will most likely be unnormalized. The first stage of the process involves removing all repeating groups and identifying the primary key. To do so, the relation needs to be broken up into two or more relations. At this point, the relations may already be of the third normal form, but it is likely more steps will be needed to transform the relations to the third normal form.

The second step involves ensuring that all nonkey attributes are fully dependent on the primary key. All partial dependencies are removed and placed in another relation.

The third step involves removing any transitive dependencies. A transitive dependency is one in which nonkey attributes are dependent on other nonkey attributes.

A Normalization Example

Figure 13.11 is a user view for the Al S. Well Hydraulic Equipment Company. The report shows the (1) SALESPERSON-NUMBER, (2) SALESPERSON-NAME, and (3) SALES-AREA. The body of the report shows the (4) CUSTOMER-NUMBER and (5) CUSTOMER-NAME. Next is the (6) WAREHOUSE-NUMBER that will service the customer, followed by the (7) WAREHOUSE-LOCATION, which is the city in which the company is located. The final information contained in the user view is the (8) SALES-AMOUNT. The rows (one for each customer) on the user view show that items 4 through 8 form a repeating group.

If the analyst were using a data flow/data dictionary approach, the same information in the user view would appear in a data structure. Figure 13.12 shows how the data structure would appear at the data dictionary stage of analysis. The repeating group is also indicated in the data structure by an asterisk (*) and indentation.

Before proceeding, note the data associations of the data elements in Figure 13.13. This type of illustration is called a *bubble diagram* or *data model diagram*. Each entity is enclosed in an ellipse, and arrows are used to show the relationships. Although it is possible to draw these relationships with an E-R diagram, it is sometimes easier to use the simpler bubble diagram to model the data.

FIGURE 13.11

A user report for the Al S. Well
Hydraulic Equipment Company.

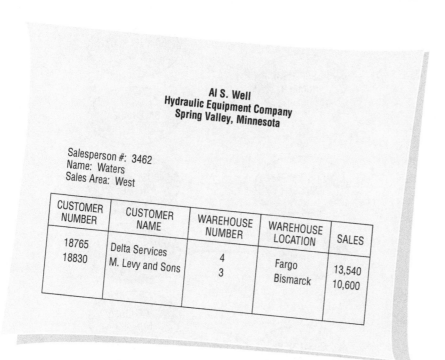

FIGURE 13.12

The analyst would find a data
structure (from a data dictionary)
useful in developing a database.

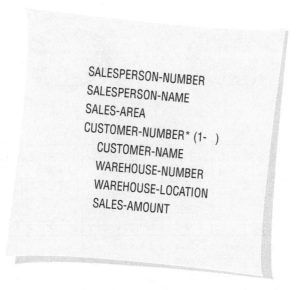

In this example, there is only one SALESPERSON-NUMBER assigned to each
SALESPERSON-NAME, and that person will cover only one SALES-AREA, but each SALES-
AREA may be assigned to many salespeople: hence, the double arrow notation from SALES-
AREA to SALESPERSON-NUMBER. For each SALESPERSON-NUMBER, there may be
many CUSTOMER-NUMBER(s).

Furthermore, there would be a one-to-one correspondence between CUSTOMER-NUMBER
and CUSTOMER-NAME; the same is true for WAREHOUSE-NUMBER and WAREHOUSE-
LOCATION. CUSTOMER-NUMBER will have only one WAREHOUSE-NUMBER and
WAREHOUSE-LOCATION, but each WAREHOUSE-NUMBER or WAREHOUSE-
LOCATION may service many CUSTOMER-NUMBER(s). Finally, to determine the SALES-
AMOUNT for one salesperson's calls to a particular company, it is necessary to know both the
SALESPERSON-NUMBER and the CUSTOMER-NUMBER.

The main objective of the normalization process is to simplify all the complex data items
that are often found in user views. For example, if the analyst were to take the user view

FIGURE 13.13

Drawing data model diagrams for data associations sometimes helps analysts appreciate the complexity of data storage.

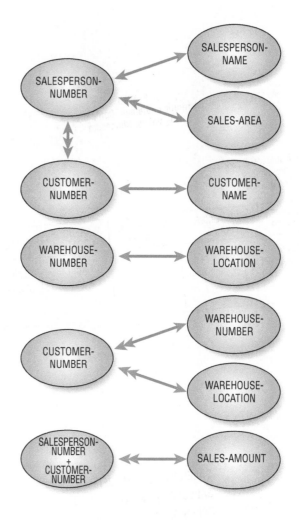

FIGURE 13.14

If the data were listed in an unnormalized table, there could be repeating groups.

SALESPERSON NUMBER	SALESPERSON NAME	SALES AREA	CUSTOMER NUMBER	CUSTOMER NAME	WAREHOUSE NUMBER	WAREHOUSE LOCATION	SALES AMOUNT
3462	Waters	West	18765	Delta Systems	4	Fargo	13540
			18830	A. Levy and Sons	3	Bismarck	10600
			19242	Ranier Company	3	Bismarck	9700
3593	Dryne	East	18841	R. W. Flood Inc.	2	Superior	11560
			18899	Seward Systems	2	Superior	2590
			19565	Stodola's Inc.	1	Plymouth	8800
etc.							

discussed previously and attempt to make a relational table out of it, the table would look like Figure 13.14. Because this relation is based on our initial user view, we refer to it as SALES-REPORT.

SALES-REPORT is an unnormalized relation because it has repeating groups. It is important to observe that a single attribute such as SALESPERSON-NUMBER cannot serve as the key. The reason is clear when one examines the relationships between SALESPERSON-NUMBER and the other attributes in Figure 13.15. Although there is a one-to-one correspondence between SALESPERSON-NUMBER and two attributes (SALESPERSON-NAME and SALES-AREA), there is a one-to-many relationship between SALESPERSON-NUMBER and the other five attributes (CUSTOMER-NUMBER, CUSTOMER-NAME, WAREHOUSE-NUMBER, WAREHOUSE-LOCATION, and SALES-AMOUNT).

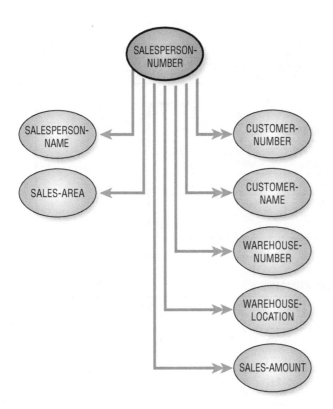

FIGURE 13.15

A data model diagram shows that in the unnormalized relation, SALESPERSON-NUMBER has a one-to-many association with some attributes.

SALES-REPORT can be expressed in the following shorthand notation:

SALES REPORT   (SALESPERSON-NUMBER,
                        SALESPERSON-NAME, SALES-AREA,
                        (CUSTOMER-NUMBER,
                        CUSTOMER-NAME,
                        WAREHOUSE-NUMBER,
                        WAREHOUSE-LOCATION,
                        SALES-AMOUNT))

where the inner set of parentheses represents the repeated group.

FIRST NORMAL FORM (1NF). The first step in normalizing a relation is to remove the repeating groups. In our example, the unnormalized relation SALES-REPORT will be broken into two separate relations. These new relations will be named SALESPERSON and SALESPERSON-CUSTOMER.

Figure 13.16 shows how the original, unnormalized relation SALES-REPORT is normalized by separating the relation into two new relations. Notice that the relation SALESPERSON contains the primary key <u>SALESPERSON-NUMBER</u> and all the attributes that were not repeating (SALESPERSON-NAME and SALES-AREA).

The second relation, SALESPERSON-CUSTOMER, contains the primary key from the relation SALESPERSON (the primary key of SALESPERSON is <u>SALESPERSON-NUMBER</u>), as well as all the attributes that were part of the repeating group (CUSTOMER-NUMBER, CUSTOMER-NAME, WAREHOUSE-NUMBER, WAREHOUSE-LOCATION, and SALES-AMOUNT). Knowing the SALESPERSON-NUMBER, however, does not automatically mean that you will know the CUSTOMER-NAME, SALES-AMOUNT, WAREHOUSE-LOCATION, and so on. In this relation, one must use a concatenated key (both <u>SALESPERSON-NUMBER</u> and <u>CUSTOMER-NUMBER</u>) to access the rest of the information. It is possible to write the relations in shorthand notation as follows:

<u>SALESPERSON</u>   (<u>SALESPERSON-NUMBER</u>,
                        SALESPERSON-NAME, SALES AREA)

and

FIGURE 13.16

The original unnormalized relation SALES-REPORT is separated into two relations, SALESPERSON (3NF) and SALESPERSON-CUSTOMER (1NF).

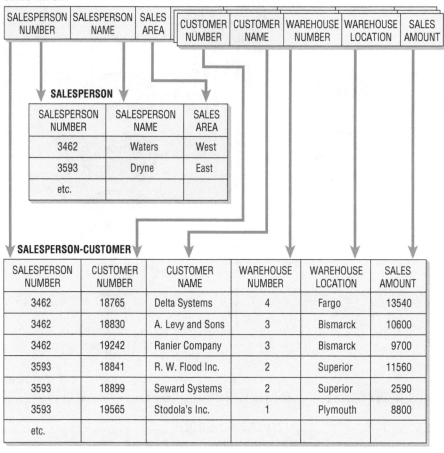

SALESPERSON-CUSTOMER   (<u>SALESPERSON-NUMBER</u>,
                                             <u>CUSTOMER-NUMBER</u>,
                                             CUSTOMER-NAME,
                                             WAREHOUSE-NUMBER,
                                             WAREHOUSE-LOCATION,
                                             SALES-AMOUNT)

The relation SALESPERSON-CUSTOMER is a first normal relation, but it is not in its ideal form. Problems arise because some of the attributes are not functionally dependent on the primary key (that is, <u>SALESPERSON-NUMBER</u>, <u>CUSTOMER-NUMBER</u>). In other words, some of the nonkey attributes are dependent only on CUSTOMER NUMBER and not on the concatenated key. The data model diagram in Figure 13.17 shows that SALES-AMOUNT is dependent on both <u>SALESPERSON-NUMBER</u> and <u>CUSTOMER-NUMBER</u>, but the other three attributes are dependent only on <u>CUSTOMER-NUMBER</u>.

SECOND NORMAL FORM (2NF). In the second normal form, all the attributes will be functionally dependent on the primary key. Therefore, the next step is to remove all the partially dependent attributes and place them in another relation. Figure 13.18 shows how the relation SALESPERSON-CUSTOMER is split into two new relations: SALES and CUSTOMER-WAREHOUSE. These relations can also be expressed as follows:

SALES   (<u>SALESPERSON-NUMBER</u>, <u>CUSTOMER-NUMBER</u>,
                                             SALES-AMOUNT)

and

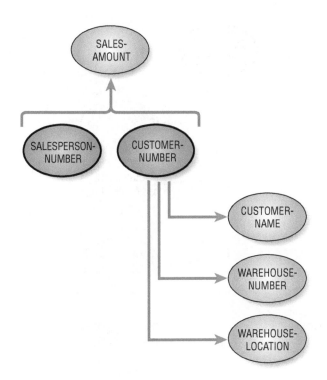

FIGURE 13.17

A data model diagram shows that three attributes are dependent on CUSTOMER-NUMBER, so the relation is not yet normalized. Both SALESPERSON-NUMBER and CUSTOMER-NUMBER are required to look up SALES-AMOUNT.

SALESPERSON-CUSTOMER

SALESPERSON NUMBER	CUSTOMER NUMBER	CUSTOMER NAME	WAREHOUSE NUMBER	WAREHOUSE LOCATION	SALES AMOUNT

FIGURE 13.18

The relation SALESPERSON-CUSTOMER is separated into a relation called CUSTOMER-WAREHOUSE (2NF) and a relation called SALES (1NF).

CUSTOMER-WAREHOUSE

CUSTOMER NUMBER	CUSTOMER NAME	WAREHOUSE NUMBER	WAREHOUSE LOCATION
18765	Delta Systems	4	Fargo
18830	A. Levy and Sons	3	Bismarck
19242	Ranier Company	3	Bismarck
18841	R. W. Flood Inc.	2	Superior
18899	Seward Systems	2	Superior
19565	Stodola's Inc.	1	Plymouth
etc.			

SALES

SALESPERSON NUMBER	CUSTOMER NUMBER	SALES AMOUNT
3462	18765	13540
3462	18830	10600
3462	19242	9700
3593	18841	11560
3593	18899	2590
3593	19565	8800
etc.		

FIGURE 13.19

This data model diagram shows that a transitive dependency exists between WAREHOUSE-NUMBER and WAREHOUSE-LOCATION.

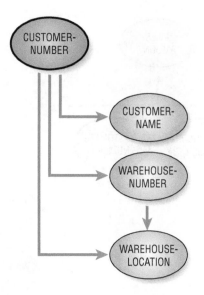

CUSTOMER WAREHOUSE   (<u>CUSTOMER-NUMBER</u>,
                             CUSTOMER-NAME,
                             WAREHOUSE-NUMBER,
                             WAREHOUSE-LOCATION)

The relation CUSTOMER-WAREHOUSE is in the second normal form. It can still be simplified further because there are additional dependencies in the relation. Some of the nonkey attributes are dependent not only on the primary key, but also on a nonkey attribute. This dependency is referred to as a transitive dependency.

Figure 13.19 shows the dependencies in the relation CUSTOMER-WAREHOUSE. For the relation to be a second normal form, all the attributes must be dependent on the primary key <u>CUSTOMER-NUMBER</u>, as shown in the diagram. WAREHOUSE-LOCATION, however, is obviously dependent on WAREHOUSE-NUMBER also. To simplify this relation, another step is required.

THIRD NORMAL FORM (3NF). A normalized relation is in the third normal form if all the nonkey attributes are fully functionally dependent on the primary key and there are no transitive (nonkey) dependencies. In a manner similar to the previous steps, it is possible to break apart the relation CUSTOMER-WAREHOUSE into two relations, as shown in Figure 13.20.

FIGURE 13.20

The relation CUSTOMER-WAREHOUSE is separated into two relations called CUSTOMER (1NF) and WAREHOUSE (1NF).

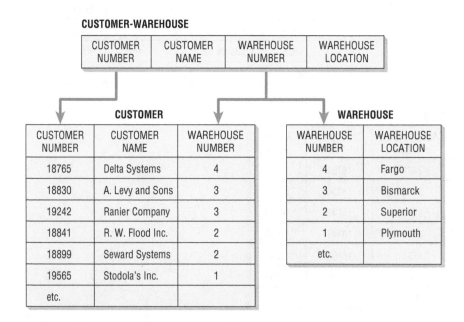

SALESPERSON

SALESPERSON NUMBER	SALESPERSON NAME	SALES AREA
3462	Waters	West
3593	Dryne	East
etc.		

SALES

SALESPERSON NUMBER	CUSTOMER NUMBER	SALES AMOUNT
3462	18765	13540
3462	18830	10600
3462	19242	9700
3593	18841	11560
3593	18899	2590
3593	19565	8800
etc.		

CUSTOMER

CUSTOMER NUMBER	CUSTOMER NAME	WAREHOUSE NUMBER
18765	Delta Systems	4
18830	A. Levy and Sons	3
19242	Ranier Company	3
18841	R. W. Flood Inc.	2
18899	Seward Systems	2
19565	Stodola's Inc.	1
etc.		

WAREHOUSE

WAREHOUSE NUMBER	WAREHOUSE LOCATION
4	Fargo
3	Bismarck
2	Superior
1	Plymouth
etc.	

FIGURE 13.21

The complete database consists of four 1NF relations called SALESPERSON, SALES, CUSTOMER, and WAREHOUSE.

The two new relations are called CUSTOMER and WAREHOUSE, and can be written as follows:

CUSTOMER   (<u>CUSTOMER-NUMBER</u>, CUSTOMER-NAME, WAREHOUSE-NUMBER)

and

WAREHOUSE   (<u>WAREHOUSE-NUMBER</u>, WAREHOUSE-LOCATION)

The primary key for the relation CUSTOMER is <u>CUSTOMER-NUMBER</u>, and the primary key for the relation WAREHOUSE is <u>WAREHOUSE-NUMBER</u>.

In addition to these primary keys, we can identify WAREHOUSE-NUMBER to be a foreign key in the relation CUSTOMER. A foreign key is any attribute that is nonkey in one relation but a primary key in another relation. We designated WAREHOUSE-NUMBER as a foreign key in the previous notation and in the figures by underscoring it with a dashed line: ---------------.

Finally, the original, unnormalized relation SALES-REPORT has been transformed into four 3NF relations. In reviewing the relations shown in Figure 13.21, we can see that the single relation SALES-REPORT was transformed into the following four relations:

SALESPERSON   (<u>SALESPERSON-NUMBER</u>, SALESPERSON-NAME, SALES-AREA)

SALES   (<u>SALESPERSON-NUMBER</u>, <u>CUSTOMER-NUMBER</u>, SALES-AMOUNT)

CUSTOMER   (<u>CUSTOMER-NUMBER</u>, CUSTOMER-NAME, WAREHOUSE-NUMBER)

FIGURE 13.22

An entity-relationship diagram for the Al S. Well Hydraulic Company database.

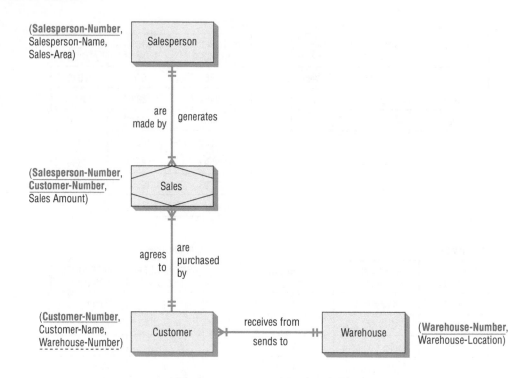

```
WAREHOUSE        (WAREHOUSE-NUMBER,
                 WAREHOUSE-LOCATION)
```

The third normal form is adequate for most database design problems. The simplification gained from transforming an unnormalized relation into a set of 3NF relations is a tremendous benefit when it comes time to insert, delete, and update information in the database.

An E-R diagram for the database is shown in Figure 13.22. One SALESPERSON serves many CUSTOMER(s), who generate SALES and receive their items from one WAREHOUSE (the closest WAREHOUSE to their location). Take the time to notice how the entities and attributes relate to the database.

Using an Entity-Relationship Diagram to Determine Record Keys

An E-R diagram can be used to determine the keys required for a record or a database relation. The first step is to construct the E-R diagram and label a unique (primary) key for each data entity. Figure 13.23 shows an E-R diagram for a customer order system. There are three data entities: CUSTOMER, with a primary key of <u>CUSTOMER-NUMBER</u>; ORDER, with a primary key of <u>ORDER-NUMBER</u>; and ITEM, with <u>ITEM-NUMBER</u> as the primary key. One CUSTOMER may place many orders, but each ORDER can be placed by one CUSTOMER only, so the relationship is one-to-many. Each ORDER may contain many ITEM(s), and each ITEM may be contained in many ORDER(s), so the ORDER-ITEM relationship is many-to-many.

A foreign key is a data field on a given file that is the primary key of a different master file. For example, a DEPARTMENT-NUMBER indicating a student's major may exist on the STUDENT MASTER table. DEPARTMENT-NUMBER could also be the unique key for the DEPARTMENT MASTER table.

FIGURE 13.23

An entity-relationship diagram for customer orders.

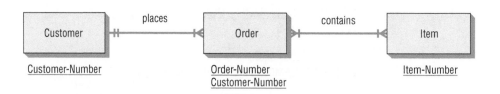

One-to-Many Relationships

A one-to-many relationship is the most common type of relationship, since all many-to-many relationships must be broken down into two or more one-to-many relationships. When a one-to-many relationship occurs, you place the primary key on the table at the one end of the relationship as a foreign key on the table on the many end of the relationship. For example, since one customer may have many orders, you place the customer number on the order record.

The design of web pages, displays, or reports that include information from only one record of the many relationship, along with information from the one end of the relationship, is easy to construct. The display will not have any repeating information. An example is an order inquiry using an order number to look up a single order. Since the order is for one customer, the result would be fields from the order and a single customer.

Designing the reverse is more complicated, since the table at the one end of the relationship may have many records for the many end. These are implemented in a variety of ways. For a simple display screen, the information from the one end is displayed with a repeating number of groups of information from the many end of the relationship. In Microsoft Access, this might be a form with a subform, such as a customer with a subform of all the customers' orders. If there were a large number of records from the many end, scroll bars would appear.

In simple situations, the relationship might also be implemented by using a drop-down list, with each record from the many end becoming one entry in the one end; an example is the display of a car along with a drop-down list containing all the models for the car. When designing websites, the information from the one end might be at the top of the page, with multiple groups of data below it or multiple links to the data. An example is one search engine topic resulting in many matching links or one genre of music and many artists that match the genre.

Many-to-Many Relationships

With a many-to-many relationship, three tables are necessary: one for each data entity and one for the relationship. The ORDER and ITEM entities in our example have a many-to-many relationship. The primary key of each data entity is stored as a foreign key of the relational table. The relational table may simply contain the primary keys for each data entity or may contain additional data, such as the grade received for a course or the quantity of an item ordered. Refer to the table layout illustrated in Figure 13.24. The ORDER ITEM table contains information about which order contains which items, and provides a link between the ORDER table and the ITEM MASTER table.

The relationship table should be indexed on each foreign key—one for each of the tables in the relationship—and may have a primary key consisting of a combination of the two foreign keys. Often corporations will use a unique key, such as sequence number, as the primary key for the relational table. To find many records from a second table given the first table, directly read the relational table for the desired key. Locate the matching record in the second *many* table. Continue to loop through the relational table until the desired key is no longer found. For example, to find records in the ITEM MASTER for a specific record in the ORDER table, you directly read the ORDER-ITEM table, using the ORDER-NUMBER as the index. Records are logically sequenced based on the data in the index, so all records for the same ORDER-NUMBER are

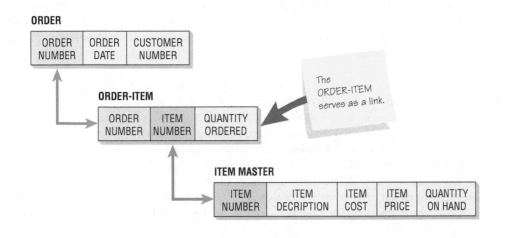

FIGURE 13.24

With a many-to-many relationship, three files are necessary.

grouped together. For each ORDER-ITEM record that matches the desired ORDER-NUMBER, directly read the ITEM MASTER table using the ITEM-NUMBER as an index.

The logic is the same for the reverse situation, such as finding all the orders for a backordered item that has been received. Use the desired ITEM-NUMBER to read the ORDER-ITEM table directly. The ORDER-ITEM index is set to the ITEM-NUMBER. For all matching ORDER ITEM records, use the ORDER-NUMBER to read the ORDER table directly. Finally, read the CUSTOMER MASTER table directly to obtain the CUSTOMER-NAME and ADDRESS using the CUSTOMER-NUMBER on the ORDER table.

Relational tables may have relationships to more tables in the database than just the two to which they directly connect. For example, there might be a relational table called Class or Section to link students and courses, since each student may take many courses, and each course may have many students. The Section table may have a relationship to the Textbook or to the Instructor for that section.

Guidelines for Master File/Database Relation Design

The following guidelines should be taken into account when designing master files or database relations:

1. Each separate data entity should create a master database table. Do not combine two distinct entities on one file. For example, items are purchased from vendors. The ITEM MASTER table should contain only item information, and the VENDOR MASTER table should contain only vendor information.
2. A specific data field should exist only on one master table. For example, CUSTOMER NAME should exist only on the CUSTOMER MASTER table, not on the ORDER table or any other master table. The exceptions to this guideline are the key or index fields, which may be on as many tables as necessary. If a report or screen needs information from many tables, the indexes should provide the linkage for obtaining the required records.
3. Each master table or database relation should have programs to *c*reate, *r*ead, *u*pdate, and *d*elete (abbreviated CRUD) the records. Ideally, only one program should add new records and only one program should delete specified records. Many programs, however, may be responsible for changing data fields in the course of normal business activities. For example, a CUSTOMER MASTER file may have a CURRENT BALANCE field that is increased by the ORDER TOTAL in the order processing program and decreased by a PAYMENT AMOUNT or an AMOUNT RETURNED from two additional programs.

Integrity Constraints

Integrity constraints are rules that govern changing and deleting records, and that help keep the data in the database accurate. Three types of integrity constraints apply to a database:

1. Entity integrity
2. Referential integrity
3. Domain integrity

Entity integrity constraints are rules that govern the composition of primary keys. The primary key cannot have a null value, and if the primary key is a composite key, none of the component fields in the key can contain a null value. Some databases allow you to define a unique constraint or a unique key. This unique key identifies only one record, which is not a primary key. The difference between a unique key and a primary key is that a unique key may contain a null value.

Referential integrity governs the nature of records in a one-to-many relationship. The table that is connected to the one end of the relationship is called the *parent*. The table connected to the many end of the relationship is called the *child table*. Referential integrity means that all foreign keys in the many table (the child table) must have a matching record in the parent table. Hence, you cannot add a record in the child (many) table without a matching record in the parent table.

A second implication is that you cannot change a primary key that has matching child table records. If you could change the parent record, the result would be a child record that would have a different parent record or an orphan record, or a child record without a parent record. Examples are a GRADE record for a student that would not be on the STUDENT MASTER table and an

MAC APPEAL

Microsoft Word, Excel, and PowerPoint are available for the Mac operating system, but the only way to run Microsoft Access on a Mac is to run Windows in virtualization mode or by booting into Windows. However, there are two other database options for the Mac: Bento and FileMaker Pro. Bento is a personal database that allows users to efficiently gather information from the address book, calendar application, Apple Mail, and Microsoft Excel, and then quickly add new fields to create a customizable database.

Some users may find Bento limiting; another option is Bento's big brother, FileMaker Pro. It is a full relational database program that has support for direct access to SQL databases. Its distinguishing characteristic is that the screens, forms, and reports that access the database are fully integrated with the database engine.

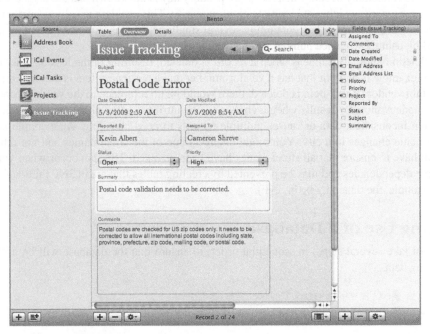

FIGURE 13.MAC

A screen from Bento, a personal database. (Screenshot from Bento, a personal database. Copyright © 2012 by Filemaker, Inc. Reprinted with permission.)

ORDER record for a CUSTOMER NUMBER that did not exist. The last implication of referential integrity is that you cannot delete a parent record that has child records. That would also lead to the orphan records mentioned earlier.

Referential integrity is implemented in two different ways. One way is to have a restricted database, in which the system can update or delete a parent record only if there are no matching child records. A cascaded database will delete or update all child records when a parent record is deleted or changed (the parent triggers the changes).

A restricted relationship is better when deleting records. You would not want to delete a customer record and have all the outstanding invoices deleted as well! The cascaded approach is better when changing records. If the primary key of a student record is changed, all the course records for that student would have their foreign keys (the STUDENT NUMBER on the COURSE MASTER) changed as well.

Domain integrity rules are used to validate the data, such as table, limit, range, and other validation checks. They are further explained in Chapter 15. The domain integrity rules are usually stored in the database structure in one of two forms. Check constraints are defined at the table level and can refer to one or more fields in the table. An example is that the DATE OF PURCHASE is always less than or equal to the current date. Rules are defined at the database level as separate objects and can be used with a number of fields. An example is a value that is greater than zero, used to validate a number of elements.

Anomalies

Four anomalies may occur when creating database tables:

1. Data redundancy
2. Insert anomaly
3. Deletion anomaly
4. Update anomaly

Data redundancy occurs when the same data are stored in more than one place in the database (except for primary keys stored as foreign keys). This problem is solved by creating tables that are in 3NF.

An insert anomaly occurs when the entire primary key is not known and the database cannot insert a new record, which would violate entity integrity. This problem usually occurs when the primary key is a composite key containing several smaller attributes. An insert anomaly may be avoided by using a sequence number for the primary key.

A deletion anomaly happens when a record is deleted, resulting in the loss of other related data. An example is an item that has a vendor number and a particular item is the only reference to a certain vendor. If that item is deleted, there would be no reference to the vendor record.

An update anomaly results when a change to one attribute value either causes the database to contain inconsistent data or causes multiple records to need changing. An example is when a street name changes in a city. You might change some of the street names and not others, or you will have to ensure that all street names have been changed. This can occur when you have transitive dependencies and may be prevented by creating tables that are in 3NF (although in the street example, the data may be in 3NF).

Making Use of a Database

You must take several steps, in sequential order, to ensure that the database will be useful for presenting data.

Steps in Retrieving and Presenting Data

There are eight steps in the retrieval and presentation of data:

1. Choose a relation from the database.
2. Join the relations together.
3. Project columns from the relation.
4. Select rows from the relation.
5. Derive new attributes.
6. Index or sort rows.
7. Calculate totals and performance measures.
8. Present the data.

The first and last steps must be done, but the six steps in between are optional, depending on how data are to be used. Figure 13.25 is a visual guide to the steps.

The final step in the retrieval of data is presentation. Presentation of the data abstracted from the database can take many forms. Sometimes the data will be presented in tabular form, sometimes in graphs, and other times as a single-word answer on a screen. Output design, as covered in Chapter 11, provides a more detailed look at presentation objectives, forms, and methods.

Denormalization

One of the main reasons for normalization is to organize data in order to reduce redundant data. If you are not required to store the same data over and over again, you can save a great deal of space. Such organization allows an analyst to reduce the amount of storage needed, which was very important when storage was expensive.

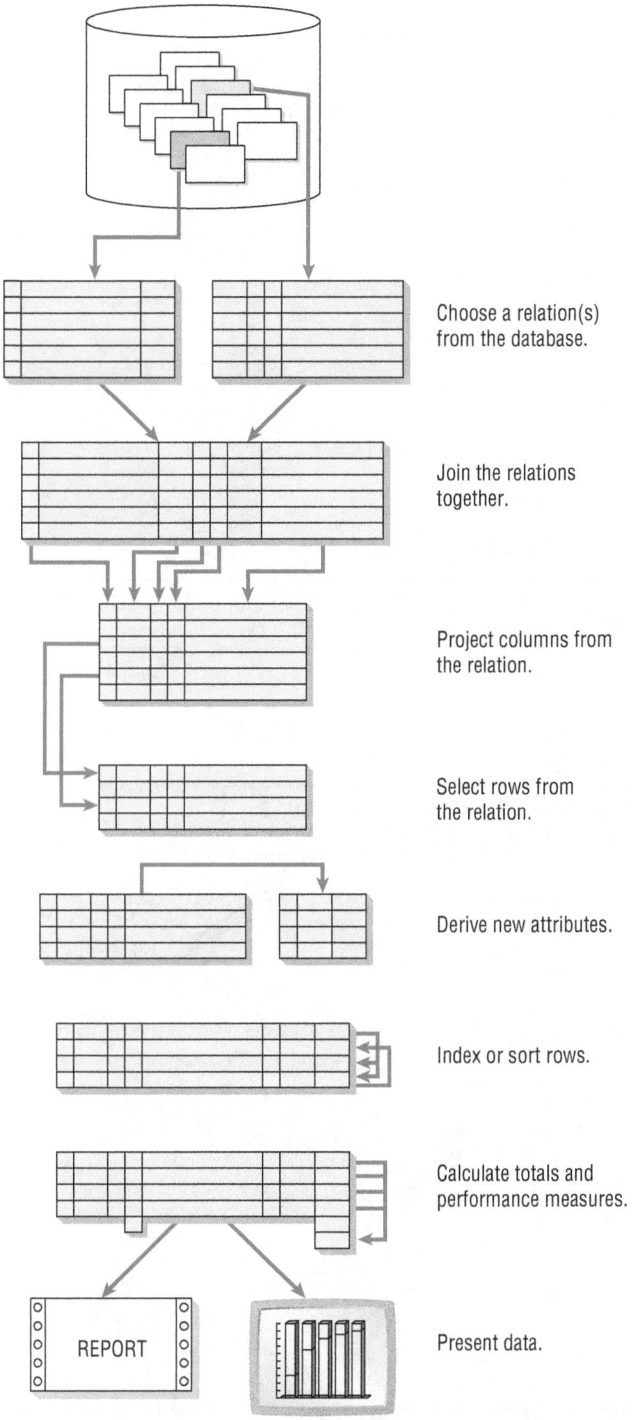

FIGURE 13.25

Data are retrieved and presented in eight distinct steps.

Choose a relation(s) from the database.

Join the relations together.

Project columns from the relation.

Select rows from the relation.

Derive new attributes.

Index or sort rows.

Calculate totals and performance measures.

REPORT

Present data.

We learned in the last section that to use normalized data, we have to progress through a series of steps that involve joining, sorting, and summarizing. When speed of querying the database (that is, asking a question and requiring a rapid response) is critical, it may be important to store data in other ways.

Denormalization is the process of transforming a logical data model into a physical model that is efficient for the most-often-needed tasks. These tasks can include report generation, but they can also mean more efficient queries. Complex queries such as online analytical processing (OLAP), as well as data mining and knowledge data discovery (KDD) processes, can also make use of databases that are denormalized.

Denormalization can be accomplished in a number of different ways. Figure 13.26 depicts some of these approaches. First, we can take a many-to-many relationship, such as that of SALESPERSON

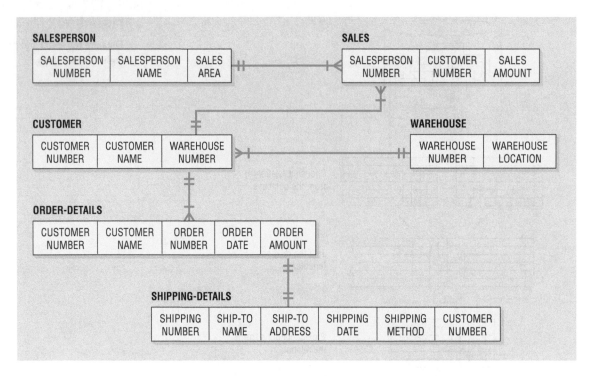

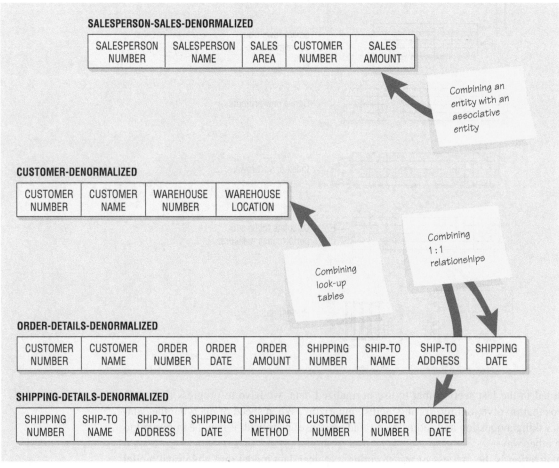

FIGURE 13.26

Three examples of denormalization in order to make access more efficient.

and CUSTOMER, which share the associative entity SALES. By combining the attributes from SALESPERSON and SALES we can avoid one of the join processes. This may result in a considerable amount of data duplication, but it makes the queries about sales patterns more efficient.

Another reason for denormalization is to avoid repeated reference to a lookup table. It may be more efficient to repeat the same information—for example, the city, state, and zip code—even though this information can usually be stored as a zip code only. Hence, in the sales example, CUSTOMER and WAREHOUSE may be combined.

Finally, we look at one-to-one relationships because they are very likely to be combined for practical reasons. If we learn that many of the queries regarding orders also are interested in how the order was shipped, it would make sense to combine, or denormalize. Hence, in the example, some of the details can appear in both ORDER-DETAILS and SHIPPING-DETAILS when we go through denormalization.

Data Warehouses

Data warehouses differ from traditional databases. The purpose of a data warehouse is to organize information for quick and effective queries. In effect, they store denormalized data, but they go one step further. They organize data around subjects. Most often, a data warehouse is more than one database processed so that data are represented in uniform ways. Therefore, the data stored in data warehouses comes from different sources, usually databases that were set up for different purposes.

The data warehouse concept is unique. Differences between data warehouses and traditional databases include the following:

1. In a data warehouse, data are organized around major subjects rather than individual transactions.
2. Data in a data warehouse are typically stored as summarized data rather than the detailed, raw data found in a transaction-oriented database.
3. Data in a data warehouse cover a much longer time frame than data in traditional transaction-oriented databases because queries usually concern longer-term decision making rather than daily transaction details.
4. Most data warehouses are organized for fast queries, whereas the more traditional databases are normalized and structured in such a way as to provide efficient storage of information.
5. Data warehouses are usually optimized for answering complex queries, known as OLAP, from managers and analysts, rather than simple, repeatedly asked queries.
6. Data warehouses allow easy access via data mining software (called siftware) that searches for patterns and is able to identify relationships not imagined by human decision makers.
7. Data warehouses include not just one but multiple databases that have been processed so that the warehouse's data are defined uniformly. These databases are referred to as clean data.
8. Data warehouses usually include data from outside sources (such as an industry report, the company's Security and Exchange Commission filing, or even information about competitors' products), as well as data generated for internal use.

Building a data warehouse is a monumental task. An analyst needs to gather data from a variety of sources and translate those data into a common form. For example, one database may store information about gender as "Male" and "Female," another may store it as "M" and "F," and a third may store it as "1" and "0." An analyst needs to set a standard and convert all the data to the same format.

Once the data are clean, the analyst has to decide how to summarize the data. Once summarized, the detail is lost, so an analyst has to predict the type of queries that might be asked.

Then, the analyst needs to design the data warehouse by logically organizing, and perhaps even physically clustering, the data by subject, requiring much analysis and design. The analyst needs to know a substantial amount about the business.

Typical data warehouses tend to be from 50 gigabytes to tens of terabytes in size. Because they are large, they are also expensive. Most data warehouses cost millions of dollars.

CONSULTING OPPORTUNITY 13.2

Storing Minerals for Health, Data for Mining

One of Marathon Vitamin Shops's employees, Esther See, approaches the owner, Bill Berry, about an observation she had. "I've noticed that our customers have different habits. Some come in regularly, and others are less predictable," Esther says. "When I see a regular customer, I pride myself on knowing what the customer will buy and maybe even suggest other vitamins they might like. I think I generate more sales that way. The customer is happier, too."

Esther continues, "I wish I could be better at helping out some of the customers who come in less frequently, though."

"That's a very nurturing attitude, Esther, and it helps out our store as well," Bill replies. "I know that we can benefit in other ways by getting a better handle on customer patterns. For instance, we can be sure that we have an item in stock."

Esther nods in agreement and adds, "It's not just the type of vitamin I'm talking about. Some customers prefer one brand over another. I don't know if it depends on their income level or the interests they have in leisure activities. Sports, for example."

"I see, Ms. See," Bill chuckles at his own joke, "but do you have anything in mind?"

"Yes, Mr. Berry," she says more formally. "We should organize the data we have about our customers using a data warehouse concept. We can merge the data we have with data from other sources. Then we can look for patterns in our data. Maybe we can identify existing patterns and predict new trends."

Think about how you would organize a data warehouse for Marathon Vitamin Shops. What other databases would you like to merge into the data warehouse? What sort of patterns should Bill Berry be looking for? Identify these patterns by type (associations, sequences, clustering, or trends) and discuss them in a page or two.

Online Analytical Processing

First introduced in 1993 by E. F. Codd, online analytical processing (OLAP) was meant to answer decision makers' complex questions. Codd concluded that a decision maker had to look at data in a number of different ways. Therefore, the database itself had to be multidimensional. Many people picture OLAP as a Rubik's Cube of data. You can look at the data from all different sides, and can also manipulate the data by twisting or turning it so that it makes sense.

This OLAP approach validated the concept of data warehouses. It then made sense for data to be organized in ways that allowed efficient queries. Of course, OLAP involves the processing of data through manipulation, summarization, and calculation, so more than a data warehouse is involved. Business intelligence (coming up in a later section) includes querying, as well as reporting, OLAP, and a variety of alerts to users.

Data Mining

Data mining can identify patterns that a human is unable to detect. Either the decision maker may not be able to see a pattern, or perhaps the decision maker is not able to think about asking whether that pattern exists. Data mining algorithms search data warehouses for patterns using algorithms. Figure 13.27 illustrates the concept of data mining.

The types of patterns decision makers try to identify include associations, sequences, clustering, and trends. Associations are patterns that occur together at the same time. For example, a person who buys cereal usually buys milk to go with the cereal. Sequences, on the other hand, are patterns of actions that take place over a period of time. For example, if a family buys a house this year, it will most likely buy durables (a refrigerator or washer and dryer) next year. Clustering is the pattern that develops among a group of people. For example, customers who live in a particular zip code may tend to buy a particular car. Finally, trends are patterns that are noticed over a period of time. For example, consumers may move from buying generic goods to premium products.

The concept of data mining came from the desire to use a database for more selective targeting of customers. Early approaches to direct mail included using zip code information as a way to determine what a family's income might be (assuming that a family must generate sufficient income to afford to live in the prestigious Beverly Hills zip code 90210 or some other affluent neighborhood). It was a way (not perfect, of course) to limit the number of catalogs sent.

Data mining takes this concept one step further. Assuming that past behavior is a good predictor of future purchases, large numbers of data are gathered on a particular person from credit

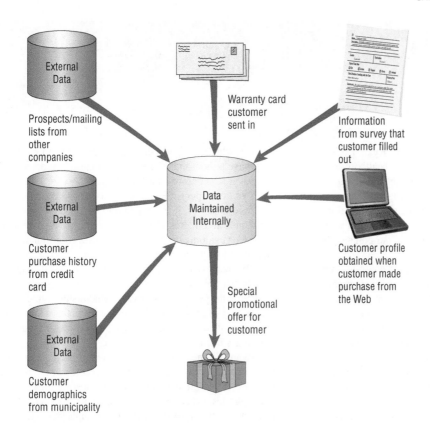

FIGURE 13.27

Data mining collects personal information about customers in an effort to be more specific in interpreting and anticipating their preferences.

card purchases. A company can identify what stores we shop in, what we have purchased, how much we paid for an item, and when and how frequently we travel. Data are also entered, stored, and used for a variety of purposes when we fill out warranties, apply for a driver's license, respond to a free offer, or apply for a membership in a fitness club. Moreover, companies share these data and often make money on the sale of them as well.

American Express has been a leader in data mining for marketing purposes. American Express will send you discount coupons for new stores or entertainment when it sends you a credit card bill, having determined that you have shopped in similar stores or attended similar events. General Motors offers a MasterCard that allows customers to accumulate bonus points toward the purchase of a new car and then sends out information about new vehicles at the most likely time that a consumer would be interested in purchasing a new car.

The data mining approach is not without problems, however. First, the costs may be too high to justify data mining, and this may be discovered only after huge setup costs have been accrued. Second, data mining has to be coordinated so that various departments or subsidiaries do not all try to reach the customer at the same time. In addition, customers may think their privacy has been invaded and resent the offers that are coming their way. Finally, customers may think profiles created solely on the basis of their credit card purchases present a highly distorted image of who they are.

Analysts should take responsibility for considering the ethical aspects of any data mining projects that are proposed. Questions about the length of time profile material is kept, its confidentiality, the privacy safeguards included, and the uses to which inferences are put should all be asked and considered with the client. The opportunities for abuse are apparent and must be guarded against. For consumers, data mining is another push technology, and if consumers do not want to be pushed, the data mining efforts will backfire.

Business Intelligence (BI)

Although not a new concept, business intelligence (BI) has grown in prominence from the late 1980s, and its uses have spread to several types of organizational employees, not just those few at the strategic level of decision making. Business intelligence is at heart a decision support system (DSS) for organizational decision makers. It is composed of features that gather and

CONSULTING OPPORTUNITY 13.3

Losing Prospects

"Market share can be a real problem," says Ryan Taylor, director of marketing systems for a large East Coast health insurer. "One of the greatest challenges we face is how to identify good leads for our salespeople. With over 50 percent market share, we must eliminate the names of most of the prospects we buy before populating our marketing database. It is critical that we get it right because our marketing database is a critical part of our company's arsenal of strategic information tools."

Ryan explains to Chandler, one of your systems analysis team members, "A marketing database, or MDB for short, is a powerful, relational database that is the heart of marketing systems. Our marketing database is used to provide information for all marketing systems. They include productivity tools, such as our Sales Force Automation and our Mass Mailing Systems, which are designed to aid our salespeople in managing the sales cycle. They also include analytical tools, such as our geographic information systems (GIS) or graphical query language tools, which are designed to provide decision support.

"The primary function of a marketing database, though, is to track information on our customers and prospects. We currently track geographic information, demographic information, and psychographic information, or, as I like to say, where they live, who they are, and how they think.

"The simplest marketing databases can be made up of just three files: Prospect Profile, Customer Profile, and Purchase and Payment History.

"Once you have designed your marketing database, the next challenge is deciding how to populate it. We currently purchase our prospect information from a list vendor. Because our company's marketing strategy is based on mass marketing, we buy every business in our area. Because of this volume, we pay less than a dime for each prospect. If, however, a company is practicing product differentiation, their prospect base will likely be more defined. This company would likely pay a premium for more detailed data that have been carefully validated," explains Ryan.

"We face a real challenge. If I had a dollar for every time a rep complained to me about the address on a prospect being wrong, I could retire and move to Florida," Ryan quips. "I'm expected to identify which prospects are bad. That's not too hard if you only have a thousand of them, but what do you do when you have over a quarter of a million?"

Ryan continues, "Because we use these data frequently for large mailings, it is very important for us to ensure that the names and addresses on that file are as accurate as possible. For example, they should conform to postal standards and should not be duplicates.

"We achieve this through a technique called data hygiene. How's that for a geeky term? Data hygiene is usually accomplished with specialized software, which is used to determine the validity of an address. This software matches the database address to its own internal database of valid streets and number ranges in a given city or zip code."

Ryan resumes, "One of the other data challenges faced by marketers is eliminating duplicate records in the marketing database. There are two types of duplicates we look for: internal duplicates, which are the existence of multiple records of the same customer or prospect, and external duplicates, which represent our inability to eliminate customers from our prospect data.

"Internal duplicates create reporting problems and increase mailing costs. External duplicates are even worse; they are both costly and embarrassing," Ryan explains. "One of the most embarrassing things for a sales representative is to make a prospecting call only to find out that the business is already our customer. The customer is generally left feeling like only a number in one of our computers. It creates a poor impression and wastes valuable time and resources."

In two paragraphs, describe some techniques Ryan could use to help identify internal and external duplicates in his company's marketing database. Describe how you would build a marketing database to minimize duplicates (use a paragraph). Are there operational methods that might cut down on this problem? List them. Who else in the organization could help with this process? Provide a brief list. In a paragraph, recommend methods to Chandler and your other systems analysis team members that can be used to help enlist and secure the assistance of other relevant organizational members.

store data, as well as using knowledge management approaches combined with analysis. This becomes input to decision makers' decision-making processes.

Business intelligence is built around the idea of processing large volumes of data. As a systems analyst, you may be asked to create systems that support BI, such as a data warehouse that is considered as the input to analytics. Or you might be involved with the creation of dashboards or even spreadsheets that convey BI to users. When data sets become too large or too complex to be handled with traditional tools or within traditional databases or data warehouses, they are often referred to as "big data."

Demand for experts possessing information management skills is ever increasing, since the growth in data far outstrips our abilities to meaningful store it, process it, or analyze it. Big data has also been called an approach to design for companies, essentially a strategy that permits organizations to cope with ever-increasing numbers of data from a myriad of sources—some human generated but many more generated via sensors of some type (for example, electronic road toll collections, satellite weather monitors).

Often you will be in charge of the interface between users and a data warehouse. As with other systems you create, you will need to understand the data generated by the organization, the business itself, and the information requirements of the analysts who will be interpreting the output of any statistical models supported by the DSS.

Five prominent methods are used for analyzing business intelligence: slice-and-dice drill-down, ad hoc queries, real-time analysis, forecasting, and scenarios. Although there has been some consolidation in the industry, there are many large and small providers of BI software. Some are moving their services to the cloud to provide on-demand services, reduce overall costs, and improve startup times.

Business analytics (BA) is a term that covers the ideas of using big data along with a variety of *quantitatively* based analytical tools (such as statistics and predictive modeling) to answer management questions about trends and what-if questions. Their output may be used as input for decision makers or as input for computerized systems.

Some of the problems inherent in BI are how systems handle semistructured and unstructured data, since most organizations still have many documents that are not available for analysis through BI because they don't fit the structured, stringent requirements of data warehouse entries. This means that a large portion of the important information and data an organization and its customers generates goes unanalyzed and, therefore, unused or underused in decision making.

One way to address the absence of these data in traditionally structured data warehouses is to develop systems that take advantage of new tools for text analytics (TA). While TA is often thought of as a way to structure the unstructured, turning qualitative material into quantitative material, a broader view is recommended here. We want to tap into *qualitative* unstructured data that can be of use to decision makers who must recommend courses of action to their organizations that are backed by data.

As discussed in Chapter 5, you can use the following five guidelines for analyzing qualitative documents you collect or observe to determine user requirements:

1. Examine documents for key or guiding metaphors.
2. Look for insiders versus outsiders, or an "us against them" mentality.
3. List terms that characterize good or evil and appear repeatedly in documents.
4. Look for the use of meaningful messages and graphics posted on common areas or on web pages.
5. Recognize a sense of humor, if present.

In the past, we effectively used qualitative analysis of organizational metaphors to determine the relationship of specific metaphors (such as family, society, machine, organism, game, journey, and war) to the success of specific types of information systems, such as MIS, DSS, cooperative systems, and competitive (strategic) systems.

While these approaches provide a fundamental start, there is now good, sophisticated software available that you can pair with your users' judgment and include in your systems designs for their use, or use them as a tool to document and reinforce your own analysis work, along with your own critical interpretations and insights.

Text Analytics

You may have the opportunity to develop systems that help organizational users to interpret and understand the unstructured data in their companies. Unstructured, qualitative, or "soft," data are generated through blogs, chat rooms, questionnaires using open-ended questions, online discussions conducted on the Web, and exchanges occurring on social media such as Facebook, Twitter, and other Web-generated dialogs between customers and an organization. Text analytics can help organizational decision makers realize valuable insights into what customers are thinking about the organization, the values and actions of the company, as well as customer or vendor motivations for beginning, maintaining, improving, or discontinuing a relationship.

Many large and small providers of text analytics software solutions are available, both proprietary and open source. If you have been active in helping an organization develop and implement an online presence, then incorporating text analytics into the new systems makes a lot of

FIGURE 13.28

Concept map showing prominence and relationships of concepts in Open Source Communities interview data.

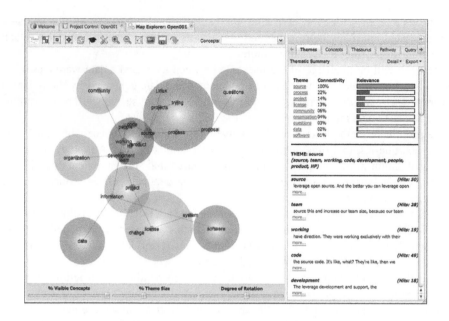

sense. As you design systems to support online presence, text analytics should be included to help the company make use of its online presence.

Text analytics provide insights for organizational members who want to have a rapid and visual yet decidedly qualitative approach to analyzing text data. A powerful text analytics software tool called Leximancer (from an Australian company) offers a very straightforward approach, since basically no set up is required. Rather, the contribution of the organizational user who will use the software is to contribute relevant knowledge about what is to be fed into the system, so that the concepts and other diagrams generated by the software are useful.

In general terms, what users do with Leximancer software is to submit documents and then omit words that they don't want to include in analysis (such as *a* and *the*); then the software performs a keyword count. The body of the text is then broken up into segments, and the concepts that result form a "thesaurus" of words that journey through the text document together.

Figure 13.28 shows a concept map generated with some interview data from the Open Communities project, creating a different visual display of the text from interviews. As mentioned earlier, submitting your documents to the software generates a concept map. Notice how different the data appear, with the user's thesaurus words, such as *Linux*, *projects*, *trying*, and *process* conceptually hanging together, and the term *community* separate but related to those concepts through *people*.

Figure 13.29 shows the ranked concepts for the Open Communities project, using Leximancer. This supplies analysis in ranked bar chart format, showing the most prominent

FIGURE 13.29

Ranked concepts for categories overview for the Open Source Communities project, using the Leximancer Insight Dashboard. (Two screenshots from Leximancer software. Copyright © by Leximancer Pty Ltd. Reprinted with permission.)

1. Ranked Concepts for Categories

Category: community

Concept	Rel Freq (%)	Strength (%)	Prominence
development	2	5	4.1
information	2	5	4.1
source	4	4	2.9
system	< 1	< 1	0.0
project	< 1	< 1	0.0
team	< 1	< 1	0.0
organization	< 1	< 1	0.0
product	< 1	< 1	0.0

Category: license

Concept	Rel Freq (%)	Strength (%)	Prominence
project	11	4	8.6
system	5	3	7.1
source	5	2	3.7
team	< 1	< 1	0.0
organization	< 1	< 1	0.0
development	< 1	< 1	0.0
product	< 1	< 1	0.0
information	< 1	< 1	0.0

HYPERCASE® EXPERIENCE 13

"I hear very good things about your team from the people in Management Systems. You even got some hard-earned praise from Tom Ketcham, who isn't easy to please these days. Even *he* is seeing some possibilities. I think you'll pull us together yet . . . unless we all go off in different directions again. I'm just teasing you. I told you to think about whether we are a family, a zoo, or a war zone. Now's the time to start designing systems that fit us. You've been here long enough now to form those opinions. I hope they're favorable. I think our famous Southern hospitality coupled with my noticeably British sense of humor should help influence you, don't you? I was so busy persuading you that we're worth the effort that I almost forgot to tell you: Tom and Snowden have agreed to think about moving toward a database of some sort. Would you have this ready in the next two weeks? Tom is at a conference in Minneapolis, but when he returns you should have some database ideas worked up for Snowden and him to discuss. Keep at it."

HYPERCASE Questions

Use the table that appears in this exercise to complete the questions below. Assume that your team members have used the Training Unit Client Characteristics Report to design this database table to store the relevant information contained on this report.

1. Apply normalization to the table your team has developed to remove repeating groups. Display your results.
2. Remove transitive dependencies from your table and show your resulting database table.

Table name: CLIENT TABLE

Column Name	Description
CLIENT ID (primary key)	Mnemonic made up by users, such as STHSP for State Hospital
CLIENT NAME	The actual, full client name
ADDRESS	The client's address
CONTACT	The name of the contact person
PHONE NUMBER	The phone number of the contact person
CLASS	The type of institution (Veterans Administration hospital, clinic, other)
STAFF-SIZE	Size of client staff (number)
TRAINING LEVEL	Minimum required expertise level of the staff (as defined by the class)
EQUIP-QTY	The number of medical machines that the client has
EQUIP TYPE	The type of medical machines (e.g., X-ray, MRI, CAT)
EQUIP MODEL-YR	The model and year of each medical machine

concepts within the designated category. Notice the prominence of the relationship between the category community and the use of the word *development*. The user can visually see the prominent relationship between the category "license" and use of the word *project*.

Your expertise in designing an interface as well as a system itself is especially critical here. Part of what you must do is to demonstrate to the organization the value of using analysis of unstructured, qualitative data to identify and predict future trends in customer, vendor, and supplier behavior. Another important element is to design the human activities surrounding the use of text analytics software. Incorporating the tools is not enough. Remember that you are not just converting qualitative data into quantitative data. Instead, you are developing systems that include text analytics software as a useful means to an end. So you should also include guidelines and suggestions for how decision makers should view the qualitative results, how they might interpret the qualitative results, and how the evidence from those results can be used to make well-reasoned, logically based recommendations which in turn can be acted upon to improve the organization.

Summary

How to store data is often an important decision in the design of an information system. There are two approaches to storing data. The first approach is to store data in individual files, one file for each application. The second approach is to develop a database that can be shared by many users for a variety of applications as the need arises.

An understanding of data storage requires a grasp of three realms: reality, data, and metadata. An entity is an object or event for which we are willing to collect and store data. Attributes are the actual characteristics of these entities. Data items can have values and can be organized into records that can be accessed by a key. Metadata describe the data and can contain restrictions about the value of a data item (such as numeric only).

Examples of conventional files include master files, table files, transaction files, work files, and report files. Databases typically are constructed with a relational structure. Legacy systems can have hierarchical or network structures, however.

Normalization is a process that transforms user views into less complex structures called normalized relations. There are three steps in the normalization process. First, all repeating groups are removed. Second, all partial dependencies are removed. Finally, the transitive dependencies are taken out. After these three steps are completed, the result is the creation of numerous relations that are of third normal form (3NF).

An entity-relationship diagram may be used to determine the keys required for a record or a database relation. The three guidelines to follow when designing master tables or database relations are that (1) each separate data entity should create a master table (do not combine two distinct entities within one table), (2) a specific data field should exist only on one master table, and (3) each master table or database relation should have programs to create, read, update, and delete records.

The process of retrieving data may involve as many as eight steps: (1) choosing a relation, (2) joining two relations together, (3) projecting (choosing) columns, (4) selecting relevant rows, (5) deriving new attributes, (6) sorting or indexing rows, (7) calculating totals and performance measures, and (8) presenting the results to the user.

Denormalization is a process that transforms the logical data model into a physical model that is efficient for tasks that are most needed. Data warehouses differ from traditional databases in many ways; one is that they store denormalized data, which are organized around subjects. Data warehouses allow easy access via data mining software, called siftware, which searches for patterns and identifies relationships not imagined by human decision makers.

Data mining involves using a database for more selective targeting of customers. Assuming that past behavior is a good predictor for future purchases, companies collect data about a person from past credit card purchases, driver's license applications, warranty cards, and so on. Data mining can be powerful, but it may be costly and it needs to be coordinated. In addition, it may infringe on consumer privacy or even a person's civil rights.

Business intelligence (BI) is composed of features that gather and store data, as well as using knowledge management approaches combined with analysis. This becomes input to decision makers' decision-making processes. Business analytics (BA) refers to statistical tools and models to quantitatively process structured data like that found in data warehouses. This in turn is input for both humans and computers. When data sets become too large or too complex to handle with traditional tools or within traditional databases or data warehouses, they are often referred to as "big data." Text analytics uses software to examine unstructured, soft data from blogs, wikis, social media sites, and other online customer interactions to support decision makers in interpreting qualitative material.

Keywords and Phrases

attribute
big data
bubble diagram
business analytics (BA)
business intelligence (BI)
clean data
concatenated key
conventional file
CRUD (create, read, update, and delete)
data element
data item
data mining
data model diagram

data storage
data warehouse
database
database administrator
database management system (DBMS)
delete anomaly
denormalization
domain integrity
entity
entity integrity constraint
entity-relationship (E-R) diagram
entity subtype
first normal form (1NF)

hierarchical data structure
information retrieval
logical view
master file
network data structure
normalization
object identifier (OID)
online analytical processing (OLAP)
partial dependencies
physical view
primary key
qualitative data
quantitative data
reality, data, and metadata
record
referential integrity

relational data structure
relationship
repeating group
report file
retrieval
second normal form (2NF)
secondary key
special characters
table file
text analytics (TA)
third normal form (3NF)
transaction file
transitive dependencies
unnormalized relation
update anomaly
work file

Review Questions

1. What are the advantages of organizing data storage as separate files?
2. What are the advantages of organizing data storage using a database approach?
3. What are the effectiveness measures of database design?
4. List some examples of entities and their attributes.
5. What is the difference between a primary key and an object identifier?
6. Define the term *metadata*. What is the purpose of metadata?
7. List types of commonly used conventional files. Which of these are temporary files?
8. Name the three main types of database organization.
9. Define the term *normalization*.
10. What is removed when a relation is converted to the first normal form?
11. What is removed when a relation is converted from 1NF to 2NF?
12. What is removed when a relation is converted from 2NF to 3NF?
13. List the three entity constraints. In a sentence, describe the meaning of each entity constraint.
14. Describe the four anomalies that may occur when creating database tables.
15. List the eight steps for retrieving, presorting, and presenting data.
16. What does a join do? What is projection? What is selection?
17. Define *denormalization*.
18. Explain the differences between traditional databases and data warehouses.
19. What is data mining?
20. What features compose business intelligence (BI)?
21. What is big data?
22. Define *business analytics*.
23. What is *text analytics*?
24. What are the sources of big data for text analytics?

Problems

1. Given the following file of renters:

Record Number	Last Name	Apartment Number	Rent	Lease Expires
41	Warkentin	102	550	4/30
42	Buffington	204	600	4/30
43	Schuldt	103	550	4/30
44	Tang	209	600	5/31
45	Cho	203	550	5/31
46	Yoo	203	550	6/30
47	Pyle	101	500	6/30

    a. Show an example of projection.

    b. Show an example of selection.

    c. Show two different examples of sorting rows.

    d. Show an example of calculating totals.

2. The following is an example of a grade report for two students at the University of Southern New Jersey:

USNJ Grade Report Spring Semester 2012

| Name: I. M. Smarte | | | Major: MIS | |
| Student: 053-6929-24 | | | Status: Senior | |

Course Number	Course Title	Professor	Professor's Department	Grade
MIS 403	Systems Analysis	Diggs, T.	MIS	A
MIS 411	Conceptual Foundations	Barre, G.	MIS	A
MIS 420	Human Factors in IS	Barre, G.	MIS	B
CIS 412	Database Design	Menzel, I.	CIS	A
DESC 353	Management Models	Murney, J.	MIS	A

USNJ Grade Report Spring Semester 2012

| Name: E. Z. Grayed | | | Major: MIS | |
| Student: 472-6124-59 | | | Status: Senior | |

Course Number	Course Title	Professor	Professor's Department	Grade
MIS 403	Systems Analysis	Diggs, T.	MIS	B
MIS 411	Conceptual Foundations	Barre, G.	MIS	A

    Draw a data model diagram with associations for the user view.

3. Convert the user view in Problem 2 to a 3NF relation. Show each step along the way.

4. What problem might arise when using a primary key of course number for the data in Problem 2? (*Hint:* Think about what would happen if the Department Name [not shown in the data] changes.)

5. Draw an entity-relationship diagram for the following situation: Many students play many different sports. One person, called the head coach, assumes the role of coaching all these sports. Each of the entities has a number and a name. (Make any assumptions necessary to complete a reasonable diagram. List your assumptions.)

6. The entity-relationship diagram you drew in Problem 5 represents the data entities that are needed to implement a system for tracking students and the sports that they play. List the tables that are needed to implement the system, along with primary, secondary, and foreign keys that are required to link the tables.

7. Draw an entity-relationship diagram for the following situation: A commercial bakery makes many different products. These products include breads, desserts, specialty cakes, and many other baked goods. Ingredients such as flour, spices, and milk are purchased from vendors. Sometimes an ingredient is purchased from a single vendor, and other times an ingredient is purchased from many vendors. The bakery has commercial customers, such as schools and restaurants, that regularly place orders for baked goods. Each baked good has a specialist that oversees the setup of the bakery operation and inspects the finished product.

8. List the tables and keys that are needed to implement the commercial bakery system.

9. Draw an E-R diagram for the ordering system in Figure 13.24.

10. Draw a data flow diagram for placing an order. Base your data flow diagram on the E-R diagram.

11. Create an entity-relationship diagram for a genealogy software package called PeopleTree to keep track of ancestors. Assume that each person will be on a Person table and that one person may have one biological father and mother as well as an adopted mother and father. The mothers and fathers must be stored on the Person table as well. Each person should have only one birthplace, stored on the Place table. Many people may be born in the same place.

12. Define the primary key used for the Person and Place tables in Problem 11.

13. GaiaOrganix is an organic food wholesale co-op that links producers and consumers. GaiaOrganix negotiates purchases by grocery and other stores from farmers who raise a variety of crops, such as fruits, vegetables, and grain. Each farmer may produce a number of crops, and each crop may be produced by a number of farmers. To provide the highest level of fresh products, the produce is shipped directly from the farm to the store. Each store may purchase from many farms, and each farm may sell to many stores. Draw an entity-relationship diagram in the third normal form showing the relationship between the producer (farms) and the retailer (stores).

14. ArticleIndex.com is a company that produces indexes of magazine and periodical articles for a given discipline. A Web user should be able to enter an article topic or authors and receive a detailed list of all the articles and periodicals in which the topic was found. Each article may have many authors, and each author may write many articles. An article may be found in only one periodical, but each periodical will usually contain many articles. Each article may have many topics, and each topic may be in many articles. Draw an entity-relationship diagram in the third normal form for the articles, authors, periodicals, and topics.

15. Identify the primary and foreign keys for the entity-relationship diagram created in Problem 14.

Group Projects

1. Gregg Baker orders tickets for two concerts over the Web. His orders are processed, exact seat locations are assigned, and the tickets are mailed separately. One of the sets of tickets gets lost in the mail. When he calls the service number, he does not remember the date or the seat numbers, but the ticket agency is able to locate his tickets quickly because the agency has denormalized the relation. Describe the ticket ordering system by listing the data elements that are kept on the order form and the shipping form. What information did Gregg give the ticket agency to retrieve the information?

Selected Bibliography

Agrawal, R., A. Ailamaki, P. A. Bernstein, E. A. Brewer, M. J. Carey, S. Chaudhuri, et al. "The Claremont Report on Database Research." *Communications of the ACM,* Vol. 52, No. 6, 2009, pp. 56–65.

Avison, D. E. *Information Systems Development: A Database Approach,* 2nd ed. London: Blackwell Scientific, 1992.

Codd, E. F. "A Relational Model of Data for Large Shared Data Banks." *Communications of the ACM,* Vol. 13, No. 6, 1970, pp. 377–387.

Davenport, T. H., and Harris, J. G. *Competing on Analytics: The New Science of Winning.* Boston: Harvard Business School Press, 2007.

Dietel, H. M., P. J. Dietel, and T. R. Nieto. *E-Business and E-Commerce: How to Program.* Upper Saddle River, NJ: Prentice Hall, 2001.

Edgington, T. M. "Introducing Text Analytics as a Graduate Business School Course." *Journal of Information Technology Education,* Vol. 10, 2011, pp. 207–234.

Evelson, Boris. "Want to Know What Forrester's Lead Data Analysts Are Thinking About BI and the Data Domain?" April 29, 2010. blogs.forrester.com/boris_evelson/10-04-29-want_know_what_forresters_lead_data_analysts_are_thinking_about_bi_and_data_domain. Last accessed August 22, 2012.

Gane, C., and T. Sarson. *Structured Systems Analysis: Tools and Techniques.* Englewood Cliffs, NJ: Prentice Hall, 1979.

Gray, P. "Data Warehousing: Three Major Applications and Their Significance." In *Emerging Information Technologies, Improving Decision, Cooperation, and Infrastructure,* edited by K. E. Kendall. Thousand Oaks, CA: Sage Publications, 1999.

Halper, B. F. "Leximancer—Concept Maps for Text Analytics and the Customer Insight Portal." October 14, 2008. fbhalper.wordpress.com/2008/10/14/leximancer-concept-maps-and-the-customer-insight-portal/. Last accessed August 22, 2012.

Hoffer, J., A. Prescott, and H. Topi. *Modern Database Management,* 9th ed. Upper Saddle River, NJ: Prentice Hall, 2009.

IBM Support Portal. "Getting Support for Your Business Analytics Product." www-01.ibm.com/software/analytics/support/. Last accessed August 22, 2012.

Leximancer. www.leximancer.com. Last accessed August 22, 2012.

Negash, S., and P. Gray. "Business Intelligence," in *Handbook of Decision Support Systems 2,* edited by F. Burstein and C. W. Holsapple, pp. 175–193. Berlin: Springer, 2008.

Sanders, G. L. *Data Modeling.* New York: International Thomson Publishing, 1995.

Shin, S. K., and G. L. Sanders. "Denormalization Strategies for Data Retrieval from Data Warehouses." *Decision Support Systems,* Vol. 42, No. 1, 2006, pp. 267–282.

The CPU Case Episode and accompanying student files are available online at www.pearsonglobaleditions.com/kendall.

CHAPTER 14

Human–Computer Interaction

LEARNING OBJECTIVES

Once you have mastered the material in this chapter you will be able to:

1. Understand human–computer interaction (HCI).
2. Design useful touch screen interfaces for smartphones and tablets.
3. Design a variety of user interfaces.
4. Design effective onscreen dialog for HCI.
5. Understand the importance of user feedback.
6. Articulate HCI implications for designing ecommerce websites.
7. Formulate queries that permit users to search the Web.

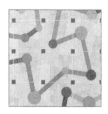

Throughout the book , your awareness of human–computer interaction (HCI) and its importance to your task as a systems analyst has grown. While awareness is important, by now you recognize that you need to master the concepts surrounding HCI as well as become proficient at assessing human information requirements and incorporating your findings into your designs. Furthermore, the European Union (EU) and the United States have created specific guidelines for usability. These guidelines mandate making websites and electronic services accessible to the able-bodied and disabled alike.

This chapter fills in some of the details about HCI and working with users. It also gives you some experience in applying HCI concepts that you have been learning to help in your design of human–computer interfaces for computers, smartphones, and tablets; feedback; ecommerce websites; and Web queries.

Understanding Human–Computer Interaction

Designing for HCI means "ensuring system functionality and usability, providing effective user interaction support, and enhancing a pleasant user experience." Furthermore, "The overarching goal is to achieve both organizational and individual user effectiveness and efficiency. To reach these goals, managers and developers need to be knowledgeable about the interplay among users, tasks, task contexts, information technology (IT), and the environments in which systems are used" (Carey et al., 2004, p. 358).

You can ensure that systems are user centered, so that they appropriately include users' needs as well as organizational needs by understanding HCI concepts, considering interfaces in the light of HCI issues, and applying standard design concepts to computers in new ways because of an HCI approach.

Knowledge about the interplay among users, tasks, task contexts, IT, and the environments in which the systems are used comprises the basis of HCI. The main tactic of HCI in systems analysis and design is to repeatedly elicit feedback from users about their experiences with prototyped designs (which could be screens, forms, interfaces, and the like), refining the design based on the suggested changes, and trying them with users again until the design is acceptable and until it is frozen by the analyst.

How Fit Affects Performance and Well-Being

Let's begin our exploration of HCI with some useful definitions that are commonly shared among those working in the field.

FIT. A good fit between the HCI elements of the human, the computer, and the task that needs to be performed leads to performance and well-being. Just as it is important that new shoes comfortably fit the shape of your foot, hold up during the activity you will be doing (such as running), and are made of a material (such as leather) that is durable and cost-effective, so too is it important that the fit among the user, computer, and task all correspond.

Analysts want the best fit in their design. You want to make the best possible use of people in designing a computerized task that is intended to meet an organizational objective. Better fit results in better performance and greater overall well-being for the human involved in the system.

Fortunately, humans' capacity to learn better ways to work also influences the fit. We would never try running a marathon with a shoe right out of the box, without first getting our foot used to it by breaking it in. Similarly, users can be trained to develop a better fit by learning their tasks and computers thoroughly. Training continues to be an important way to improve fit.

TASK. In the foregoing chapters you have learned many methods to help you understand, document, and graphically depict the tasks that people currently perform in the organization. You have also learned methods to help you design new tasks that will help people reach their objectives with the new systems you are creating. As you recall, tasks can be structured and routine, or they can be ill defined and without apparent structure. Complex tasks that require human, system, and task interaction are supported by ecommerce and Web systems, ERP systems, and wireless systems inside and outside the organization.

PERFORMANCE. The definition of the word *performance* in the HCI context is key. In this case, the term *performance* refers to a combination of the efficiency involved in performing a task and the quality of the work that is produced by the task. For example, if analysts are using high-level software or a CASE tool to create data flow diagrams in which they are proficient, we would predict that the quality of the data flow diagrams produced would be high. The performance is also efficient because analysts are using an automated tool with which they are familiar. They can work rapidly, with good results. The task fits the objective, which is to create high-quality data flow diagrams to document a system. The efficiency of producing such diagrams with a CASE tool, which can then be used to store, retrieve, communicate, and modify the UML diagrams, is excellent, compared to alternatives such as using a drawing tool unrelated to a data dictionary or drafting diagrams by hand, neither of which offer such features.

WELL-BEING. At this point, we can introduce the concept of well-being, which is a concern for a human's overall comfort, safety, and health; in sum, it is a human's physical as well as psychological state. Does using a CASE tool for producing UML diagrams or DFDs on a computer serve the analyst's well-being? Yes, because the task fits well with the analyst, the software, the objective, and the computer. Notice that analysts are working in an environment where they are physically comfortable, are psychologically stimulated to be creative, and can be productive; also, each analyst's work is valued by peers and clients, as well as valued monetarily by the employing organization.

Psychological attitudes (the affective component) are also important. How users feel about themselves, their identities, their work life, and performance can all be gauged through assessing their attitudes. As an analyst taking an HCI perspective, you are concerned about how humans' attitudes color the way they feel about technology and their tasks, and whether their attitudes hinder or enhance their experience.

Usability

Usability is a term that is defined differently depending on which branch of science you are investigating. For our purposes in exploring usability through an HCI lens, we will try to focus on usability as a way for designers to evaluate the systems and interfaces they create with an eye toward addressing as many HCI concerns as we can as thoroughly as possible. Usability studies (according to www.useit.com) are all about finding out what works in the world and what doesn't. The International Organization for Standardization (ISO) has created usability standards that you can explore at www.usabilitynet.org/tools/r_international.htm. The standards cover the use of a product (effectiveness, efficiency, and satisfaction in a particular context of use), the user interface and interaction, the process used to develop the product, and the capability of an organization to apply user-centered design.

Nielsen and Mack (1994) and Nielsen, Molich, Snyder, and Farrell (2001) have published usability heuristics (or rules of thumb) based on thousands of usability tests of interfaces and, later, tests of ecommerce websites. They include visibility of system status, match between the system and the real world, user control and freedom, consistency and standards, error prevention, reconnection rather than recall, flexibility and efficiency of use, aesthetic and minimalist design, help that users recognize, diagnosis and recovery from errors, and help and documentation. Some of these are already familiar to you from Chapter 11 on output design and Chapter 12 on input design.

Figure 14.1 shows a usability survey to administer directly to users who have personally interacted with a prototype. It asks users outright about some important usability and ergonomic dimensions. Another approach is to write up use case scenarios for the system. These are helpful in examining usability concerns.

Designing for the Cognitive Styles of Individual Users

One important consideration is that data, particularly data used for decision making, need to be made available in different forms so that users with different cognitive abilities can make sense of them. Some users may prefer to examine tables and make decisions, some prefer graphs, and others want to read narrative text.

It is also possible that the same person might want different types of presentations at different times. For example, suppose a manager wants to compare inventory held at different stores in a region. A graph can present the data very effectively. A column chart can use colors to show when a store is near its stockout level, and it can also show the relative amount of stock by allowing the user to visually compare the height of the bars directly.

Suppose now that the same decision maker wants information about a particular store in a given month. The graphical depiction may have been set up to show the stores from highest to lowest inventory on a month-by-month basis. The user may prefer to return to the table that lists stores alphabetically, with the months listed chronologically. As you can see, the same person may want to see the same data in very different ways.

PIVOT TABLES. Pivot tables allow users to arrange data in a table in any way they choose. An example of a pivot table template created in Microsoft Excel is shown in Figure 14.2. The user would take an item from the pop-up box called "Pivot Table Field List," such as **Product,** drag it over to the table template, and drop it in one of the blank areas. In this example, the user drags and drops **Product** into the area on the left titled "Drop Row Fields Here." The user drops **Sales** into the largest area that says "Drop Data Items Here."

Usability Survey

Please fill this out after you complete your interaction with the prototype. Circle a number as you respond to each question. Please hand your survey to the analyst when you have completed it. Thank you for this important feedback.

Prototype being evaluated _____   Version _____   Date ____/____/_____

Human–Computer Interaction Factors

	Very Poor		Average		Very Good

Physical/Safety Concerns

1. How well were you able to read the display or form?
2. If audio was used, were you able to hear it?
3. Did you consider the system safe to use?

1	2	3	4	5
1	2	3	4	5
1	2	3	4	5

Usability Concerns   How well did the system:

4. Help you cut down on making errors?
5. Allow you to recover from an error if you made one?
6. Help you use it easily?
7. Help you remember how to use it?
8. Make it easy to learn how to use it?

1	2	3	4	5
1	2	3	4	5
1	2	3	4	5
1	2	3	4	5
1	2	3	4	5

Pleasing and Enjoyable Attributes

9. Was the system attractive?
10. Was the system engaging (you wanted to use it)?
11. Do you trust it as a system?
12. Was it satisfying to use?
13. Was it enjoyable to use?
14. Was the system entertaining?
15. Was the system fun to use?

1	2	3	4	5
1	2	3	4	5
1	2	3	4	5
1	2	3	4	5
1	2	3	4	5
1	2	3	4	5
1	2	3	4	5

Usefulness Attributes   How well did the system:

16. Support your individual task or tasks?
17. Help you to extend your capabilities?
18. Make itself rewarding to use?
19. Permit you to do tasks that the other system would not allow you to do?

1	2	3	4	5
1	2	3	4	5
1	2	3	4	5
1	2	3	4	5
1	2	3	4	5

FIGURE 14.1

A form may be used to survey users of prototypes on key usability and ergonomic factors. (Categories based on Zhang, Carey, Te'eni, and Tremaine, 2005, table of HCI concerns, p. 522.)

Finally the user takes the item called **Quarter** and drops it into the area called "Drop Column Fields Here." The result is a table that shows each of the products in alphabetical order and its sales for each of the four quarters we have data for, followed by the grand total for the year. This table is shown in Figure 14.3.

Of course, the user could have dragged the item **Quarter** to the leftmost column and **Product** to the area that says "Drop Column Fields Here." That operation, however, would have produced a table with many columns (one for each product) and only five rows (one for each quarter plus a row for the total). The resulting table would have been difficult to read.

FIGURE 14.2

A pivot table template can make it easier for users to see information displayed in different ways.

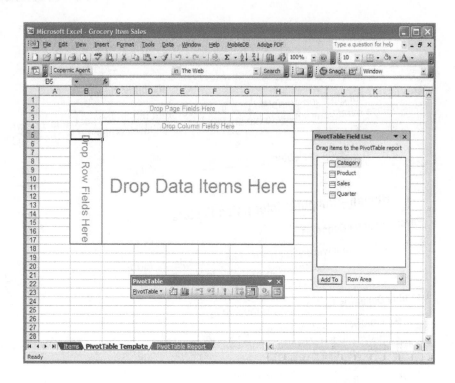

Many different tables are possible just by rearranging these four variables. If the user dragged the variable **Category** over to the area that says "Drop Column Fields Here," the columns would have been categories of products, rather than the quarters, and the resulting table would have clearly shown which of the items belonged in each category and produced subtotals for each category. If **Category** was dragged to the area at the very top of the template that says "Drop Page Fields Here," then each category would have its own table beginning on a separate page.

Pivot tables are useful because they grant users greater control over how they look at data in different ways within a table. We can examine this same concept for graphs in the next section.

Sum of Sales	Quarter				
Product	Qtr 1	Qtr 2	Qtr 3	Qtr 4	Grand Total
Alice Mutton	2667.6	4013.1	3900	6000.15	16580.85
Aniseed Syrup	544	600	140	440	1724
Boston Crab Meat	1474.41	2272	3887.92	2162	9796.33
Camembert Pierrot	3329.28	3989.9	10273.1	3060	20652.28
Carnarvon Tigers	1500	2362.5	7100	4987.5	15950
Chai	705.6	878.4	1174.5	2128.5	4887
Chang	2720.8	228	2061.5	2028.25	7038.55
Chartreuse verte	590.4	360	1100.7	2424.6	4475.7
Chef Anton's Cajun Seasoning	225.28	2970	1337.6	682	5214.88
Chef Anton's Gumbo Mix			288.22	85.4	373.62
Chocolade	744.6	162.56	68.85	306	1282.01
Côte de Blaye	25127.36	12806.1	7312.12	1317.5	46563.08
Escargots de Bourgogne		265	1393.9	417.38	2076.28
Filo Mix	187.6	742	226.8	911.75	2068.15
Fløtemysost	4454.8	174.15	2541.29	2472.5	9642.74
Geitost	294	242.5	99.5	150	786
Genen Shouyu		176.7	1298.12		1474.82
Gnocchi di nonna Alice	6931.2	9868.6	6771.6	9032.6	32604
Gorgonzola Telino	487	2993.12	1458.75	2681.87	7620.74
Grandma's Boysenberry Spread			1750	750	2500
Gravad lax	208	421.2			629.2
Guaraná Fantástica	529.2	467.55	219.37	337.5	1553.62
Gudbrandsdalsost	2649.6	1267.2	4473	5652	14041.8

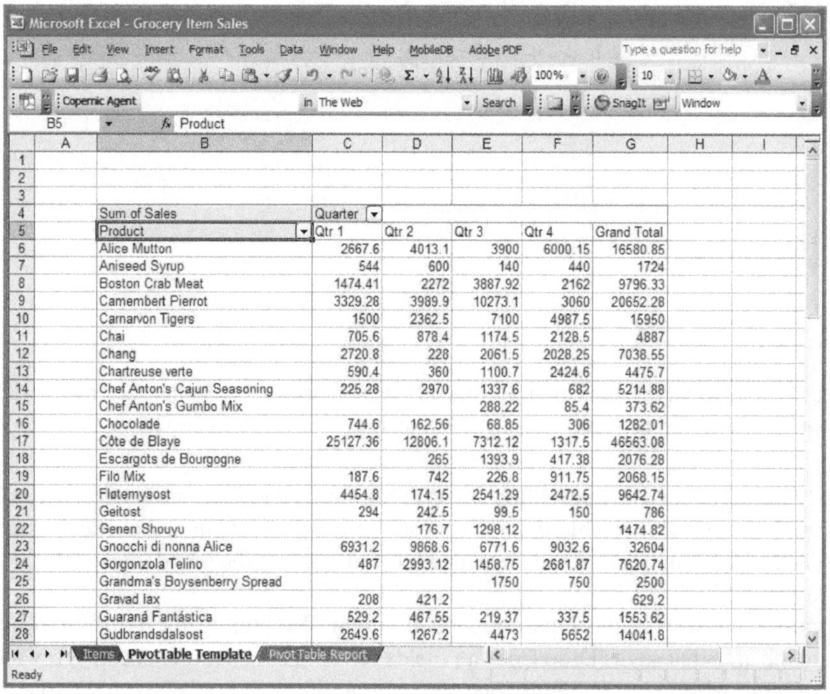

FIGURE 14.3

After the user drags the items Product, Quarter, and Sales to the template, the table looks like this.

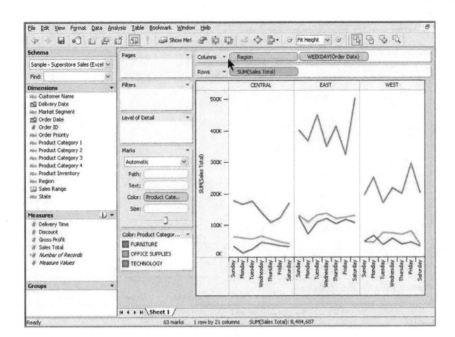

FIGURE 14.4

This graph, showing the daily sales by category and by region, was produced using Tableau. (Screenshots courtesy of www. tableausoftware.com. Reprinted with permission.)

VISUAL ANALYSIS OF DATABASES. Innovative visual displays of data have existed for quite some time, even as early as the eighteenth century. Barriers to widespread use of visual displays included lack of imagination, the inability to draw graphs and charts in a cost-effective manner, and a lack of appreciation for such displays. The consumer of information must be able to interpret the information in the diagram or it adds little value.

Software that enables the user to visually examine a database or spreadsheet is available. One example is Tableau Software's product (www.tableausoftware.com). Using an approach similar to the pivot tables we saw in Microsoft Excel, Tableau allows the user to drag and drop variables onto either a row or a column, and they appear on a graph. In Figure 14.4, **Region** and **Weekday** were designated as columns, and **SUM (Sales Total)** was designated as a row. Each **Product Category** was then graphed (with "furniture" in blue, "office supplies" in orange, and "technology" in green). The graph demonstrates that technology sales were higher than the other categories, but in particular, technology sales were much higher than either furniture or office supplies in the East. The user was easily able to see this because **Region** was dragged to the area as a column and singled out as one of the separators.

Tableau is a well-designed software package because it goes much further than other applications in extending user capabilities to perform their tasks through the use of pivot table techniques. The developers also realized that users might want to cluster the data into what they consider a meaningful group. Users may then continue analysis by examining one of the groups further.

Figure 14.5 examines **SUM (Gross Profit)** from each **Product Category** from our example. This graph uses color to indicate a profit (green) or a loss (red). In fact, the intensity of the color indicates the amount of profit or loss. (A user or designer might select different graph colors based on whether the user knows they experience color vision deficiency.)

This graph can be used to explore the situation more deeply by selecting the three clusters of circles that are bright red, isolating them, and looking at the data for those observations in more detail. Users can examine graphs or simply look at the observations in a table. Once again, they have control over how the information is presented and thus control their task for best cognitive fit.

Another example from Tableau, presented in Figure 14.6, shows that this software can also create a dashboard (explained in Chapter 11). Here a table, a scatter plot, and a column chart are all shown on the same page. Visual analysis tools like this support visual thinking and extend the user's cognitive capabilities to do so. An appropriate visual display will increase the chances of making an appropriate decision.

FIGURE 14.5

Products yielding losses are highlighted in bright red on this scatter plot, created using Tableau. (Screenshots courtesy of www.tableausoftware.com. Reprinted with permission.)

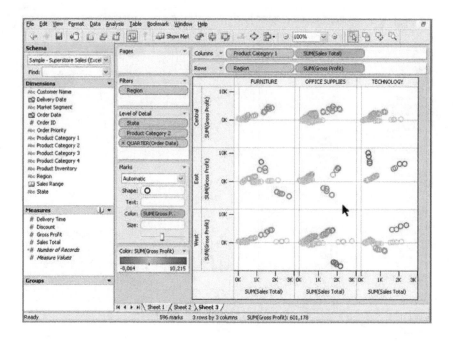

Physical Considerations in HCI Design

In Chapters 11, 12, and 13 you learned the basis for sound design of screens, forms, websites, and databases. This included the special use of fonts, color, and layout design to communicate to users and to help them do the right thing with the input and output they encountered. To examine the underlying reasons for much of the design you learned, it is useful to look at human sensory capabilities and limitations that will inform our design. In keeping with the HCI philosophy, an analyst should be able to compensate, overcome, or replace human senses to a varying extent.

VISION. As you become a systems analyst, you are becoming accustomed to designing screens and reports for sighted people. The use of color, fonts, graphics, software, and PowerPoint presentations for displays and printed reports as input and output were detailed in Chapters 11 and 12. However, from an HCI perspective, you will also want to think in terms of limitations on human vision. Factors such as length of the distance from display to the person performing a task; the angle of the display in relation to the person viewing it; the size and uniformity of

FIGURE 14.6

When different graphs or tables can be displayed on the same page, the page resembles a dashboard. (Screenshots courtesy of www.tableausoftware.com. Reprinted with permission.)

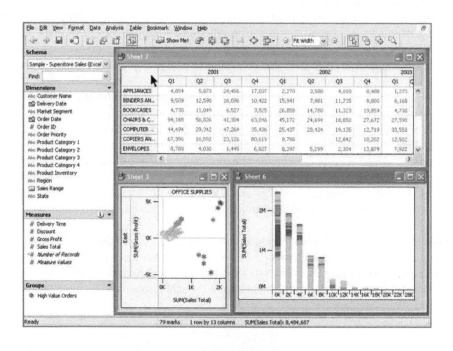

the characters; the brightness, contrast, balance, and glare of the screen; and whether a display is blinking or stable can all be designed to standards established through ISO and other national and international groups.

HEARING. Humans have limits on the amount of stress their senses can withstand. Noisy laser printers, phone conversations, and shredders can lead to overload on human hearing. Office workers can wear noise-canceling headphones or get a personal music player like an iPod, but these solutions may have the effect of isolating a person from the organizational setting and may even diminish their capability to perform tasks. As an analyst, you will need to consider noise when you design office systems.

TOUCH. When using an HCI perspective to evaluate the usefulness of keyboards and other input devices, you can rate the human–computer fit as well as the dimensions examining the human–computer–task fit. Keyboards have been ergonomically designed to provide the correct feedback for a person doing data entry. Users know by the firmness of the key under their finger that the keystroke has been entered. Although keyboards can be silenced, they are often designed with a click of feedback that is emitted when a key is pressed. Keyboards also include slightly raised bumps on what are called home keys, often the *f* and the *j* keys, which orients users to where their fingers are positioned on the keyboard, enabling them to look at the screen or type from a printed page on their desk without continually glancing at the keyboard.

Although the popular QWERTY keyboard that we most often use with computers today was originally designed to slow down typists so that mechanical keys of the day would not become entangled, using this layout has proved to be quite an efficient way to enter data. In fact, since users do so well with this familiar interface, it is difficult to conduct experiments comparing the efficiency of QWERTY keyboards with other innovative keyboards.

Designing for data entry using numeric keypads as the human entry device also provides a decision point for designers. Notice that numbers on your mobile phone are ordered differently than numbers on a numeric keypad or calculator. Your phone may be arranged with the numbers 1, 2, and 3 on the top row. When you look at a calculator layout or a numeric keypad on your keyboard, you will see 7, 8, and 9 on the top row instead. Research now points to the superiority of the calculator layout when the user is doing a lot of data entry. However, the phone digit layout is supposed to be better for locating a number. As a designer, you are constantly examining the fit between the human, the computer, and the tasks set by the organization.

Considering Human Limitations, Disabilities, and Design

All humans have limitations in their physical capabilities. Some are immediately visible, others are not. When designing from an HCI perspective, you start realizing that limitations are often discussed in terms of disabilities. The application of HCI to supporting and enhancing the physical capabilities of humans is one of the most promising application areas. Strides in biomedical engineering mean that there is research to support the blind or those with low vision, those who are deaf or have impaired hearing, and people with limited mobility.

There are also improvements in the technical supports available to those who face difficulties in cognitive processing, including persons suffering with symptoms of autism, dyslexia, and attention-deficit disorder. As a systems analyst you will be subject to the legal provisions of the country in which you are working. For instance, if you are designing for workplaces in the United States, you may want to check on the obligations of an employer on the Americans with Disabilities Act website. There you will find definitions of who is considered disabled, including the following: "An individual with a disability is a person who: has a physical or mental impairment that substantially limits one or more major life activities; has a record of such impairment; or is regarded as having such an impairment" (www.ada.gov).

An employer in the United States is expected to make reasonable accommodation to employ a disabled person, which includes "making existing facilities used by employees readily accessible to and usable by persons with disabilities; job restructuring, modifying work schedules, reassignment to a vacant position; acquiring or modifying equipment or devices, adjusting or modifying examinations, training materials, or policies, and providing qualified readers or interpreters" (www.ada.gov).

A qualified employee or applicant is an individual who, "with or without reasonable accommodation, can perform the essential functions of the job in question." An employer is required to make reasonable accommodation to the known disability of a qualified applicant or employee if it would not impose an undue hardship on the operation of the business. Undue hardship is defined as "an action requiring significant difficulty or expense when considered in light of factors such as an employer's size, financial resources, and the nature and structure of its operation. An employer is not required to lower quality or production standards to make an accommodation" (www.ada.gov).

One of the best ways to ensure the broadest possible accommodation is to begin designing from an HCI perspective. That way, your foremost concern will always be assisting a user in accomplishing a task, set by the organization, with the use of technology. When accommodations for disabled people are necessary, there are many sources to examine and many assistive devices to consider.

For people who are blind or who have low vision, there are Braille keyboards as well as special speech software that reads web pages and other documents aloud. Both the Apple iPad and the Kindle 2 reader have text-to-voice features that enable them to read electronic books out loud. There are also screen magnifiers that fit over a display to magnify the entire screen.

For people who lack certain perceptual sensitivity (incorrectly called color blindness), you can work at testing the colors you are choosing for screens or forms to make certain that they can be easily distinguished from each other. Particular problems occur telling the difference between red and green, for instance. Always design the screen or form with alternative cues, such as icons, written text, or audio cues that reinforce the content. For instance, if a hyperlink that has been clicked on turns blue to show it has been followed, you can also add another icon to the display to indicate that it has been followed or create a separate sidebar list that shows which websites have been visited. These are better alternatives than relying solely on color to convey your message.

For users who experience impaired hearing, you can make sure that the documents and screens you design include access to written versions of the audio material. Alternatively, you might design tasks where headphones can be successfully used.

If you are designing computer tasks for those with limited mobility, you can think of speech input rather than keyboarding. In addition, new advances in biomedical engineering permit mobility-impaired users to move the cursor on the screen by breathing into a tube or by directing the cursor to the desired spot on the screen by looking at that spot or even, in some highly specialized interfaces, by thinking about where the cursor should move.

Implementing Good HCI Practices

The ideal is to invite a usability specialist to serve on the systems development team with the other team members. However, many systems groups are quite small, and not many professionals are available who are involved in the practice of usability per se; so even if you make this recommended change to your project, the odds are that the position will go unstaffed or understaffed. However, don't let that discourage you. You can take some simple steps that will positively influence the outcome of your systems project. Figure 14.7 provides a list of guidelines for taking an HCI approach to systems design.

FIGURE 14.7

The HCI approach to systems design emphasizes the fit among the human, computer, and task.

Guidelines for the HCI Approach to Systems Design

- Examine the task to be done and consider the fit among the human, computer, and task.
- Identify what obstacles exist for users in their attempts to accomplish their assigned tasks.
- Keep in mind perceived usefulness and perceived ease of use of technology.
- Consider usability. Examine the usage environment by creating use case scenarios that depict what is going on between users and the technology.
- Use the information you have gained beforehand to figure out the physical and organizational environmental characteristics. Design with prototyping to accommodate diverse users and users with disabilities.

School Spirit Comes in Many Sizes

Matt Scott manages the student-alumni clothing department for a large bookstore in Saratoga Springs.

"Our clothing sales depend not only on whether our sports teams win or lose, but the overall well-being of our students and alums. If they are proud of their university and want to show their school spirit, they'll buy up everything on our racks!" exclaims Matt. "But don't underestimate the weather as a factor," he adds. "If the weather turns cold in October, you'll see a surge in people buying warm sweaters, hoodies, pullovers, and gloves."

"Our store serves the major three universities in our area," Matt goes on to state. "First, there is Hyde Park, what we call 'the football school.' They have about 17,000 students going there. They have high demand for school-branded clothing, particularly in the fall. Then, of course, there's Pierce University. Pierce thinks it's part of the Ivy League, so the students like to buy crew and Lacrosse shirts. They have about 7,500 students. Then there is St. David's, with about 3,000 students. They are devoted to their basketball team. They really have faith in them. You'll see sales pick up in the second semester, particularly during 'March Madness.'"

Mr. Scott continues, admitting, "I thought about asking the students what to stock, but an email survey is out of the question. I get spammed a lot, so I mostly don't bother with email. Unfortunately, the lead-time for getting official branded sportswear into the store is really long, and we run the risk of stocking out. But we try to never run out."

You've been asked to design a set of tables and graphs that will help analyze the sales of Matt Scott's school clothing. Start by listing about 20 different items of school-branded clothing for men

and women fans, including items such as hooded sweatshirts, T-shirts, baseball caps, sweatbands, running shorts, and so on. Many of them feature fanciful embroidered designs depicting their mascots in menacing or endearing poses. Hyde Park has their Golden Retrievers; Pierce has their much beloved bird, the Puffins; and St. David's cheers with their Dragons.

Put the items into categories. Then think about what the data would look like. Does it make sense for Matt to look at the data weekly, monthly, or by semester? Will he want to look back five years to see if there were any trends? Set up tables identifying the rows and columns and the content of the main cells. Suggest several tables so that Matt can analyze them in different ways.

Now construct graphs that analyze the same data. Using some of the examples found in this book, suggest the type of graphs and show the data so that different users with different styles can make some decisions regarding the trend of sales over the last few years. Remember to compare the schools as well. Suggest the appropriate graphs from column, line, scatter plots, or even pie charts.

Also suggest three or four specific changes you would make to allow someone who has low vision to be able to read the graphs more easily. Magnification is one way to change a graph, but it may not be the best approach.

Consider the size of the schools since this may become the most important factor when determining how Matt Scott should adjust his ordering for St. David's, Hyde, and Pierce.

Although we have been discussing the system in the abstract, it is important to recognize that the interface *is* the system for most users. However well or poorly designed, it stands as the representation of the system and, by reflection, your competence as a systems analyst. A well-designed interface improves the fit among the task, the technology, and the user.

Your goal must be to design interfaces that help users and businesses get the information they need in and out of the system by addressing the following objectives:

1. Matching the user interface to the task
2. Making the user interface efficient
3. Providing appropriate feedback to users
4. Generating usable queries
5. Improving the productivity of computer users

Types of User Interface

In this section, several different kinds of user interfaces are described, including natural-language interfaces, question-and-answer interfaces, menus, form-fill interfaces, command-language interfaces, graphical user interfaces (GUIs), and a variety of Web interfaces for use on the Internet. The user interface has two main components: presentation language, which is the computer-to-human part of the transaction, and action language, which characterizes the human-to-computer portion. Together, both concepts cover the form and content of the term *user interface*.

CONSULTING OPPORTUNITY 14.2

I'd Rather Do It Myself

"**I** can get Mickey to download any data I need from the Web or our server to my PC," DeWitt Miwaye, an upper-level manager for Yumtime Foods (a Midwest food wholesaler), tells you. "Getting data is no problem. What I don't want are a lot of reports. I'd rather get into the data myself."

Miwaye goes on to tell you that as an executive, he doesn't use his PC as often as he'd like, maybe only three times a month, but he has some very specific ideas about what he'd like to do with it.

"I'd like to be able to make some comparisons myself. I could compare the turnover rate for all 12 of our warehouses. I'd also like

to see how effectively the capacity of each of our warehouses is being used. Sometimes I'd like to be able to graph the comparisons or see a chart of them over time."

In three paragraphs, compare three different types of interfaces that Miwaye could use. Then recommend one interface for his use that takes into account his infrequent use of the PC, his enjoyment of working with raw data, and his desire to see data displayed in a variety of ways.

Natural-Language Interfaces

Natural-language interfaces are perhaps the dream and ideal of inexperienced users because they permit them to interact with the computer in their everyday, or natural, language. No special skills are required of the user, who interfaces with the computer by typing a sentence such as "Schedule an appointment on Wednesday at 1 PM with Karla Salguero in marketing." On an Apple iPhone, Siri is a voice-enabled interface that uses natural language.

Question-and-Answer Interfaces

In a question-and-answer interface, the computer displays a question to the user on the display. To interact, the user enters an answer (via a keyboard stroke or a mouse click), and the computer then acts on that input information in a preprogrammed manner, typically by moving to the next question. Wizards used to install software are a common example of a question-and-answer interface. The user responds to questions about the installation process, such as where to install the software or features.

Menus

In using a menu, a user is limited to the options displayed. The user need not know the system but does need to know what task should be accomplished. For example, with a typical word processing menu, users can choose from the Edit, Copy, or Print options. To utilize the menu best, however, users must know which task they want to perform.

Menus can be nested within one another to lead a user through options in a program. Nested menus allow the screen to appear less cluttered, which is consistent with good design. They also allow users to avoid seeing menu options in which they have no interest. Nested menus can also move users quickly through the program.

GUI menus are used to control software and have the following guidelines:

1. The main menu bar is always displayed either at the top of the application (Windows) or the top of the screen (Mac OS).
2. The main menu should have secondary options grouped into similar sets of features. For example, on a **Format** menu you can format a **Font, Paragraph, Document**, etc. Each of these will be a secondary menu.
3. Menu items that are unavailable at this time should be grayed out to signify that they cannot be used at this time.

An object menu, also called a pop-up menu or context menu, is displayed when the user clicks on a GUI object with the right mouse button. These menus contain items specific for the current activity, and most of them duplicate functions of main menu items.

Form-Fill Interfaces

Form-fill interfaces consist of onscreen forms or Web-based forms displaying fields that contain data items or parameters that need to be communicated to the user. Figure 14.8 shows widgets that can be useful in designing forms for iPads and iPhones using an app called FormEntry by Widget Press, Inc. In this instance, the designer is creating a patient admit form using widgets to add a date picker.

Forms for display screens are set up to show what information should be input and where. Blank fields requiring information can be highlighted. The user moves the cursor from field to field by using a single stroke of an arrow or Tab key. This provides the user good control over data entry. Web-based forms afford an opportunity to include hyperlinks to examples of correctly filled-out forms or to further help and provide examples.

Form input for displays can be simplified by supplying default values for fields and then allowing users to modify default information if necessary. For example, a database management system designed to show a form for inputting checks may supply the next sequential check number as a default when a new check form is exhibited. If checks are missing, the user changes the check number to reflect the actual check being input.

Input for display screen fields can be alphanumerically restricted so that, for example, users can enter only numbers in a field requesting a Social Security number, or they can input only letters where a person's name is required. If numbers are input where only letters are allowed, the computer may alert the user via audio output that the field was filled out incorrectly.

Web forms can return incomplete forms to the user with an explanation of what data must be entered to complete the transaction. Often, fields with missing data are marked in red, and the user is required to complete the entry before moving on to the next web page.

Web-based documents can be sent directly to billing if a transaction is involved, or they can go directly to a real-time database if a survey is being submitted. Web-based forms push the responsibility for accuracy to the user and make the form available for completion and submission on a 24-hour, 7-day-a-week, worldwide basis.

Choosing and Evaluating Interfaces

When choosing and evaluating interfaces, keep some standards in mind:

1. The necessary training period for users should be acceptably short.
2. Early in their training, users should be able to enter commands without thinking about them or without referring to a help menu or manual. Keeping interfaces consistent throughout applications can help in this regard.
3. The interface should be seamless so that errors are few and those that do occur are not occurring because of poor design.
4. The time that users and the system need to bounce back from errors should be short.
5. Infrequent users should be able to relearn the system quickly.

Many different interfaces are available, and it is important to realize that an effective interface goes a long way toward addressing key HCI concerns. Users should want to use the system, and they should find it attractive, effective, and pleasing to use.

FIGURE 14.8

Many useful widgets for designing forms for iPads and iPhones are immediately available to the designer in an app is called FormEntry by Widget Press, Inc. (Screenshot from "FormEntry." Copyright © 2012 by Widget Press, Inc. Reprinted by permission.)

CONSULTING OPPORTUNITY 14.3

Don't Slow Me Down

"**I**'ve seen 'em all," Carrie Moore tells you. "I was here when they got their first computer system. I guess I've sort of made a career of this," she says cheerfully, pointing to the large stack of medical insurance claim forms she has been entering into the computer system. As a systems analyst, you are interviewing Carrie, a data entry operator for AbundaCare (a large medical insurance company), about changes being contemplated in the computer system.

"I'm really fast compared with the others," she states as she nods toward the six other operators in the room. "I know, because we have little contests all the time to see who's the fastest, with the fewest errors. See that chart on the wall? That shows how much we enter and how quickly. The gold stars show who's the best each week. Performance measures are my friends."

"I don't really mind if you change computers. Like I say, 'I've seen 'em all.'" She resumes typing on her keyboard as she continues the interview. "Whatever you do, though, don't slow me down. One of the things I'm most proud of is that I can still beat the other operators. They're good too, though," Carrie adds.

Based on this partial interview with Carrie Moore, what type of user interface will you design for her and the other operators? Assume that even though the new system is improved, it will still require massive amounts of data entry from a variety of paper medical insurance forms sent in by claimants.

Compare and contrast interfaces such as natural language, question and answer, menus, and Web-based form-fill interfaces in two paragraphs. Then choose and defend one alternative. What qualities possessed by Carrie and the other operators—and the data they will be entering—shaped your choice? Make a list of them. Is there more than one feasible choice? Why or why not? Respond in a paragraph.

Designing Interfaces for Smartphones And Tablets

Smartphones and tablets open up a new creative approach for designers. Touch-sensitive screens allow a user to use a finger to activate the display. The operating systems for these small devices use multitouch gestures (capacitive sensing featuring screens controlled with a human finger or a conductive stylus) for moving from one screen to another or from one state to another on the same screen. An example of the latter is enlarging a photo by flicking your fingers outward.

Gestures

Some gestures are basic. All humans start poking at objects before they know what a computer is; nudging and poking are inherent. It is also intuitive to swipe your finger from right to left when you need to turn a page when reading English, for example.

Using a pinching gesture to zoom in and out to shrink or enlarge a photo or to zoom in and out on a map is not intuitive. Once people learn it, however, they remember it. Somehow it makes sense to people.

Designers can take advantage of the fact that people easily adapt to new interfaces if they make sense. When you choose existing gestures or design new gestures for an app you are developing, make sure you observe how other people use the gesture. Chances are that they won't think exactly like you. You have thought about the gesture; they have not. They may not discover a designed gesture on their own.

If a designer is fearful that a user will not read the manual or talk to someone who knows how to pinch the screen, they can use something more easily discovered through trial and error. A typical user will tap the screen once and, if the desired result doesn't happen, will try tapping the screen twice. The designer may decide this is an alternative to pinching or may include both options so that one way or another, the user will be able to zoom in.

On a desktop, arrow keys are used to go from one screen to the next, and using a scroll bar is the common way to get to the bottom of a news story on the Web. On smartphones and tablets, a swipe may be more logical when going from left to right and a tug may be the preferred way to look down further at an article or a list.

When a designer creates new nonstandard gestures, problems can occur. Suppose that a user has other apps that use a swipe from left to right as a copy command, and you develop a calculator app where a swipe from left to right clears the display. A user who owns an app that regularly requires the copy gesture will likely be upset by your adoption of the gesture for something different (and potentially destructive of data). You just lost a customer.

Then there is the temptation to take advantage of the latest techniques without really thinking through what the user would like to actually do. One example of this is the use of a motion detector that detects motion in 3 dimensions, called an accelerometer, to change something in the app by shaking the smartphone. Shaking is incorporated into apps for games, seismometers, pedometers, and apps that give biofeedback. In those contexts the designer has made a choice that makes sense to the user. Remember, however, that shaking is not an obvious or innate gesture. Although many users shook their iPhone, fewer users wanted to shake their iPad. The size of the platform and the context of the app mattered. Also, shaking the iPhone may look odd at the office or on public transportation. If a designer still feels that shaking is highly useful for some apps, the designer should add an option for users uncomfortable with it. Pressing part of a screen or a button is subtle. Shaking is not.

Designers need to think about the metaphor before programming a gesture simply because it exists. Urbanspoon (an app that locates and recommends restaurants) introduced an app that allows the user to randomly select a restaurant, using the metaphor of spinning a slot machine; however, some users might not think the gambling metaphor adequately reflects their quest for fine dining.

Remember, too, that every gesture needs to provide the user with feedback. If the user tugs and nothing more is in the list, it needs to appear obvious to the user that he or she has reached the end of the list. If the user turns a page, the designer should show motion so that the user understands that the page is turning. Feedback is discussed in the section "Feedback for Users."

In Chapter 11, we discussed designing for both portrait and landscape modes. Make sure that the gestures are the same for both orientations.

Alerts, Notices, and Queries

Alerts, notices, and queries are forms of output on smartphones and tablets. An alert is useful if a severe thunderstorm is approaching, but if the user's phone is on during a theatre performance, the user may be asked to leave. Alerts can include a sound, but remember that they do not need to do so to still be helpful.

Alerts are for critical information that the user needs to know in a timely manner. It is not a good idea to use an alert to tell the user that signal strength is low, for instance. That's what the signal bars are for.

Notifications convey information to a user. Some of that information, such as the notification that an update to the app is ready, may not be urgently important. Since in this context it breaks the user's concentration, a notification such as an update might be best kept as part of the app launch process but not used as an alert.

Some developers like to include queries such as "Would you like to rate this app today?" This makes the designer seem less professional, and by including it you could annoy the very people you are trying to please.

Always give the user a chance to opt out. Many users just do not want to be notified with information or alerts. Users can change their notification settings in the "Settings" area at any time, but it's nice if they have a chance to opt out during the first launch of the app.

Badges

Another feature of smartphones and tablets is the ability to put "badges" on home screen icons. On the iPhone and iPad, these are little red circles. A badge for the App Store signifies how many updates are waiting for the user to download and install. Some weather apps use a badge to display the current temperature.

Badges are good because they are an unobtrusive way to send a message to the user. They are quiet and passive, not loud and active like alerts and notifications. One problem is that they are ignored; another is that they may be outdated. (For example, the temperature displayed as a badge on your weather app may indicate the last time you opened the app sometime last week, not the temperature for today.) So unless you have something meaningful to convey, try to avoid using a badge.

Voice Recognition

In 2012 Apple introduced Siri, billed as your personal assistant that "helps you get things done, just by asking." A user can speak to Siri in natural language, just as he or she would talk to a friend. Unlike older voice recognition systems, Siri does not need to be taught to respond to your voice commands.

Siri may be a good interface, but in its first generation, it is not yet an intelligent agent or assistant. It doesn't yet understand you and your preferences. Siri relies on apps such as Yelp and Wikipedia to obtain information, rather than using Google searches. Siri can book appointments for you, remind you of activities you need to take care of, and call a friend. In that way, Siri is a personal assistant, but not an intelligent one. You will need to consider Siri and future versions of Siri when you develop apps for smartphones.

Some additional interfaces are changing the ways smartphone developers think of the user interface. The Microsoft Windows phones were designed with the Microsoft Design Language as was the interface for the Xbox 360. Windows 8 was also released with the tiled, colorful interface. Microsoft's new design language is based on easy-to-read fonts found on Swiss packaging and transportation signs. The idea is that following them will get the user somewhere or let the user accomplish something. The emphasis is on users' content, not on stand-alone icons or on frivolous add-ons needed to access content. The shift is toward letting the content become the user interface. The Windows 8 user interface is accessed through a series of specially designed gestures. Transitions between content are handled through consistent animations or motions.

Guidelines for Dialog Design

A dialog is the communication between a computer and a person. Well-designed dialog makes it easier for people to use a computer and lessens their frustration with a computer system. There are several key points for designing good dialog. They include the following:

1. Meaningful communication, so that the computer understands what people are entering and people understand what the computer is presenting or requesting
2. Minimal user action
3. Standard operation and consistency

Meaningful Communication

A system should clearly present information to the user. This means having an appropriate title for each display, minimizing the use of abbreviations, and providing clear user feedback. Inquiry programs should display code meanings as well as data in an edited format, such as displaying slashes between the month, day, and year in a date field or commas and decimal points in an amount field. User instructions should be supplied regarding details, such as available function key assignments. In a graphical user interface, the cursor may change shape, depending on the work being performed.

Users with less skill in using the computer or doing their tasks with a computer require more communication. Websites must display more text and instructions to guide the user through the site. Intranet sites may have less dialog because there is a measure of control over how well trained users are. Internet graphics should have pop-up text or rollover descriptions when images are used as hyperlinks, because there may be uncertainty in interpreting their meaning, especially if the site is used internationally. Notice that EU guidelines for the display of Web graphics requires that all images be labeled, so that visually impaired users will be able to hear written descriptions announced through special software. Status line information for GUI screens is another way of providing instructions for users.

Easy-to-use help screens should be provided. Many PC help screens have additional topics that may be directly selected using highlighted text displayed on the first help screen. These hyperlinks are usually in a different color, which makes them stand out in contrast to the rest of the help text. Remember to use icons or text in addition to color coding in order to reach the largest number of users. Many GUIs incorporate tool tip help, displaying a small help message identifying the function of a command button when the cursor is placed over it. The other side of communication is that the computer should understand what the user has entered. Hence, all data entered on the screen should be checked for validity.

Minimal User Action

Keying is often the slowest part of a computer system, and good dialog will minimize the number of keystrokes required. You can accomplish this goal in a number of different ways:

1. ***Keying codes, such as airport codes when making a flight reservation, instead of whole words on entry screens.*** Codes are also keyed when using a command-language interface, such as a two-letter state postal abbreviation. On a GUI screen, the codes may be entered by selecting descriptions of the codes from a pull-down list of available options. This helps to ensure accuracy, since the code is stored as a value of the drop-down list, as well as helping to provide meaningful communication since descriptions that are familiar to the user are selected. An example would be selecting a Canadian province and having the two-character postal code stored.

2. ***Entering only data that are not already stored on files.*** For example, when changing or deleting item records, only the item number should be entered. The computer responds by displaying descriptive information that is currently stored on the item file. Another example is when a user logs on to a website, the user ID is used to find related records, such as a customer record, outstanding bills, orders, and so on.

3. ***Supplying the editing characters (for example, slashes as date field separators).*** Users should not have to enter formatting characters such as leading zeros, commas, or a decimal point when entering a dollar amount; nor should they have to enter slashes or hyphens when entering a date. In general, websites are an exception to this rule, since Web forms do not include slashes or decimal points. Some Web forms use a series of entry fields with editing characters between them, such as parentheses around an area code.

4. ***Using default values for fields on entry screens.*** Defaults are used when a user enters the same value in a screen field for the majority of the records being processed. The value is displayed, and the user may press the Enter key to accept the default or overtype the default value with a new one. GUIs may contain check boxes and radio buttons that are selected when a Web form or dialog box opens. Context-sensitive menus appear when an object is clicked with the right mouse button. These menus contain options specific to the object under the mouse.

5. ***Designing an inquiry (or change or delete) program so that the user needs to enter only the first few characters of a name or item description.*** The program displays a list of all matching names, and, when the user chooses one, the matching record is displayed.

6. ***Providing keystrokes for selecting pull-down menu options.*** Often, these options are selected using a mouse, followed by keying. Users must move their hands from the keyboard to the mouse and back. As users become familiar with the system, shortcut keystrokes provide a faster method for manipulating the pull-down menus because both hands remain on the keyboard. This helps users become efficient at their tasks. On a PC or Mac, keystrokes usually involve pressing a function key or the Alt key followed by a letter or another key. Figure 14.9 shows an example of nested pull-down menus.

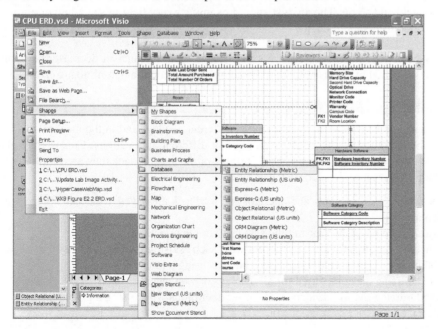

FIGURE 14.9

Example of nested pull-down menus with shortcut keys from Microsoft Visio Professional.

CONSULTING OPPORTUNITY 14.4

Waiting to Be Fed

"Yeah, we were sold a package all right. This one right here. Don't get me wrong, it gets the work done. We just don't know when."

You are talking with Owen Itt, who is telling you about the sales unit's recent purchase of new software for its networked PCs that allows the input of sales data for each of its 16 salespeople, provides output showing comparison data for them, and projects future sales based on past sales records.

"We've had some odd experiences with this program, though," Owen continues. "It seems slow or something. For instance, we're never sure when it's done. I type in a command to get a file and nothing happens. About half a minute later, if I'm lucky, the display I want might come up, but I'm never sure. If I ask it to save sales data, I just get a whirring sound. If it works, I'm returned to where I was before. If it doesn't save data, I'm still returned to where I was before. It's confusing, and I never know what to do. There's nothing on the display screen that tells me what to do next. See the little manual that came with it? It's dog-eared because we have to keep thumbing through it trying to figure out what to do next. Or we go online to try to get some help, but their technical assistance is just about nonexistent. It takes way too much time, too."

Based on what you've heard in the interview, take this opportunity to supplement the software by designing some onscreen feedback for Owen and his sales team. The feedback should address all of Owen's concerns, and follow the guidelines for giving feedback to users, and the guidelines for designing good displays. Draw a prototype of the displays you think are necessary to address the problems Owen lists.

7. *Use radio buttons and drop-down lists to control displays of new web pages or to change Web forms.* For example, when a radio button is clicked, a drop-down list may change to reflect the radio button choice. A radio button may be clicked, and a form may change according to the choice. A drop-down list may change or a radio button may be clicked to move to a new web page. Drop-down lists are often provided on a web page for quick navigation; selecting a new web page from the drop-down list takes the viewer to that page.

8. *Provide cursor control for Web forms and other displays so that the cursor moves to the next field when the right number of characters has been entered.* An example would be when a user enters an area code for a U.S. telephone number, and, following the entry of three characters, the cursor moves to the local phone number field. Entering software registration key codes is another example. The codes are often in groups of four or five letters and, when the first field is filled, the cursor moves to the next field and so on. The analyst should examine every field to see whether automatic cursor control should occur.

Any combination of these eight approaches can help the analyst decrease the number of keystrokes required by the user, thereby speeding up data entry and minimizing errors.

Standard Operation and Consistency

A system should be consistent throughout its different displays and in the mechanisms for controlling the operation of the displays throughout different applications. Consistency makes it easier for users to learn how to use new portions of the system once they are familiar with one component. You can achieve consistency by:

1. Locating titles, date, time, and operator and feedback messages in the same places on all displays
2. Exiting each program by using the same key or menu option
3. Canceling a transaction in a consistent way, such as using the **Esc** key
4. Obtaining help in a standardized way, such as using a function key
5. Standardizing the colors used for all displays or web pages
6. Standardizing the use of icons for similar operations when using a graphical user interface

7. Using consistent terminology in a display screen or website
8. Providing a consistent way to navigate through the dialog
9. Using consistent font alignment, size, and color on web pages

Feedback for Users

All systems require feedback to monitor and change behavior. Feedback usually compares current behavior with predetermined goals and gives back information describing the gap between actual and intended performance.

Because humans themselves are complex systems, they require feedback from others to meet psychological and cognitive processing needs discussed earlier in this chapter. Feedback also increases human confidence. How much feedback is required is an individual characteristic.

When users interface with machines, they still need feedback about how their work is progressing. As designers of user interfaces, systems analysts need to be aware of the human need for feedback and build it into the system. In addition to text messages, icons can often be used. For example, displaying an hourglass while the system is processing encourages the user to wait a while rather than repeatedly pressing keys to get a response.

Types of Feedback

Feedback to a user from a system is necessary in seven distinct situations. Feedback that is ill timed or too plentiful is not helpful because humans possess a limited capacity to process information. Websites should display a status message or some other way of notifying the user that the site is responding and that input is either correct or in need of further information. Figure 14.10 provides some examples for each type of feedback.

ACKNOWLEDGING ACCEPTANCE OF INPUT. The first situation in which users need feedback is to learn that the computer has accepted the input. For example, when a user enters a name on a line, the computer provides feedback to the user by advancing the cursor one character at a time when the letters are entered correctly. A Web example would be a web page displaying the message "Your payment has been processed. Your confirmation number is 1234567. Thank you for using our services."

RECOGNIZING THAT INPUT IS IN THE CORRECT FORM. Users need feedback to tell them that the input is in the correct form. For example, a user inputs a command, and the feedback states "READY" as the program progresses to a new point. A poor example of feedback that tells the user that input is in the correct form is the message "INPUT OK," because that message takes extra space, is cryptic, and does nothing to encourage the input of more data. When placing an order on the Web or making a payment, a confirmation page often displays,

Type of Feedback	Sample Feedback Response
Acknowledging acceptance of input	"Your payment has been processed."
Recognizing that input is in the correct form	An LED-like display turns from red to green
Notifying that input is not in the correct form	"The telephone number should be separated by dashes (-)"
Explaining a delay in the processing	An hourglass or a rotating circle
Acknowledging that a request is completed	"You successfully renewed your membership. Thank you."
Notifying that a request was not completed	"Your attempt was not successful. Please try at a later time."
Offering the user more detailed feedback	"Your credit card was not charged. You must enter a valid expiry date. Press help if you need more information."

FIGURE 14.10

Feedback can take many forms.

requesting that the user review the information and click a button or image to confirm the order or payment.

NOTIFYING THAT INPUT IS NOT IN THE CORRECT FORM. Feedback is necessary to warn users that input is not in the correct form. If a user enters a phone number incorrectly, the software or website needs to let the user know that. Caution must be used so that the error messages are bold enough for the user to notice. A small red line of text may go unnoticed. An analyst must decide whether to detect and report errors when a Submit button or link is clicked, called batch validation, or detect errors one at a time, such as when a user enters a month of 14 and leaves the field. The second method is a riskier approach since poor coding may put the browser into a loop, and the user will have to shut down the browser.

Additional feedback that tells the user that the input is not in the correct form might be to not permit the user to advance to the next field or screen, along with an appropriate popup message explaining what is wrong and supplying instructions on how to correct it.

Audio has also been used to inform a user that something was entered incorrectly, but audio feedback alone is not very descriptive, so it is not as helpful to users as onscreen instructions. Use audio feedback sparingly, perhaps to denote urgent situations. The same advice also applies to the design of websites, which may be viewed in an open office, where sounds carry and a coworker's desktop speakers are within earshot of several other people.

EXPLAINING A DELAY IN PROCESSING. One of the most important kinds of feedback informs the user that there will be a delay in processing his or her request. Delays longer than 10 seconds or so require feedback so that the user knows the system is still working.

Sometimes during delays, while new software is being installed, a short tutorial on the new application is run, which is meant to serve as a distraction rather than feedback about the installation. Often, a list of files that are being copied and a status bar are used to reassure the user that the system is functioning properly. Web browsers usually display the web pages that are being loaded and the time remaining.

ACKNOWLEDGING THAT A REQUEST IS COMPLETED. Users need to know when their request has been completed and new requests may be input. Often a specific feedback message is displayed when an action has been completed by a user, such as "Employee record has been added," "Customer record has been changed," or "Item number 12345 has been deleted."

NOTIFYING THAT A REQUEST WAS NOT COMPLETED. Feedback is needed to let the user know that the computer is unable to complete a request. If the display reads "Unable to process request. Check request again," the user can then go back and check to see if the request has been input correctly rather than continue to enter commands that cannot be executed.

OFFERING THE USER MORE DETAILED FEEDBACK. Users need to be reassured that more detailed feedback is available, and they should be shown how they can get it. Commands such as Assist, Instruct, Explain, and More may be employed. Or the user may type a question mark or click on an appropriate icon to get more feedback. Using the command Help as a way to obtain further information has been questioned, because users may feel helpless or caught in a trap from which they must escape. This convention is in use, and its familiarity to users may overcome this concern.

When designing Web interfaces, hyperlinks can be embedded to allow the user to jump to the relevant help screens or to view more information. Hyperlinks are typically highlighted with underlining, italics, or a different color. Hyperlinks can be graphics, text, or icons.

Including Feedback in Design

If used correctly, feedback can be a powerful reinforcer of users' learning processes, serve to improve user performance with the system, increase motivation to produce, and improve the fit among the user, the task, and the technology.

A VARIETY OF HELP OPTIONS. Feedback on personal computers has developed over the years. "Help" originally started as a response to the user who pressed a function key, such as F1;

the GUI alternative is the pull-down help menu. This approach was cumbersome because end users had to navigate through a table of contents or search via an index. Next came context-sensitive help. Users could simply click on the right mouse button, and topics or explanations about the current screen or area of the screen would be revealed. A third type of help on personal computers occurs when the user places the arrow over an icon and leaves it there for a couple of seconds. At this point, some programs pop up a balloon similar to those found in comic strips. This balloon explains a little bit about the icon function.

The fourth type of help is a wizard, which asks the user a series of questions and then takes action accordingly. Wizards help users through complicated or unfamiliar processes such as setting up network connections or booking an airline seat online. Most users are familiar with wizards through creating a PowerPoint presentation or choosing a style for a word processing memo.

Besides building help into an application, software manufacturers offer online help (either automated or personalized with live chat) or help lines (most customer service telephone lines are not toll free, however). Some COTS software manufacturers offer a fax-back system. A user can request a catalog of various help documents to be sent by fax and then can order from the catalog by entering the item number with a touch-tone phone.

Finally, users can seek and find support from other users through software forums. This type of support is, of course, unofficial, and the information thus obtained may be true, partially true, or misleading. The principles regarding the use of software forums are the same for those mentioned later on in Chapter 16, where folklore and recommendation systems are discussed. Approach any software fixes posted on bulletin boards, blogs, discussion groups, or chat rooms with wariness and skepticism.

Besides informal help on software, vendor websites are extremely useful for updating drivers, viewers, and the software itself. Most online computer publications have some sort of "driver watch" or "bug report" that monitors the bulletin boards and websites for useful programs that can be downloaded. Programs will forage vendor websites for the latest updates, inform the user of them, assist with downloads, and actually upgrade user applications.

Special Design Considerations for Ecommerce

Many of the user interface design principles you have learned concerning feedback also extend to designing ecommerce websites. A few extra considerations shown in this section can give your Web interface designs improved functionality.

Soliciting Feedback from Ecommerce Website Customers

Not only do you need to give users feedback about what is happening with an order, but you need to elicit feedback as well. Most ecommerce websites have a **Feedback** button. There are two standard ways to design what users will experience when they click on the **Feedback** button.

The first way is to launch the user's email program with the email address of the company's contact automatically entered into the **To:** field. This method prevents typing errors and facilitates ease in contacting the organization. The user does not need to leave the site to communicate with it.

These messages, however, raise expectations that they will be answered just as regular mail or phone calls are. Research indicates that 60 percent of organizations with this type of email contact feature on their sites do not have anyone assigned to reply to the email messages received. These businesses lose valuable feedback, allow customers to harbor the impression that they are communicating, and engender ill will when they provide no response. If you design this type of feedback opportunity, you also need to design procedures for the organization to reply to email from the website. Some designers handle this problem by creating systems to automatically return an email reply, which generates a unique case or incident number, provides further instructions on how to proceed (hyperlinks to FAQ pages perhaps), or offers phone numbers to help lines that are unavailable to the general public.

The second type of design for garnering feedback from customers using an ecommerce website is to take users to a blank message template when they click on **Feedback.** Some

CONSULTING OPPORTUNITY 14.5

When You Run a Marathon, It Helps to Know Where You're Going

Marathon Vitamin Shops was successful in getting its website up and running. The Web developers put the company's entire catalog online and included a choice of skins (or personas, as they are called in the Firefox browser) so that each type of customer would enjoy using the website. (See Consulting Opportunity 1.1 for more details.)

The analysts are meeting with owner Bill Berry and some employees to evaluate customer feedback as well as give their own reactions to the new website. They are meeting in a large conference room, where they have a computer with Internet access and a projector. As they sit down at the table, the entry screen for the website is projected at the front of the room. "The website has attracted lots of attention, but we want to give the customers even more so that they keep coming back," says Bill, gesturing to the screen.

He continues, "It's not like we're closing our retail stores or anything. In fact, it's just the opposite. When customers notice we're on the Web, they're eager to locate the store in their community. They want to be able to walk into a store and talk to a trained expert rather than buying everything on the Web. We need to tell people how to get there."

"We think we can improve the site by adding special enhancements and features," says Al Falfa, a member of the systems team who originally developed and implemented the ecommerce website.

"Yes," says Ginger Rute, one of the other members of the systems development team, as she nods in agreement. "The university uses a mapping facility from MapQuest, and Home Depot uses maps from Microsoft."

Vita Ming, another member of the original systems development team, speaks up enthusiastically, saying, "We know of a couple good message board services and chat rooms we can build into our website. We think they can improve the stickiness of the site, making people stay on the site longer and also making them want to return."

"That's a great idea," says Jin Singh, one of the technologically savvy Marathon employees. "We can let customers talk with one another, tell each other about a product they liked, and so on. We could even let them start their own blogs."

Vita continues by moving to the computer keyboard and saying, "Let me show you some good sites." As she types in the first URL, the group sees the site projected. "They use chat systems from iChat and Multicity.com," she continues.

"Customers also need to search for more information about a product or manufacturer," Al adds. "Let's make it easier for them. Let's look at www.cincinnati.com for an example. They use Google to search for information."

After listening intently, Bill speaks up. "Medical information could also be useful" he says. "I've noticed that some websites show medical news from Acquire Media. I've seen people on the treadmills at my health and fitness center watching the financial channels while they exercise."

"While we're at it, why don't we add news and financial information to the website?" Ginger asks. "I notice that a company called Moreover.com appears often as a provider."

Think about the conversation between the systems development team and the people from Marathon Vitamin Shops. Some of the enhancement suggestions involved taking advantage of free services; others required payments ranging from $1,000 to $5,000 annually. Although some were good ideas, others may not have been practical or feasible. Perhaps some of the ideas just do not make sense for the company.

For each of the following, review what you know about the mission and business activities of Marathon Vitamin Shops. Then make a recommendation regarding each option the analysts and clients have made and defend it:

- Mashups using Google Maps
- Chat rooms and message boards
- Blogs
- Search engines
- Medical information
- News feeds and financial markets information

Web creation tools permit you to create and insert a feedback form into your site easily. This form might begin with a header that states "Company X Feedback" and then "You can use the form below to send suggestions, comments, and questions about the X site to our Customer Service team."

Fields can include First Name, Last Name, Email Address, Regarding (a subject field that supplies a drop-down menu of the company's product or service selections, asking the user to "Please make a selection"), an "Enter Your Message Here:" section (a free-form space where users can type in their message), and the standard **Submit** and **Clear** buttons at the bottom of the form. Using this type of form permits the analyst to have the user data already formatted correctly for storage in a database. Consequently, it makes the data entered into a feedback form easier to analyze in the aggregate.

Thus, an analyst does more than just design a response to individual email. The analyst helps the organization capture, store, process, and analyze valuable customer information in a

MAC APPEAL

Megasearch engines that obtain results from multiple search engines, aggregate the results, and display them in a more useful way than any single search engine can have been available for a long time. There is a unique application on the Mac platform that goes one step further.

That application is DEVONagent, software that uses both general and specialized search engines to get results and then gives the user the option to view the results in a graphical topic map. Another option is to view the results in a relevance-ranked list.

Analysts will find DEVONagent to be useful if they understand and make use of the graphical topic map. It is also useful if complex searches are required (that is, if standard searches do not dig deep enough to find the exact information needed). It is also useful for searches that need to be repeated often.

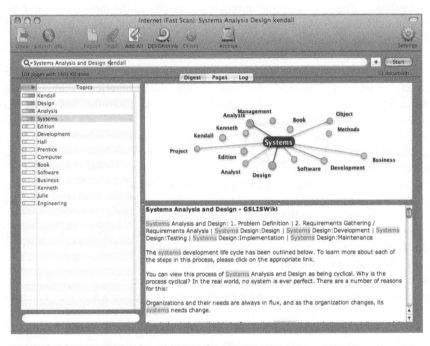

FIGURE 14.MAC

DEVONagent from DEVONtechnologies. (Screenshot from DEVONAGENT. Copyright © by DEVONtechnologies. Reprinted by permission.)

manner that makes it more likely that the company will be capable of spotting important trends in customer response rather than simply reacting to individual queries.

Easy Navigation for Ecommerce Websites

Many authors speak of "intuitive navigation" for ecommerce websites. Users need to know how to navigate a site without having to learn a new interface and without having to explore every inch of the website before they can find what they want. The standard for this type of navigational approach is called one-click navigation.

There are four enhancements you can add to design easy, one-click navigation for an ecommerce site: (1) create a rollover menu, (2) build a collection of hierarchical links so that the home page becomes an outline of the key topic headings associated with the website, (3) place a site map on the home page and emphasize the link to it (this would also be placed on every other page on the site), and (4) place a navigational bar on every inside page (usually at the top or on the left side of the page) that repeats the categories used on the entry screen.

ROLLOVER MENUS. A rollover menu (or rollover button) can be created with cascading style sheets (CSS) with JavaScript and HTML divisions. The rollover menu appears when the customer using the website moves the cursor over a link.

HIERARCHICAL LINKS. Creating an outline of the content of the site through the presentation of a table of contents on the home page is another way to speed navigation of the site. This design, however, imposes severe constraints on the designer's creativity, and sometimes simply presenting a list of topics does not adequately convey the strategic mission of the organization to the user.

SITE MAP. Designing and then prominently displaying the link to a site map is a third way to improve navigational efficiency. Remember to include the link to the site map on the home page and on every other page as well.

NAVIGATION BAR. You can design navigation bars that are consistently displayed on the home page as well as at the top and on the left of all other pages that comprise the site. Once you have established (during the information requirements phase) the most useful and most used categories (usually categories such as "Our Company," "Our Products," "Buy Now," "Contact Us," "Site Map," and "Search"), remember to include them on all pages.

OTHER NAVIGATION OPTIONS. Including a search function is another option. Consider adding a search engine such as Google to your site. Simple search functions work well for small, manageable sites, but as a site grows large, advanced search functions that include Boolean logic (discussed later in this chapter) are needed.

Creating flexibility in the way users navigate the Web is also important. An expert website designer would try to incorporate many different ways to look up information on a particular subject. Figure 14.11 shows a web page from DinoTech. For example, a user interested in an international IT career can find out information from the DinoTech website in three different ways. If they are interested in working in Argentina, they can click on the Argentine flag, click on the name of the country, or click on the map representing Argentina.

Designing a website with navigation for users with different cognitive processing or interests is desirable. It is even possible that the same user may use all three of these methods at different times. They all add to the usability of a website.

The main priority in navigation, however, is to make it extremely easy for users to return to a previous page and make it somewhat easy to return to the place where they entered the site.

FIGURE 14.11

An example of a web page that allows users to navigate to the desired page in different ways.

Your main concern is keeping customers on the website. The longer customers are on the site, the greater the chance that they will purchase something. So make sure that, if users navigate to a link in your client's website, they can easily find their way back. Doing these things will ensure the stickiness of the website. Do not create any barriers to the customer who wants to return to the client's website.

Mashups

An application programming interface (API) is a set of small programs and protocols used like building blocks for building software applications. When two or more APIs are used together, they form a mashup. Many mashups are open source, so developers can use an API from a site like Google Maps and combine it with an API that contains other data, resulting in a new website that creates an entirely new application.

A large corporation that has many retail outlets in a region may want to make it easier for customers to find their retail stores. They may want to hire a company like Blipstar, which provides a service that allows companies to upload information about retail stores. Blipstar geocodes them and places them on a Google map. The company then puts a link to this information on its own website, so customers can simply enter their zip or postal code and let the mashup display the location of the nearest retail store. Blipstar works on mobile devices.

Mashups are becoming a popular way to present information. Expect to see many useful mashup applications soon. Look for them at www.programmableweb.com.

Designing Queries

When users ask questions of, or communicate with, a database, they are said to *query* it. Six different types of queries are among the most common. Your careful attention to query design can help reduce users' time spent in querying the database, help them find the data they want, and result in a smoother user experience overall.

Query Types

The questions we pose concerning data from our database are referred to as queries. There are six basic query types. Each query involves three items: an entity, an attribute, and a value. In each case, two of these are given, and the intent of the query is to find the remaining item. Figure 14.12 will be used to illustrate all the query examples.

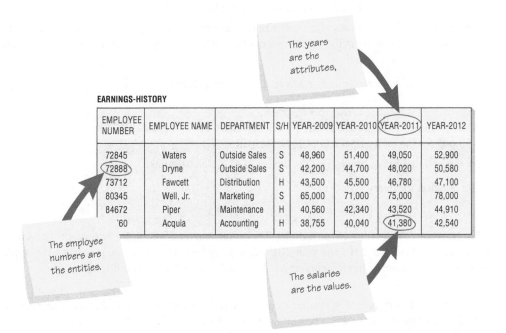

FIGURE 14.12

It is possible to perform six basic types of queries on a table that contains entities, attributes, and values.

QUERY TYPE 1. The entity and one of the entity's attributes are given. The purpose of the query is to find the value. The query can be expressed as follows:

What is the value of a specified attribute for a particular entity?

Sometimes it is more convenient to use notation to formulate the query. This query can be written as

$$V \longleftarrow (E, A)$$

where V stands for the value, E for entity, and A for attribute, and the variables in parentheses are given.

The question

What did employee number 73712 make in year 2012?

can be stated more specifically as

What is the value of the attribute YEAR-2012 for the entity EMPLOYEE NUMBER 73712?

The record containing employee number 73712 will be found, and the answer to the query will be $47,100.

QUERY TYPE 2. The intent of the second query type is to find an entity or entities when an attribute and value are given. Query type 2 can be stated as follows:

What entity has a specified value for a particular attribute?

Because values can also be numeric, it is possible to search for a value equal to, greater than, less than, not equal to, greater than or equal to, and so on. An example of this type of query is as follows:

What employee(s) earned more than $50,000 in 2012?

The notation for query type 2 is

$$E \longleftarrow (V, A)$$

In this case, three employees made more than $50,000, so the response will be a list of the employee numbers for the three employees: 72845, 72888, and 80345.

QUERY TYPE 3. The purpose of this query type is to determine which attributes fit the description provided when the entity and value are given. Query type 3 can be stated as follows:

What attribute(s) has a specified value for a particular entity?

This query is useful when many similar attributes have the same property. The following example has similar attributes (specific years) that contain the annual salaries for the employees of the company:

What years did employee number 72845 make over $50,000?

or, more precisely,

What attributes {YEAR-2009, YEAR-2010, YEAR-2011, YEAR-2012} have a value > 50,000 for the entity EMPLOYEE-NUMBER = 72845?

where the optional list in braces ({ }) is the set of eligible attributes.

The notation for query type 3 is

$$A \longleftarrow (V, E)$$

In this example, Waters (employee number 72845) made over $50,000 for two years. Therefore, the response will be year 2010 and year 2012. Query type 3 is rarer than the preceding two types due to the requirement of having similar attributes exhibiting the same properties.

QUERY TYPE 4. Query type 4 is similar to query type 1. The difference is that the values of all attributes are desired. Query 4 can be expressed as follows:

List all the values for all the attributes for a particular entity.

An example of query type 4 is:

List all the details in the earnings history file for employee number 72888.

The notation for query type 4 is

$$\text{all } V \longleftarrow (E, \text{all } A)$$

The response for this query will be the entire record for the employee named Dryne (employee number 72888).

QUERY TYPE 5. The fifth type of query is another global query, but it is similar in form to query type 2. Query type 5 can be stated as follows:

List all entities that have a specified value for all attributes.

An example of query type 5 is:

List all the employees whose earnings exceeded $50,000 in any of the years available.

The notation for query type 5 is

$$\text{all } E \longleftarrow (V, \text{all } A)$$

The response to this query will be 72845, 72888, and 80345.

QUERY TYPE 6. The sixth query type is similar to query type 3. The difference is that query type 6 requests a listing of the attributes for all entities rather than a particular entity. Query type 6 can be stated as follows:

List all the attributes that have a specified value for all entities.

The following is an example of query type 6:

List all the years for which earnings exceeded $40,000 for all employees in the company.

The notation for query type 6 is

$$\text{all } A \longleftarrow (V, \text{all } E)$$

The response will be YEAR-2010, YEAR-2011, and YEAR-2012. As with query type 3, query type 6 is not used as much as other types.

BUILDING MORE COMPLEX QUERIES. The preceding six query types are building blocks for more complex queries. Expressions, referred to as Boolean expressions, can be formed for queries. An example of a Boolean expression is:

List all the customers who have zip codes greater than or equal to 60001 and less than 70000, and who have ordered more than $500 from our catalogs or have ordered at least five times in the past year.

One difficulty with this statement is determining which operator (for example, AND) belongs with which condition; it is also difficult to determine the sequence in which the parts of the expression should be carried out. The following may help to clarify this problem:

LIST ALL CUSTOMERS HAVING (ZIP-CODE GE 60001 AND ZIP-CODE LT 70000) AND (AMOUNT-ORDERED GT 500 OR TIMES-ORDERED GE 5)

Now some of the confusion is eliminated. The first improvement is that the operators are expressed more clearly as GE, GT, and LT than as English phrases, such as "at least." Second, the attributes are given distinct names, such as AMOUNT-ORDERED and TIMES-ORDERED. In the earlier sentence, these attributes were both referred to as "have ordered." Finally, parentheses are used to indicate the order in which the logic is to be performed. Whatever is in parentheses is done first.

Operations are generally performed in a predetermined order of precedence. Arithmetic operations are usually performed first (exponentiation, then either multiplication or division, and then addition or subtraction). Next, comparative operations are performed. These operations are GT (greater than), LT (less than), and others. Finally, the Boolean operations are performed

FIGURE 14.13

Arithmetic, comparative, and Boolean operators are processed in a hierarchical order of precedence unless parentheses are used.

Type	Level	Symbol
Arithmetic Operators	1	* *
	2	* /
	3	+ −
Comparative Operators	4	GT LT
		EQ NE
		GE LE
Boolean Operators	5	AND
	6	OR

(first AND and then OR). Within the same level, the order generally goes from left to right. The precedence is summarized in Figure 14.13.

Query Methods

Two popular query languages are Query by Example and Structured Query Language.

QUERY BY EXAMPLE. Query by Example (QBE) is a simple but powerful method for implementing queries in database systems, such as Microsoft Access. The database fields are selected and displayed in a grid, and the requested query values are entered either in the field area or below the field. The query should be able to select both rows from the table that match conditions as well as specific columns (fields). Complex conditions may be set to select records, and the user may easily specify the columns to be sorted. Figure 14.14 is an example of a query using Microsoft Access. The query design screen is divided into two portions. The top portion contains the tables selected for the query and their relationships, and the bottom portion contains the query selection grid. Fields from the database tables are dragged to the grid.

The first two rows contain the field and the table in which the field is located. The next row contains sorting information. In this example, the results will be sorted by CUSTOMER NAME. A check mark in the Show box (fourth row down) indicates that the field is to be displayed in the results. Notice that the CUSTOMER NUMBER, CUSTOMER NAME, and STATUS CODE MEANING are selected for the resulting display (other fields are displayed as well, but they do not show in the display). Notice that the ACCOUNT STATUS CODE and ACCOUNT TYPE CODE are not checked and therefore will not be in the final results. In the criteria rows, there is a *1* in the ACCOUNT STATUS CODE (indicating an active record) and a *C* and *D* (selecting

FIGURE 14.14

Query by example using Microsoft Access.

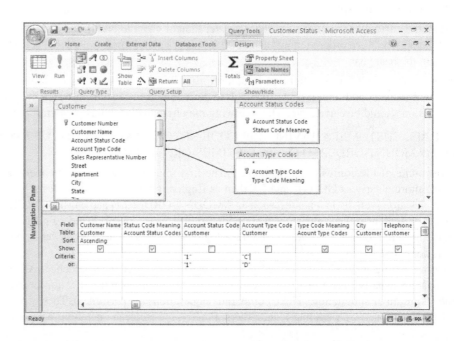

CONSULTING OPPORTUNITY 14.6

Hey, Look Me Over (Reprise)

Y ou have been called back to take another look at Merman's Costume Rentals. Here is part of the database created for Annie Oaklea of Merman's (with whom you last worked in Consulting Opportunities 7.1 and 8.1). The database contains information, such as the cost of the rental; the date checked out, the date due back, and the number of days the costume has been rented since the beginning of the year (YTD DAYS OUT) (see Figure 14.C1).

Analyzing Annie's typical day in the costume rental business, you realize there are several requests she must make of the database so that she can make decisions on when to replace frequently used costumes or even when to buy more costumes of a particular

type. She also needs to remember to keep in the good graces of customers she has previously turned down for a particular costume rental, to know when to recall an overdue costume, and so on.

Formulate several queries that will help her get the information she needs from the database. (*Hint:* Make any assumptions necessary about the types of information she needs to make decisions and use as many of the different query types discussed in this chapter as you can.) In a paragraph, describe how Annie's queries would be different if she were working with a Web-based or hyperlinked system.

COSTUME-RENTAL

COSTUME NUMBER	DESCRIPTION	SUIT NUMBER	COLOR	COST OF	DATE CHECKED OUT	DUE DATE	YTD DAYS OUT	TYPE OF COSTUME	REQUESTS TURNED DOWN
0003	Lady MacBeth F, SM	01	Blue	15.00	10/15	11/30	150	Standard	2
1342	Bear F, MED	01	Dk. Brown	12.50	10/24	11/09	26	Standard	0
1344	Bear F, MED	02	Dk. Brown	12.50	10/24	11/09	115	Standard	0
1347	Bear F, LG	01	Black	12.50	10/24	11/09	22	Standard	0
1348	Bear F, LG	02	Black	12.50	11/01	11/08	10	Standard	0
1400	Goldilocks F, MED	01	Light Blue	7.00	10/24	11/09	140	Standard	0
1402	Goldilocks F, MED	02	Light Blue	7.00	10/28	11/09	10	Standard	0
1852	Hamlet M, MED	01	Dark Green	15.00	11/02	11/23	115	Standard	3
1853	Ophelia F, SM	01	Light Blue	15.00	11/02	11/23	22	Standard	0
4715	Prince M, LG	01	White/purple	10.00	11/04	11/21	145	Standard	5
4730	Frog M, SM	01	Green	7.00	11/04	11/21	175	Standard	2
7822	Jester M, MED	01	Multi	7.50	11/10	12/08	12	Standard	0
7824	Jester M, MED	02	Multi	7.50	11/09	11/15	10	Standard	0
7823	Executioner M, LG	01	Black	7.00	11/19	12/05	21	Standard	0
8645	Mr. Spock N, LG	01	Orange	18.00	09/07	09/12	150	Trendy	4
9000	Pantomime F, LG	01	Red	7.00	08/25	09/15	56	Standard	0
9001	Pantomime M, MED	01	Blue	7.00	08/25	09/15	72	Standard	0
9121	Juggler M, MED	01	Multi	7.00	11/05	11/19	14	Standard	0
9156	Napoleon M, SM	01	Blue/white	15.00	10/26	11/23	56	Standard	1

FIGURE 14.C1

A portion of the database from Merman's Costume Rental shop.

a General Customer or a Discount Customer) in the ACCOUNT TYPE CODE columns. Two conditions in the same row indicate an AND condition, and two conditions in different rows represent an OR condition. This query specifies that the user should select both an Active Customer and either a General or Discount Customer.

The results of a query are displayed in a table, illustrated in Figure 14.15. Notice that the ACCOUNT STATUS CODE and ACCOUNT TYPE CODE do not display. They are not checked and are included in the query for selection purposes only. Instead, the code meanings are displayed, which are more useful to the user. The customer names are sequenced alphabetically.

FIGURE 14.15

A query by example for
CUSTOMER STATUS yields
these results.

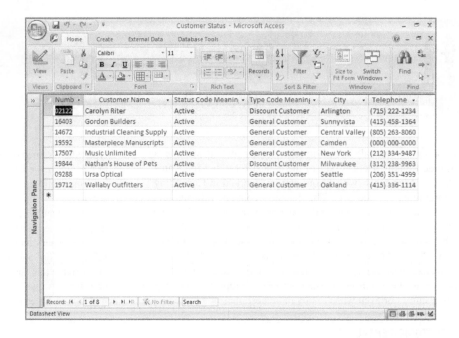

STRUCTURED QUERY LANGUAGE. Structured Query Language (SQL) is another popular language for implementing queries. It uses a series of words and commands to select the rows and columns that should be displayed in the resulting table. Figure 14.16 contains SQL code. The SELECT DISTINCT ROW keyword determines which rows are to be selected. The WHERE keyword specifies the condition that the CUSTOMER NAME should be used to select the data entered in the LIKE parameter.

FIGURE 14.16

Structured Query Language (SQL)
for the CUSTOMER NAME
parameter query.

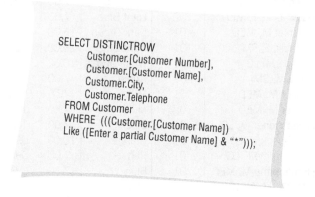

```
SELECT DISTINCTROW
        Customer.[Customer Number],
        Customer.[Customer Name],
        Customer.City,
        Customer.Telephone
FROM Customer
WHERE  (((Customer.[Customer Name])
Like ([Enter a partial Customer Name] & "*")));
```

Summary

In this chapter, we examined human–computer interaction (HCI), a variety of interfaces, designing the user interface, developing multitouch gestures to interface with smartphones and tablets, designing user feedback, and designing ecommerce website feedback and navigation. We focused on understanding HCI to ensure the functionality and usability of computer systems we design. When analysts create a proper fit among the HCI elements of the human, the computer, and the task, it leads to improved performance and overall psychological and physical well being of the individual. Designs focus on developing a proper fit.

Usability identifies what works for users and what does not. Physical considerations of HCI design include vision, hearing, and touch. Physical disabilities and limitations should be taken into consideration during task and interface design. A variety of user interfaces and input devices are possible. Some interfaces are particularly well suited to inexperienced users, whereas others are better suited to experienced users. Combine interfaces such as drop-down menus and graphical interfaces for increased effectiveness. The Web has posed new challenges for designers because the user is not known.

HYPERCASE® EXPERIENCE 14

"I have no problem with using a mouse or any other rodent you throw my way. Really, though, whatever Snowden needs is what I try to do. Everyone is different, however. I've seen people here go out of their way to avoid using a computer altogether. Other people would prefer not to talk with a human. In fact, they would be as happy as a puppy chewing on a new bedroom slipper if they could use simple commands to interact. Most of the folks we have here are open to new things. Otherwise, they wouldn't be here at MRE. We do pride ourselves on our creativity. I have you signed up for a meeting with people from the training group, including Tom Ketchem, Melissa Smith, and Kathy Blandford. I've included Ashley Heatherford who is our usability expert consultant. Melanie Corazón might be there since she wants to see if what you're doing could be applicable to ecommerce systems. Snowden may sit in as well, if he has time. They'll be very curious to see what kind of interface you are suggesting for them on the new project reporting system."

HYPERCASE Questions

1. Write a short proposal describing what type of user interface would be appropriate for the users of the project reporting system who are in the training group. Include reasons for your decision.

2. Design a user interface using a CASE tool, such as Visible Analyst, a software package such as Microsoft Access, or paper layout forms for the training group. What are the key features that address the needs of the people in the training group?

3. Write a short proposal describing what type of user interface would be appropriate for the users of the B2B ecommerce systems Melanie Corazón described in her interview. Include reasons for your decision.

4. Design a user interface using a CASE tool, such as Visible Analyst, a software package such as Microsoft Access, or paper layout forms for the users of the B2B ecommerce systems Melanie Corazón described. What are the key features that address the needs of the people who use B2B ecommerce systems on the Web?

5. Make a bulleted list of any usability concerns Ashley might have with the interfaces you designed.

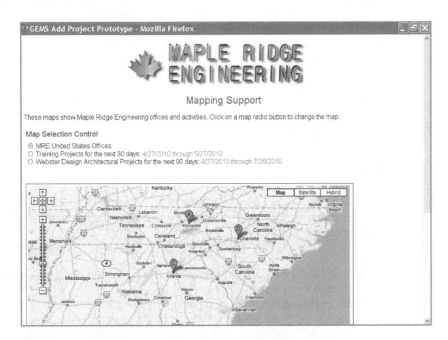

FIGURE 14.HC1

In HyperCase, you can see how users process information in order to create a more effective user interface.

Creative opportunities for developers to design multitouch gestures for interfacing with touch-sensitive (capacitive sensing) smartphones and tablets are increasing. Some conventions should be observed, users should be able to learn the interface quickly, and gestures should make sense to users. Developers can include alerts, notices, and queries for smartphones and tablets but should always offer an opt-out option to them. Using badges is an unobtrusive way to provide additional information to users, but badges should be used with caution so that they don't rely on user interaction to display accurate information.

Users' need for feedback from a system is an important consideration. Feedback is most often visual, with text, graphics, or icons the most common. Audio feedback can also be effective.

Improve functionality of websites by eliciting customer feedback through automatic email feedback buttons or by including blank feedback forms on the website. Four important navigation design strategies improve the stickiness of ecommerce websites: (1) rollover menus, (2) hierarchical displays of links on the entry screen, (3) site maps, and (4) navigation bars that provide one-click navigation.

Queries are designed to allow users to extract meaningful data from the database. There are six basic types of queries, and they can be combined using Boolean logic to form more complex queries. Query by Example and SQL are two common languages for querying database systems.

Keywords and Phrases

accelerometer
alerts
application programming interface (API)
badges
Boolean operators
capacitive sensing
cognitive considerations of HCI
continuous speech system
dialog box
disabilities and design
feedback
feedback for users
fit
form-fill interfaces
gestures
graphical user interface (GUI)
intuitive navigation
mashup
menu
multitouch gestures
natural-language interface
navigation bar
nested menus
notices

one-click navigation
performance
physical considerations of HCI
psychological considerations of HCI
pull-down menu
query
Query by Example (QBE)
question-and-answer interface
rollover menu
site map
smartphone
speech recognition and synthesis
stickiness
Structured Query Language (SQL)
stylus
tablet
task
template
touch screen
touch-sensitive screen
usability
Web-based form-fill interface
voice recognition

Review Questions

1. Define *HCI*.
2. Explain how fit among the HCI elements of the human, the computer, and the tasks to be performed leads to performance and well-being.
3. What are the components of the term *performance* in the HCI context?
4. What is meant by the word *well-being* when used in an HCI context?
5. List 5 of the 11 usability heuristics for judging the usability of computer systems and ecommerce websites provided by Nielsen and others.
6. Describe some of the ways that a pivot table allows a user to arrange data.
7. List three physical considerations that HCI design addresses.
8. List three ways that analysts can improve task or interface design to help, respectively, a person who is visually impaired, hearing impaired, or mobility impaired.
9. What are the five objectives for designing user interfaces?
10. Define *natural-language interface*. What is the major drawback of these interfaces?
11. Explain what is meant by question-and-answer interfaces. To what kind of users are they best suited?

12. Describe how users use onscreen menus.
13. What is a nested menu? What are its advantages?
14. Define *onscreen input/output form*. What is the chief advantage of these forms?
15. What are the advantages of Web-based fill-in forms?
16. What are the drawbacks of Web-based form-fill interfaces?
17. Define *graphical user interface*. What is the key difficulty these interfaces present for programmers?
18. For what type of user is a GUI particularly effective?
19. What is a synonym for the term "capacitive sensing?"
20. List three gestures that can be used to interface with touch-sensitive smartphones and tablets.
21. Describe a problem that can occur if a designer uses nonstandard gestures to create an interface for smartphones or tablets.
22. Why is the metaphor important in designing a user interface for an app?
23. What do alerts do in a smartphone or tablet interface
24. What do badges do in a smartphone or tablet interface?
25. Why is it important to include an opt-out option on all notices and alerts for smartphone and tablet users?
26. Why is it a good idea to avoid the use of a badge on apps?
27. Designs for the iPhone can take advantage of using voice recognition, called Siri. Is Siri an intelligent agent? Why or why not?
28. What are the three guidelines for designing good screen dialog?
29. What are the roles of icons, graphics, and colors in providing feedback?
30. List eight ways to achieve the goal of minimal operator action when designing a user interface.
31. List five standards that can aid in evaluating user interfaces.
32. What are the seven situations that require feedback for users?
33. What is a suitable way of telling a user that input was accepted?
34. When a user is informed that his or her input is not in the correct form, what additional feedback should be given at the same time?
35. List four ways to notify a Web user that input is not in the correct form.
36. Why is it unacceptable to notify the user that input is not correct solely through the use of audible beeping or buzzing?
37. When a request is not completed, what feedback should be provided to the user?
38. Describe two types of website designs for eliciting feedback from customers.
39. List four practical ways that an analyst can improve the ease of user navigation and the stickiness of an ecommerce website.
40. What are hypertext links? Where should they be used?
41. Describe what a mashup is.
42. List in shorthand notation the six basic query types.

Problems

1. Manu Narayan owns several first-class hotels worldwide, including properties in Manhattan, Mumbai, and even some in suburbia. He wants to make sure that the human–computer interface is appropriate to each culture but also wants to be able to share the software among all of his hotel reservation departments. Design a nested menus interface for a check-in and checkout hotel reservation system that can be used internationally. Use numbers to select a menu item. Show how each menu would look on a standard computer display.
2. Stefan Lano needs displays that will show the musical instrument inventory in his chain of music stores that caters to musicians playing in world-class symphony orchestras in Basel, Switzerland; Buenos Aires, Argentina; and Philadelphia and New York in the US. Design a form-fill interface for the inventory control of musical instruments in all four stores that could be used on a computer display screen. Assume that English will be the interface language.
3. Design a Web-based form-fill interface to accomplish the same task as in Problem 2.

   a. What difficulties did you encounter? Discuss them in a paragraph.
   b. Of the two designs you did, which would you say is better suited to Mr. Lano's task? Why? List three reasons for your choice. How would you test their usability?

4. A UK-based travel agent, Euan Morton, LLC, would like your systems team to design a smartphone and tablet interface he can give away free to his clients for booking seats for airlines with which his firm has solid business ties, such as British Air, RyanAir, and Virgin Atlantic.

   a. Show what the interface would look like on a smartphone display.

b. Make a list of gestures that you should include for users to activate the display to book an airline seat and write down what each gesture does.

5. An IT executive, Felicia Finley, from Jersey IT Innovators, Inc., has asked that you design a graphical user interface for an executive desktop to help her in her work. Use icons for file cabinets, a wastebasket, a telephone, and so on. Show how they would appear on the computer display.

6. Josh and Colleen, a celebrity chef couple and restaurant owners from Cherry Hill, New Jersey, want to be able to receive clear feedback on the systems used to manage food inventory at their many "farm to table" restaurants in Philadelphia and New York City. Design a display that provides appropriate feedback for a user whose daily update report on locally available produce has been successfully sent to the chefs.

7. Design a screen for a payroll software package that displays information telling Josh and Colleen from Problem 6 how to get more detailed feedback.

8. Design a Web-based display that shows an acceptable way to tell Josh and Colleen that input to their system was accepted.

9. Design a feedback form for Josh and Colleen's devoted restaurant customers using an ecommerce website to purchase their cookbook.

10. Write six different queries for the file in Problem 1 in Chapter 13.

11. Write six different queries for the 3NF relation in Problem 5 in Chapter 13.

12. Design a search that will find potential competitors of a company such as World's Trend on the Web. Assume that you are the customer.

13. Search for World's Trend's potential competitors on the Web. (You won't find World's Trend itself on the Web. It is a fictional company.) Make a list of those you've found.

Group Projects

1. With your group members, create a pull-down menu for an employment agency that matches professional candidates to position openings. Include a list of keystrokes that would directly invoke the menu options using the Alt-X format. The menu has the following options:

    Add employee
    Change employee
    Delete employee
    Employee inquiry
    Position inquiry
    Employer inquiry
    Add employer
    Change employer
    Delete employer
    Match employee to opening
    Print open positions report
    Print successful matches report
    Add position
    Change position
    Delete position

2. In a paragraph, describe the problems your group faced in creating this menu.

3. The drag-and-drop feature is used in GUIs and allows the user to move sentences around in a word processing package. As a group, suggest how drag and drop can be used to its fullest potential in the following applications:

    a. Project management software (Chapter 3)
    b. Relational database program (Chapter 13)
    c. Display or forms designer (Chapter 12)
    d. Spreadsheet program
    e. CASE tool for drawing data flow diagrams (Chapter 7)
    f. Smartphone calendar (Chapter 3)
    g. Illustration in a drawing package
    h. CASE tool for developing data dictionaries (Chapter 8)
    i. Decision tree drawing program (Chapter 9)

j. Website for collecting consumer opinions on new products (Chapter 11)

k. Organizing bookmarks for websites

For each solution your group designs, draw the display and show movement by using an arrow.

4. Ask each of the members of your group to request a search based on their leisure activities (such as running, dancing, going to the movies). If there are four people in your group, there will be four unique searches to perform. Now go ahead and do all the searches. Compare your results. Does the person who is involved with the activity have an advantage over the people who know less about it? Explain.

5. Look at the following mashup websites and describe how each of them adds value by providing a service.

a. Airport parking, www.aboutairportparking.com

b. Global Incident Map, www.globalincidentmap.com

c. Hawkee Application Developer Network, www.hawkee.com

d. Realtors on Homethinking, www.homethinking.com

e. Streeteasy, www.streeteasy.com

6. The following mashup has a political agenda. Suggest three other mashups that attempt to change something by appealing to the public.

Health Care That Works, www.healthcarethatworks.org/maps/nyc

7. Try these mashups just for fun.

a. The Geography of Seinfeld, www.stolasgeospatial.com/seinfeld.htm

b. PlotShot, www.plotshot.com

c. Flickr Sudoku, http://flickrsudoku.com

d. Liveplasma, www.liveplasma.com

Selected Bibliography

Adam, A., and D. Kreps. "Web Accessibility: A Digital Divide for Disabled People?" In IFIP International Federation for Information Processing, Vol. 208, *Societal and Organizational Implications for Information Systems*. Edited by E. Trauth, D. Howcraft, T. Butler, B. Fitzgerald, and J. DeGross, pp. 217–228. Boston: Springer, 2006.

Barki, H., and J. Hartwick. "Measuring User Participation, User Involvement, and User Attitude." *MIS Quarterly,* Vol. 18, No. 1, 1994, pp. 59–82.

Berstel, J., S. C. Reghizzi, G. Roussel, and P. San Pietro. "A Scalable Formal Method for Design and Automatic Checking of User Interfaces." *ACM Transactions on Software Engineering and Methodology,* Vol. 14, No. 2, April 2005, pp. 124–167.

Carey, J., D. Galletta, J. Kim, D. Te'eni, B. Wildemuth, and P. Zhang. "The Role of Human–Computer Interaction in Management Information Systems Curricula: A Call to Action." *Communications of the Association for Information Systems,* Vol. 13, 2004, pp. 357–379.

Davis, G. B., and M. H. Olson. *Management Information Systems: Conceptual Foundations, Structure, and Development.* New York: McGraw-Hill, 1985.

Galleta, D., and P. Zhang (Eds.). *Human–Computer Interaction and Management Information Systems: Applications.* Armonk, NY: M.E. Sharpe, 2006.

Greene, K. "A Better, Cheaper Multitouch Interface." *Technology Review,* March 2009. www.technolo-gyreview.com/computing/22358/?a=f. Last accessed April 19, 2009.

Hornbaek, K., and E. Frokjaer. "Comparing Usability Problems and Redesign Proposals as Input to Practical Systems Development." *CHI 2005,* April 2–7, 2005, pp. 391–400.

Mantei, M. M., and T. J. Teorey. "Incorporating Behavioral Techniques in the System Development Lifecycle." *MIS Quarterly,* Vol. 13, No. 3, September 1989, pp. 257–267.

Microsoft. "Metro Design Language of Windows Phone 7." www.microsoft.com/design/toolbox/tutorials/windows-phone-7/metro/. Last accessed August 31, 2012.

Nielsen, J., and R. L. Mack. *Usability Inspection Methods.* New York: John Wiley, 1994.

Nielsen, J., R. Molich, C. Snyder, and S. Farrell. *E-Commerce User Experience.* Fremont, CA: Norman Nielsen Group, 2001.

Rubin, J. *Handbook of Usability Testing.* New York: John Wiley, 1994.

Schneiderman, B., and C. Plaisant. *Designing the User Interface: Strategies for Effective Human–Computer Interaction.* New York: Addison-Wesley, 2005.

Te'eni, D., J. Carey, and P. Zhang. *Human–Computer Interaction: Developing Effective Organizational Systems.* New York: John Wiley, 2007.

U.S. Department of Health and Human Services. "Usability Guide" (for developing websites). www.usability.gov. Last accessed May 30, 2012.

U.S. Equal Employment Opportunity Commission. http://www.eeoc.gov/employers/index.cfm. Last accessed August 31, 2012.

UsabilityNet. "Overview of the User Centered Design Process." www.usabilitynet.org/management/ b_overview.htm. Last accessed May 30, 2012.

Zhang, P., J. Carey, D. Te'eni, and M. Tremaine. "Integrating Human–Computer Interaction Development into the Systems Development Life Cycle: A Methodology." *Communications of the Association for Information Systems,* Vol. 15, 2005, pp. 512–543.

The CPU Case Episode and accompanying student files are available online at www.pearsonglobaleditions.com/kendall.

Designing Accurate Data Entry Procedures

LEARNING OBJECTIVES

Once you have mastered the material in this chapter you will be able to:

1. Understand the uses of effective coding to support users in accomplishing their tasks.

2. Design effective and efficient data capture approaches for people and systems.

3. Recognize how to ensure data quality through validation.

4. Articulate accuracy advantages of user input on ecommerce websites.

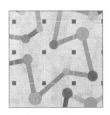

Making sure that users are able to enter data into a system accurately is of utmost importance. It is by now axiomatic that the quality of data input determines the quality of information output. A systems analyst can support accurate data entry through the achievement of four broad objectives: (1) creating meaningful coding for data, (2) designing efficient data capture approaches, (3) assuring complete and effective data capture, and (4) assuring data quality through validation.

The quality of data is a measurement of how consistently correct the data are within certain preset limits. Effectively coded data facilitate accurate human data entry by cutting down on the sheer quantity of data and thus the time required to enter the information.

When users enter data efficiently, data entry meets predetermined performance measures that give the relationship between the time spent on entry and the number of data items entered. Effective coding, effective and efficient data capture and entry, and assurance of data quality through validation procedures are all data entry objectives covered in this chapter.

Effective Coding

One of the ways that data can be entered more accurately and efficiently is through the knowledgeable use of various codes. The process of putting ambiguous or cumbersome data into short, easily entered digits or letters is called *coding* (not to be confused with program coding).

Coding aids a systems analyst in reaching the objective of efficiency because data that are coded require less time for people to enter and thus reduce the number of items entered. Coding can also help in the appropriate sorting of data at a later point in the data transformation process. In addition, coded data can save valuable memory and storage space. Coding is a way of being eloquent but succinct in capturing data. Besides providing accuracy and efficiency, codes should have a purpose that supports users. Specific types of codes allow us to treat data in a particular manner. Human purposes for coding include the following:

1. Keeping track of something
2. Classifying information
3. Concealing information
4. Revealing information
5. Requesting appropriate action

Each of these purposes for coding is discussed in the following sections, along with some examples of codes.

Keeping Track of Something

Sometimes we want merely to identify a person, place, or thing just to keep track of it. For example, a shop that manufactures custom-made upholstered furniture needs to assign a job number to a project. The salesperson needs to know the name and address of the customer, but the job shop manager or the workers who assemble the furniture need not know who the customer is. Consequently, an arbitrary number is assigned to the job. The number can be either random or sequential, as described in the following subsection.

SIMPLE SEQUENCE CODES. The simple sequence code is a number that is assigned to something if it needs to be numbered. It therefore has no relation to the data themselves. Figure 15.1 shows how a furniture manufacturer's orders are assigned an order number. With this easy reference number, the company can keep track of the order in process. It is more efficient to enter job "5676" than "that brown and black rocking chair with the leather seat for Arthur Hook, Jr."

Using a sequence code rather than a random number has some advantages. First, it eliminates the possibility of assigning the same number multiple times. Second, it gives users an approximation of when the order was received.

Sequence codes should be used when the order of processing requires knowledge of the sequence in which items enter the system or the order in which events unfold. For example, consider a bank running a special promotion that makes it important to know when a person applied for a special, low-interest home loan because (all other things being equal) the special mortgage loans will be granted on a first-come, first-served basis. In this case, assigning a correct sequence code to each applicant is important.

ALPHABETIC DERIVATION CODES. At times it is undesirable to use sequence codes. The most obvious instance is when you do *not* wish to have someone read the code to figure out how many numbers have been assigned. Another situation in which sequence codes may not be useful is when a more complex code is desirable to avoid a costly mistake. One possible error would be to add a payment to account 223 when you meant to add it to account 224 because you entered an incorrect digit.

FIGURE 15.1

Using a simple sequence code to indicate the sequence in which orders enter a custom furniture shop.

Order #	Product	Customer
5676	Rocking Chair/with Leather	Arthur Hook, Jr.
5677	Dining Room Chair/Upholstered	Millie Monice
5678	Love Seat/Upholstered	J. & D. Pare
5679	Child's Rocking Chair/Decals	Lucinda Morely

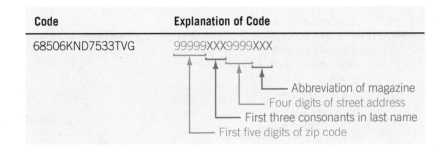

FIGURE 15.2

Identifying the account of a magazine subscriber with an alphabetic derivation code.

The alphabetic derivation code is a commonly used approach in identifying an account number. The example in Figure 15.2 comes from a mailing label for a magazine. The code becomes the account number. The first five digits come from the first five digits of the subscriber's zip code, the next three are the first three consonants in the subscriber's name, the next four numbers are from the street address, and the last three make up the code for the magazine. The main purpose of this code is to identify an account.

A secondary purpose is to print mailing labels. When designing this code, the zip code is the first part of the account number. The subscriber records are usually updated only once a year, but the primary purpose of the records is to print mailing labels once a month or once per week. Having the zip code as the first part of a primary key field means that the records do not have to be sorted by zip code for bulk mailing, because records on a file are stored in primary key sequence. Notice that the expiration date is not part of the account number, because that number can change more frequently than the other data.

One disadvantage of an alphabetic derivation code occurs when the alphabetic portion is small (for example, the name Po) or when the name contains fewer consonants than the code requires. The name Roe has only one consonant and would have to be derived as RXX, or derived using some other scheme. Another disadvantage is that some of the data may change. Changing one's address or name would change the primary key for the file.

Classifying Information

Classification affords the ability to distinguish among classes of items. Classifications are necessary for many purposes, such as reflecting what parts of a medical insurance plan an employee carries, or showing which student has completed the core requirements of his or her coursework.

To be useful, classes must be mutually exclusive. For example, if a student is in class F, meaning freshman, having completed 0 to 36 credit hours, he or she should not also be classifiable as a sophomore (S). Overlapping classes would be F = 0 – 36 credit hours, S = 32 – 64 credit hours, and so on. Data are unclear and not as readily interpretable when coding classes are not mutually exclusive.

CLASSIFICATION CODES. Classification codes are used to distinguish one group of data with special characteristics from another. Classification codes can consist of either a single letter or a number. Using them is a shorthand way of describing a person, place, thing, or event.

Classification codes are listed in manuals or posted so that users can locate them easily. Many times, users become so familiar with frequently used codes that they memorize them. A user classifies an item and then enters its code directly into an online system.

An example of classification coding is the way you may wish to group tax-deductible items for the purpose of completing your income taxes. Figure 15.3 shows how codes are developed

Code	Tax-Deductible Item
I	Interest Payments
M	Medical Payments
T	Taxes
C	Contributions
D	Dues
S	Supplies

FIGURE 15.3

Grouping tax-deductible items through the use of a one-letter classification code.

FIGURE 15.4

Problems in using a one-letter classification code occur when categories share the same letter.

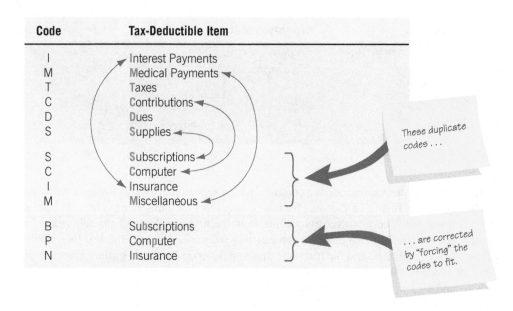

for items such as interest, medical payments, contributions, and so on. The coding system is simple: Take the first letter of each of the categories; contributions are C, interest payments are I, and supplies are S.

All goes well until we get to other categories (such as computer items, insurance payments, and subscriptions) that begin with the same letters we used previously. Figure 15.4 demonstrates what happens in this case. The coding was stretched so that we could use P for "comPuter," N for "iNsurance," and B for "suBscriptions." Obviously, this situation is far from perfect. One way to avoid this type of confusion is to allow for codes longer than one letter, discussed later in this chapter under the subheading of mnemonic codes. Pull-down menus in a GUI system often use classification codes as a shortcut for running menu features, such as **Alt-F** for the **File** menu.

BLOCK SEQUENCE CODES. Earlier we discussed sequence codes. The block sequence code is an extension of the sequence code. Figure 15.5 shows how a business user assigns numbers to computer software. Main categories of software are browsers, database packages, and Web design. These were assigned sequential numbers in the following "blocks," or ranges: browser, 100–199; database, 200–299; and so forth. The advantage of the block sequence code is that the data are grouped according to common characteristics, but still take advantage of the simplicity of assigning the next available number (within the block, of course) to the next item needing identification.

FIGURE 15.5

Using a block sequence code to group similar software.

Code	Name of Software Package	Type
100	Apple Safari	Browser
101	Mozilla Firefox	
102	Microsoft Internet Explorer	
103	Google Chrome	
.		
.		
.		
200	Microsoft Access	Database
201	MySQL	
202	Oracle	
.		
.		
.		
300	Adobe Dreamweaver	Web design
301	Freeway Pro	
302	SiteGrinder	

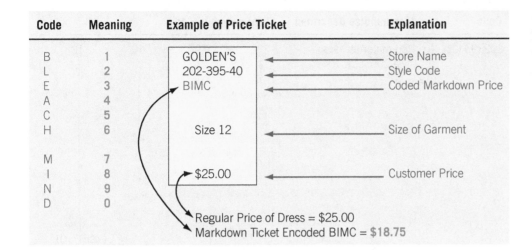

FIGURE 15.6

Encoding markdown prices with a cipher code is a way of concealing price information from customers.

Concealing Information

You can use codes to conceal or disguise information you do not wish others to know. There are many reasons a business user may want to do this. For example, a corporation may not want information in a personnel file to be accessed by data entry workers. A store may want its salespeople to know the wholesale price to show them how low a price they can negotiate, but they may encode it on price tickets to prevent customers from finding that out. A restaurant may want to capture information about the service without letting the customer know the name of the server. Concealing information and security have become very important in the last few years. Corporations have started to allow vendors and customers to access their databases directly, and handling business transactions over the Internet has made it necessary to develop tight encryption schemes. The following subsection describes an example of concealing information through codes.

CIPHER CODES. Perhaps the simplest coding method is the direct substitution of one letter for another, one number for another, or one letter for a number. A popular type of puzzle called a cryptogram is an example of letter substitution. Figure 15.6 is an example of a cipher code taken from a Buffalo, New York, department store that coded all markdown prices with the words BLEACH MIND. No one really remembered why those words were chosen, but all the employees knew them by heart, and so the cipher code was successful. Notice in this figure that an item with a retail price of $25.00 would have a markdown price of BIMC, or $18.75, when decoded letter-by-letter.

Revealing Information

Sometimes it is desirable to reveal information to specific users through a code. In a clothing store, information about the department, product, color, and size is printed along with the price on the ticket for each item. This information helps the salespeople and stock people locate the place for the merchandise.

Another reason for revealing information through codes is to make the data entry more meaningful for humans. A familiar part number, name, or description supports more accurate data entry. The examples of codes in the following subsection explain how these concepts can be realized.

SIGNIFICANT-DIGIT SUBSET CODES. When it is possible to describe a product by virtue of its membership in many subgroups, we can use a significant-digit subset code to help describe it. The clothing store price ticket example in Figure 15.7 is an example of an effective significant-digit subset code.

To the casual observer or customer, the item description appears to be one long number. A salesperson knows, however, that the number is made up of a few smaller numbers, each one having a meaning of its own. The first three digits represent the department, the next three the product, the next two the color, and the last two the size.

Significant-digit subset codes may consist of either information that actually describes the product (for example, the number 10 means size 10) or numbers that are arbitrarily assigned (for

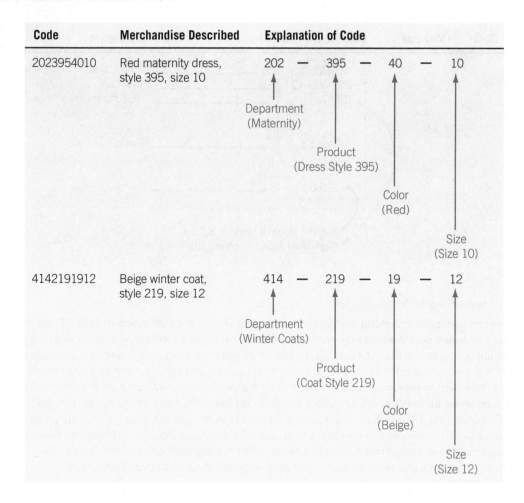

instance, 202 is assigned to mean the maternity department). In this case, the advantage of using a significant-digit subset code is that it makes it possible to locate items that belong to a certain group or class. For example, if the store's manager decided to mark down all winter merchandise for an upcoming sale, salespeople could locate all items belonging to departments 310 through 449, the block of codes used to designate "winter" in general.

MNEMONIC CODES. A mnemonic (pronounced nî-môn'-ĭk) is a human memory aid. Any code that helps either a data entry person remember how to enter the data or a user remember how to use the information can be considered a mnemonic. Using a combination of letters and symbols affords a strikingly clear way to code a product so that the code is easily seen and understood.

The city hospital codes formerly used by the Buffalo Regional Blood Center were mnemonic, as shown in Figure 15.8. The simple codes were invented precisely because the blood center administrators and systems analysts wanted to ensure that hospital codes were easy to memorize and recall. Mnemonic codes for the hospitals helped lessen the possibility of blood being shipped to the wrong hospital.

Code	City Hospitals
BGH	**B**uffalo **G**eneral **H**ospital
ROS	**Ro**swell Park Memorial Institute
KEN	**Ken**more Mercy
DEA	**Dea**coness Hospital
SIS	**Sis**ters of Charity
STF	**S**aint **F**rancis Hospital
STJ	**S**aint **J**oseph's Hospital
OLV	**O**ur **L**ady of **V**ictory Hospital

Code	Function
1	Delivered
2	Sold
3	Spoiled
4	Lost or Stolen
5	Returned
6	Transferred Out
7	Transferred In
8	Journal Entry (Add)
9	Journal Entry (Subtract)

FIGURE 15.9

Function codes compactly capture functions that a computer must perform.

UNICODE. Codes allow us to reveal characters that we normally cannot input or view. Traditional keyboards support character sets that are familiar to people using Western alphabetic characters (referred to as Latin characters), but many languages, such as Greek, Japanese, Chinese, and Hebrew, do not use the Western alphabet. These languages may use Greek letters or glyphs or symbols representing syllables or whole words. The International Organization for Standardization (ISO) has defined the Unicode character set, which includes all standard language symbols and has room for 65,535 characters. You can display web pages written in other alphabets by downloading an input method editor from Microsoft.

Glyph symbols are represented using an "&#xnnnn;" notation, in which nnnn represents a specific letter or symbol and x means that hexadecimal notation, or base 16 numbering, is used to represent the Unicode characters. For example, &#30B3 represents the Japanese Katakana symbol *ko*. The code used for the Japanese word for hello, *konichiwa*, is こにち&#x308F. In Japanese, the word looks like this:

こ に ち わ
ko ni chi wa
hello

The Unicode characters are grouped by language and may be found at www.unicode.org.

Requesting Appropriate Action

Codes are often needed to instruct either a computer or a decision maker about what action to take. Such codes are generally referred to as function codes, and they typically take the form of either sequence or mnemonic codes.

FUNCTION CODES. The functions that an analyst or a programmer desires a computer to perform with data are captured in function codes. Spelling out precisely what activities are to be accomplished is translated into a short numeric or alphanumeric code.

Figure 15.9 shows examples of a function code for updating inventory. Suppose you managed a dairy department; if a case of yogurt spoiled, you would use the code 3 to indicate this event. Of course, data required for input vary, depending on what function is needed. For example, appending or updating a record would require only the record key and function code, whereas adding a new record would require all data elements to be input, including the function code.

General Guidelines for Coding

In the previous sections, we examined the purposes for using different types of codes when humans and machines enter and store data. Next, we examine a few heuristics for establishing a coding system. These rules are highlighted in Figure 15.10.

BE CONCISE. Codes should be concise. Overly long codes mean more keystrokes and consequently more errors. Long codes also mean that storing the information in a database will require more memory.

Short codes are easier for people to remember and easier to enter than long codes. If codes must be long, they should be broken up into subcodes. For example, 5678923453127 could be broken up with hyphens as follows: 5678-923-453-127. This approach is much more manageable and takes advantage of the way people are known to process information in short chunks. Sometimes codes are made longer than necessary for a reason. Credit card numbers are often

CONSULTING OPPORTUNITY 15.1

It's a Wilderness in Here

"I can't stand this. I've been looking for this hat for the last 45 minutes," complains Davey, as he swings a coonskin cap by its tail above his head. He is one of the new warehouse workers for Crockett's, a large retail catalog and ecommerce firm. "The catalog slip calls it a 'Coo m5–9w/tl.' Good thing you told me 'Coo' stands for coonskin. Then, of course, I thought about caps and looked over here. I found it here in this bin labeled BOYS/CAP. Wouldn't it be easier if the catalog and web pages matched the bins? To me, this invoice says, 'Cookware, metallic, 5–9-piece set with Teflon.' I've been stranded in the cookware sets the whole time."

Daniel, Davey's coworker, barely listens as he hurriedly pulls items out of bins to fill another order. "You'll get used to it. They've got to have it this way so that the computers can understand the bill later. Mostly, I look at the catalog page number on the invoice, then I look it up in the book and sort of translate it to back here . . . unless I remember it from finding it before," Daniel explains.

Davey persists, saying, "Computers are smart, though, and we have to fill so many orders. We should tell the people up in billing the names we've got on our bins."

Daniel replies cynically, "Oh, sure. They're dying to know what we think." Then he continues in a quieter tone. "You know, we used to have it like that, but when they got all the new computers

and went to 24-hour phone and Internet orders, it all changed. They said the operators (and users) had to know more about what they were selling (or buying), so they changed their codes to be more like a story."

Davey, surprised at Daniel's revelation, asks, "What's the story for the one I was working on?"

Inspecting the code on the cap's invoice, Daniel replies, "The one you were working on was 'Coo m5–9w/tl.' After looking it up real fast on the computer the operator can tell the customer, 'It's a coonskin (Coo) cap for boys (m for male) ages 5–9 with a real tail (w/tl).' We can't see the forest for the trees because of their codes, but you know Crockett's. They've got to make the sale."

How important is it that the warehouse bins and invoices are coded inconsistently? Respond in a paragraph. What are some of the problems created when a code appears to be mnemonic but employees are never given an appropriate "key" to decode it? Discuss your response in two paragraphs. What changes would you make to invoice/warehouse coding for Crockett's? Document your changes, identify the type of code you would use, and use the code in an example of a product that Crockett's might sell. Remember to decipher it as well.

---

long to prevent people from guessing a credit card number. Visa and MasterCard use 16-digit numbers, which would accommodate 9 trillion customers. Because the numbers are not assigned sequentially, chances of guessing a credit card number are very slight.

KEEP THE CODES STABLE. Stability means that the identification code for a customer should not change each time new data are received. Earlier, we presented an alphabetic derivation code for a magazine subscription list. The expiration date was not part of the subscriber identification code because it was likely to change.

Don't change the code abbreviations in a mnemonic system. Once you have chosen the code abbreviations, do not try to revise them, because that makes it extremely difficult for data entry personnel to adapt.

ENSURE THAT CODES ARE UNIQUE. For codes to work, they must be unique. Make a note of all codes used in the system to ensure that you are not assigning the same code number or name to the same items. Code numbers and names are an essential part of the entries in data dictionaries, discussed in Chapter 8.

FIGURE 15.10

There are eight general guidelines for establishing a coding system.

In Establishing a Coding System, the Analyst Should:
Keep codes concise
Keep codes stable
Make codes that are unique
Allow codes to be sortable
Avoid confusing codes
Keep codes uniform
Allow for modification of codes
Make codes meaningful

Incorrect Sorting Using MMM-DD-YYYY	Incorrect Sorting Using MM-DD-YYYY	Incorrect Sorting (Year 2000 Problem) YY-MM-DD	Correct Sorting Using YYYY-MM-DD
Dec-25-1998	06-04-1998	00-06-11	1997-06-12
Dec-31-1997	06-11-2000	97-06-12	1997-12-31
Jul-04-1999	06-12-1997	97-12-31	1998-06-04
Jun-04-1998	07-04-1999	98-06-04	1998-10-24
Jun-11-2000	10-24-1998	98-10-24	1998-12-25
Jun-12-1997	12-25-1998	98-12-25	1999-07-04
Oct-24-1998	12-31-1997	99-07-04	2000-06-11

FIGURE 15.11

Plan ahead in order to be able to do something useful with data that have been entered. In this example, the person creating the codes did not realize the data would have to be sorted.

ALLOW CODES TO BE SORTABLE. If you are going to manipulate the data usefully, the codes must be sortable. For example, if you were to perform a text search on the months of the year in ascending order, the "J" months would be out of order (January, July, and then June). Dictionaries are sorted in this way, one letter at a time from left to right. So, if you sorted MMMDDYYYY where the MMM stood for the abbreviation for the month, DD for the day, and YYYY for the year, the result would be in error.

Figure 15.11 shows what would happen if a text search were performed on different forms of the date. The third column shows a problem that was part of the year 2000 (Y2K) crisis that caused some alarm and even made the cover of *Time* magazine.

It is important to make sure that users can do what you intend them to do with the codes you create. Numeric codes are much easier to sort than alphanumerics; therefore, consider converting to numerics wherever practical.

AVOID CONFUSING CODES. Try to avoid using coding characters that look or sound alike. The characters O (the letter oh) and 0 (the number zero) are easily confused, as are the letter I and the number 1, and the letter Z and the number 2. Therefore, codes such as B1C and 280Z are unsatisfactory.

One example of a potentially confusing code is the Canadian Postal Code, as shown in Figure 15.12. The code format is X9X 9X9, where X stands for a letter and 9 stands for a number. One advantage to using letters in the code is to allow more data in a six-digit code (there are 26 letters, but only 10 numbers). Because the code is used on a regular basis by Canadians, the code makes perfectly good sense to them. To foreigners sending mail to Canada, however, it may be difficult to tell if the second-to-last symbol is a Z or a 2.

KEEP THE CODES UNIFORM. To be effective and efficient for humans, codes need to follow readily perceived forms most of the time. Codes used together, such as BUF-234 and KU-3456, are poor because the first contains three letters and three numbers, whereas the second has only two letters followed by four numbers.

When you are required to add dates, try to avoid using the codes MMDDYYYY in one application, YYYYDDMM in a second, and MMDDYY in a third. It is important to keep codes uniform among, as well as within, programs.

ALLOW FOR MODIFICATION OF CODES. Adaptability is a key feature of a good code. An analyst must keep in mind that the system will evolve over time, and the coding system should be able to encompass change. The number of customers should grow, customers will change names, and suppliers will modify the way they number their products. An analyst needs to be able to forecast the predictable changes that business users will desire and anticipate a wide range of future needs when designing codes.

	Code Format for Canadian Postal Code X9X 9X9		
Handwritten Code	**Actual Code**	**City, Province**	**Problem**
L8S 4M4	L8S 4M4	Hamilton, Ontario	S looks like a 5
T3A ZE5	T3A 2E5	Calgary, Alberta	2 looks like a Z 5 looks like an S
L0S 1J0	L0S 1J0	Niagara-on-the-Lake, Ontario	Zero and Oh look alike S looks like a 5 1 looks like an I

FIGURE 15.12

Combining look-alike characters in codes can result in errors.

Catching a Summer Code

Vicky takes her fingers off her keyboard and bends over her workstation to verify the letters on the invoices stacked in front of her. "What on earth?" Vicky asks aloud as she further scrutinizes the letters that encode cities where orders are to be shipped.

Shelly Overseer, her supervisor, who usually sits a couple of workstations away, is passing by and sees Vicky's consternation. "What's the matter? Did the sales rep forget to write in the city code again?"

Vicky swings around in her chair to face Shelly. "No, there are codes here, but they're weird. We usually use a three-letter code, right? Like CIN for Cincinnati, SEA for Seattle, MIN for Minneapolis, BUF for Buffalo. They're all *five*-letter codes here, though."

"Look," Vicky says, lifting the invoice to show Shelly. "CINNC, SEATT, MINNE. It'll take me all day to enter these. No kidding, it's really slowing me down. Maybe there's a mistake. Can't I just use the standard?"

Shelly backs away from Vicky's workstation as if the problem were contagious. Excusing herself apologetically, Shelly says, "It's the part-timers. They are learning sales now, and management was worried that they'd get messed up on their cities. I think it has something to do with mixing up Newark and New Orleans on the last orders. So, a committee decided to make the cities more recognizable by having them add two letters. Those kids can't learn everything we know overnight, even though they try. It's just until August 19, though, when the part-timers go back to school."

As Vicky glumly turns back to her keyboard, Shelly shakes her head sympathetically and says, "I know it's a strain and it's making you feel miserable, but don't worry. You'll get over it. It's just a summer code."

What general guidelines of coding has management overlooked in its decision to use a summer code for cities? Make a list of them. What is the effect on full-time data entry personnel of changing codes for the ease of temporary help? Respond in two paragraphs. What future impact could the temporary change in codes have on sorting and retrieving data entered during the summer period? Take two paragraphs to discuss these implications. What changes can you suggest so that the part-timers don't get mixed up on codes in the short term? In a memo to the supervisor of this work group, make a list of five to seven changes in the data capture or data entry procedures that can be made to accommodate short-term hires without disrupting normal business. In a paragraph, indicate how this goal can be accomplished without marring the productivity of data entry personnel.

MAKE CODES MEANINGFUL. Unless the analyst wants to hide information intentionally, codes should be meaningful. Effective codes not only contain information, but they also make sense to the people using them. Meaningful codes are easy to understand, work with, and recall. The job of data entry becomes more interesting when working with meaningful codes instead of just entering a series of meaningless numbers.

USING CODES. Codes are used in a number of ways. In validation programs, input data is checked against a list of codes to ensure that only valid codes have been entered. In report and inquiry programs, a code stored on a file is transformed into the meaning of the code. Reports and displays should not show or print the actual code. If they did, the user would have to memorize code meanings or look them up in a manual. Codes are used in GUI programs to create drop-down lists.

Effective and Efficient Data Capture

To ensure the quality of data that users enter into a system, it is important to capture data effectively. Data capture has received increasing attention as the point in information processing at which excellent productivity gains can be made. Great progress in improving data capture has been made in the last four decades, as we have moved from multiple-step, slow, and error-prone systems such as keypunching to using sophisticated systems including such things as optical character recognition (OCR), bar codes, and point-of-sale terminals.

Deciding What to Capture

The decision about what to capture precedes user interaction with the system. Indeed, it is vital in making the eventual interface worthwhile, for the adage "garbage in, garbage out" is still true.

Decisions about what data to capture for system input are made among systems analysts and systems users. Much of what will be captured is specific to the particular business. Capturing data, inputting them, storing them, and retrieving them are all costly endeavors. With all these factors in mind, determining what to capture becomes an important decision.

There are two types of data to enter: data that *change* or *vary* with every transaction, and data that concisely *differentiate* the particular item being processed from all other items.

An example of changeable data is the quantity of supplies purchased each time an advertising firm places an order with the office supply wholesaler. Because quantities change depending on the number of employees at the advertising firm and on how many accounts they are servicing, quantity data must be entered each time an order is placed.

An example of differentiation data is the inclusion on a patient record of the patient's Social Security number and the first three letters of his or her last name. In this way, the patient is uniquely differentiated from other patients in the same system.

Letting the Computer Do the Rest

When considering what data to capture for each transaction and what data to leave to the system to enter, a systems analyst must take advantage of what computers do best. In the preceding example of the advertising agency ordering office supplies, it is not necessary for the operator entering the stationary order to reenter each item description each time an order is received. The computer can store and access this information easily.

Computers can automatically handle repetitive tasks, such as recording the time of the transaction, calculating new values from input, and storing and retrieving data on demand. By employing the best features of computers, efficient data capture design avoids needless data entry, which in turn alleviates much human error and boredom, and permits people to focus on higher-level or creative tasks. Software can be written to ask the user to enter today's date or capture the date from the computer's internal clock. Once entered, the system proceeds to use that date on all transactions processed in that data entry session.

A prime example of reusing data entered once is the Online Computer Library Center (OCLC), used by thousands of libraries in the United States. OCLC was built on the idea that each item bought by a library should have to be cataloged only once for all time. Once an item is entered, cataloging information goes into the huge OCLC database and is shared with participating libraries. In this case, implementation of the simple concept of entering data only once has saved enormous data entry time.

The calculating power of the computer should also be taken into account when deciding what *not* to reenter. Computers are adept at long calculations, using data already entered.

For example, a person doing data entry may enter the flight numbers and account number of an air trip taken by a customer belonging to a frequent-flyer incentive program. The computer then calculates the number of miles accrued for each flight, adds it to the miles already in the customer's account, and updates the total miles accrued to the account. The computer may also flag an account that, by virtue of the large number of miles flown, is now eligible for an award. Although all this information may appear on the customer's updated account, the only new data entered were the flight numbers of the flights flown.

In systems that use a graphical user interface (GUI), codes are often stored either as a function or as a separate table in the database. There is a trade-off on creating too many tables because the software must find matching records from each table, which may lead to slow access. If the codes are relatively stable and rarely change, they may be stored as a database function. If the codes change frequently, they are stored on a table so that they may be easily updated.

Figure 15.13 shows how a drop-down list is used to select the codes for adding or changing a record in the CUSTOMER table. Notice that the code is stored, but the drop-down list displays both the code and the code meaning. This method helps to ensure accuracy, because the user does not have to guess at the meaning of the code and there is no chance of typing an invalid code.

Avoiding Bottlenecks and Extra Steps

A bottleneck in data entry is an apt allusion to the physical appearance of a bottle. Data are poured rapidly into the wide mouth of the system only to be slowed in its "neck" because of an artificially created instance of insufficient processing for the volume or detail of the data being entered. One way a bottleneck can be avoided is by ensuring that there is enough capacity to handle the data that are being entered.

FIGURE 15.13

A table of codes used in a drop-down list shown in Microsoft Access. This list is used to select a code for adding or changing an item in a record.

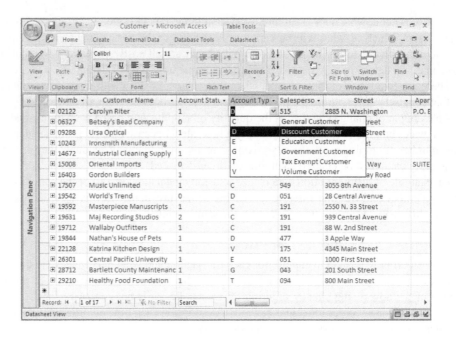

Ways to avoid extra steps are determined not only at the time of analysis, but also when users begin to interact with prototypes of the system. The fewer steps involved in inputting data, the fewer chances there are for the introduction of errors. So, beyond the obvious consideration of saved labor, avoiding extra steps is also a way to preserve the quality of data. Once again, use of an online, real-time system that captures customer data without necessitating the completion of a form is an excellent example of saving steps in data entry.

Starting with a Good Form

Effective data capture is achievable only if prior thought is given to what the source document should contain. A data entry operator inputs data from the source document (usually some kind of form); this document is the source of a large amount of all system data. Online systems (or special data entry methods such as bar codes) may circumvent the need for a source document, but often some kind of paper form, such as a receipt, is created anyway.

With effective forms, it is not necessary to reenter information that the computer has already stored, or data such as time or date of entry that the computer can determine automatically. Chapter 12 discussed in detail how a form or source document should be designed to maximize its usefulness for capturing data and to minimize the time users need to spend entering data from it.

Choosing a Data Entry Method

Several efficient data entry methods are available, and choosing one of them is shaped by many factors, including the need for speed, accuracy, and user training; the cost of the data entry method (whether it is materials or labor intensive); and the methods currently in use in the organization.

KEYBOARDS. Keyboarding is the oldest method of data entry, and certainly it is the one with which organizational members are the most familiar. Some improvements have been made over the years to improve keyboards. Features include special function keys to open programs, keys used to scroll and explore the Web, and keys that can be programmed with macros to reduce the number of keystrokes required. Ergonomic keyboards and infrared or Bluetooth-enabled keyboards and mice are big improvements, as well.

BAR CODES. Bar codes typically appear on product labels, and they also appear on patient identification bracelets in hospitals and in almost any context in which a person or object needs to be checked into and out of any kind of inventory system. A bar-coded label, such as the one shown in Figure 15.14, includes the following elements of coding for a particular product: the manufacturer identification number, the product identification number, a code to verify the scan's accuracy, and codes to mark the beginning and end of the scan.

ISBN 4324234329

FIGURE 15.14

One dimensional bar codes found on most products allow quick and highly accurate data entry. (Pastushenko Taras / Shutterstock. com)

Bar codes can be thought of as metacodes, or codes encoding codes, because they appear as a series of bars or objects that encodes numbers or letters. These symbols in turn have access to product data stored in computer memory. Until recently, scanners were needed to read linear codes, or one-dimensional codes, but today the camera in mobile phones or the camera built into your iMac has become capable of reading UPC codes. An inexpensive app such as Delicious Library (www.delicious-monster.com) can be used to scan in one-dimensional bar codes and create a database of all of your books, movies, music, software, or games.

Instead of one-dimensional or linear bar codes, matrix bar code squares are two-dimensional bar codes. They are still optical, machine-readable codes, and they take up much less space than the older linear bar codes. They are cheaper than RFID tags (discussed in an upcoming section) and can appear in print. There are more than 30 different types of these bar codes in existence. These different codes are known generally as 2D bar codes.

Examples of the two most popular codes, the QR code and the Microsoft Tag, can be seen in Figure 15.15. Users can get apps for their mobile phones that will read these codes and direct them to their respective web pages.

QR CODES. QR codes were first created in 1994 by DENSO WAVE, which was a Toyota subsidiary at that time. QR codes have been used in Japan and South Korea for some time. They are beginning to show up in the United States now and have the potential of becoming the dominant 2D bar code in the world.

A QR (Quick Response) code is easily identifiable by the position marker (which looks like three nested squares) that appears in three of its corners. QR codes are completely free, since the license holder has chosen not to exercise their rights

2D code readers feature a couple of codes, but all (with the exception of the Microsoft Tag reader, explained in a later section) are likely to be able to read a QR code. You can easily find more QR creators available.

Rutgers School of Law–Camden is a leading national law school. Its Web and Information Technology development team is generating QR codes for all school events and electronically sending them to the school's digital signage system. Each code provides the user with a link to the event's web page, giving users a way to easily view event data, register for an event, and add the event to a mobile calendar.

FIGURE 15.15

Two popular 2D bar codes.

QR codes are also being placed outside student study rooms, allowing students to self-schedule reservations and activate video recording capabilities. Study room recordings of practice sessions are then sent to the student's mobile device and can be viewed instantly.

MICROSOFT TAGS. Microsoft Tags, formerly called High Capacity Color Barcodes (HCCBs), now use four colors, making the codes more aesthetically appealing then the black-and–white QR codes. Microsoft tags work if they are printed monochromatically, but Microsoft discourages doing so. So far, creating, publishing, and using Microsoft Tags is free.

Microsoft Tags differ from other 2D codes in another way. Suppose that you wanted to create a code directing users to your website. The Microsoft Tag system requires that a code be read by a server. Therefore, the code is really just a Web link. Once the code is read and the reader app reaches the server, it is redirected to the intended website. This redirection has the advantage of collecting and storing data that can be subjected to sophisticated analytics. It is possible to get information about a Microsoft Tag's usage from the Microsoft Tag site. A drawback to this redirection is that there is only one reader app—the official one from Microsoft. Of course, some may point out that then there is no confusion about the quality of the reader.

Recently one of the leaders in quality winemaking in the United States, Shafer Vineyards of Napa, California, began to place Microsoft Tags in its customer literature. The marketing design team liked the smaller size of the Microsoft Tag and they praised the easy signup and creation processes, as well as the basic tracking of responses that it supports. Shafer Vineyards has since changed to the more widely used QR codes and has started to print the QR code on their bottles. Two-dimensional codes are perfect tools to create a bridge between their print and Web-based efforts. The 2D codes also support links to the company's online store.

Two-dimensional bar codes can be used in imaginative ways. Figure 15.16 lists some ideas.

There are many advantages to incorporating 2D codes into your designs. Almost everyone in business already owns and uses a device (such as a smartphone) that can read different types of bar codes. The 2D codes are either nonproprietary or free. Bar coding affords an extraordinarily high degree of accuracy for data entry. Furthermore, 2D codes can be generated easily by using a Web-based creator or a stand-alone app.

One of the disadvantages of 2D codes is the ease with which a printed code can be stickered over or tampered with. There is growing concern that users of smartphones are vulnerable to malicious attacks launched when an unsuspecting user scans QR codes indiscriminately without checking whether the QR code is a genuine, authorized one. Some security software that is in beta testing will permit Android or iOS users to scan a code and find out instantly if it is malicious before it is loaded into the user's browser. However, at this point, many users and even designers are unaware of the potential risks of bar codes.

RFID. Commonly known as RFID, radio frequency identification allows the automatic collection of data. It uses RFID tags or transponders, each of which contains a chip and an antenna. An RFID tag may or may not have its own power source. If it does not have its own power, the antenna provides just enough power from an incoming signal to power the chip and transmit a response. RFID tags can be attached to products, packages, animals, and even humans so that the item or person can be identified using a radio frequency.

FIGURE 15.16

Two dimensional bar codes can be used in many ways.

QR Codes and Microsoft Tags Can

Take you to a specific web page

Find the item you want for purchase at a lower price

Initiate an email, SMS text message, or make a phone call

Find and capture a digital coupon

Use the digital coupon

Tell you more about the art gallery painting you are gazing at

Get you information about the time and place of an event

Show you a promotional video for a movie, play, or concert

Allow you to view or extend the amount of money on a transit card

RFID tags, also called proximity cards because of their limited range, can be passive or active. Passive RFID tags have no internal power source; active tags do. Passive tags are inexpensive (less than 5¢ per tag) and are typically the size of a postage stamp. They are used in large retail stores, including Wal-Mart and Target. Wal-Mart has been actively pursuing RFID technology in improving its inventory management and supply chain processes.

Active tags are much more reliable than passive ones because they have their own power supply. The U.S. Department of Defense has used these tags to minimize the costs related to logistics and increase supply chain visibility. Active tags cost a few dollars each.

To capture the data on an RFID tag, a reader is required. The reader activates the tag so that it can be read. The reader decodes the data and the unique product code on the chip inside the tag and then passes it along to a host computer that processes the data.

One example is an electronic toll pass used in vehicles traversing toll roads. An RFID transponder can be attached to the windshield and read every time the vehicle passes through a toll booth. The toll booth's RFID reader can also act as a writer to store a balance on the RFID chip.

The Moscow Metro was the first transportation system to use RFID smartcards in 1998. Other applications include the tracking of cattle to identify the herd of origin, which enables better tracking of mad cow disease, as well as RFID tracking in bookstores, airline baggage services, pharmaceuticals, and even tracking of patients or inmates.

RFID tags have found common use in most shipping applications. The technology will soon be used in general electronic cash transactions. They may even replace UPC codes since their advantages include security (by reducing the number of items stolen) and not requiring scanning (they can simply pass through the reader zone).

RFID is not without controversy. Privacy is a concern. An individual who pays for a tagged item by using a credit card or a shoppers' card could be identifiable with such a system.

A systems analyst needs to think of the users involved and their rights when considering whether this technology is suited for an application.

NEAR FIELD COMMUNICATION (NFC). Another approach to inputting data is near field communication (NFC), which builds upon RFID. NFC allows two-way communication and is used in contactless payment systems at checkout counters and other venues. Customers simply use a smartphone to touch or come close to the point-of-sale device in order to establish radio communications.

NFC can also be used for payment on transportation systems, to exchange information (such as a schedule or map), receive custom coupons, or exchange business cards. It is considered secure because of its short range of communication.

OPTICAL CHARACTER RECOGNITION. Optical character recognition (OCR) lets a user read input from a source document with an optical scanner rather than off the magnetic media we have been discussing so far. Using OCR devices can speed data input from 60 to 90 percent over some keying methods.

The increased speed of OCR comes through not having to encode or key in data from source documents. It eliminates many of the time-consuming and error-fraught steps of other input devices. In doing so, OCR demands few employee skills and commensurately less training, resulting in fewer errors and less time spent by employees in redundant efforts. It also decentralizes responsibility for quality data directly to the unit that is generating it. OCR, which has become available to all, has one additional, highly practical use: the transformation of faxes into documents that can be edited.

MAGNETIC INK CHARACTER RECOGNITION. Magnetic ink characters are found on the bottom of bank checks and some credit card bills. This method is akin to OCR in that special characters are read, but its use is limited. Data entry through magnetic ink character recognition (MICR) is done through a machine that reads and interprets a single line of material encoded with ink that is made up of magnetic particles.

Some advantages of using MICR are (1) it is a reliable and high-speed method that is not susceptible to accepting stray marks (because they are not encoded magnetically); (2) if it is required on all withdrawal checks, it serves as a security measure against bad checks; and (3) data entry personnel can see the numbers making up the code if it is necessary to verify it.

MARK-SENSE FORMS. Mark-sense forms allow data entry through the use of a scanner that senses where marks have been made on special forms. A common usage is for scoring answer sheets

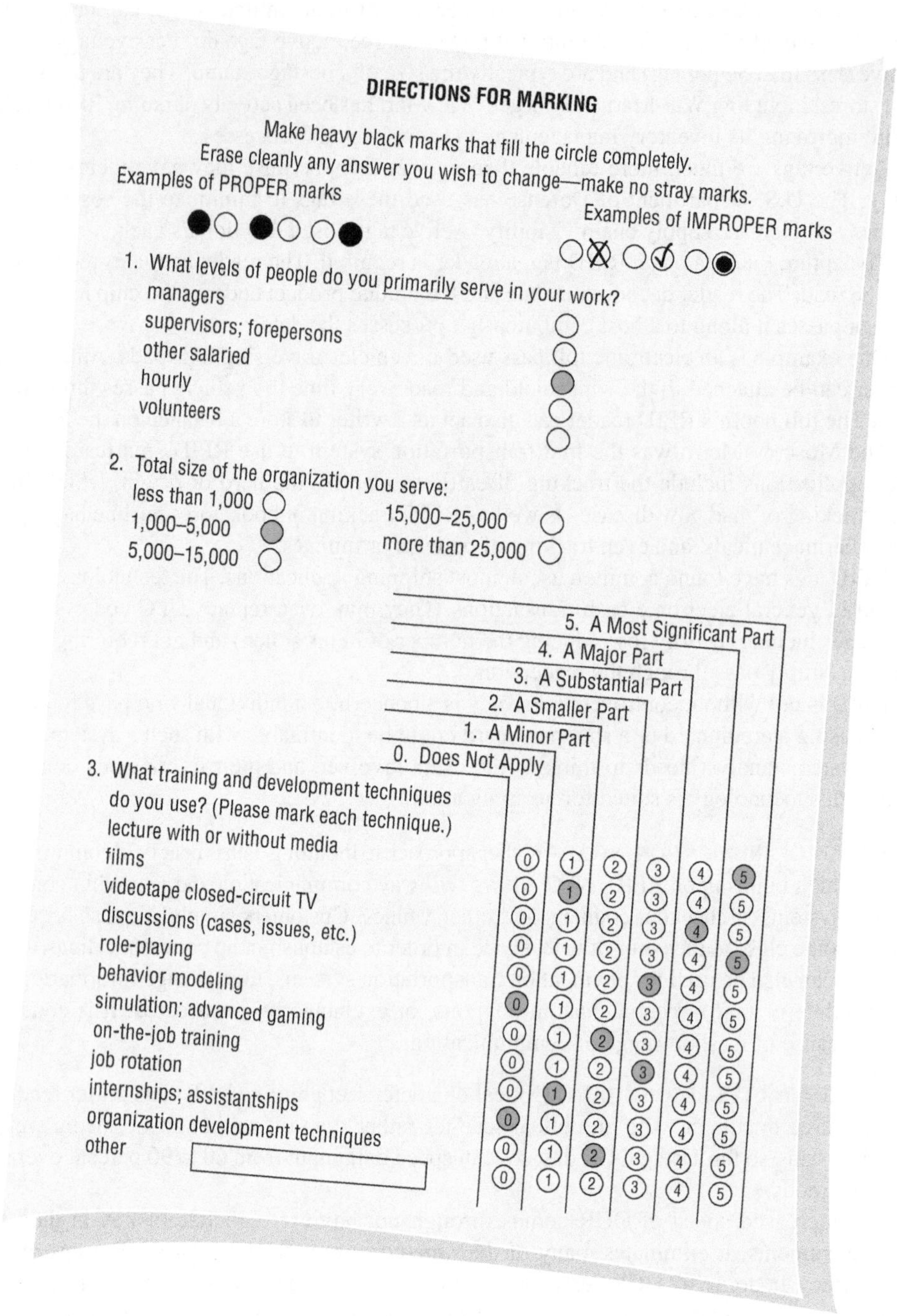
DIRECTIONS FOR MARKING

Make heavy black marks that fill the circle completely.
Erase cleanly any answer you wish to change—make no stray marks.

Examples of PROPER marks

Examples of IMPROPER marks

1. What levels of people do you primarily serve in your work?
managers
supervisors; forepersons
other salaried
hourly
volunteers

2. Total size of the organization you serve:
less than 1,000
1,000–5,000
5,000–15,000
15,000–25,000
more than 25,000

5. A Most Significant Part
4. A Major Part
3. A Substantial Part
2. A Smaller Part
1. A Minor Part
0. Does Not Apply

3. What training and development techniques do you use? (Please mark each technique.)
lecture with or without media
films
videotape closed-circuit TV
discussions (cases, issues, etc.)
role-playing
behavior modeling
simulation; advanced gaming
on-the-job training
job rotation
internships; assistantships
organization development techniques
other

FIGURE 15.17

A mark-sense form that can be read by a scanner speeds data entry.

for questionnaires, as shown in Figure 15.17. Little training of entry personnel is necessary, and a high volume of forms can be processed quickly.

One drawback of mark-sense forms is that although the readers can determine whether a mark has been made, they cannot interpret the mark in the way that optical character readers do. Stray marks on forms can thus be entered as incorrect data. In addition, choices are limited to the answers provided on the mark-sense form, forms have difficulty in capturing alphanumeric data because of the space required for a complete set of letters and numbers, and it is easy for those filling out mark-sense forms to get confused and put a mark in an incorrect position.

Ensuring Data Quality Through Input Validation

So far, we have discussed ensuring the effective capture of data onto source documents and the data's efficient entry into the system through various input devices. Although these conditions are necessary for ensuring quality data, they alone are not sufficient.

CONSULTING OPPORTUNITY 15.3

To Enter or Not to Enter: That Is the Question

"I've just taken on the presidency of Elsinore Industries," says Rose N. Krantz. "We're actually part of a small cottage industry that manufactures toy villages for children seven years old and up. Our tiny hamlets consist of various kits that will build what children want from interlocking plastic cubes, essentials such as city hall, the police station, the gas station, and a hot dog stand. Each kit has a unique part number from 200 to 800, but not every number is used. The wholesale prices vary from $54.95 for the city hall to $1.79 for a hot dog stand.

"I've been melancholy over what I've found out since signing on at Elsinore. 'Something is rotten' here, to quote a famous playwright. In fact, the invoicing system was so out of control that I've been working around the clock with our bookkeeper, Gilda Stern," Krantz soliloquizes.

"I would like you to help straighten things out," Rose continues. "We ship to 12 distribution warehouses around the country. Each invoice we write out includes the warehouse number, 1 through 12, its street address, and the U.S. postal code (zip code). We also put on each invoice the date we fill the order, code

numbers for the hamlet kits they order, a description of each kit, the price per item, and the quantity of each kit ordered. Of course, we also include the subtotals of kit charges, shipping charges, and the total that the warehouse owes us. No sales tax is added, because they resell what we send them to toy stores in all 50 states. I want you to help us design a computerized order entry system that will be part of the invoicing system for Elsinore Industries."

For your design of a data entry system for Elsinore, take into consideration all the objectives for data entry discussed throughout this chapter. Draw any displays necessary to illustrate your design. How can you make the order entry system efficient? Respond in a paragraph. Specify what data can be stored and retrieved and what data must be entered anew for each order. How can unnecessary work be avoided? Write a paragraph to explain why the system you propose is more efficient than the old one. How can data accuracy be ensured? List three strategies that will work with the type of data that are being entered for Elsinore Industries.

Errors cannot be ruled out entirely, and the critical importance of catching errors during input, *prior* to processing and storage, cannot be overemphasized. The snarl of problems created by incorrect input can be a nightmare, not the least of which is that many problems take a long time to surface. The systems analyst must assume that errors in data *will* occur and must work with users to design input validation tests to prevent erroneous data from being processed and stored, because initial errors that go undiscovered for long periods are expensive and time-consuming to correct.

You cannot imagine everything that will go awry with input, but you must cover the kinds of errors that give rise to the largest percentage of problems. A summary of potential problems that must be considered when validating input is given in Figure 15.18.

This Type of Validation	Can Prevent These Problems
Validating Input Transactions	Submitting the wrong data Data submitted by an unauthorized person Asking the system to perform an unacceptable function
Validating Input Data	Missing data Incorrect field length Data have unacceptable composition Data are out of range Data are invalid Data do not match with stored data

FIGURE 15.18

Validating input is important to ensure that most potential problems with data are eliminated early.

Validating Input Transactions

Validating input transactions is largely done through software, which is a programmer's responsibility, but it is important that a systems analyst know what common problems might invalidate a transaction. Businesses committed to quality will include validity checks as part of their routine software.

Three main problems can occur with input transactions: submitting the wrong data to the system, an unauthorized person submitting data, or asking the system to perform an unacceptable function.

SUBMITTING THE WRONG DATA. An example of submitting the wrong data to the system is the attempt to input a patient's Social Security number into a hospital's payroll system. This error is usually an accidental one, but it should be flagged before data are processed.

AN UNAUTHORIZED PERSON SUBMITTING DATA. A system should be able to discover if otherwise correct data are submitted by an unauthorized person. For instance, only the supervising pharmacist should be able to enter inventory totals for controlled substances in the pharmacy. Invalidation of transactions submitted by an unauthorized individual applies to privacy and security concerns surrounding payroll systems and employee evaluation records that determine pay levels, promotions, or discipline; files containing trade secrets; and files holding classified information, such as national defense data.

ASKING THE SYSTEM TO PERFORM AN UNACCEPTABLE FUNCTION. The third error that invalidates input transactions is asking the system to perform an unacceptable function. For instance, it would be logical for a human resources manager to update the existing record of a current employee, but it would be invalid to ask the system to create a new file rather than merely to update an existing record.

Validating Input Data

It is essential that input data themselves, along with the transactions requested, be valid. Several tests can be incorporated into software to ensure this validity. In the following subsections, we consider a number of possible ways to validate input.

TEST FOR MISSING DATA. The first kind of validity test examines data to see if there are any missing items. For some situations, *all* data items must be present. For example, a Social Security file for paying out retirement or disability benefits would be invalid if it did not include the payee's Social Security number.

In addition, the record should include both the key data that distinguish one record from all others and the function code telling the computer what to do with the data. A systems analyst needs to interact with users to determine what data items are essential and to find out whether exceptional cases ever occur that would allow data to be considered valid even if some data items were missing. For example, a second address line containing an apartment number or a person's middle initial may not be a required entry.

TEST FOR CORRECT FIELD LENGTH. A second kind of validity test checks input to ensure it is of the correct length for the field. For example, if the Omaha, Nebraska, weather station reports into the national weather service computer but mistakenly provides a two-letter city code (OM) instead of the national three-letter city code (OMA), the input data might be deemed invalid, and hence would not be processed.

TEST FOR CLASS OR COMPOSITION. The test for class or composition validity checks to see that data fields that are supposed to be exclusively composed of numbers do not include letters, and vice versa. For example, a credit card account number for American Express should not include any letters. Using a composition test, the program should not accept an American Express account number that includes both letters and numbers.

TEST FOR RANGE OR REASONABLENESS. Validity tests for range or reasonableness are really common-sense measures of input that answer the question of whether data fall within an acceptable range or whether they are reasonable within predetermined parameters. For instance, if a user were trying to verify a proposed shipment date, the range test would neither permit a shipping date on the 32nd day of October nor accept shipment in the 13th month, the respective ranges being 1 to 31 days and 1 to 12 months.

A reasonableness test ascertains whether the item makes sense for the transaction. For example, when adding a new employee to the payroll, entering an age of 120 years would not be reasonable. Reasonableness tests are used for data that are continuous, that is, data that have a smooth range of values. These tests can include a lower limit, an upper limit, or both a lower and an upper limit.

TEST FOR INVALID VALUES. Checking input for invalid values works if there are only a few valid values. This test is not feasible for situations in which values are neither restricted nor predictable. This kind of test is useful for checking responses where data are divided into a limited number of classes. For example, a brokerage firm divides accounts into three classes only: class 1 = active account, class 2 = inactive account, and class 3 = closed account. If data are assigned to any other class through an error, the values are invalid. Value checks are usually performed for discrete data, which are data that have only certain values. If there are many values, they are usually stored in a table of codes file. Having the values in a file provides an easy way to add or change values.

CROSS-REFERENCE CHECKS. Cross-reference checks are used when one element has a relationship with another one. To perform a cross-reference check, each field must be correct in itself. For example, the price for which an item is sold should be greater than the cost paid for the item. Price must be entered, numeric, and greater than zero. The same criterion is used to validate cost. When both price and cost are valid, they may be compared.

A geographical check is another type of cross-reference check. In the United States, the state abbreviation may be used to ensure that a telephone area code is valid for that state and that the first two digits of the zip code are valid for the state.

TEST FOR COMPARISON WITH STORED DATA. The next test for validity of input data that we consider is one comparing it with data that the computer has already stored. For example, a newly entered part number can be compared with the complete parts inventory to ensure that the number exists and is being entered correctly.

SETTING UP SELF-VALIDATING CODES (CHECK DIGITS). Another method for ensuring the accuracy of data, particularly identification numbers, is to use a check digit in the code itself. Suppose we had the part number 53411 that needed to be typed into the system. While that is being entered, different types of errors can occur. One possible error is a mis-keying a single digit; for example, the clerk types in 54411 instead of 53411. A second type of error is transposed digits; it commonly occurs when the intended number 53411 gets typed in as number 54311 instead, just because two keys are pressed in reverse order. Transposition errors are also difficult for humans to detect.

The procedure for using a check digit involves beginning with an original numeric code, performing some mathematics to arrive at a derived check digit, and then adding the check digit to the original code. The mathematical process involves multiplying each of the digits in the original code by some predetermined weights, summing these results, and then dividing this sum by a modulus number. The modulus number is needed because the sum usually is a large number, and we need to reduce the result to a single digit. Finally, the remainder is subtracted from the modulus number, giving us the check digit. A well-known formula for check digits is the Luhn formula created in the 1960s. It is used by credit card companies as described below.

VERIFYING CREDIT CARDS. When credit cards are entered into a website or computer program, the first check is the length of the number. Credit card companies designed their cards to include a different number of digits. For example, Visa cards are 16 digits long, while American Express card numbers are 15 digits long.

Another test is to match the credit card company and bank to verify that it is indeed a card issued by that company. The first four digits usually signify the type of card. The middle digits usually represent the bank and the customer. The last digit is a check digit.

In addition to these verification methods, the Luhn formula can be used. Suppose we are given a number 7-7-7-8-8-8, where the first five numbers represent a bank account number and the last digit is a check digit. Let's apply the Luhn formula to see whether this is a valid number:

1. Double the second-last digit, then double every other digit (that is, skip a digit, double the next, skip a digit, double the next, etc.). For example, the number 7-7-7-8-8-8 becomes 14-7-14-8-16-8.
2. If doubling any digit results in a number that is larger than 10, reduce this two-digit number to a single digit by adding the numbers together. In our example, the 14 becomes $1 + 4 = 5$ and the 16 becomes $1 + 6 = 7$. In doing so, our original number, 7-7-7-8-8-8 has been transformed into a new number, 5-7-5-8-7-8.
3. Now add all the digits in the new number together. So, $5 + 7 + 5 + 8 + 7 + 8 = 40$.
4. Look at the total. If it ends in zero, the number is valid according to the Luhn formula. Since 40 ends in zero, we can say that it passes the Luhn formula test.

The Luhn formula can be used to identify mistakes in entering an incorrect credit card. For example, the credit card number 1334-1334-1334-1334 is assumed to be valid because the digits of the transformed number 2364-2364-2364-2364 will add up to 60, a number ending in zero. If a user incorrectly enters a wrong digit, the total would not be a multiple of zero.

The Luhn formula does not catch every error, however. If a user makes mistakes in entering more than one digit, for example entering 1334-1334-1334-3314, the total of the transformed number, 2364-2364-2364-6324, is still 60. This transposition error (flipping the second-last and fourth-last digits) will not be caught.

Credit card companies also use the expiration date and a three- or four-digit verification code, often written on the reverse side of the card for more security.

The tests for checking on validity of input can go a long way toward protecting a system from the entry and storage of erroneous data. As a systems analyst, you should always assume that human errors in input are more likely than not to occur. It is your responsibility to understand which errors will invalidate data and how to use a computer to guard against those human errors and thus limit their intrusion into system data.

The Process of Validation

It is important to validate each field until it is either valid or an error has been detected. The order of testing data is to first check for missing data. Then a syntax test can check the length of the data entered and check for proper class and composition. Only after the syntax is correct are the semantics, or meaning, of the data validated. This includes a range, reasonable, or value test, followed by a check digit test.

GUI screens help reduce the number of human input errors when they incorporate radio buttons, check boxes, and drop-down lists. When radio buttons are used, one should be set as the default, and the only way it can be unchecked is if the user clicks a different radio button. In the case of drop-down lists, the first choice should contain a message informing the user to change the list. If the first choice is still selected when the form is submitted, a message should inform the user to select a different option.

Usually validating a single field is done with a series of IF . . . ELSE statements, but there are also pattern validation methods. Usually these patterns are found in the database design (as in Microsoft Access) but may be included in programming languages, such as Perl, JavaScript, and XML schemas. The patterns are called regular expressions and contain symbols that represent the type of data that must be present in a field. Figure 15.19 illustrates characters used in JavaScript regular expressions.

An example of pattern validation used to test an email address is

[A-Za-z0-9]\w{2,}@[A-Za-z0-9]{3,}\.[A-Za-z]{3}/

The meaning of this pattern is as follows: The first letter must be any uppercase letter, lowercase letter, or number ([A-Za-z0-9]). This is followed by two or more characters that are any letter, number, or an underscore (\w{2,}). There must then be an @ symbol, followed by at least three letters or numbers, a period, and exactly three characters after the period.

A cross-reference check assumes that the validity of one field may depend on the value of another field. An example of a cross-reference check is checking for a valid date. In one very special case, the validity of the day of the month depends on the year. That is, February 29 is valid only during leap years. Once single fields have been checked, you can perform cross-reference checks. Obviously, if one of the fields is incorrect, the cross-reference check is meaningless and should not be performed.

Character Code	Meaning Used in Regular Expression Validation
\d	Any digit 0–9
\D	Any nondigit character
\w	Any letter, number, or underscore
\W	Any character other than a letter, number, or underscore
.	Matches any character
[characters]	Matches the characters in the brackets
[char-char]	Matches the range of characters
[a–z][A–Z][0–9]	Will accept any letter or digit
[^characters]	Match anything other than the characters
[^char-char]	Match anything outside the range of characters
[^a–z]	Will accept anything except lowercase letters
{n}	Match exactly n occurrences of the preceding character
{n,}	Match at least n occurrences of the character
\s	Any white space formatting character (tab, new line, return, etc.)
\S	Any nonwhite space character

FIGURE 15.19

These characters are used in regular expression (pattern) validation.

XML documents may be validated by comparing them to a document type definition (DTD) or a schema (refer to Chapter 8). The DTD will check to see whether the format of the document is valid, but a schema is much more powerful and will check the type of data, such as a short or long integer, a decimal number, or date. A schema will also check a range of values, the number of digits to the left and right of a decimal point, and the values of codes. There are free tools to validate a DTD or schema. IEXMLTLS is a Microsoft extension to Internet Explorer that adds new menu options when the user right-clicks in an XML document.

Data Accuracy Advantages in Ecommerce Environments

One of the many bonuses of ecommerce transactions is increased accuracy of data, due to four reasons:

1. Customers generally key or enter data themselves.
2. Data entered by customers are stored for later use.
3. Data entered at the point of sale are reused throughout the order fulfillment process.
4. Information is used as feedback to customers.

An analyst needs to be aware of the advantages that have resulted from ecommerce and the electronic capture and use of information.

Customers Keying Their Own Data

First, customers know their own information better than anyone else. They know how to spell their street address, they know whether they live on a "Drive" or a "Street," and they know their own area code. If this information is transmitted by phone, it is easier to make a mistake spelling the address; if it is entered by using a faxed paper form, mistakes can occur if the fax transmission is difficult to read. If users enter their own information, however, accuracy increases.

Storing Data for Later Use

After customers enter information at an ecommerce site, it may be stored on their own personal computers. If they visit ecommerce sites and fill out similar forms to complete a transaction, they will experience the advantage of storing this information. As they begin to type their name, dropdown lists will prompt them with their full name even though only a couple of characters were

HYPERCASE® EXPERIENCE 15

"Sometimes I think I'm the luckiest person on earth. Even though I've been here five years, and away from the UK all this time, I still enjoy the people I meet and what I do. Yes, I know Snowden's demanding. You've experienced some of that, haven't you? He, for one, loves codes. I think they are a pain. I always forget them or try to make up new ones or something. Some of the physicians, though, think they're great. It must be all those Latin abbreviations they studied in med school. I hear that your most pressing assignment this week has to do with actually getting the information into the project reporting system. The Training group wants your ideas, and it wants them fast. Good luck with it. Oh, and when Snowden gets back from Thailand, I'm certain he'll want to take a peek at what your team has been up to."

HYPERCASE Questions

1. Using a CASE tool, a software package such as Microsoft Access, or a paper layout form, design a data entry procedure for the proposed project reporting system for the Training group. Assume that we are particularly concerned about the consulting physicians staff, who don't want to spend a great deal of time keying in large amounts of data when using the system.
2. Test your data entry procedure on three teammates. Ask for feedback concerning the appropriateness of the procedure, given the type of users the system will have.
3. Redesign the data entry procedure to include the feedback you have received. Explain in a paragraph how your changes reflect the comments you were given.

entered. By clicking on this prompt, the full name is entered, and no further typing is necessary for this field. This is called autocomplete or auto-suggestion.

Companies that want to store information to enable faster and more accurate transactions do so in small files called cookies. Once information is stored in a cookie, a website can suggest matches for credit card and password information as well, and this information is encrypted so that websites cannot read the information stored on the user's computer. Personal information can only be accessed by the company that placed the cookie on the user's computer.

Using Data through the Order Fulfillment Process

When companies capture information from a customer order, they can use and reuse that information throughout the order fulfillment process. Hence, the information gathered to complete an order can also be used to send an invoice to a customer, obtain the product from the warehouse, ship the product, send feedback to the customer, and restock the product by notifying the manufacturer. It can also be used again to send a paper catalog to the customer or send a special offer by email.

These ecommerce enhancements replace the traditional approach of using a paper-based procurement process with purchase orders sent via fax or mail. This electronic process not only speeds up the delivery of the product but also increases accuracy so that the product is delivered to the correct address. Rather than reading a fax or a mailed-in form, a shipper uses the more accurate electronic version of the data. Electronic information allows better supply chain management, including checking product and resource availability electronically, and automating planning, scheduling, and forecasting.

Providing Feedback to Customers

Confirmations and order status updates are ways to enhance feedback to customers. If a customer receives confirmation of a mistake in an order just placed, the order can be corrected immediately. For example, suppose a customer mistakenly submits an order for two copies of a DVD rather than one. After submitting the order, the customer receives an email confirming the order. The customer notices the mistake, immediately contacts the company, and has the order corrected, thereby avoiding having to return the extra copy of the DVD. Accuracy is improved by better feedback.

Summary

Ensuring the quality of the data input to an information system is critical to ensuring quality output. The quality of data entered can be improved through effective coding, effective and efficient data capture, and the validation of data.

Data entry by humans can be speeded up through effective use of coding, which puts data into short sequences of digits and/or letters. Both simple sequence codes and alphabetic derivation codes can be used to follow the progress of a given item. Classification codes and block sequence codes are useful for distinguishing classes of items from each other. Cipher codes are also useful because they can conceal information that is sensitive or is restricted to employees.

Codes are also worthwhile for revealing information to users, since they can enable employees to locate items in stock and also make data entry more meaningful. Significant-digit subset codes use subgroups of digits to describe a product. Mnemonic codes also reveal information by serving as human memory aids that can help a data entry operator enter data correctly or help the user. The Unicode character set includes all standard language symbols. You can display web pages written in other alphabets by downloading an input method editor from Microsoft. Function codes are useful shortcuts for informing computers or people about what functions to perform or what actions to take.

Effective data entry should also consider input devices. Creating a well-designed, effective form that serves as a source document is the first step. Data can be input through many different methods, each with varying speed and reliability. Keyboards have been redesigned for efficiency and improved ergonomics. Optical character recognition (OCR), magnetic ink character recognition (MICR), and mark-sense forms each have special capacities for improving efficiency. Bar codes also speed data entry, improve data accuracy, and increase reliability. Two dimensional codes such as QR codes and Microsoft Tags are becoming popular and efficient ways to direct users to information or promotional websites. Many QR code scanner apps are available for smartphones, allowing users to interact easily with QR codes. Microsoft Tags support one official tag reader. RFID allows the automatic collection of data using RFID tags on products, people, or animals. They can improve inventory management and supply chain processes.

Accurate data entry also can be enhanced through the use of input validation. Analysts must work with users to design input validation tests to prevent erroneous data from being processed and stored, which is costly and potentially detrimental.

Input transactions should be checked to ensure the request is acceptable, authorized, and correct. Input data can be validated through software using several types of tests that check for missing data, length of data items, range and reasonableness of data, and invalid values for data. Input data can also be compared with stored data for validation purposes. Once numerical data are input, they can be checked and corrected automatically through the use of check digits and the Luhn formula.

There is a set order for the testing of data to validate each field. There are also pattern validation methods found in the database design or included in programming languages. The patterns are called regular expressions and contain symbols that represent the type of data that must be present in a field.

Ecommerce environments afford the opportunity for increasing accuracy of data. With proper emphasis on user-centered design elements, customers can enter their own data, store data for later use, use the same stored data throughout the order fulfillment process, and receive feedback regarding order confirmations and updates.

Keywords and Phrases

alphabetic derivation code	Luhn formula
autocomplete feature	magnetic ink character recognition (MICR)
bar code	mark-sense form
block sequence code	Microsoft Tags
bottleneck	mnemonic code
changeable	optical character recognition (OCR)
check digit	Quick Response (QR) code
cipher code	radio frequency identification (RFID)
classification code	redundancy in input data
coding	regular expression
cookies	self-validating code
cross-reference test	significant-digit subset code
differentiated	simple sequence code
function code	supply chain management
keyboarding	test for class or composition

test for comparison with stored data   test for range or reasonableness
test for correct field length   two dimensional bar code
test for invalid values   Unicode
test for missing data   validating input

Review Questions

1. What are the four primary objectives of data entry?
2. List the five general purposes for coding data.
3. Define *simple sequence code.*
4. When is an alphabetic derivation code useful?
5. Explain what is accomplished with a classification code.
6. Define *block sequence code.*
7. What is the simplest type of code for concealing information?
8. What are the benefits of using a significant-digit subset code?
9. What is the purpose of using a mnemonic code for data?
10. Define *function code.*
11. List eight general guidelines for proper coding.
12. What are changeable data?
13. What are differentiation data?
14. What is one specific way to reduce the redundancy of data being entered?
15. Define the term *bottleneck* as it applies to data entry.
16. What three repetitive functions of data entry can be done more efficiently by a computer than by a data entry operator?
17. List six data entry methods.
18. List the three main problems that can occur with input transactions.
19. List two types of two dimensional (2D) bar codes.
20. What are the main functions of two dimensional bar codes?
21. List two ways developers can include QR codes into their designs.
22. List two ways developers can include Microsoft Tags into their designs.
23. Define *RFID*. What are the differences between active and passive RFID tags?
24. Give two examples of the use of RFID tags in process or inventory management in retail or health care environments.
25. What are eight tests for validating input data?
26. Which test checks to see whether data fields are correctly filled in with either numbers or letters?
27. What common error is missed by the Luhn formula?
28. Which test would not permit a user to input a date such as October 32?
29. Which test ensures data accuracy by the incorporation of a number in the code itself?
30. List four improvements to data accuracy that transactions conducted over ecommerce websites can offer.
31. What is Unicode, and how is it used?
32. What is the process for validating data entered into fields?
33. What is a regular expression?

Problems

1. A small, private university specializing in graduate programs needs to keep track of the list of students who (a) apply, (b) are accepted, and (c) actually enroll in the university. For security purposes the university also must send a report to the government with a list of foreign students who enroll but fail to register. Suggest a kind of code for this purpose, and give an example of its use in the university that demonstrates its appropriateness. What are its advantages?
2. The Central Pacific University Chipmunks have been using a simple sequence code to keep track of season ticket holders and fans who are not season ticket holders for all of its sports programs. There have been some upsetting mix-ups.

    In a paragraph, suggest a different coding scheme that will help uniquely identify each ticket holder and explain how it will prevent mix-ups.
3. A code used by an ice cream store to order its products is 12DRM215-220. This code is deciphered in this manner: 12 stands for the count of items in the box, DRM stands for Dreamcicles (a particular kind of ice cream novelty), and 215-220 indicates the entire class of low-fat products carried by the distributor.

   a. What kind of code is used? Describe the purpose behind each part of the code (12, DRM, 215-220).

   b. Construct a coded entry using the same format and logic for an ice cream novelty called Pigeon Bars, which come in a 6-count package and are *not* low-fat.

   c. Construct a coded entry using the same format and logic for an ice cream novelty called Airwhips, which come in a 24-count package and are low-fat.

4. The data entry operators at Michael Mulheren Construction have been making errors when entering the codes for residential siding products, which are as follows: U = stUcco, A = Aluminum, R = bRick, M = Masonite, EZ = EZ color-lok enameled masonite, N = Natural wood siding, AI = pAInted finish, SH = SHake SHingles. Only one code per address is permitted.

   a. List the possible problems with the coding system that could be contributing to erroneous entries. (*Hint:* Are the classes mutually exclusive?)

   b. Devise a mnemonic code that will help the operators understand what they are entering and subsequently help their accuracy.

   c. How would you redesign the classes for siding materials? Respond in a paragraph.

5. The following is a code for one product in an extensive cosmetic line: L02002Z621289. L means that it is a lipstick, 0 means it was introduced without matching nail polish, 2002 is a sequence code indicating in what order it was produced, Z is a classification code indicating that the product is hypoallergenic, and 621289 is the number of the plant (there are 15 plants) where the product is produced.

   a. Critique the code by listing the features that might lead to inaccurate data entry.

   b. Designer Brian d'Arcy James owns the cosmetic firm that uses this coding scheme. Always interested in new design, Brian is willing to look at a more elegant code that encodes the *same* information in a better way. Redesign the coding scheme and provide a key for your work.

   c. Write a sentence for each change you have suggested, indicating what data entry problem (from Problem 5a) the change will eliminate.

   d. Brian is delighted with your work. He says the firm would like to hire you to help branch out into selling theatrical makeup. (Shows such as *Wicked* and *Shrek* with eight performances a week use a lot of green greasepaint.) Add any necessary new codes for the coding scheme you suggested in 5b, and provide a key for your work.

6. The d'Arcy James cosmetics firm requires its salespeople to use laptops to enter orders from retail department stores (their biggest customers). This information is then relayed to warehouses, and orders are shipped on a first-come, first-served basis. Unfortunately, the stores are aware of this policy and are extremely competitive about which one of them will offer a new d'Arcy James product first. Many retailers have taken the low road and persuaded salespeople to falsify their order dates on sales forms by making them earlier than they actually were.

   a. This problem is creating havoc at the warehouse. Disciplining any of the personnel involved is not feasible. How can the warehouse computer be used to certify when orders are actually placed? Explain in a paragraph.

   b. Salespeople are complaining that they have to ignore their true job of selling so that they can key in order data. List the data items relating to sales of cosmetics to retailers that should be stored in and retrieved from the central computer rather than keyed in for every order.

   c. Describe in a paragraph or two how bar coding might help solve the problem in Problem 6b.

7. List the best data entry method and your reason for choosing it for each of the following six situations:

   a. Turnaround document for a utility company that wants notification of a change in the customer address.

   b. Data retrieval allowed only if there is positive machine identification of the party requesting data.

   c. Not enough trained personnel available to interpret long, written responses; many forms submitted that capture answers to multiple-choice examinations; high reliability necessary; fast turnaround not required.

   d. Warehouse set up for a discount compact disc operation; bins are labeled with price information, but individual discs are not; and few skilled operators are available to enter price data.

   e. Poison control center that maintains a large database of poisons and antidotes; needs a way to enter data on the poison taken; also enter weight, age, and general physical condition of the victim when a person calls the center's toll-free number for emergency advice.

   f. Download of a movie by a consumer with a credit card.

8. Ben Coleman, one of your systems analysis team members, surprises you by asserting that when a system uses a test for correct field length, it is redundant also to include a test for range or reasonableness. In a paragraph, give an example that demonstrates that Ben is mistaken on this one.

9. Several retailers have gotten together and begun issuing a "state" credit card that is good only in stores in their state. As a courtesy, salesclerks are permitted to transcribe the 15-digit account number by hand (after getting it from the accounting office) if the customer is not carrying the card. The only

problem with accounts that retailers have noticed so far is that sometimes erroneous account numbers are accepted into the computer system, resulting in a bill being issued to a nonexistent account.

   a. What sort of validity test would clear up the problem? How? Respond in a paragraph.

   b. Suggest an alternative data entry method that might alleviate this problem altogether.

10. Provide an example of submitting the wrong data to a payroll system. Also state at what time in the input process this error should be flagged.

11. Describe a good composition test for a program that accepts first and last names of users. What should the input be composed of? What input should it flag as unacceptable?

12. Recommend an appropriate range or reasonableness test for entering the birth date of a college student enrolling in a four-year college into the college registration system.

13. Explain how a test for comparison with stored data would work for a retailer that was adding new models of smart phones by manufacturers already represented in their database.

14. Define a regular expression for validating each of the following:

   a. A U.S. zip code, which must have five digits, followed by an optional hyphen and four digits.

   b. A telephone number in the format (aaa) nnn-nnnn, where aaa represents the area code and the *n*s represent digits.

   c. A date in the form of day-month-year, where the month is a three-letter code and the year is four digits. A hyphen must separate the day and month and year and month.

   d. The alphabetic derivation code illustrated in this chapter for a magazine subscriber. The format is 99999XXX9999XXX, where X represents a letter and 9 represents a number.

15. For the following codes, define the validation criteria (there may be multiple checks for each field) and the order in which you would test each of the conditions.

   a. **A credit card number entered on a Web form:** The customer has selected the type of credit card from a drop-down list.

   b. **A part number in a hardware store:** The part number is a complex code, where the first digit represents the department (such as housewares, automotive, and so on), and the number should be self-checking. There are seven different departments.

   c. **The date that a book was postmarked when returned to an online bookstore:** A copy of the customer receipt must be included with the book. Returned books must be postmarked within 30 days of the purchase date.

   d. **A language spoken code used on a website:** *Hint:* Search the Web for standard language codes.

   e. **A driver's license number, composed of several parts:** The person's birth month, the birth day, and birth year, not necessarily together; a code representing eye color; and a sequence number. The driver's license contains the date of birth, the eye and hair color, as well as the person's name and address.

   f. **The Canadian postal code:** The format is X9X 9X9 (X is any letter, 9 is any number).

   g. **Airport codes:** Such as LAX for Los Angeles or DUB for Dublin.

   h. **A product key used to unlock purchased software:** The key consists of four groups of five characters each. The first group must have two letters followed by three numbers; the second group must contain two numbers followed by three letters; the third group must contain two letters, each from A through G, followed by three numbers from one through four; and the last group must contain a letter, either an E, G, or C, two digits with values from four through seven, and two letters, either an A, B, or C. *Hint:* Using a pattern may be the best way to validate the product key.

Group Projects

1. Along with your group members, read Consulting Opportunity 15.3, "To Enter or Not to Enter: That Is the Question," presented earlier in this chapter. Design an appropriate data entry system for Elsinore Industries. Your group's design should emphasize efficiency and accuracy. In addition, distinguish between data that are changeable and data that differentiate an item being entered from all others. Draw prototypes of any screen necessary to explain what you are recommending.

2. Divide your group into analysts and Elsinore Industries employees to role-play. The analysts should present the new data entry system, complete with prototype displays. Ask for feedback on the design from Elsinore employees.

3. Write a brief paragraph describing how to improve the original data entry design based on the comments received.

Selected Bibliography

Davis, G. B., and M. H. Olson. *Management Information Systems, Conceptual Foundations, Structure, and Development,* 2nd ed. New York: McGraw-Hill, 1985.

Kendall. J. E. "2D or Not 2D: That Is the Barcode Question." *Decision Line*, Vol. 42, Issue 5., October, 2011. Last accessed August 31, 2012.

Lamming, M. G., P. Brown, K. Carter, M. Eldridge, M. Flynn, G. Louie, P. Robinson, and A. Sellan. "The Design of a Human Memory Prosthesis." *Computer Journal,* Vol. 37, 1994, pp. 153–163.

Lee, Y. M., F. Cheng, and Y. T. Leung. "Exploring the Impact of RFID on Supply Chain Dynamics." In *Proceedings of the 2004 Winter Simulation Conference.* Edited by R. G. Ingalls, M. D. Rossetti, J. S. Smith, and B. A. Peters, pp. 1145–1152.

MacKay, D. J. *Information Theory, Inference and Learning Algorithms.* Cambridge, UK: Cambridge University Press, 2004.

Miller, G. A. "The Magical Number Seven, Plus or Minus Two: Some Limits on Our Capability for Processing Information." *Psychological Review,* Vol. 63, No. 2, March 1956, pp. 81–97.

Newman, W. N., M. G. Lamming, and M. Lamming. *Interactive System Design.* Reading, MA: Addison-Wesley Longman Publishing Co., 1995.

Niederman, F., R. G. Mathieu, R. Morley, and I. Kwon. "Examining RFID Applications in Supply Chain Management." *Communications of the ACM,* Vol. 50, No. 7, July 2007, pp. 92–101.

Owsowitz, S., and A. Sweetland. "Factors Affecting Coding Errors." *Rand Memorandum RM-4346-PR.* Santa Monica, CA: Rand Corporation, 1965.

Robey, D., and W. Taggart. "Human Processing in Information and Decision Support Systems." *MIS Quarterly,* Vol. 6, No. 2, June 1982, pp. 61–73.

Ryder, J. "Credit Card Validation Using LUHN Formula." www.freevbcode.com. Last accessed May 30, 2012.

The CPU Case Episode and accompanying student files are available online at www.pearsonglobaleditions.com/kendall.

CHAPTER 16

Quality Assurance and Implementation

LEARNING OBJECTIVES

Once you have mastered the material in this chapter you will be able to:

1. Recognize the importance of users and analysts taking a total quality approach to improve the quality of software design and maintenance.

2. Realize the importance of documentation, testing, maintenance, and auditing.

3. Understand how service-oriented architecture and cloud computing are changing the nature of information system design.

4. Design appropriate training programs for users of a new system.

5. Recognize the differences among physical conversion strategies and be able to recommend an appropriate one to a client.

6. Address security, disaster preparedness, and disaster recovery concerns for traditional and Web-based systems.

7. Understand the importance of evaluating a new system and be able to recommend a suitable evaluation technique to a client.

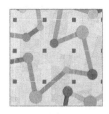

Quality has long been a concern of businesses, as it should be for systems analysts in the analysis and design of information systems. The user of an information system is the single most important factor in establishing and evaluating its quality. It is far less costly to correct problems in their early stages than it is to wait until a problem is articulated through user complaints or crises. The three approaches to quality assurance through software engineering are (1) securing total quality assurance by designing systems and software with a top-down, modular approach; (2) documenting software with appropriate tools; and (3) testing, maintaining, and auditing software.

The process of ensuring that an information system is operational and then allowing users to take over its operation for use and evaluation is called implementation. Implementation concerns moving computer power to individual users by shifting computer power and responsibility to groups and individuals throughout the business with the help of distributed computing, cloud computing, and service-oriented architecture; training users and making sure that each user understands any new roles they must take on because of the new information system; choosing a conversion strategy; providing proper security, privacy, and disaster plans; and evaluating the new or modified information system.

The Total Quality Management Approach

Total quality management (TQM) is essential throughout all the systems development steps. According to Evans and Lindsay (2004), the primary elements of TQM are meaningful only when they occur in an organizational context that supports a comprehensive quality effort. It is in this context that the elements of customer focus, strategic planning and leadership, continuous improvement, empowerment, and teamwork are united to change employees' behavior and, ultimately, the organization's course. Notice that the concept of quality has broadened over the years to reflect an organizational, rather than an exclusively production, approach. Instead of conceiving of quality as controlling the number of defective products produced, quality is now thought of as an evolutionary process toward perfection that is referred to as total quality management.

Systems analysts must be aware of the factors that are driving the interest in quality. It is important to realize that the increasing commitment of businesses to TQM fits extraordinarily well into the overall objectives for systems analysis and design.

Six Sigma

The advent of Six Sigma has changed the approach to quality management. Systems analysts and systems users need to be aware of Six Sigma and apply some of the principles to their systems analysis projects. Originally developed by Motorola in the 1980s, Six Sigma is more than a methodology; it is a culture built on quality. The goal of Six Sigma is to eliminate all defects. This applies to any product, service, or process. In operations management textbooks from the 1970s to the end of the century, quality control was expressed in terms of three standard deviations from the mean, or three sigma, which equals about 67,000 defects per 1 million opportunities. Six Sigma implies a goal of no more than 3.4 defects per 1 million opportunities.

Six Sigma is a top-down approach. It requires a CEO to adopt the philosophy and an executive to serve as project champion. A Six Sigma project leader is called a Black Belt. (The metaphor of the Black Belt comes from the ranking system of capabilities in martial arts.) Black Belts are certified after they have successfully led projects. Other project members are called Green Belts. Master Black Belts are Black Belts who have worked on many projects and are available as a resource to project teams.

Six Sigma can be summarized as a methodology. The steps of Six Sigma are shown in Figure 16.1. Six Sigma, however, is much more than a methodology; it is a philosophy and a culture. For more information on Six Sigma and quality management, visit the website for the Juran Center at the Carlson School of Management, University of Minnesota, Twin Cities (www.csom. umn.edu). In 2002 the Juran Center issued a proclamation to support and encourage quality. The authors of this book signed the charter at that time, and we agree wholeheartedly with its principles.

The late Joseph M. Juran said, "All quality improvement occurs on a project-by-project basis and in no other way" (Juran, 1964). Systems analysts, project managers, and users should take that to heart.

Responsibility for Total Quality Management

Practically speaking, a large portion of the responsibility for the quality of information systems rests with systems users and management. Two things must happen for TQM to become a reality with systems projects. First, the full organizational support of management must exist, which is a departure from merely endorsing the newest management gimmick. Such support means establishing a context for management people to consider seriously how the quality of information systems and information itself affects their work.

Early commitment to quality from analysts and business users is necessary to achieve the goal of quality. This commitment results in exerting an evenly paced effort toward quality throughout the systems development life cycle, and it stands in stark contrast to having to pour huge amounts of effort into ironing out problems at the end of the project.

Organizational support for quality in management information systems can be achieved by providing on-the-job time for IS quality circles, which consist of six to eight organizational peers specifically charged with considering both how to improve information systems and how to implement improvements.

Through work in IS quality circles or through other mechanisms already in place, management and users must develop guidelines for quality standards of information systems. Preferably,

FIGURE 16.1

Every systems analyst should
understand the methodology and
philosophy of Six Sigma.

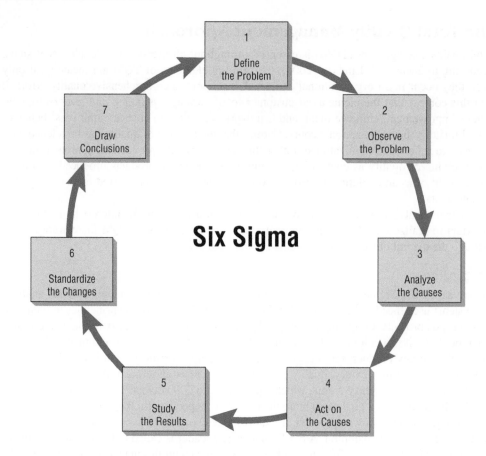

standards will be reshaped every time a new system or major modification is to be formally proposed by the systems analysis team.

Hammering out quality standards is not easy, but it is possible and has been done. Part of a systems analyst's job is encouraging users to crystallize their expectations about information systems and their interactions with them.

Departmental quality standards must then be communicated through feedback to the systems analysis team. The team is often surprised at what has developed. Expectations typically are less complex than what experienced analysts know can be done with a system. In addition, human issues that have been overlooked or underrated by the analysis team may be designated as extremely pressing in users' quality standards. Getting users involved in spelling out quality standards for information systems will help an analyst avoid expensive mistakes in unwanted or unnecessary systems development.

Structured Walkthrough

One of the strongest quality management actions a systems analysis team can take is to do structured walkthroughs routinely. Conducting structured walkthroughs is a way of using peer reviewers to monitor a system's programming and overall development, point out problems, and allow the programmer or analyst who is responsible for that portion of the system to make suitable changes.

Structured walkthroughs involve at least four people: the person responsible for the part of the system or subsystem being reviewed (a programmer or analyst), a walkthrough coordinator, a programmer or analyst peer, and a peer who takes notes about suggestions.

Each person attending a walkthrough has a special role to play. The coordinator is there to ensure that the others adhere to any roles assigned to them and to ensure that any activities scheduled are accomplished. The programmer or analyst is there to listen, not to defend his or her thinking, rationalize a problem, or argue. The programmer or analyst peer is present to point out errors or potential problems, not to specify how the problems should be remedied. The note taker records what is said so that the others present can interact without encumbrance.

Structured walkthroughs fit well in a total quality management approach when performed throughout the systems development life cycle. The time they take should be short—half an hour

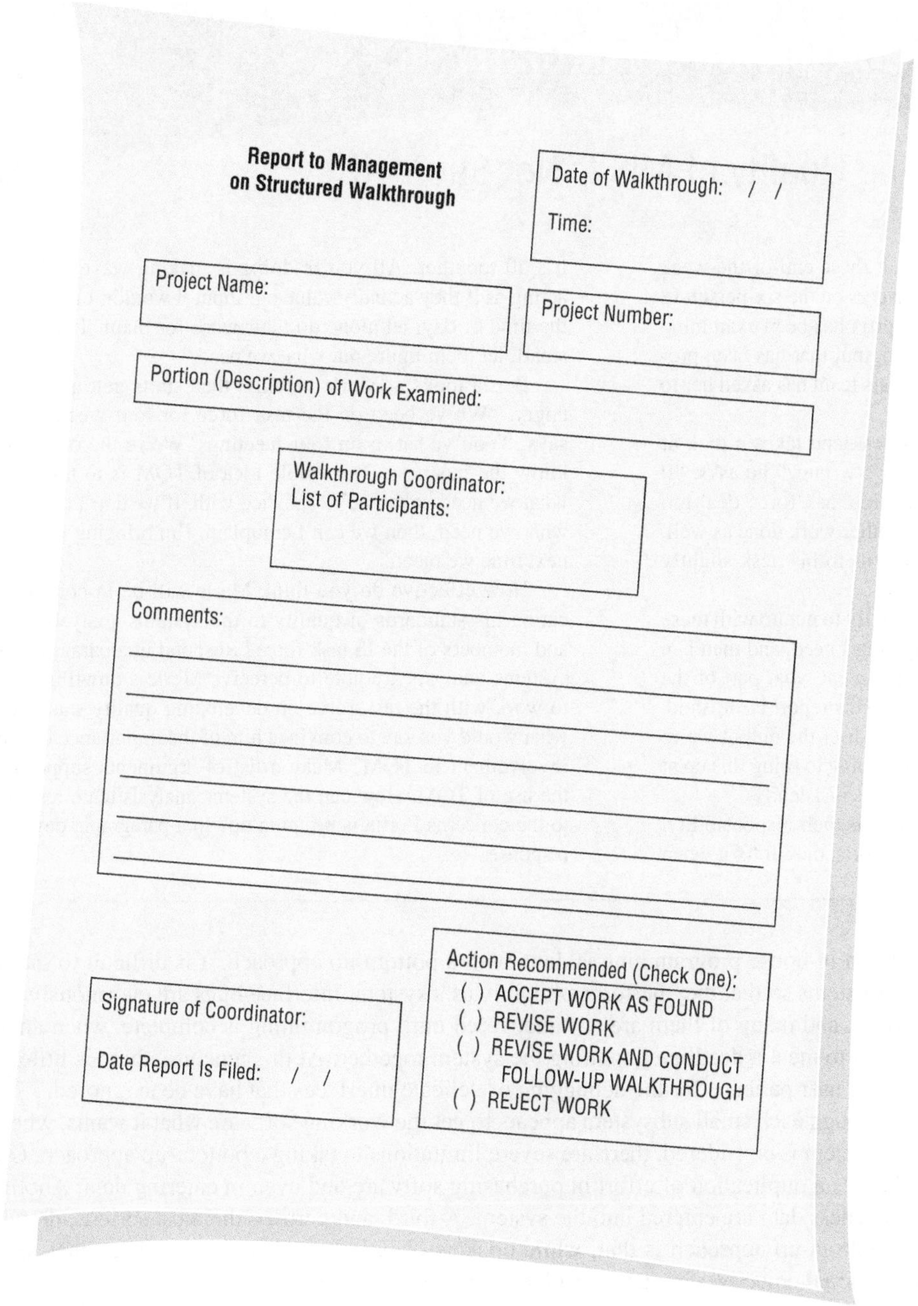

FIGURE 16.2

**Report to Management
on Structured Walkthrough**

Date of Walkthrough: / /

Time:

Project Name:

Project Number:

Portion (Description) of Work Examined:

Walkthrough Coordinator:
List of Participants:

Comments:

Signature of Coordinator:

Date Report Is Filed: / /

Action Recommended (Check One):
( ) ACCEPT WORK AS FOUND
( ) REVISE WORK
( ) REVISE WORK AND CONDUCT
     FOLLOW-UP WALKTHROUGH
( ) REJECT WORK

A form to document structured walkthroughs; walkthroughs can be done whenever a portion of coding, a system, or a subsystem is complete.

to an hour at most—which means that they must be well coordinated. Figure 16.2 shows a form that is useful in organizing the structured walkthrough and reporting its results. Because walkthroughs take time, it is important not to overuse them.

You can use structured walkthroughs as a way to obtain (and then act on) valuable feedback from a perspective that you lack. As with all other quality assurance measures, the point of walkthroughs is to evaluate a product systematically on an ongoing basis rather than wait until completion of the system.

Top-Down Systems Design and Development

Many companies first introduced computer systems on the lowest level of the organization. This is where the immediate benefits to computerization are most observable and cost-effective. Businesses often take this approach to systems development by going out and acquiring, for example, COTS software for accounting, a different package for production scheduling, and another one for marketing.

CONSULTING OPPORTUNITY 16.1

The Quality of MIS Is Not Strained

"Merle, come here and take a look at these end-of-the-week reports," Portia pleads. As one of the managers on the six-person IS task force/quality assurance committee, Portia has been examining for her marketing department the system output that has been produced by the prototype. The systems analysis team has asked her to review the output.

Merle Chant walks over to Portia's desk and takes a look at the prospectus she's holding. "Why, what's wrong?" he asks. "It looks okay to me. I think you're taking this task force deal too much to heart. We're supposed to get our other work done as well, you know." Merle turns to leave and returns to his desk slightly perturbed at being interrupted.

"Merle, have a little mercy. It is really silly to put up with these reports the way they are. I can't find anything I need, and then I'm supposed to tell everyone else in the department what part of the report to read. I, for one, am disappointed. This report is slipshod. It doesn't make any sense to me. It's a rehash of the output we're getting now. Actually, it looks worse. I am going to bring this up at the next task force meeting," Portia proclaims insistently.

Merle turns to face her, saying, "Quality is their responsibility, Portia. If the system isn't giving us good reports, they'll fix it when

it's all together. All you're doing is making waves. You're acting as if they actually value our input. I wouldn't give them the time of day, let alone do their work for them. They're so smart, let them figure out what we need."

Portia looks at Merle blankly, then starts getting a little angry. "We've been on the task force for four weeks," she says. "You've sat in on four meetings. We're the ones who know the business. The whole idea of TQM is to tell them what we need, what we're satisfied with. If we don't tell them what we need, then we can't complain. I'm bringing it up the next time we meet."

How effective do you think Merle will be in communicating his standards of quality to the systems analysis team and members of the IS task force? Respond in a paragraph. If systems analysts are able to perceive Merle's unwillingness to work with the task force on developing quality standards, what would you say to convince him of the importance of user involvement in TQM? Make a list of arguments supporting the use of TQM. How can the systems analysis team respond to the concerns Portia is bringing up? In a paragraph, devise a response.

When in-house programming is done with a bottom-up approach, it is difficult to interface the subsystems so that they perform smoothly as a system. Interface bugs are enormously costly to correct, and many of them are not uncovered until programming is complete, when analysts are trying to meet a deadline in putting the system together. At this juncture, there is little time, budget, or user patience for the debugging of delicate interfaces that have been ignored.

Although each small subsystem appears to get the working software what it wants, when the overall system is considered, there are severe limitations to taking a bottom-up approach. One is that there is a duplication of effort in purchasing software and even in entering data. Another is that worthless data are entered into the system. A third, and perhaps the most serious, drawback of the bottom-up approach is that, while pockets of users' needs may have been met, overall organizational objectives are not considered and hence cannot be met.

Top-down design allows a systems analyst to ascertain overall organizational objectives first, as well as to ascertain how they are best met in an overall system. Then the analyst divides that system into subsystems and their requirements.

Top-down design is compatible with the general systems thinking that was discussed in Chapter 2. When systems analysts employ a top-down approach, they are thinking about the interrelationships and interdependencies of subsystems as they fit into the existing organization. The top-down approach also provides desirable emphasis on synergy or the interfaces that systems and their subsystems require, which is lacking in the bottom-up approach. It helps to answer the question of how teams must work together to accomplish their goals.

The advantages of using a top-down approach to systems design include avoiding the chaos of attempting to design a system all at once. As we have seen, planning and implementing management information systems is incredibly complex. Attempting to get all subsystems in place and running at once is agreeing to fail.

A second advantage of taking a top-down approach to design is that it enables separate systems analysis teams to work in parallel on different but necessary subsystems, which can save a great deal of time. The use of teams for subsystems design is particularly well suited to a total quality assurance approach.

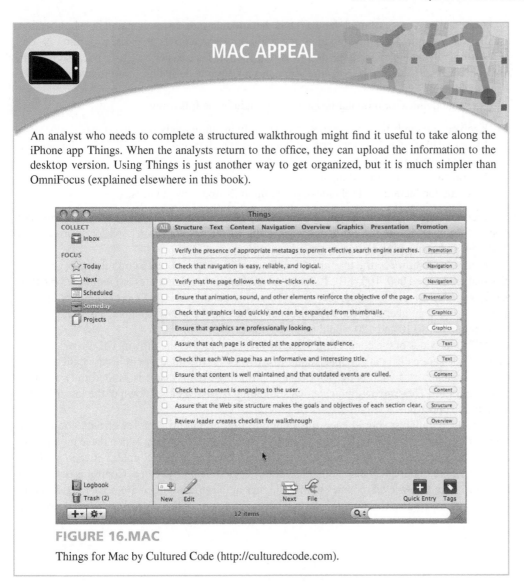

MAC APPEAL

An analyst who needs to complete a structured walkthrough might find it useful to take along the iPhone app Things. When the analysts return to the office, they can upload the information to the desktop version. Using Things is just another way to get organized, but it is much simpler than OmniFocus (explained elsewhere in this book).

FIGURE 16.MAC

Things for Mac by Cultured Code (http://culturedcode.com).

A third advantage is that a top-down approach avoids a major problem associated with a bottom-up approach: It prevents systems analysts from getting so mired in detail that they lose sight of what the system is supposed to do.

Total quality management and the top-down approach to design can go hand-in-hand. The top-down approach provides the systems group with a ready-made division of users into task forces (specialized teams of users) for subsystems. Task forces set up in this manner can then serve a dual function as quality circles for the management information system. The necessary structure for quality assurance is then in place, as is proper motivation for getting the subsystem to accomplish the departmental goals that are important to the users involved.

Using Structure Charts to Design Modular Systems

With the top-down design approach, the modular approach is useful in programming. This approach involves breaking the programming into logical, manageable portions, or modules. This kind of programming works well with top-down design because it emphasizes the interfaces between modules and does not neglect them until later in systems development. Ideally, each individual module should be functionally cohesive so that it is charged with accomplishing only one function.

Modular program design has three main advantages. First, modules are easier to write and debug because they are virtually self-contained. Tracing an error in a module is less complicated, because a problem in one module should not cause problems in others.

A second advantage of modular design is that modules are easier to maintain. Modifications usually will be limited to a few modules and will not be spread over an entire program.

A third advantage of modular design is that modules are easier to grasp, because they are self-contained subsystems. Hence, a reader can pick up a code listing of a module and understand its function.

Some guidelines for modular programming include the following:

1. Keep each module to a manageable size (ideally including only one function).
2. Pay particular attention to the critical interfaces (the data and control variables that are passed to other modules).
3. Minimize the number of modules the user must modify when making changes.
4. Maintain the hierarchical relationships set up in the top-down phases.

The recommended tool for designing a modular, top-down system is called a structure chart. A structure chart is simply a diagram consisting of rectangular boxes, which represent the modules, and connecting arrows.

Figure 16.3 shows a structure chart for changing a customer record, involving seven modules that are labeled 000, 100, 110, 120, and so on. Higher-level modules are numbered by 100s, and lower-level modules are numbered by 10s. This numbering allows programmers to insert modules using a number between the adjacent module numbers. For example, a module inserted between modules 110 and 120 would receive number 115.

Off to the sides of the connecting lines, two types of arrows are drawn. The arrows with the empty circles are called data couples, and the arrows with the filled-in circles are called control flags or switches. A switch is the same as a control flag except that it is limited to two values: either yes or no. These arrows indicate that something is passed either down to the lower module or up to the upper one.

Ideally, an analyst should keep this coupling to a minimum. The fewer data couples and control flags one has in the system, the easier it is to change the system. When these modules are actually programmed, it is important to pass the least number of data couples between modules.

Even more important is that numerous control flags should be avoided. Control is designed to be passed from lower-level modules to those higher in the structure. On rare occasions, however, it will be necessary to pass control downward in the structure. When control is passed downward,

FIGURE 16.3

A structure chart encourages
top-down design using modules.

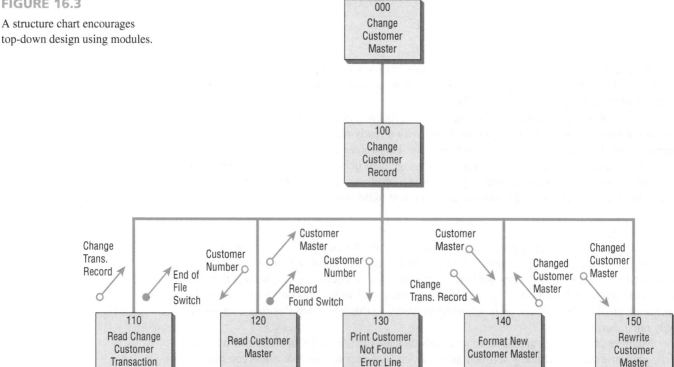

a low-level module is allowed to make a decision, and the result is a module that performs two different tasks. This result violates the ideal of a functional module: It should perform only one task.

Even when a structure chart accomplishes all the purposes for which it was drawn, the structure chart cannot stand alone as the sole design and documentation technique. First, it doesn't show the order in which the modules should be executed (a data flow diagram will accomplish that). Second, it doesn't show enough detail (Structured English will accomplish that).

Service-Oriented Architecture (SOA)

Modular development has led to a concept called service-oriented architecture (SOA), which is very different from the modules in the structure chart. Instead of being hierarchical like the top-down approach found in structure charts, the SOA approach is to make individual SOA services that are unassociated or only loosely coupled to one another.

Each service executes one action. One service may return the number of days in this month; another may tell us if this is a leap year; a third service may reserve five nights in a hotel room from the end of February to the beginning of March. Although the third service needs to know the values obtained from the first and second services, they are independent of one another. Each service can be used in other applications within the organization or even in other organizations.

We can say that service-oriented architecture is simply a group of services that can be called upon to provide specific functions. Rather than including calls to other services, a service can use certain defined protocols so that it can communicate with other services.

Figure 16.4 shows how services are called upon throughout the system. Services can be general in nature and can be outsourced or even be available on the Web. Other services are more specialized and oriented toward the business itself. These enterprise-based services provide business rules and can also differentiate one business from another. Services can be called upon at a time and can be called on repeatedly in many application modules.

The burden of connecting services in a useful fashion, a process called orchestration, is placed upon a systems designer. This can even be accomplished by selecting services from a menu of services and monitoring them by setting up an SOA dashboard.

In order to set up an SOA, the services must be:

1. Modular
2. Reusable
3. Able to work together with other modules (interoperable)
4. Able to be categorized and identified
5. Able to be monitored
6. Compliant with industry-specific standards

While the advantages of reusability and interoperability are obvious, SOA is not without challenges. First, industry standards must be agreed upon. Next, a library must be maintained so

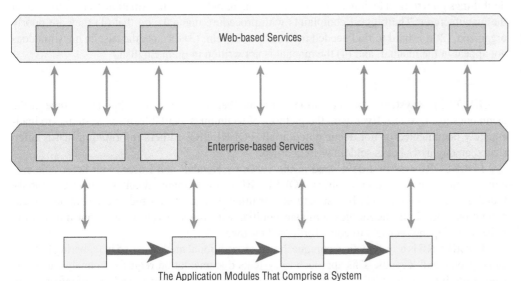

FIGURE 16.4

Modules in service-oriented architecture are independent and can be ubiquitous.

that developers can find the services they need. Finally, security and privacy can be issues when using software developed by someone else. Advocates of SOA claim that it has made many of the features found in Web 2.0 possible. Others have called cloud computing an offspring of SOA.

Documentation Approaches

A total quality assurance effort requires that programs be documented properly. Software, systems, and formal and informal procedures need to be documented so that systems can be maintained and improved. Documentation allows users, programmers, and analysts to "see" the system, its software, and procedures without having to interact with it.

Turnover of information service personnel has traditionally been high in comparison with other departments, so chances are that the people who conceived of and installed the original system will not be the same ones who maintain it. Consistent, well-updated documentation will shorten the number of hours required for new people to learn the system before performing maintenance.

There are many reasons why systems and programs are undocumented or underdocumented. Some of the problems reside with the systems and programs themselves, others with systems analysts and programmers.

Systems analysts may fail to document systems properly because they do not have the time or are not rewarded for time spent documenting. Some analysts do not document because they dread doing so or think it is not their real work. Furthermore, many analysts are reticent about documenting systems that are not their own, perhaps fearing reprisals if they include incorrect material about someone else's system. Defenders of the SDLC approach remind us that documentation accomplished by means of a CASE tool during the analysis phases can address many of these problems.

Procedure Manuals

Procedure manuals are common organizational documents that most people have seen. They are the English-language component of documentation, although they may also contain program codes, flowcharts, and so on. Manuals are intended to communicate to those who use them. They may contain background comments, steps required to accomplish different transactions, instructions on how to recover from problems, and what to do next if something isn't working (troubleshooting). Many manuals are now available online, with hypertext capability that facilitates use.

A straightforward, standardized approach to creating user support documentation is also desirable. To be useful, user documentation must be kept up to date. Use of the Web has revolutionized the speed with which assistance can be obtained by users. Many software developers have moved user support—complete with manuals, FAQ pages, online chat, and user communities—to the Web.

Key sections of a manual should include an introduction, how to use the software, what to do if things go wrong, a technical reference section, an index, and information on how to contact the manufacturer. The biggest complaints with procedure manuals are that (1) they are poorly organized, (2) it is hard to find needed information in them, (3) the specific case in question does not appear in the manual, and (4) the manual is not written in plain English.

The FOLKLORE Method

FOLKLORE is a systems documentation technique that was created to supplement some of the techniques just covered. Even with the plethora of techniques available, many systems are inadequately documented or not documented at all. FOLKLORE gathers information that is often shared among users but is seldom written down.

FOLKLORE was first developed in the 1980s by Kendall and Losee, well before the creation of blogs and user communities. FOLKLORE has two main advantages over commonly found user communities: (1) It is structured, resulting in more organized, more complete documentation, and (2) it encourages someone familiar with the software to seek out information rather than depend on users to come forth on their own.

FOLKLORE is a systematic technique, based on traditional methods used in gathering folklore about people and legends. This approach to systems documentation requires an analyst to interview users, investigate existing documentation in files, and observe the processing of information.

CONSULTING OPPORTUNITY 16.2

Write Is Right

"It's so easy to understand. I say if everybody uses pseudocode, we won't have trouble, you know, with things not being standardized," says Al Gorithm, a new programmer who will be working with your systems analysis team. Al is speaking at an informal meeting among three members of the systems analysis team, a six-person MIS task force from the advertising department, and two programmers, who were all working to develop an information system for advertising personnel.

Philip, an advertising account executive and one of the members of the MIS task force, looks up in surprise. "What is this method called?" The two programmers reply at the same time: "Pseudocode." Philip looks unimpressed and says, "That doesn't say anything to me."

Neeva Phail, one of the systems analysts, begins explaining: "It probably won't matter one way or the other what we use, if—"

Flo Chart, another systems analyst, breaks in saying, "I hate pseudocode." She looks hopefully at the programmers. "I'm sure we can agree on a better technique."

David, an older advertising executive, seems slightly upset, stating, "I learned about flowcharting from the first systems analysts we had years ago. Don't you people do that anymore? I think they work best."

What was at first a friendly meeting suddenly seems to have reached an impasse. The participants are looking at each other warily. As a systems analyst who has worked on many different projects with many different kinds of people, you realize that the group is looking to you to make some reasonable suggestions.

Based on what you know about the various documentation techniques, what technique(s) would you propose to the members of the group? How will the technique(s) you proposed overcome some of the concerns they have voiced? What process will you use to decide on appropriate techniques? Compose your answer in one page.

The objective is to gather information corresponding to one of four categories: customs, tales, sayings, and art forms. Figure 16.5 suggests how each category relates to the documentation of information systems.

When documenting customs, an analyst (or other folklorist) tries to capture in writing what users are currently doing to get all programs to run without problems. An example of a custom is: "Usually, we take two days to update the monthly records because the task is quite large. We run commercial accounts on day one and save the others for the next day."

Tales are stories that users tell regarding how the system worked. The accuracy of the tale, of course, depends on the user's memory and is at best an opinion about how the program worked. Tales normally have a beginning, a middle, and an end. So we would have a story about a problem (the beginning), a description of the effects (the middle), and the solution (the end).

Sayings are brief statements representing generalizations or advice. We have many sayings in everyday life, such as "April showers bring May flowers" or "A stitch in time saves nine."

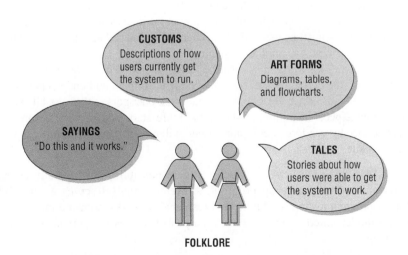

FOLKLORE

FIGURE 16.5

Customs, tales, sayings, and art forms used in the FOLKLORE method of documentation apply to information systems.

In systems documentation, we have many sayings, such as "Omit this section of code, and the program will bomb" or "Always back up frequently." Users like to give advice, and an analyst should try to capture this advice and include it in the FOLKLORE documentation.

Gathering art forms is another important activity of traditional folklorists, and a systems analyst should understand its importance, too. Flowcharts, diagrams, and tables that users draw sometimes may be better or more useful than flowcharts drawn by the original system author. Analysts will often find such art posted on bulletin boards, or they may ask the users to clean out their files and retrieve any useful diagrams.

Contributors to a FOLKLORE document do not have to document the entire system, only the parts they know about. Just like with Web-based user communities, the danger of relying on FOLKLORE is that the information gathered from users may be correct, partially correct, or incorrect.

Choosing a Design and Documentation Technique

The techniques discussed in this chapter are extremely valuable as design tools, memory aids, productivity tools, and as a means of reducing dependencies on key staff members. A systems analyst, however, is faced with a difficult decision regarding which method to adopt. The following is a set of guidelines to help an analyst use the appropriate technique. A systems analyst should choose a technique that:

1. Is compatible with existing documentation
2. Is understood by others in the organization
3. Allows you to return to working on the system after you have been away from it for a period of time
4. Is suitable for the size of the system on which you are working
5. Allows for a structured design approach if that is considered to be more important than other factors
6. Allows for easy modification

Testing, Maintenance, and Auditing

Once an analyst has designed and coded the system, testing, maintenance, and auditing of it are prime considerations.

The Testing Process

All of a system's newly written or modified application programs—as well as new procedural manuals, new hardware, and all system interfaces—must be tested thoroughly. Haphazard, trial-and-error testing will not suffice. Testing is done throughout systems development, not just at the end. It is meant to turn up previously unknown problems, not to demonstrate the perfection of programs, manuals, or equipment.

Although testing is tedious, it is an essential series of steps that helps ensure the quality of the eventual system. It is far less disruptive to test beforehand than to have a poorly tested system fail after installation. Testing is accomplished on subsystems or program modules as work progresses. Testing is done on many different levels at various intervals. Before a system is put into production, all programs must be desk checked, checked with test data, and checked to see if the modules work together with one another as planned.

A system as a working whole must also be tested. Included here are testing the interfaces between subsystems, the correctness of output, and the usefulness and understandability of systems documentation and output. Programmers, analysts, operators, and users all play different roles in the various aspects of testing, as shown in Figure 16.6. Testing of hardware is typically provided as a service by vendors of equipment, who will run their own tests on equipment when it is delivered on-site.

PROGRAM TESTING WITH TEST DATA. Much of the responsibility for program testing resides with the original author(s) of each program. A systems analyst serves as an advisor and coordinator for program testing. In this capacity, an analyst works to ensure that correct testing techniques are implemented by programmers but probably does not personally carry out this level of checking.

HYPERCASE® EXPERIENCE 16.1

"This is a fascinating place to work. I'm sure you agree, now that you've had a chance to observe us. Sometimes I think it must be fun to be an outsider . . . don't you feel like an anthropologist discovering a new culture? I remember when I first came here. Everything was so new, so strange. Why, even the language was different. It wasn't a 'customer,' it was a 'client.' We didn't have 'departments,' we had 'units.' It's not an employee cafeteria, it's the 'canteen.' That goes for the way we work, too. We all have our different ways to approach things. I think I'm getting the hang of what Snowden expects, but every once in a while I make a mistake, too. For instance, if I can give him work online, he'd just as soon see it that way than get a printed report. That's why I have two computers on my desk, too! I always see you taking so many notes. . . . I guess it makes sense, though. You're supposed to document what

we do with our systems and information as well as what your team is doing, aren't you?"

HYPERCASE Questions

1. Use the FOLKLORE method to complete the documentation of the Management Information Systems Unit GEMS system. Be sure to include customs, tales, sayings, and art forms.
2. In two paragraphs, suggest a PC-based approach for capturing the elements of FOLKLORE so that it is not necessary to use a paper-based log. Make sure that your suggested solution can accommodate graphics as well as text.
3. Design input and output screens for FOLKLORE that facilitate easy entry, and provide prompting so that recall of FOLKLORE elements is immediate.

FIGURE 16.HC1

In HyperCase, use FOLKLORE to document art forms that users have created or collected to make sense of their systems.

At this stage, programmers must first desk check their programs to verify the way the system will work. In desk checking, the programmer follows each step in the program on paper to check whether the routine works as it is written.

Next, programmers must create both valid and invalid test data. These data are then run to see if base routines work and also to catch errors. If output from main modules is satisfactory, you can add more test data so as to check other modules. Created test data should test possible

FIGURE 16.6

Programmers, analysts, operators, and users all play different roles in testing software and systems.

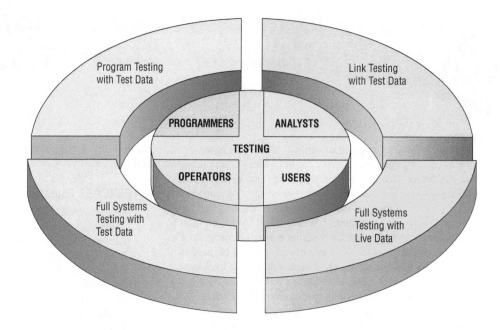

minimum and maximum values as well as all possible variations in format and codes. File output from test data must be carefully verified. It should never be assumed that data contained in a file are correct just because a file was created and accessed.

Throughout this process, a systems analyst checks output for errors, advising the programmer of any needed corrections. An analyst will usually not recommend or create test data for program testing but might point out to the programmer omissions of data types to be added in later tests.

LINK TESTING WITH TEST DATA. When programs pass desk checking and checking with test data, they must go through link testing, which is also referred to as string testing. Link testing checks to see if programs that are interdependent actually work together as planned.

An analyst creates special test data that cover a variety of processing situations for link testing. First, typical test data are processed to see if the system can handle normal transactions, those that would make up the bulk of its load. If the system works with normal transactions, variations are added, including invalid data used to ensure that the system can properly detect errors.

FULL SYSTEMS TESTING WITH TEST DATA. When link tests are satisfactorily concluded, a system as a complete entity must be tested. At this stage, operators and end users become actively involved in testing. Test data, created by the systems analysis team for the express purpose of testing system objectives, are used.

As can be expected, there are a number of factors to consider when systems testing with test data:

1. Examining whether operators have adequate documentation in procedure manuals (hard copy or online) to afford correct and efficient operation
2. Checking whether procedure manuals are clear enough in communicating how data should be prepared for input
3. Ascertaining whether work flows necessitated by the new or modified system actually "flow"
4. Determining whether output is correct and whether users understand that this output is, in all likelihood, as it will look in its final form

Remember to schedule adequate time for system testing. Unfortunately, this step often gets dropped if system installation is lagging behind the target date.

Systems testing includes reaffirming the quality standards for system performance that were set up when the initial system specifications were made. Everyone involved should once again agree on how to determine whether the system is doing what it is supposed to do. This step will include measures of error, timeliness, ease of use, proper ordering of transactions, acceptable down time, and understandable procedure manuals.

Cramming for Your Systems Test

"We're strapped for time. Just look at this projection," says Lou Scuntroll, the newest member of your systems analysis team, showing you the PERT diagram that the team has been using to project when the new system would be up and running. "We can't possibly make the July target date for testing with live data. We're running three weeks behind because of that slow equipment shipment."

As one of the systems analysts who has seen deadlines come and go on other projects, you try to remain calm and to size up the situation carefully before you speak. Slowly, you question Lou about the possibility of delaying testing.

Lou replies, "If we try to push the testing off until the first weeks of August, there are two key people from accounting who are going to be out on vacation." Lou is visibly upset at the possibility of missing the deadline.

Stan Dards, another junior member of your systems analysis team, enters Lou's office. "You two look terrible. Things are going okay, aren't they? I'm not reassigned to program a payroll application, am I?"

Lou looks up, obviously neither appreciating Stan's sense of humor nor what seems like his single-minded self-concern. "Good thing you came in when you did. We've got some big decisions to make about scheduling." Lou holds up the PERT diagram for Stan's inspection. "Notice the July test date. Notice that there is no way we can make it. Any bright ideas?"

Stan contemplates the chart momentarily, then states, "Something's got to go. Let's see here . . . maybe move testing of the accounting module to—"

Lou interrupts, saying bluntly, "Nope, already thought of that, but Stanford and Binet from accounting are out of town in August. Maybe we can skip that portion of the testing. They've been really cooperative. I don't think they'd object if we just 'do it for real' and test as we actually go into production."

"I think that's a good idea, Lou," Stan agrees, trying to make up for his earlier jokes. "We haven't had any real trouble with that, and the programmers sure are confident. That way we could stay on schedule with everything else. I vote for *not* testing the accounting portion, but just sort of winging it when it starts up."

As the most senior member of the team present, what can you do to convince Lou and Stan about the importance of testing the accounting module with live data? What can systems analysts do in planning their time to allow adequate time for testing with test and live data? What are some of the possible problems the team members may encounter if they do not test the system completely with live data before putting the system into production? Realistically, are there steps in the systems analysis and design process that can be collapsed to bring a delayed project in on time? Respond to these questions in two double-spaced pages.

FULL SYSTEMS TESTING WITH LIVE DATA. When systems tests using test data prove satisfactory, it is a good idea to try the new system with several passes on what is called live data, data that have been successfully processed through the existing system. This step allows an accurate comparison of the new system's output with what you know to be correctly processed output, as well as a good idea for testing how actual data will be handled. Obviously, this step is not possible when creating entirely new outputs (for instance, output from an ecommerce transaction from a new corporate website). As with test data, only small amounts of live data are used in this kind of system testing.

Although much thought is given to user–system interaction (see Chapter 14), you can never fully predict the wide range of differences in the way users will actually interact with a system. It is not enough to interview users about how they are interacting with the system; you must observe them firsthand.

Items to watch for are ease of learning the system and user reaction to system feedback, including what happens when an error message is received, and what happens when the user is informed that the system is executing his or her commands. Be particularly sensitive to how users react to system response time and to the language of responses. Also listen to what users say about the system as they encounter it. Any real problems need to be addressed before the system is put into production, not just glossed over as adjustments to the system that users and operators ought to make on their own.

Procedure manuals, just like computer software, also need to be tested. Although manuals can be proofread by support staff and checked for technical accuracy by the systems analysis team, the only real way to test them is to have users and operators try them, preferably during full systems testing with live data. Consider user suggestions, and incorporate them into the final versions of web pages, printed manuals, and other documentation.

Maintenance Practices

Your objective as a systems analyst should be to install or modify systems that have a reasonably useful life. You want to create a system whose design is comprehensive and farsighted enough to serve current and projected user needs for several years to come. Part of your expertise should be used to project what those needs might be and then build flexibility and adaptability into the system. The better the system design, the easier it will be to maintain and the less money the business will have to spend on maintenance.

Reducing maintenance costs is a major concern, and software maintenance alone can devour upward of 50 percent of the total data processing budget for a business. Excessive maintenance costs reflect directly back on the system's designer, because approximately 70 percent of software errors have been attributed to inappropriate software design. From a systems perspective, it makes sense that detecting and correcting software design errors early on is less costly than letting errors remain unnoticed until maintenance is necessary.

Maintenance is performed most often to improve the existing software rather than to respond to a crisis or system failure. Maintenance is also done to update software in response to the changing organization. This work is not as substantial as enhancing the software, but it must be done. Emergency and adaptive maintenance comprises less than half of all system maintenance.

Part of a systems analyst's job is to ensure that there are adequate channels and procedures in place to permit feedback about—and subsequent response to—maintenance needs. Users must be able to communicate problems and suggestions easily to those who will be maintaining the system. Solutions are to provide users email access to technical support, as well as to allow them to download product updates or patches from the Web.

Auditing

Auditing is yet another way of ensuring the quality of the information contained in a system. Broadly defined, auditing refers to having an expert who is not involved in setting up or using a system examine information in order to ascertain its reliability. Whether or not information is found to be reliable, the finding on its reliability is communicated to others for the purpose of making the system's information more useful to them.

For information systems, there are generally two kinds of auditors: internal and external. Whether both are necessary for the system you design depends on what kind of system it is. Internal auditors work for the same organization that owns the information system, whereas external (also called independent) auditors are hired from the outside.

External auditors are used when the information system processes data that influences a company's financial statements. External auditors audit the system to ensure the fairness of the financial statements being produced. They may also be brought in if there is something out of the ordinary occurring that involves company employees, such as suspected computer fraud or embezzlement.

Internal auditors study the controls used in the information system to make sure that they are adequate and that they are doing what they are purported to be doing. They also test the adequacy of security controls. Although they work for the same organization, internal auditors do not report to the people responsible for the system they are auditing. The work of internal auditors is often more in-depth than that of external auditors.

Implementing Distributed Systems

If the reliability of a telecommunications network is high, it is possible to have distributed systems for businesses, a setup that can be conceived of as an application of telecommunications. The concept of distributed systems is used in many different ways. Here it will be taken in a broad sense, including workstations that can communicate with each other and data processors, as well as different hierarchical architectural configurations of data processors that communicate with each other and that have differing data storage capabilities.

In this model, the processing functions are delegated either to clients (users) or to servers, depending on which machines are most suitable for executing the work. In this type of architecture, the client portion of a network application will run on the client system, with the server part of the application running on the file server. With a client/server model, users interact with limited parts of the application, including the user interface, data input, database queries, and report generation. Controlling user access to centralized databases, retrieving or processing data, and other functions (such as managing peripheral devices) are handled by the server.

Client/Server Technology

The client/server model, client/server computing, client/server technology, and client/server architecture all refer to a design model that can be thought of as applications running on a network. In very basic terms, you can picture the client requesting—and the server executing or in some way fulfilling—the request. This would be considered two-tiered client/server architecture.

A more involved configuration uses three sets of computers to accomplish retrieval, processing, storage, and receiving of data. Figure 16.7 shows a three-tiered client/server model. In this figure, client computers access three different tiers of servers; Web servers, which handle Web-based exchange of information; application servers, which process data to and from the client computers and the database server; and the database server, which stores and receives data. The computers on the network are programmed to perform work efficiently by dividing up processing tasks among clients and servers.

When you think of the client/server model, you should think of a system that accentuates the users as the center of the work, with their interaction with data being the key concept. Although there are two elements working—the client and the server—it is the intent of the client/server model that users view it as one system. Indeed, the hope is that users are unaware of how the client/server network is performing its distributed processing, because it should have the look and feel of a unified system. In a peer-to-peer network, PCs can act as either the server or the client, depending on the requirements of the application.

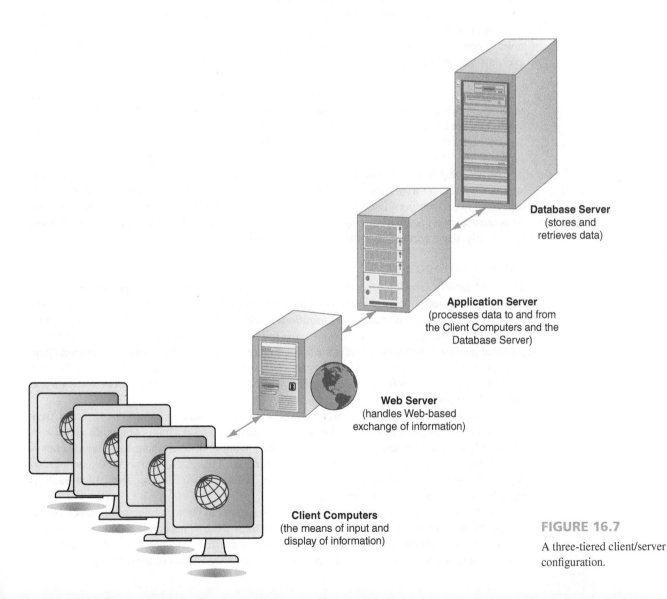

Database Server
(stores and retrieves data)

Application Server
(processes data to and from the Client Computers and the Database Server)

Web Server
(handles Web-based exchange of information)

Client Computers
(the means of input and display of information)

FIGURE 16.7

A three-tiered client/server configuration.

CLIENTS AS PART OF THE CLIENT/SERVER MODEL. When you see the term *client*, you might be tempted to think of people or users; for example, we speak of "clients of our consulting practice." In the client/server model, however, the term *client* refers not to people but to networked machines that are typical points of entry to the client/server system that is used by humans. Therefore, clients could be networked desktop computers, a workstation, or notebook computers, or any other way in which the user can enter the system.

Using a graphical user interface (GUI), individuals typically interface directly only with the client part. Client workstations use smaller programs that reside in the client to do front-end processing (as opposed to the back-end processing, mentioned later), including communicating with the user. If an application is called a client-based application, the application resides in a client computer and cannot be accessed by other users on the network.

WEIGHING THE ADVANTAGES AND DISADVANTAGES OF THE CLIENT/SERVER MODEL. Early adopters of the client/server model found that this model was not always the best solution to an organization's computing problems. Often, the systems designer is asked to endorse a client/server model that is already in the works. Just as with any other corporate computing proposal that you did not have an active part in creating, you must review the plan carefully. Will the organization's culture support a client/server model? What kinds of changes must be made in the informal culture and in the formal work procedures before a client/server model can be used to its full potential? What should your role as a systems analyst be in this situation?

Although lower processing costs are cited as a benefit of the client/server model, there is very little actual data available to prove it (even though there is some anecdotal evidence to support this claim). There are well-documented high start-up or switch-over costs associated with a movement to a client/server architecture. Applications for the client/server model must be written as two separate software components, each running on separate machines, but they must appear as if they are operating as one application. Using the client/server model, however, affords greater computer power and greater opportunity to customize applications than other options.

Although networks can be characterized by their shape or topology, they are also discussed in terms of their geographic coverage and the kinds of services they offer. Standard types of networks include a wide area network (WAN) and a local area network (LAN). LANs are standard for linking local computers or terminals within a department, building, or several buildings of an organization. WANs can serve users over several miles or across entire continents.

Networking is now technically, economically, and operationally feasible for small offices as well, and it provides a solution that analysts must consider for small businesses. One of the costly aspects of implementing a LAN is that each time it is moved, it must be rewired. Some organizations are coping with this by setting up a high-speed, wireless local area network (WLAN). More specifically, these wireless networks are called Wi-Fi.

Cloud Computing

The most rapidly growing type of computing is cloud computing. Cloud computing has been described as a metaphor for the Internet, since the Internet is often drawn as a cloud in network diagrams. Using cloud computing, organizations and individual users can use Web services, database services, and application services over the Internet, without having to invest in corporate or personal hardware, software, or software tools. Figure 16.8 depicts the exchanges between client computers and services in the cloud. Businesses use Web browsers such as Microsoft Internet Explorer or Mozilla Firefox to access applications. As you can see, servers store software and data for businesses.

Many large, well-established hardware, software, and consulting companies such as Cisco, Dell, IBM, HP, Microsoft, SAP, and others are creating massive cloud computing endeavors, often with what are termed "virtualized resources." What is distinct about these approaches is their ability to grow and adapt to changing business needs. That is, they are scalable to suit growing (or changing) demand by users. The model of Software as a Service, also called SaaS, is included in the concept of cloud computing.

Users do not need to understand, control, or be experts in the technology infrastructure that composes the complex cloud infrastructure that enables them to accomplish their work. Often, organizations do not need to keep IT staff to scale up or down, even when a contract or company budget changes upward or downward because of the lessened impact of these changes.

FIGURE 16.8

Cloud computing offers many services.

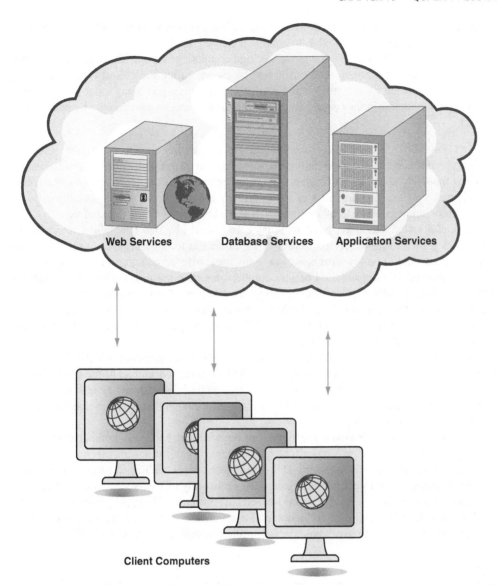

Web Services   Database Services   Application Services

Client Computers

Often organizations that are using cloud computing do not find it necessary to make up-front capital expenditures on IT infrastructure, so it enables smaller companies with smaller and less predictable budgets to make advances in processing more quickly. It also allows larger corporations to invest in strategic projects rather than IT infrastructure.

Sharing IT resources via cloud computing means that a large collection of corporate users share Web services, but also then jointly bear the lowered cost of them; realize increases in peak-load capacity, and ensure that underutilized systems are used more efficiently and widely.

Companies also hope to improve their ability to perform disaster recovery by using cloud computing that provides many redundant sites. While cloud computing is not immune to outages, it can spread the risk to multiple servers.

Organizations try to improve security via cloud computing by using services that are sold as possessing a security focus. However, there are concerns that centralization of this sort also can translate into loss of control over mission-critical data. Users might benefit from the mobility afforded by being liberated from a single computer installation or single interface. Rather, Web browsers and Web-based services made available through cloud computing free users to access applications from anywhere at any time without regard for location or the device they are using.

Many large software companies (some of which are called "pure players" since they have never existed as brick-and-mortar enterprises) are offering applications using cloud computing, where users can use their Web browser to access applications. These include Google Apps (for spreadsheets and calendars), Amazon Web Services, Akami, and CRM software by Salesforce.com that is now available on the iPhone as well. These software purveyors state that they are attempting to lower the cost to the user, as well as to provide increased flexibility.

Some observers believe that the move to cloud computing is a way for older, larger companies to solidify and retain their core businesses by incorporating SaaS (Software as a Service), SOA (service-oriented architecture), virtualization, open source, and many other trends from the past decade into their offerings of software, software tools, services, and computing power via the Internet.

ERP SYSTEMS AND THE CLOUD. Researchers are suggesting that many of the issues and challenges associated with implementing ERP systems can be mitigated by new, lower-cost ERP offerings that take advantage of cloud computing. Organizations do not have to purchase costly new IT; new versions and upgrades of ERP software are maintained and installed by the ERP cloud vendor, and switching to other modules may be less cumbersome than before.

A vendor called Workday, www.workday.com (created by two PeopleSoft veterans), provides SaaS (Software-as-a-Service) solutions for midrange to large organizations. These solutions are advertised to be low cost and easy to upgrade, offer a reduction in data storage costs, provide a greener IT footprint, and be less cumbersome than traditional ERP implementations using a predictable subscription pricing model. In addition, Workday asserts that its systems address some of the long-standing problems experienced with legacy ERP implementations, including unpredictable costs of on-premises ERP systems, piecemeal solutions to IT business problems, and variable level of service.

Another prominent cloud ERP vendor is NetSuite, www.netsuite.com, which offers accounting software real-time inventory management, CRM, ecommerce, and real-time global business management. NetSuite differentiates its services by emphasizing reduced IT cost and real-time visibility and anywhere, anytime access to business information that decision makers find critical to running an organization.

As with any other endeavor in the cloud, but especially an ERP system that can store sensitive corporate data and run mission-critical applications, organizations are creating new tools to assess the potential value to furthering strategic objectives, as well as to assess the risk to the enterprise, examine the security issues, gauge the probability of user acceptance, and gauge the likelihood of success in integrating a cloud-based ERP into their IT supply chain. As an analyst, you have developed several data collection and analytical tools that can facilitate the ability of the organization to come to reasonable decisions about adopting cloud-based ERP.

Network Modeling

Because networking has become so important, a systems designer needs to consider network design. Whether a systems designer gets involved with decisions about the configurations of networks—or whether he or she worries about hardware such as routers and bridges that must be in place when networks meet—the systems designer must always consider the logical design of networks.

An analyst should adopt a set of symbols such as the ones in Figure 16.9 to model a network. It is useful to have distinct symbols to distinguish among hubs, external networks, and workstations. It is also useful to adopt a convention for illustrating multiple networks and workstations. The first step is to draw a network decomposition diagram that provides an overview of the system. Next, draw a hub connectivity diagram. Finally, explode the hub connectivity diagram to show the various workstations and how they are to be connected.

DRAWING A NETWORK DECOMPOSITION DIAGRAM. We can illustrate drawing a network decomposition model by referring once again to the World's Trend Catalog Division example from earlier chapters. Start by drawing a circle and labeling it "World's Trend Network." Now draw a number of circles below the first one, as shown in Figure 16.10. These circles represent hubs for the Marketing Division and each of the three order-entry and distribution centers (the U.S. Division, the Canadian Division, and the Mexican Division).

Then you can extend this drawing further by drawing another level. This time, you can add the workstations. For example, the Marketing Division has two workstations connected to it, whereas the U.S. Division has 33 workstations on its LAN (Administration, the Warehouse, the Order-Entry Manager, and 30 Order-Entry Clerks). This network is simplified for the purpose of providing a readily understandable example.

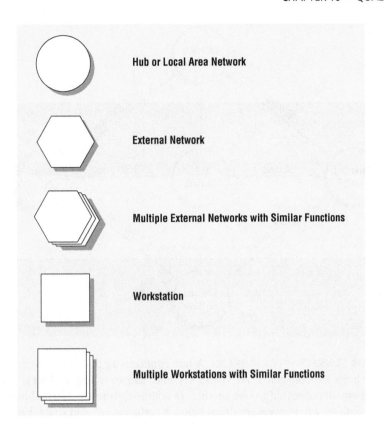

FIGURE 16.9

Use special symbols when drawing network decomposition and hub connectivity diagrams.

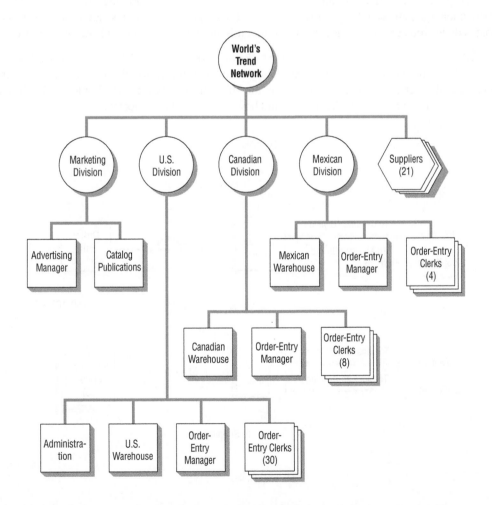

FIGURE 16.10

A network decomposition diagram for World's Trend.

FIGURE 16.11

A hub connectivity diagram for World's Trend.

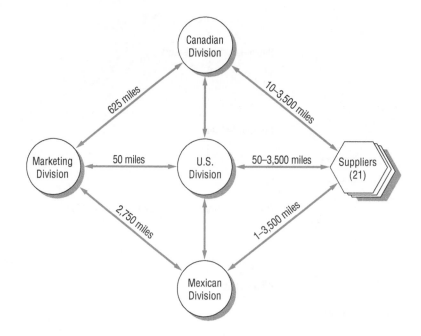

CREATING A HUB CONNECTIVITY DIAGRAM. A hub connectivity diagram is useful for showing how the major hubs are connected. At World's Trend, shown in Figure 16.11, there are four major hubs that are all connected to one another. In addition, there are external hubs (suppliers) that need to be notified when inventory drops below a certain point, and so on. Each of the three country divisions are connected to the 21 suppliers; the Marketing Division, however, does not need to be connected to suppliers.

To produce an effective hub connectivity diagram, start by drawing all the hubs. Then experiment (perhaps sketching it first on a sheet of paper) to see which links are necessary. Once that is done, you can redraw the diagram so that it is attractive and communicates well to users.

EXPLODING THE HUB CONNECTIVITY DIAGRAM INTO A WORKSTATION CONNECTIVITY DIAGRAM. The purpose of network modeling is to show the connectivity of workstations in some detail. Therefore, you need to explode the hub connectivity diagram. Figure 16.12 shows the 33 workstations for the U.S. Division and how they are to be connected.

FIGURE 16.12

A workstation connectivity diagram for World's Trend.

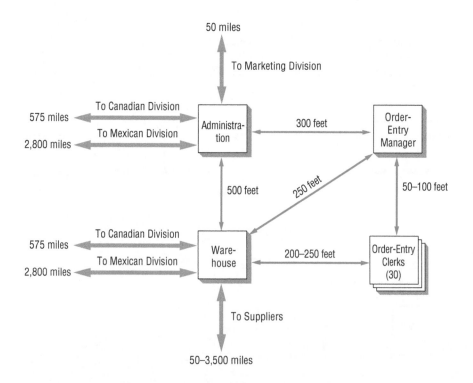

You draw the diagrams for this level by examining the third level of the network decomposition diagram. You group items such as Order-Entry Manager and Order-Entry Clerks together because you already recognize that they must be connected. Use a special symbol to show multiple workstations and indicate in parentheses the number of similar workstations. In our example, there are 30 Order-Entry Clerks.

On the perimeter of the diagram, you place workstations that must be connected to other hubs. In this way, it will be easier to represent these connections using arrows. You draw the external connections in a different color or use thicker arrows. External connections are usually long distance. For example, Administration is connected to the Marketing Division, which is 50 miles away, and also to the Canadian and Mexican Divisions. The Warehouse needs to communicate directly with the Canadian and Mexican warehouses in case it is possible to obtain the merchandise from another warehouse. The Order-Entry Manager and Order-Entry Clerks do not have to be connected to anyone outside their LAN.

Training Users

Systems analysts engage in an educational process with users that is called training. Users have been involved throughout the systems development life cycle, so by now an analyst should have an accurate assessment of the users who must be trained.

In the implementation of large projects, an analyst will often be managing the training rather than be personally involved in it. One of the most prized assets an analyst can bring to any training situation is the ability to see the system from the user's viewpoint. An analyst must never forget what it is like to face a new system. Those recollections can help analysts empathize with users and facilitate their training.

Training Strategies

Training strategies are determined by who is being trained and who will train them. An analyst will want to ensure that anyone whose work is affected by the new information system is properly trained by the appropriate trainer.

WHOM TO TRAIN. All people who will have primary or secondary use of a system must be trained. They include everyone from data entry personnel to those who will use output to make decisions without personally using a computer. The amount of training a system requires depends on how much someone's job will change because of the new interactions required by the revised system.

You must ensure that users of different skill levels and job interests are separated. It is certain trouble to include novices in the same training sessions as experts, because novices are quickly lost and experts are rapidly bored with basics. Both groups are then lost.

PEOPLE WHO TRAIN USERS. For a large project, many different trainers may be used, depending on how many users must be trained and who they are. Possible training sources include the following:

1. Vendors
2. Systems analysts
3. External paid trainers
4. In-house trainers
5. Other system users

This list gives just a few of the options an analyst has in planning for and providing training.

Large vendors often provide off-site one- or two-day training sessions on their equipment as part of the service benefits offered when corporations purchase expensive COTS software. These sessions include both lectures and hands-on training in a focused environment. They may also extend the experience with online user groups, dedicated blogs, or annual user conferences.

Because systems analysts know the organization's people and the system, they can often provide good training. The use of analysts for training purposes depends on their availability, because they also are expected to oversee the complete implementation process.

External paid trainers are sometimes brought into an organization to help with training. They may have broad experience in teaching people how to use a variety of computers, but they may not give the hands-on training that is needed for some users. In addition, they may not be able to custom-tailor their presentations enough to make them meaningful to users.

Full-time, in-house trainers are usually familiar with the skills and learning preferences of personnel and can tailor materials to their needs. One of the drawbacks of in-house trainers is that they may possess expertise in areas other than information systems and may therefore lack the depth of technical expertise that users require.

It is also possible to have any of these trainers train a small group of people from each functional area that will be using the new information system. They can, in turn, train the remaining users. This approach can work well if the original trainees still have access to materials and trainers as resources when they themselves are providing training. Otherwise, it might degenerate into a trial-and-error situation rather than a structured one.

Guidelines for Training

An analyst has four major guidelines for setting up training: (1) Establish measurable objectives, (2) use appropriate training methods, (3) select suitable training sites, and (4) employ understandable training materials.

TRAINING OBJECTIVES. Who is being trained in large part dictates the training objectives. Training objectives for each group must be spelled out clearly. Well-defined objectives are of enormous help in letting trainees know what is expected of them. In addition, objectives allow evaluation of training when it is complete. For example, operators must know such basics as turning on the machine, what to do when common errors occur, basic troubleshooting, and how to end an entry.

TRAINING METHODS. Each user and operator will need slightly different training. To some extent, their jobs determine what they need to know, and their personalities, experience, and backgrounds determine how they learn best. Some users learn best by seeing, others by hearing, and still others by doing. Because it is often not possible to customize training for an individual, using a combination of methods is often the best way to proceed. That way, most users are reached through one method or another.

Methods for those who learn best by seeing include demonstrations of equipment and exposure to training manuals. Those who learn best by hearing will benefit from lectures about procedures, discussions, and question-and-answer sessions among trainers and trainees. Those who learn best by doing need hands-on experience with new equipment. For jobs such as that of computer operator, hands-on experience is essential, whereas a quality assurance manager for a production line may only need to see output, learn how to interpret it, and know when it is scheduled to arrive.

TRAINING SITES. Training takes place in many different locations, some of which are more conducive to learning than others. Large computer vendors provide special off-site locations at which operable equipment is maintained free of charge. Their trainers offer hands-on experience as well as seminars in settings that allow users to concentrate on learning the new system. One of the disadvantages of off-site training is that users are away from the organizational context in which they must eventually perform.

On-site training in the users' organization is also possible with several different kinds of trainers. The advantage is that users see the equipment placed as it will be when it is fully operational in the organizational context. A serious disadvantage is that trainees often feel guilty about not fulfilling their regular job duties if they remain on-site for training. Thus, full concentration on training may not be possible.

Off-site training sites are also available for a fee through consultants and vendors. Training sites can be set up in places with rented meeting space, such as a hotel, or may even be permanent facilities maintained by the trainers. These arrangements allow workers to be free from regular job demands, but they may not provide equipment for hands-on training.

CONSULTING OPPORTUNITY 16.4

You Can Lead a Fish to Water . . . But You Can't Make It Drink

"Sam Monroe, Belle Uga, Wally Ide, and you make up a four-member systems analysis team that is developing an information system to help managers monitor and control water temperature, the number of fish released, and other factors at a large commercial fish hatchery. (They were last seen in "To Hatch a Fish," Consulting Opportunity 6.3, when they asked you, as their fourth member, to help solve a problem involving the timely delivery of a system prototype.)

With your input, the team successfully turned the tide of the earlier dilemma, and the project has continued. Now you are discussing the training that you have begun to undertake for managers and other systems users. Due to some scheduling difficulties, you have decided to cut down on the number of different training sessions offered, which has resulted in users at a variety of levels of management and computer expertise being in the same training sessions in some instances.

Laurie Hook, one of the operators who is being trained, has been in the same training "tank" with Wade Boot, one of the managers with whom you have been working. Both Laurie and Wade have come to the team privately with different concerns.

Wade told you, "I'm mad that I have to type in my own data in the sessions. The Mississippi will freeze solid before I ever do that on my job. I've got to know *when* to expect output and how to interpret it when it comes. I'm not spending time in training sessions if I can't get that."

Laurie, who shares training sessions with Wade, also complained to your group. "We should be getting more hands-on training. All we hear is a bunch of lectures. It's like school. Not only that, but the managers in the group like to spin these 'fish stories' about what happened to them with the old system. It's boring. I want to know how to operate the thing. It's bait and switch, if you ask me. I'm not learning what you said I would, and besides, with all those bosses in there, I feel like a fish out of water."

What problems are occurring with the training sessions? How can they be addressed, given the scheduling constraints mentioned? What basic advice on setting up training sessions did your team ignore? Write a one-page response to these questions.

TRAINING MATERIALS. In planning for the training of users, systems analysts must realize the importance of well-prepared training materials. These materials include training manuals; training cases, in which users are assigned to work through a case that incorporates most of the commonly encountered interactions with the system; and prototypes and mock-ups of output. Users of larger systems will sometimes be able to train on elaborate Web-based simulations or software that is identical to what is being written or purchased. Most COTS software vendors provide online tutorials that illustrate basic functions, and vendors may maintain websites that feature pages devoted to FAQs, which can be downloaded and printed. Changes to manuals can also be gleaned from many vendors' websites.

The user's understanding of a system depends on training materials being clearly written for the correct audience with a minimum of jargon. Training materials should also be well indexed and available to everyone who needs them. A summary of considerations for training objectives, methods, sites, and materials is provided in Figure 16.13.

Elements	Relevant Factors
Training Objectives	Depend on requirements of user's job
Training Methods	Depend on user's job, personality, background, and experience; use combination of lecture, demonstration, hands-on, and study
Training Sites	Depend on training objectives, cost, availability; free vendor sites with operable equipment; in-house installation; rented facilities
Training Materials	Depend on user's needs; operating manuals, cases, prototypes of equipments and output; online tutorials

FIGURE 16.13

Appropriate training objectives, methods, sites, and materials depend on many factors.

Conversion to a New System

A third approach to implementing a system is to physically convert an old information system to a new or modified one. There are many conversion strategies available to analysts, and also a contingency approach that takes into account several user and organizational variables in deciding which conversion strategy to use. There is no single best way to proceed with conversion. The importance of adequate planning and scheduling of conversion with the strategic involvement of users (which often takes many weeks), file backup, and adequate security cannot be overemphasized.

Conversion Strategies

The five strategies for converting from the old system to the new are given in Figure 16.14 and are as follows:

1. Direct changeover
2. Parallel conversion
3. Gradual, or phased, conversion
4. Modular conversion
5. Distributed conversion

Each of the five conversion approaches is described separately in the following subsections.

DIRECT CHANGEOVER. Conversion by direct changeover means that, on a specified date, users stop using the old system, and the new system is put into use. Direct changeover can be successful only if extensive testing is done beforehand, and it works best when some delays in processing can be tolerated. Direct changeover is considered a risky approach to conversion. Disruption to the work environment may occur if users resent being forced to use an unfamiliar system, without recourse. Finally, there is no adequate way to compare new results with old.

PARALLEL CONVERSION. Parallel conversion refers to running an old system and a new system at the same time, in parallel. When the same results can be gained over time, the new system continues to be used and the old one is stopped. One advantage of running both systems in parallel is the possibility of checking new data against old data to catch any errors in processing in the new system. The main disadvantages include the cost of running two systems at the same time and the burden on employees by virtually doubling their workload during conversion.

GRADUAL CONVERSION. Gradual, or phased, conversion attempts to combine the best features of the two previously mentioned plans, without incurring all the risks. In this plan, the volume of transactions handled by the new system is gradually increased as the system is phased in. The advantages of this approach include allowing users to get used to the system gradually, the possibility of detecting and recovering from errors without a lot of down time, and the ability to add features one-by-one. Agile methodologies tend to use this conversion approach.

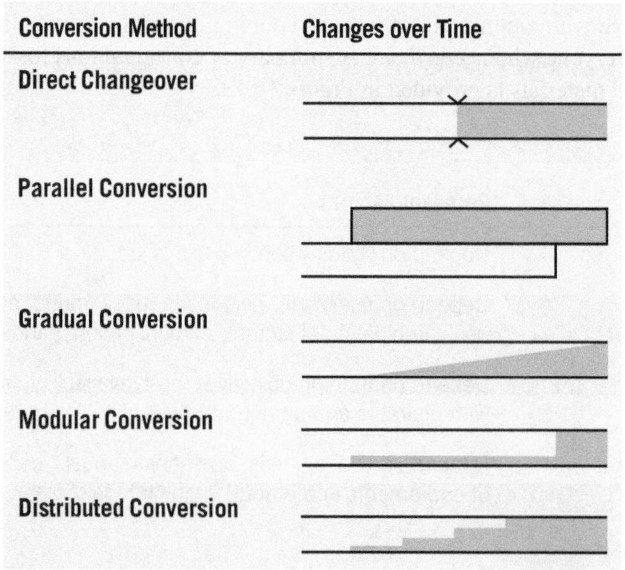

FIGURE 16.14

Five conversion strategies for information systems.

MODULAR CONVERSION. Modular conversion uses the building of self-contained, operational subsystems to change from old systems to new in a gradual manner. As each module is modified and accepted, it is put into use. One advantage is that each module is thoroughly tested before being used. Another advantage is that users are familiar with each module as it becomes operational. Their feedback has helped determine the final attributes of the system. Object-oriented methodologies often use this approach.

DISTRIBUTED CONVERSION. Distributed conversion refers to a situation in which many installations of the same system are contemplated, as is the case in banking or in franchises such as restaurants or clothing stores. One entire conversion is done (with any of the four approaches considered previously) at one site. When that conversion is successfully completed, other conversions are done for other sites. An advantage of distributed conversion is that problems can be detected and contained rather than inflicted simultaneously on all sites. A disadvantage is that even when one conversion is successful, each site will have its own people and culture, along with regional and local peculiarities to work through, and they must be handled accordingly.

Other Conversion Considerations

Conversion entails other details for an analyst, including the following:

1. Ordering equipment (up to three months ahead of planned conversion)
2. Ordering any necessary materials that are externally supplied to the information system, such as toner cartridges, paper, preprinted forms, and magnetic media
3. Appointing a manager to supervise, or personally supervising, the preparation of the installation site
4. Planning, scheduling, and supervising programmers and data entry personnel who must convert all relevant files and databases

For many implementations, your chief role will be accurately estimating the time needed for each activity, appointing people to manage each subproject, and coordinating their work. For smaller projects, you will do much of the conversion work on your own. Many of the project management techniques discussed in Chapter 3, such as Gantt charts, PERT, function point analysis, and successfully communicating with team members, are useful for planning and controlling implementation.

Organizational Metaphors and Their Relationship to Successful Systems

Be aware of organizational metaphors when you attempt to implement a system you have just developed. Our research has suggested that the success or failure of a system may be related to the metaphors used by organizational members.

When people in an organization describe the company as a zoo, you can infer that the atmosphere is chaotic; if they describe it as a machine, everything is working in an orderly fashion. When the predominant metaphor is war, journey, or jungle, the environment is chaotic, as with the zoo. The war and journey metaphors are oriented toward an organization goal, however, whereas the zoo and jungle metaphors are not.

In addition to the machine, metaphors such as society, family, and the game all signify order and rules. Although the machine and game metaphors are goal oriented, the society and zoo metaphors do not stress the company's goal but instead allow individuals in the corporation to set their own standards and rewards. Another metaphor, organism, appears balanced between order and chaos, corporate and individual goals.

Our research suggests that the success or failure of a system may have something to do with the predominant metaphor. Figure 16.15 shows that a traditional MIS will tend to succeed when the predominant metaphor is society, machine, or family, but it might not succeed if the metaphor is war or jungle (two chaotic metaphors). Notice, however, that competitive systems will most likely succeed if the metaphor is war.

Positive metaphors appear to be game, organism, and machine. Negative metaphors appear to be jungle and zoo. The others (journey, war, society, and family) show mixed success, depending on the type of information system being developed. More research needs to be done in this

FIGURE 16.15

Organizational metaphors may contribute to the success or failure of an information system.

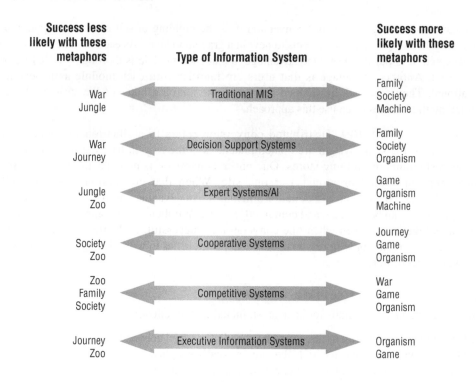

Success less likely with these metaphors	Type of Information System	Success more likely with these metaphors
War Jungle	Traditional MIS	Family Society Machine
War Journey	Decision Support Systems	Family Society Organism
Jungle Zoo	Expert Systems/AI	Game Organism Machine
Society Zoo	Cooperative Systems	Journey Game Organism
Zoo Family Society	Competitive Systems	War Game Organism
Journey Zoo	Executive Information Systems	Organism Game

area. In the meantime, as a systems analyst, you should be aware that metaphors communicated in interviews could be meaningful and may even be a factor that contributes to the success of the information system implementation.

Security Concerns for Traditional and Web-Based Systems

Security of computer facilities, stored data, and the information generated is part of a successful conversion. Recognition of the need for security is a natural outgrowth of the belief that information is a key organizational resource. With increasingly complex transactions and many innovative exchanges, the Web has brought heightened security concerns to the IS professional's world.

It is useful to think of security of systems, data, and information on an imaginary continuum from totally secure to totally open. Although there is no such thing as a totally secure system, the actions analysts and users take are meant to move systems toward the secure end of the continuum by lessening the system's vulnerability. It should be noted that as more people in the organization gain greater computer power, gain access to the Web, or connect to intranets and extranets, security becomes increasingly difficult and complex. Sometimes, organizations will hire a security consultant to work with a systems analyst when security is crucial to successful operations.

Security is the responsibility of all those who come into contact with the system and is only as good as the most lax behavior or policy in the organization. Security has three interrelated aspects: physical, logical, and behavioral. All three must work together if the quality of security is to remain high.

Physical Security

Physical security refers to securing the computer facility, its equipment, and software through physical means. It can include controlling access to the computer room by means of machine-readable badges, biometric systems, or a human sign-in/sign-out system, as well as using closed-circuit television cameras to monitor computer areas, backing up data frequently, and storing backups in a fireproof, waterproof area, often at a secure off-site location, or using data storage provided by cloud vendors.

In addition, small computer equipment should be secured so that a typical user cannot move it, and it should be guaranteed uninterrupted power. Alarms that notify appropriate people of fire, flood, or unauthorized human intrusion must be in working order at all times.

Decisions about physical security should be made along with users when an analyst is planning for computer facilities and equipment purchases. Obviously, physical security can be much tighter if anticipated in advance of actual installation and if computer rooms are specially equipped for security when they are constructed rather than outfitted as an afterthought.

Logical Security

Logical security refers to logical controls in the software itself. The logical controls familiar to most users are passwords or authorization codes of some sort. When used, they permit the user with the correct password to enter the system or a particular part of a database.

Passwords, however, are treated cavalierly in many organizations. Employees have been overheard yelling a password across crowded offices, taping passwords to their display screens, and sharing personal passwords with authorized employees who have forgotten their own.

Special encryption software has been developed to protect commercial transactions on the Web, and business transactions are proliferating. Internet fraud is also up sharply, however, with few authorities trained in catching Internet criminals and a "wild west" or "last frontier" mentality clearly evidenced in those instances when authorities have been able to apprehend Web criminals.

One way for networks to cut down on the risk of exposure to security challenges from the outside world is to build a firewall or firewall system. A firewall constructs a barricade between an internal organization's network and an external (inter)network, such as the Internet. The internal network is assumed to be trustworthy and secure, whereas the Internet is not. Firewalls are intended to prevent communication into or out of the network that has not been authorized and that is not wanted. A firewall system is not a perfect remedy for organizational and Internet security; it is, however, an additional layer of security that is now widely endorsed. There is still no fully integrated way to address security problems with internal and external networks, but they do deserve analysts' attention when planning any new or improved systems.

Logical and physical controls are important but clearly not enough to provide adequate security. Behavioral changes are also necessary.

Behavioral Security

The behavioral expectations of an organization are implicit in its policy manuals and even on signs posted in workrooms and lunch rooms, as we saw in Chapter 5. The behavior that organization members internalize, however, is also critical to the success of security efforts. (One reason firewalls are not attack proof is that many attacks to information systems come from within the organization.)

Security can begin with the screening of employees who will eventually have access to computers, data, and information, to ensure that their interests are consistent with the organization's interests and that they fully understand the importance of carrying through on security procedures. Policies regarding security must be written, distributed, and updated so that employees are fully aware of expectations and responsibilities. It is typical that a systems analyst will first have contact with the behavioral aspects of security. Some organizations have written rules or policies prohibiting employees from surfing the Web during work hours, or even prohibiting Web surfing altogether, if company equipment is involved. Other corporations use software locks to limit access to websites that are judged to be objectionable in the workplace, such as game, gambling, or pornographic sites.

Part of the behavioral facet of security is monitoring behavior at irregular intervals to ascertain that proper procedures are being followed and to correct any behaviors that may have eroded with time. Having a system log the number of unsuccessful sign-on attempts of users is one way to monitor whether unauthorized users are attempting to sign on to the system. Periodic and frequent inventorying of equipment and software is desirable. In addition, unusually long sessions or atypical after-hours access to the system should be examined.

Employees should clearly understand what is expected of them, what is prohibited, and the extent of their rights and responsibilities. In the United States and the European Union, employers are legally obligated to disclose all monitoring that is being done or that is being contemplated, and they must supply the rationale behind it. Such disclosure should include the use of video cameras, software, and phone monitoring.

Output generated by a system must be recognized for its potential to put an organization at risk in some circumstances. Controls for output include displays that can only be accessed via password, the classification of information (that is, to whom it can be distributed and when), and secure storage of printed and stored documents, no matter what their format.

In some cases, provisions must be made for shredding documents that are classified or proprietary. Shredding or pulverization services can be contracted from an outside firm that, for a fee, will shred magnetic media, printer cartridges, and paper. A large corporation may shred upward of 76,000 pounds of output in a variety of media annually.

Special Security Considerations for Ecommerce

It is well known that intruders can violate the integrity of any computer system. As an analyst, you need to take a series of precautions to protect the computer network from both internal and external Web security threats. A number of actions and products can help you:

1. Software that detects and removes malware.
2. Email filtering products that provide policy-based email and email attachment scanning and filtering to protect companies against both incoming and outgoing email. Incoming scanning protects against spam (unsolicited email such as advertising) attacks, and outgoing scanning protects against the loss of proprietary information.
3. URL filtering products that provide employees with access to the Web by user, by groups of users, by computers, by the time, or by the day of the week.
4. Firewalls, gateways, and virtual private networks that prevent hackers from gaining backdoor access to a corporate network.
5. Intrusion detection and antiphishing products that continually monitor usage, provide messages and reports, and suggest actions to take.
6. Vulnerability management products that assess the potential risks in a system and discover and report vulnerabilities. Some products correlate the vulnerabilities to make it easier to find the root cause of the security breach. Risk cannot be eliminated, but this software can help manage the risk by balancing security risk to the financial bottom line.
7. Security technologies such as Secure Sockets Layer (SSL) for authentication.
8. Encryption technologies such as Secure Electronic Transaction (SET).
9. Public Key Infrastructure (PKI) and digital certificates (obtained from a company such as VeriSign). Use of digital certificates ensures that the reported sender of the message is really the company that sent the message.

Privacy Considerations for Ecommerce

The other side of security is privacy. To make your website more secure, you must ask the user or customer to give up some privacy.

As a website designer, you will recognize that the company for which you design exercises a great deal of power over the data its customers are providing. The same tenets of ethical and legal behavior apply to website design as to the design of any traditional application that accepts personal data from customers. The Web, however, allows the data to be collected faster and allows different data to be collected (such as the browsing habits of the customer). In general, information technology makes it possible to store more data in data warehouses, process that data, and distribute the data more widely.

Every company for which you design an ecommerce application should adopt a privacy policy. Here are some guidelines:

1. Start with a corporate policy on privacy. Make sure it is prominently displayed on the website so that all customers can access the policy whenever they complete a transaction.
2. Ask for only information the application requires to complete the transaction at hand. For example, is it necessary to the transaction to ask a person's age or gender?
3. Make it optional for customers to fill out personal information on the website. Some customers do not mind receiving targeted messages, but you should always give customers an opportunity to maintain the confidentiality of their personal data by not responding.
4. Use sources that allow you to obtain anonymous information about classes of customers. There are companies that offer audience profiling technology and technology solutions for management of advertisements, their targeting, and their delivery. They do so by

maintaining a dynamic database of consumer profiles without linking them to individuals, thereby respecting customers' rights to privacy.

5. Be ethical. Avoid the latest cheap trick that permits your client to gather information about the customer in highly suspect ways. Tricks such as screen scraping (capturing remotely what is on a customer's screen) and email cookie grabbing are clear violations of privacy, and may prove to be illegal as well.

A coordinated policy of security and privacy is essential. It is essential to establish these policies and adhere to them when implementing an ecommerce application.

Disaster Recovery Planning

No matter how diligently you and your organizational colleagues work to ensure the security and stability of systems, all employees and systems are inevitably vulnerable to some kind of natural or human-caused disaster that threatens security as well as the functioning of the business. Some disasters are quite common, such as power outages. We can assess the probability of some disasters occurring, such as a hurricane or an earthquake. However, many disasters are unexpected in their timing or their severity, perhaps even causing loss of life, creating chaos for people and the organization itself.

The fields of disaster preparedness and disaster recovery are interdependent, and they build on each other. Disaster preparedness includes what a company should do if it encounters a crisis. The field of disaster recovery is focused on how a business can continue in the aftermath of a disaster and how it can restore essential systems in the IT infrastructure. This section focuses on disaster recovery as it relates to information systems. The traditional disaster recovery process consists of planning, a walkthrough, practice drills, and recovery from the disaster.

When hit with a disaster, a company stands to lose people, money, reputation, and its own assets, as well as those of their clients. It is important to do the right things to minimize potential losses. Analysts should determine what the organization's level of disaster planning is and how well articulated the role of information systems is in their disaster response and recovery plans. The key questions that analysts must ask early on are (1) whether employees know where to go and (2) what to do in the face of a disaster. The answers to these questions will guide your further planning. Conventional wisdom provides seven elements to consider during and after a disaster. As you will see, many of them involve information systems and relate specifically to the planning required of you as a systems analyst:

1. Identify the teams responsible for managing a crisis.
2. Eliminate single points of failure.
3. Determine data replication technologies that match the organization's timetable for getting systems up and running.
4. Create detailed relocation and transportation plans.
5. Establish multiple communication channels among employees and consultants who are on-site, such as analysis teams.
6. Provide recovery solutions that include an off-site location.
7. Ensure the physical and psychological well-being of employees and others who may be physically present at the work site when a disaster hits.

The disaster preparedness plan should identify who, in the event of a disaster, is responsible for making several pivotal decisions. These include decisions about whether business operations will continue, how to support communications (both computer and voice), where people will be sent if the business is uninhabitable, where personnel will go in an emergency, seeing to the personal and psychological needs of the people present in the business and those who might be working virtually, and restoring the main computing and work environments.

Redundancy of data provides the key for eliminating single points of failure for servers running Web applications. As an analyst, you can be especially helpful in setting up this type of backup and redundancy.

Some businesses are moving to storage area networks (SANs) to get away from some of the unreliability associated with physical tape backups and storage. Synchronous remote replication, also called data mirroring, for nearly real-time backup is also gaining favor. However, if companies are farther than 100 miles away from the site, the data mirroring process can be affected. Asynchronous remote replication sends data to the secondary storage location at designated time intervals. Online options are available for small businesses, too.

An organization should develop and distribute a one-page memo that contains evacuation routes and employee assembly points. This should be distributed to everyone in the organization. The three common choices are either to send employees home, to have them remain on-site, or to relocate them to a recovery facility that is set up to continue operations. The entire gamut of transportation options should be considered when developing this memo.

Organizational and analysis team members must be able to communicate in the event that their typical email is disrupted. If email is unavailable for broadcasting an emergency message, an emergency information web page or emergency hotline can serve as a viable alternative. Recently, some software companies have started offering a suite of software tools that permits ad hoc communication by emergency response agencies that allows them to rapidly set up secure VoIP, Web connectivity, and Wi-Fi hot spot capabilities. Wider availability and lower prices will undoubtedly bring these important communication capabilities to other types of organizations in the future.

To better protect an organization's backup systems and to ensure the continued, uninterrupted flow of banking transactions in the event of a disaster, new regulations in the United States stipulate that bank off-site locations must be at least 100 miles away from the original site. Since paper files and backups also present a monumental problem and are highly vulnerable to natural and human-caused disasters, organizations are strongly encouraged to create a plan that helps them move toward a digital documentation project that is meant to convert all of their paper documents to electronic formats within three to five years of inception (Stephens, 2003).

Support for humans working at an organization that experiences a disaster is paramount. There must be plentiful and easily available water, especially if employees are unable to leave the site for a number of days due to outside weather conditions or partial building collapses. While food is important, water is more so. Employees should also be issued a safety kit containing water, a dust mask, a flashlight, glow sticks, and a whistle. One way to learn what should comprise a personal workspace disaster supplies kit is to go to the American Red Cross website (www.redcross.org), which provides details for supporting humans during disasters and providing for them in the aftermath.

Evaluation

Throughout the systems development life cycle, an analyst, management, and users have been evaluating the evolving information systems and networks to give feedback for their eventual improvement. Evaluation is also called for following system implementation.

Evaluation Techniques

In recognition that the ongoing evaluation of information systems and networks is important, many evaluation techniques have been devised. These techniques include cost-benefit analysis (as discussed in Chapter 3); models that attempt to estimate the value of a decision based on the effects of revised information using information theory, simulation, or Bayesian statistics; user evaluations that emphasize implementation problems and user involvement; and information system utility approaches that examine the properties of information.

Each type of evaluation serves a different purpose and has inherent drawbacks. Cost-benefit analysis may be difficult to apply because information systems provide information about objectives for the first time, making it impossible to compare performance before and after implementation of the system or distributed network. The revised decision evaluation approach presents difficulty because all variables involved with the design, development, and implementation of the information system cannot be calculated or quantified. The user involvement approach yields some insight for new projects by providing a checklist of potentially dysfunctional behavior by various organizational members, but it stresses implementation over other aspects of IS design. The information system utility approach to evaluation can be more comprehensive than the others if it is expanded and systematically applied.

The Information System Utility Approach

The information system utility approach for evaluating information systems can be a comprehensive and fruitful technique for measuring the success of a developed system. It also can serve as a guide in the development of any future projects an analyst might undertake.

Utilities of information include possession, form, place, and time. To evaluate the information system comprehensively, these utilities must be expanded to include actualization utility and goal utility. Then the utilities can be seen to address adequately the questions of who (possession), what (form), where (place), when (time), how (actualization), and why (goal). An example of this information utility approach can be seen in the evaluation of a blood inventory system in Figure 16.16.

An information system can be evaluated as successful if it possesses all six of these utilities. If the system module is judged as "poor" in providing one of the utilities, the entire module will be destined to fail. A partial or "fair" attainment of a utility will result in a partially successful module. If the information system module is judged as "good" in providing every utility, the module is a success.

Information Systems Modules	Form Utility	Time Utility	Place Utility	Possession Utility	Actualization Utility	Goal Utility
Inventory Lists **Success**	Good. Acronyms used were the same as shipping codes. As systems grew, too much information was presented; this overload called for summary information.	Good. Reports were received at least one hour before scheduled shipments on a daily basis.	Good. Inventory lists were printed at the regional blood center. Lists were delivered to hospitals with the current shipments.	Good. The same people who originally kept manual records received these reports.	Good. Implementation was easy because hospitals found the inventory lists to be extremely useful.	Good. Information about the location of particular units was made available.
Management Summary Reports **Success**	Good. Summary report was designed to exact format specifications of manual summary reports developed by the blood administrator for city hospitals.	Good. Same as listings.	Good. Summary reports were printed at the center where they were needed.	Good. Blood administrators who originally kept manual reports received these reports.	Good. Blood administrators participated in the design of the reports.	Good. Summary reports helped reduce outdating and prevent shortages.
Short-Term Forecasting **Success**	Good. A forecast was issued for each blood type.	Good. Forecasts were updated daily.	Good. Printed at blood center.	Good. Administrators concerned with distribution and collections received the report.	Good. Output design could have been more participative.	Good. Shortages were prevented by calling in more donors.
Heuristic Allocation **Failure**	Poor. The people who allocated blood mistrusted the mysterious numbers produced by the computer.	Good. Reports were provided one hour before allocation decisions were made.	Good. Printed at blood center.	Fair. Administrators responsible for daily blood allocation received the original.	Poor. Too many people were involved with blood inventories to be able to participate in the design of the system.	Poor. This was not an immediate goal of the blood region. Shipping costs were passed on to patients.

FIGURE 16.16

Evaluating a blood inventory information and decision support system using the information system utility approach.

POSSESSION UTILITY. Possession utility answers the question of who should receive output, or, in other words, who should be responsible for making decisions. Information has no value in the hands of someone who lacks the power to make improvements in the system or someone who lacks the ability to use the information productively.

FORM UTILITY. Form utility answers the question of what kind of output is distributed to the decision maker. The documents must be useful for a particular decision maker in terms of the document's format and the jargon used. Acronyms and column headings must be meaningful to the user. Furthermore, information itself must be in an appropriate form. For example, the user should not have to divide one number by another to obtain a ratio. Instead, a ratio should be calculated and prominently displayed. At the other extreme is the presentation of too much irrelevant data. Information overload certainly decreases the value of an information system.

PLACE UTILITY. Place utility answers the question of where the information is distributed. Information must be delivered to the location where the decision is made. More detailed reports or previous management reports should be filed or stored to facilitate future access.

TIME UTILITY. Time utility answers the question of when information is delivered. Information must arrive before a decision is made. Late information has no utility. At the other extreme is the delivery of information too far in advance of the decision. Reports may become inaccurate or may be forgotten if delivered prematurely.

ACTUALIZATION UTILITY. Actualization utility involves how the information is introduced and used by the decision maker. First, the information system has value if it possesses the ability to be implemented. Second, actualization utility implies that an information system has value if it is maintained after its designers depart, or if a one-time use of the information system obtains satisfactory and long-lasting results.

GOAL UTILITY. Goal utility answers the "why" of information systems by asking whether the output has value in helping the organization obtain its objectives. The goal of the information system must not only be in line with the goals of decision makers, but it must also reflect their priorities.

Evaluating Corporate Websites

Evaluating the corporate website that you are developing or maintaining is an important part of any successful implementation effort. Analysts can use the information system utility approach previously described to assess the aesthetic qualities, content, and delivery of the site. As an analyst or a Webmaster, you should go one step further and analyze Web traffic.

A visitor to your website can generate a large amount of useful information for you to analyze. This information can be gathered automatically by capturing information about the source, including the previous website the user visited and the keywords used to find the site; the information can also be obtained through using cookies (files left on a user's computer about when they last were on the site).

A leading Web activity monitoring package is Webtrends. Figure 16.17 is a sample report showing the most-downloaded files on the website by day of the week. The graph displays the top five downloaded files, and the table at the bottom is a sorted list of all downloads.

An analyst or a Webmaster can gain valuable information by using a service such as Webtrends. (Although some services are free, the pay services usually provide the detail needed to evaluate a site in depth. The cost is an ongoing budget item for maintaining the website.) Information to help you evaluate your client's site and make improvements is plentiful and easy to obtain. The seven essential items are described next:

1. *Know how often your client's website is visited.* The number of hits a website had in the past few days, the number of visitor sessions, and the number of pages visited are a few of the general things you need to know.
2. *Learn details about specific pages on the site.* It is possible to get statistics on the most requested pages, most requested topics, top paths a visitor takes through the client's website, or even the most downloaded files. If the website is a commercial one, shopping cart reports can show how many visitors were converted into buyers and how many abandoned their carts or failed to complete the checkout process.

CONSULTING OPPORTUNITY 16.6

Mopping Up with the New System

"I don't know what happened. When the new system was installed, the systems analysts made a clean getaway, as far as I can tell," says Marc Schnieder, waxing philosophic. Recall that he is owner of the Marc Schnieder Janitorial Supply Company. (You last met Marc in Consulting Opportunity 13.1, in which you helped him with his data storage needs. In the interim, he has had a new information system installed.)

"The systems analysis team asked us some questions about how we liked the new system," Marc supplies eagerly. "We didn't really know how to tell them that the output wasn't as spotless as we'd like. I mean, it's confusing. It isn't getting to the right people

at the right time or anything. We never really did get into the nitty-gritty about the finished system with that consulting team. I feel as if we had to hire your group just to mop up after what they left."

After further discussions with Stan Lessink and Jill Oh, the company's chief programmers, you realize that the team that did the initial installation had no evaluation mechanism. Suggest a suitable framework for evaluating the kinds of concerns that Mr. Schnieder raised about the system. What are the problems that can occur when a system is not evaluated systematically? Respond in a paragraph.

3. *Find out more about the website's visitors.* Visitor demographics and information such as the number of visits by a particular visitor in a period of time, whether the visitor is a new or a returning one, and who the top visitors are—all are valuable information when evaluating a website. The display in Figure 16.18 shows the number of unique visitors (top graph), the number of first-time visitors (middle graph), and the average length of visits over time (bottom graph).

4. *Discover whether visitors can properly fill out the forms you designed.* If the error rate is high, redesign the form and see what happens. Analysis of the statistics will reveal whether bad form design was to blame for errors in response.

5. *Find out who is referring website visitors to the client's site.* Find out which sites are responsible for referring visitors to the client's website. Get statistics on the top referring site, the top search engines leading to the site, and the keywords visitors used to locate your client's website. After promoting a site, you can use Web traffic analysis to track whether the site promotion really made a difference.

6. *Determine what browsers visitors are using.* By knowing what browsers are being used, you can add browser-specific features that improve the look and feel of the site and

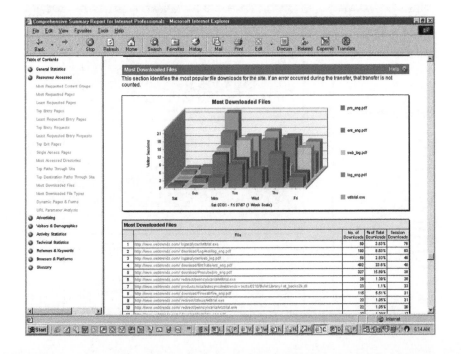

FIGURE 16.17

A sample report from Webtrends Corporation showing the most downloaded files on the corporate website.

HYPERCASE® EXPERIENCE 16.2

"As you know, Snowden is determined to implement some kind of computerized tracking for the Training people. Even after having you and your team here at MRE for all this time, though, it isn't clear to me how that will ever come about. You've probably noticed by now that people such as Tom Ketcham are pretty set in their ways, but so is Snowden, and he definitely has the upper hand. I'm not telling you anything you don't know already, am I?

"Also, please check with Jack O'Malley and Kate Eckert to make sure that what you're planning will be covered by their latest disaster recovery plan. I think when Snowden comes back from Poland, you should be ready to show him how we can implement an automated tracking system for the Training group, but it really has to be acceptable to the new users. After all, they're the ones who have to live with it. I'll pencil you in for a meeting with Snowden two weeks from today."

HYPERCASE Questions

1. Develop an implementation plan that would be useful to the Training group in changing to a computerized project tracking system. Use a paragraph to explain your approach. Be sure that what you are doing also meets Snowden's expectations.
2. In two paragraphs, discuss what *conversion* approach is appropriate for adopting a new automated project tracking system for the Training group.
3. Create a bulleted list of the measures you would take to secure and back up the new project tracking system you are proposing for the Training group.

encourage visitors to stay longer, thereby improving the stickiness of the site. It helps to know whether visitors are using current or outdated browsers.

7. ***Find out whether the client's website visitors are interested in advertising.*** Finally, find out whether visitors to the site are interested in the ad campaigns you have on your site, such as offering a product on sale for a specific period.

Web activity services can be helpful in evaluating whether a site is meeting its stated objectives in terms of traffic, advertising effectiveness, employee productivity, and return on investment. It is one of the ways an analyst can evaluate whether the corporate Web presence is meeting management goals and whether it accurately portrays the organization's vision.

FIGURE 16.18

A report comparing statistics on visitors generated by Commerce Trends (from Webtrends Corporation).

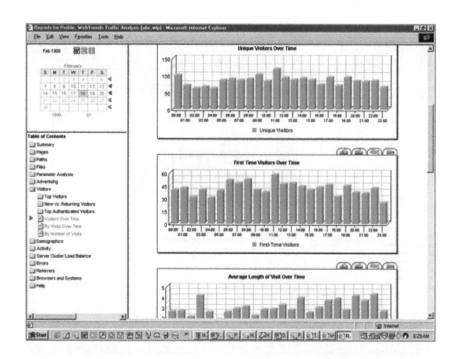

Summary

A systems analyst can ensure total quality management (TQM) for analyzing and designing information systems in many ways. Six Sigma is a culture, philosophy, methodology, and approach to quality that has as its goal the elimination of all defects. A tool for designing a top-down, modular system is called a structure chart. Service-oriented architecture is an approach that uses independent services to perform various functions. Two of the structured techniques that can aid a systems analyst are procedure manuals and FOLKLORE. Systems analysts must choose a technique that fits in well with what was previously used in the organization and that allows flexibility and easy modification.

Testing of specific programs, subsystems, and total systems is essential to quality. System maintenance is an important consideration. Both internal and external auditors are used to determine the reliability of the system's information. They communicate their audit findings to others so as to improve the usefulness of the system's information.

Implementation is the process of ensuring that information systems and networks are operational and then involving well-trained users in their operation. In large systems projects, the primary role of an analyst is overseeing implementation by correctly estimating the time needed and then supervising the installation of equipment for information systems.

Distributed systems take advantage of telecommunications technology and database management to interconnect people manipulating some of the same data in meaningful but different ways. As hardware and software are evaluated, a systems analyst also needs to consider the costs and benefits of employing a distributed system to fulfill user requirements. One of the most popular ways to approach distributed systems is through the use of a client/server model. Cloud computing allows commerce, applications, and data storage to be served using the Internet. The move to cloud computing is a potential way for older, larger companies to solidify and retain their core businesses by incorporating SaaS (Software as a Service), SOA (service-oriented architecture), virtualization, open source, and many other trends from the past decade into their offerings of software, software tools, services, and computing power via the Internet.

Enterprise systems (ERP) are also developed for use via the cloud. Standard types of organizational networks include the local area network (LAN) and the wide area network (WAN). Using a top-down approach, analysts can use five symbols to help draw network decomposition and hub connectivity diagrams.

Training users and personnel to interact with the information system is an important part of implementation, because users must usually be able to run the system without the intervention of an analyst. Conversion is the process of changing from an old information system to a new one. The five conversion strategies are direct changeover, parallel conversion, phased or gradual conversion, modular conversion, and distributed conversion. Research suggests that systems analysts can improve the chances that newly implemented systems will be accepted if they develop systems with predominant organizational metaphors in mind.

Security of data and systems has taken on increased importance for analysts who are designing more ecommerce applications. Security has several facets—physical, logical, and behavioral—that must all work together. Analysts can take a number of precautions, such as virus protection software, email filtering, URL filters, firewalls, gateways, virtual private networks, intrusion detection products, secure socket layering, secure electronic translation, and the use of a public key infrastructure, to improve privacy, confidentiality, and the security of systems, networks, data, individuals, and organizations. In addition, every company for which you design an ecommerce application should adopt a privacy policy following five guidelines.

Even when you take all possible measures to ensure system security, privacy, and stability, all employees and systems are vulnerable to natural or human-caused disaster. Disaster recovery is focused on how a business can continue after a disaster has hit and how it can restore essential IT infrastructure.

Many different evaluation approaches are available, including cost-benefit analysis, the revised decision evaluation approach, and user involvement evaluations. The information system utility framework is a direct way to evaluate a new system based on the six utilities of possession, form, place, time, actualization, and goal.

Keywords and Phrases

behavioral security
bottom-up design
client/server model
cloud computing
control flag (switch)
corporate privacy policy
data couple
desk check

direct changeover
disaster preparedness
disaster recovery
distributed conversion
distributed processing
enterprise system
enterprise resource planning (ERP) system
email filtering products

encryption software
firewall or firewall system
FOLKLORE
full systems testing with live data
full systems testing with test data
gradual or phased conversion
hub connectivity
information system utility
internal auditor
IS quality circle
link testing with test data (string testing)
local area network (LAN)
logical security
malware
modular conversion
modular design and development
network decomposition
network modeling
organizational metaphors

parallel conversion
program testing with test data
Public Key Infrastructure (PKI)
Service Oriented Architecture (SOA)
Six Sigma
Software as a Service (SaaS)
software documentation
software maintenance
storage area networks (SANs)
structure chart
structured walkthrough
top-down design
total quality management (TQM)
URL filtering products
virus protection software
Web traffic analysis
wide area network (WAN)
wireless local area network (WLAN)

Review Questions

1. What are the three broad approaches available to a systems analyst for attaining quality in newly developed systems?
2. Who or what is the most important factor in establishing and evaluating the quality of information systems or decision support systems? Why?
3. Define the total quality management (TQM) approach as it applies to the analysis and design of information systems.
4. What is meant by the term *Six Sigma?*
5. What is an IS quality circle?
6. Define what is meant by doing a structured walkthrough. Who should be involved? When should structured walkthroughs be done?
7. List the advantages of taking a top-down approach to design.
8. Define modular development.
9. List four guidelines for correct modular programming.
10. Name the two types of arrows used in structure charts.
11. What is service-oriented architecture (SOA)?
12. Give two reasons that support the necessity of well-developed systems and software documentation.
13. In what four categories does the FOLKLORE documentation method collect information?
14. List six guidelines for choosing a design and documentation technique.
15. Whose primary responsibility is it to test computer programs?
16. What is the difference between test data and live data?
17. What are the two types of systems auditors?
18. List the four approaches to system implementation.
19. Describe what is meant by a distributed system.
20. What is the client/server model?
21. Describe how a client is different from a user.
22. What is distributed computing?
23. What are the advantages of using a client/server approach?
24. What are the disadvantages of using a client/server approach?
25. What are the benefits of implementing systems and software using cloud computing?
26. What does the term Software as a Service (SaaS) mean?
27. Explain why ERP served with a cloud platform might be more affordable to small and medium businesses than traditional ERP installations.
28. Who should be trained to use the new or modified information system?
29. List the five possible sources of training for users of information systems.
30. List the five conversion strategies for converting old information systems to new ones.
31. List the nine organizational metaphors and the hypothesized success of each type of system given their presence.
32. Define the terms *physical, logical,* and *behavioral security* and give an example of each one that illustrates the differences among them.

33. Define what encryption software means.
34. What is a firewall or firewall system?
35. List five of the measures an analyst can take to improve the security, privacy, and confidentiality of data, systems, networks, individuals, and organizations that use ecommerce Web applications.
36. List five guidelines for designing a corporate privacy policy for ecommerce applications.
37. Briefly describe the differences between disaster preparedness and disaster recovery.
38. List and describe the utilities of information systems that can be used to evaluate the information system.
39. What are seven essential items that an analyst should include in performing a website traffic analysis?

Problems

1. One of your systems analysis team members has been discouraging user input on quality standards, arguing that because you are the experts, you are really the only ones who know what constitutes a quality system. In a paragraph, explain to your team member why getting user input is critical to system quality. Use an example.

2. Write a detailed table of contents for a procedure manual that explains to users how to log onto your school's computer network, as well as the school's network policies (who is an authorized user, and so on). Make sure that the manual is written with the user in mind.

3. Your systems analysis team is close to completing a system for Meecham Feeds. Roger is quite confident that the programs that he has written for Meecham's inventory system will perform as necessary, because they are similar to programs he has done before. Your team has been very busy and would ideally like to begin full systems testing as soon as possible. Two of your junior team members have proposed the following:
   a. kip desk checking of the programs (because similar programs were checked in other installations; Roger has agreed).
   b. Do link testing with large amounts of data to prove that the system will work.
   c. Do full systems testing with large amounts of live data to show that the system is working.
      Respond to each of the three steps in their proposed test schedule. Use a paragraph to explain your response.

4. Propose a revised testing plan for Meecham Feeds (Problem 3). Break down your plan into a sequence of detailed steps.

5. Draw a local area network or some other configuration of distributed processing using the client/server approach to solve some of the data sharing problems that Bakerloo Brothers is having. It wants to be able to allow teams of architects to work on blueprints at headquarters, let the construction supervisor enter last-minute changes to plans under construction from the field, and permit clients to view plans from almost anywhere. Currently, the company has a LAN for the architects who are in one city (Philadelphia) that lets them share some drawing tools and any updates that team members make with architects in other cities (New York, Terre Haute, Milwaukee, Lincoln, and Vancouver). The supervisor uses a notebook computer, cannot make any changes, and is not connected to a database. Clients view plans on displays, but sales representatives are not able to enter modifications to show them what would happen if a wall were moved or a roof line altered. (*Hint*: List the problems that the company is encountering, analyze the symptoms, think of a solution, and then start drawing.) More than one network may be necessary, and not all problems will be amenable to a systems solution.

6. Create a disaster recovery plan for one of the networks you recommended to Bakerloo Brothers in Problem 5.

7. Cramtrack, the regional commuter train system, is trying to train users of its newly installed computer system. For the users to get the proper training, the systems analysts involved with the project sent a memo to the heads of the four departments that include both primary and secondary users. The memo said in part, "Only people who feel as if they require training need to make reservations for off-site training; all others should learn the system as they work with it on the job." Only 3 of a possible 42 users signed up. The analysts were satisfied that the memo effectively screened people who needed training from those who did not.
   a. In a paragraph, explain how the systems analysts got off the track in their approach to training.
   b. Outline the steps you would take to ensure that the right people at Cramtrack are trained.
   c. Suggest in a paragraph how the Web might be used to assist in training for Cramtrack.

8. A beautiful, full-color brochure arrived on Bill Cornwell's desk, describing the Benny Company's off-site training program and facilities in glowing terms; it showed happy users at computers and professional-looking trainers leaning over them with concerned looks. Bill ran excitedly into Roseann's office and told her, "We've got to use these people. This place looks terrific!" Roseann was not persuaded by the brochure, but didn't know what to say in defense of the on-site training for users that she had already authorized.

    a. In a few sentences, help Roseann argue the usefulness of on-site training with in-house trainers in contrast to off-site training with externally hired trainers.

    b. If Bill does decide on Benny Company training, what should he do to verify that this company is indeed the right place to train the company's information system users? Make a list of actions he should take.

9. "Just a little longer … I want to be sure this is working correctly before I change over," says Buffy, the owner of three bathroom accessories boutiques called Tub 'n Stuff. Her accountant, who helped her set up a new accounting information system, is desperately trying to persuade Buffy to change over completely to the new system. Buffy has insisted on running the old and new systems in parallel for an entire year.

    a. Briefly describe the general problems involved in using a parallel conversion strategy for implementing a new information system.

    b. In a paragraph, try to convince the owner of Tub 'n Stuff that a year of running a system in parallel is too long. Suggest a way to end Tub 'n Stuff's dual systems that will provide enough reassurance to Buffy. (Assume that the new system is reliable.)

10. Draft a plan to perform Web traffic analysis for the ecommerce application developed for Marathon Vitamin Shops. (See Consulting Opportunities 1.1, 13.2, and 14.5 for more information about the organization, their products, and their goals.) Your plan should take the form of a written report to the owner of the chain, Bill Berry. Be sure to indicate what statistics you will monitor and why they are important for Marathon Vitamin Shops to know.

11. Ayman's Office Supplies Company recently had a new information system installed to help its managers with inventory. In speaking with the managers, you notice that they seem disgruntled with the system output, which is a series of displays that show current inventory, customer and supplier addresses, and so on. All screens need to be accessed through several special commands and the use of a password. The managers had several opinions about the system but had no systematic way to evaluate it.

    a. Devise a checklist or form that helps Ayman's managers evaluate the utilities of an information system.

    b. Suggest a second way to evaluate the information system.

Group Projects

1. Nicholas Ventola is the General Manager for the world-famous Le Corked restaurant. His information system was developed over time and, as it exists now, it consists of two computer systems that do not communicate with one another. One system handles reservations and maintains a database of customers' preferences (likes, dislikes, food allergies), birthdates and anniversaries, and other information. The other system assigns each party to a table on a given night. In your group use a top-down approach to identify the modules that would be necessary to accomplish everything Nicholas wants to do using only one computer system, from making reservations to ordering food. From your own experience, determine what systems are necessary to operate a fine dining establishment, then describe the modules and how and when you would use them.

2. Divide your group into two subgroups. One subgroup should interview the members of the other subgroup about their experiences encountered in registering for a class. Questions should be designed to elicit information on customs, tales, sayings, and art forms that will help document the registration process at your school.

3. Reunite your group to develop a web page for a short excerpt for a FOLKLORE manual that documents the process of registering for a class, one based on the FOLKLORE passed on in the interviews in Group Project 2. Remember to include examples of customs, tales, sayings, and art forms.

Selected Bibliography

Evans, J. R., and W. M. Lindsay. *An Introduction to Six Sigma and Process Improvement.* Florence, KY: Cengage Learning, 2004.

Hecht, J. A. "Business Continuity Management." *Communications of the AIS,* Vol. 8, Article 30, 2002.

Juran, J. M. *Managerial Breakthrough.* New York: McGraw-Hill, 1964.

Kendall, J. E., and K. E. Kendall. "Metaphors and Methodologies: Living Beyond the Systems Machine." *MIS Quarterly,* Vol. 17, No. 2, June 1993, pp. 149–171.

Kendall, K. E. "Evaluation of a Regional Blood Distribution Information System." *International Journal of Physical Distribution and Materials Management,* Vol. 10, No. 7, 1980.

Kendall, K. E., J. E. Kendall, and K. C. Lee. "Understanding Disaster Recovery Planning through a Theatre Metaphor: Rehearsing for a Show That Might Never Open." *Communications of AIS,* Vol. 16, 2005, pp. 1001–1012.

Kendall, K. E., and R. Losee. "Information System FOLKLORE: A New Technique for System Documentation." *Information and Management,* Vol. 10, No. 2, 1986, pp. 103–111.

Stephens, D. O. "Protecting Records in the Face of Chaos, Calamity, and Cataclysm," *The Information Management Journal,* January/February 2003, pp. 33–40.

Warkentin, W., R. S. Morse, E. Bekkering, and A. C. Johnston. "Analysis of Systems Development Project Risks: An Integrative Framework." *The DATA BASE for Advances in Information Systems,* Volume 40, Number 2, May, 2009, pp. 8–21.

Wikipedia. "Cloud Computing." http://en.wikipedia.org/wiki/Cloud_computing. Last accessed August 31, 2012.

Zmud, R. W., and J. F. Cox. "The Implementation Process: A Change Approach." *MIS Quarterly,* Vol. 3, No. 2, 1979, pp. 35–44.

The CPU Case Episode and accompanying student files are available online at www.pearsonglobaleditions.com/kendall.

GLOSSARY

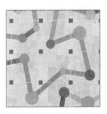

Numbers in parentheses refer to the chapters in which the terms are defined.

ACTOR   In UML, a particular role of a user of the system. The actor exists outside the system and interacts with the system in a specific way. An actor can be a human, another system, or a device such as a keyboard. (10) *See also* use case.

AGGREGATION   Often described as a "has a" relationship when using UML for an object-oriented approach. Aggregations provide a means of showing that the whole object is composed of the sum of its parts (other objects). (10)

AGILE APPROACH (OR AGILE MODELING)   A systems development approach that has values, principles, and practices useful for systems analysts who desire a flexible, interactive, and participative approach. (6) *See also* Extreme Programming.

AJAX   A method using JavaScript and XML to dynamically change Web pages without displaying a new page by obtaining small amounts of data from the server. (12)

ALIAS   Alternative name for a data element used by different users. Recorded in a data dictionary. (8)

APPLICATION SERVICE PROVIDER (ASP)   A company that hosts application software, which is leased by other organizations for use on the Web. Applications include traditional ones as well as collaboration and data management. (16)

ASSOCIATIVE ENTITY   An entity type that associates the instances of one or more entity types and contains attributes that are peculiar to the relationship between those entity instances. (2)

ATTRIBUTE   Some characteristic of an entity. There can be many attributes for each entity. (13) *See also* data item.

ATTRIBUTIVE ENTITY   One of the types of entities used in entity-relationship diagrams. Something useful in describing attributes, especially repeating groups. (2)

BEHAVIOR   How an object acts and reacts. (10)

BESPOKE SOFTWARE   Another term for custom software, it is the opposite of commercial-off-the-shelf (COTS) software. It is software developed to serve a particular function, or to support a unique organizational feature. (1)

BIPOLAR QUESTION   A subset of closed questions that can be answered in two ways only, such as yes or no, true or false, and agree or disagree. (4) *See also* closed question, open-ended question.

BRING YOUR OWN DEVICE (BYOD)   A corporate policy that permits employees to use their preferred personal mobile technology (for example, their mobile phone or tablet) for work purposes once the technology undergoes customization by the corporate IT department, which uploads work software, encryption for security purposes, and so on to prepare the device for work. (3)

BRING YOUR OWN TECHNOLOGY (BYOT)   A special instance of BYOD that often refers to bringing one's own computer or tablet into (often) an educational setting or classroom. (3)

BROWSER   Special software that runs on an Internet-connected computer enabling users to view hypertext-based Web pages on the Internet. Microsoft Internet Explorer and Mozilla Firefox are examples of graphical browsers. (11)

BUBBLE DIAGRAM   A simple diagram that shows data associations of data elements. Each entity is enclosed in an ellipse, and arrows are used to show the relationships. Also called a data model diagram. (13)

BUSINESS INTELLIGENCE (BI)   Information that is useful to an organization through the processing of large amounts of data with BI technologies to help an organization meet its strategic goals. (13)

BUSINESS RULES   Statements specific to an organization's functioning that provide a logical description of business activities. Used to help create data flow diagrams. (7)

CAPACITIVE DEVICE   An input device (a human finger or conductive stylus) for providing input to a touch sensitive screen that has capacitive sensing such as those used on smartphones and tablets. (14)

CASE TOOLS Computer-aided software engineering tools that include computer-based automated diagramming, analyzing, and modeling capabilities. (1)

CHILD DIAGRAM The diagram that results from exploding the process on Diagram 0 (called the parent process). (7)

CLASS A common template for a group of individual objects with common attributes and common behavior in object-oriented analysis and design and UML. (10)

CLASS DIAGRAM A diagram that is used to graphically model the static structural design view of a system; illustrates the functional requirements of the system gathered by way of analysis, as well as the physical design of the system. (10)

CLIENT/SERVER ARCHITECTURE A design model that features applications running on a local area network (LAN). Computers on the network divide processing tasks among servers and clients. Clients are networked machines that are points of entry into the client/server system. (16)

CLOSED QUESTION A type of question used in interviews or on surveys that closes the possible response set available to respondents. (4) *See also* bipolar question, open-ended question.

CLOSED SYSTEM Part of general systems theory; a system that does not receive information, energy, people, or raw materials as input. Systems are never totally closed or totally open, but exist on a continuum from more closed to more open. (2) *See also* open system.

CLOUD COMPUTING A system in which organizations and individual users use Web services, database services, and application services over the Internet (the cloud), without having to invest in corporate or personal hardware, software, or software tools beyond the Web. Businesses use Web browsers to access applications and servers store software and data for businesses. (16)

COMPUTER-AIDED SOFTWARE ENGINEERING (CASE) (1) *See* CASE tools.

CONCATENATED KEY A composite key created when it is not possible to identify a record uniquely by using one of the data items found in a record; a key can be constructed by choosing two or more data items and combining them. (13)

CONTEXT-LEVEL DATA FLOW DIAGRAM The most basic data flow diagram of an organization showing how processes transform incoming data into outgoing information. Also called an environmental model. (2) *See also* data flow diagram.

CONTROL FLAG A flag that is used in structure charts to govern which portion of a module is to be executed, associated with IF, THEN, ELSE, and other similar types of statements. (16)

CONVERSION The physical switch from an old information system to a new one. There are five conversion

strategies: direct changeover, parallel conversion, phased or gradual conversion, modular prototype conversion, and distributed conversion. (16)

CRC CARDS Class, Responsibilities, and Collaborators cards that represent the responsibilities of classes and the interaction between the classes when beginning to model the system from an object-oriented perspective. Analysts create these cards based on scenarios that outline system requirements. (10)

CRITICAL PATH The longest path calculated using the PERT scheduling technique; the path that will cause the whole systems project to fall behind if even one day's delay is encountered on it. (3)

DASHBOARD A screen for decision makers that includes a variety of displays of relevant performance measurements. (11)

DATA COUPLE Depiction of the passing of data between two modules on a structure chart. (16)

DATA DICTIONARY A reference work of data about data (metadata) created by a systems analyst based on data flow diagrams; collects and coordinates specific data terms, confirming what each term means to different people in the organization. (8)

DATA ELEMENT A simple piece of data that can be base or derived; should be defined in the data dictionary. (8)

DATA FLOW Data that move in a system from one place to another; input and output are depicted using an arrow with an arrowhead in data flow diagrams. (7)

DATA FLOW DIAGRAM (DFD) A graphical depiction of data processes, data flows, and data stores in a business system. (7)

DATA ITEM The smallest unit in a file or database. Used interchangeably with the word *attribute*. (13)

DATA MINING Techniques that apply algorithms for extracting patterns from data stored in data warehouses that are typically not apparent to human decision makers. Also known as knowledge data discovery (KDD). (14)

DATA REPOSITORY A centralized database that contains all diagrams, form and report definitions, data structures, data definitions, process flows and logic, and definitions of other organizational and system components; provides a set of mechanisms and structures to achieve seamless data-to-tool and data-to-data integration. (8)

DATA STORE Data that are at rest in a system; depicted using an open-ended rectangle in data flow diagrams. (7)

DATA STRUCTURE A structure composed of data elements, typically described using algebraic notation to produce a view of the elements. An analyst begins with the logical design and then designs the physical data structures. (8)

DATA WAREHOUSE A collection of data in support of management decision processes that is subject oriented, integrated, time variant, and nonvolatile. (13) *See also* data mining.

DATABASE A formally defined and centrally controlled store of electronic data intended for use in many different applications. (13)

DATABASE MANAGEMENT SYSTEM (DBMS) Software that organizes data in a database, providing information storage, organization, and retrieval capacities. (13)

DECISION SUPPORT SYSTEM (DSS) An interactive information system that supports the decision-making process through the presentation of information designed specifically for the decision maker's problem-solving approach and application needs. It does not make a decision for the user. (10)

DECISION TABLE A tool for examining, describing, and documenting structured decisions. Four quadrants are drawn to describe the conditions, identify possible decision alternatives, indicate which actions should be performed, and describe the actions. (9)

DECISION TREE A method of decision analysis for structured decisions; an appropriate approach when actions must be accomplished in a certain sequence. (9)

DEFAULT VALUE A value that a field will assume unless an explicit value is entered for it. (14)

DELIVERABLES Any of the software, documentation, procedures, user manuals, or training sessions that a systems analyst delivers to a client based on specific contractual promises. (3)

DENORMALIZATION Definition of physical records not in third or higher normal forms; includes joining attributes from several relations together to avoid the cost of accessing several files. Partitioning is an intentional form of denormalization. (13)

DIGITAL SUBSCRIBER LINE (DSL) Protocols that allow high-speed data transmission over regular telephone wire. (16)

DISASTER RECOVERY PLANNING Strategic and tactical plans to aid people and systems to recover in the face of natural and human-made disasters. (16)

DISPLAY Any one of a number of display alternatives that users employ to view computer software, including monitors and liquid plasma screens. (11)

DISTRIBUTED SYSTEMS Computer systems that are distributed geographically, as well as having their processing, data, and databases distributed. One common architecture for distributed systems is a LAN-based client/server system. (16)

DOCUMENTATION Written material created by the analyst that describes how to run the software, gives an overview of the system, or details the program code used. Analysts can use a CASE tool to facilitate documentation. (16)

DROP-DOWN LIST A GUI design element that permits users to click on a box that appears to drop down on the screen and list a number of alternatives, which can be subsequently chosen. (11)

ECOMMERCE Business conducted electronically, including via email, Web technologies, BBS, smartcards, EFT, and EDI, among suppliers, customers, governmental agencies, and other businesses to conduct and execute transactions in business, administrative, and consumer activities. (1)

ENCAPSULATION In object-oriented analysis and design, the behavior of an object that defines the object. An object maintains data about the real-world things it represents in a true sense. An object must be asked or told to change its own data with a message. (10)

ENCRYPTION The process of converting a message into an encrypted message by using a key so that the message cannot be read by a person. The intended receiver of the message can then use a key to decode and read the encrypted message. (15)

END USERS In an organization, non-information system professionals who specify the business requirements for and use software applications. End users often request new or modified applications, test and approve applications, and may serve on project teams as business experts. (3)

ENTERPRISE SYSTEMS Information systems that are integrated organization-wide (enterprise-wide) that help companies in the coordination of critical organizational processes. (1)

ENTITY A person, group, department, or system that either receives or originates information or data. One of the primary symbols on a data flow diagram. (2) *See also* data flow diagram, external entity.

ENTITY-RELATIONSHIP (E-R) DIAGRAM A graphical representation of an E-R model. (8)

ENTITY TYPE A collection of entities that share common properties or characteristics. (8)

ENVIRONMENT Anything external to an organization. Multiple environments exist, such as the physical, economic, legal, and social environments. (2)

EXECUTIVE SUPPORT SYSTEM (ESS) A computer system that helps executives organize their interactions with the external environment by providing graphical and communication support. (1)

EXPLORATION PHASE The beginning phase of agile development, where an analyst asserts his or her conviction that the agile approach is the correct one and then assembles a development team and assesses their skills. This phase can last for a few weeks or up to a few months. (1)

EXTERNAL ENTITY A source or destination of data considered to be external to the system being described. Also called an entity. (7) *See also* data flow diagram.

EXTREME PROGRAMMING (XP) A systems development approach that accepts what we know as good systems development practices and takes them to the extreme. The genesis of agile approaches. (6)

FAVICON   A small icon displayed next to a bookmarked address in a browser. Copying the bookmarked link to a desktop results in a larger version of the icon being placed there. Unique favicons can be generated with a Java icon generator or with other graphics programs. (11)

FIELD   A physical part of a database that can be packed with several data items; the smallest unit of named application data recognized by system software. (13)

FIREWALL   Computer security software used to erect a barrier between an organization's LAN and the Internet. Although it prevents hackers from getting into an internal network, it also stops organizational members from getting direct access to the Internet. (16)

FIRST NORMAL FORM (1NF)   The first step in normalizing a relation in data used in a database so that it contains no repeating groups. (13) *See also* second normal form, third normal form.

FIT   The way that HCI elements of a human, a computer, and a task that needs to be performed work together to improve performance and well-being. (14)

FOLKLORE   A system documentation technique based on traditional methods used in gathering information about people and legends. (16)

FORECASTING   System analysts' predictions about certain key variables before the systems proposal is submitted to the client. Forecasting is the art and science of predicting key variables, often assisted with mathematical forecasting models. (3)

FORM-FILL INTERFACE   Part of GUI design elements that automatically prompt the user to fill in a standard form. Useful for ecommerce applications. (14)

FUNCTION POINT ANALYSIS   A way to estimate project size, considering five main components of computer systems: external inputs, external outputs, external log queries, internal logical files, and external interface files. (3)

GANTT CHART   A graphical representation of a project that shows each task activity as a horizontal bar, the length of which is proportional to its time for completion. (3)

GESTURES   A variety of movements for a finger or a capacitive device such as a stylus across a touch-sensitive screen on a smartphone or tablet that enables the user to change screens, enlarge or shrink content with a pinch, zoom in or out of a page, or rotate or scroll content. Used with apps for smartphones and tablets. (14)

GRAPHICAL USER INTERFACE (GUI)   An icon-based user interface with features such as pull-down menus, drop-down lists, and radio buttons. (14)

HUMAN–COMPUTER INTERACTION (HCI)   The aspect of a computer that enables communications and interactions between humans and the computer; the layer of the computer between humans and the computer. (14)

HYPERLINK   A highlighted word in a hypertext system that will display another document when clicked on by the user. (11)

ICON   A small picture that represents an activity and function available to users when they activate it, often with a mouse click. Frequently used in GUI design. (14)

IMPLEMENTATION   The last phase of the systems development life cycle, in which an analyst ensures that the system is in operation and then allows users to take over its operation and evaluation. (16)

INDEXED FILE ORGANIZATION   A type of file organization that uses separate index files to locate records. (13)

INHERITANCE   The capacity of a derived class to inherit all the attributes and behaviors of the base class in object-oriented analysis and design. Classes can have children; the parent class is known as the base class, and the child class is called a derived class. (10)

INPUT   Any data, either text or numbers, entered into an information system for storage or processing via forms, screens, voice, or interactive Web fill-in forms. (12)

INTANGIBLE BENEFITS   Benefits that accrue to an organization as a result of a new information system and that are difficult to measure, such as improving decision making, enhancing accuracy, and becoming more competitive. (3) *See also* intangible costs, tangible benefits, tangible costs.

INTANGIBLE COSTS   Costs that are difficult to estimate and may not be known, including losing a competitive edge, losing a reputation for innovation, and declining company image, due to untimely or inaccessible information. (3) *See also* intangible benefits, tangible benefits, tangible costs.

INTEGRATED SERVICES DIGITAL NETWORK (ISDN)   A switched network service that provides end-to-end digital connectivity for transmitting voice, data, and video simultaneously over a single line versus multiple lines. (16)

INTERNET SERVICE PROVIDER (ISP)   A company that provides access to the Internet and that may provide other services, such as Web hosting and Web traffic analysis, for a fee. (12)

IP (INTERNET PROTOCOL) ADDRESS   The number used to represent an individual computer on a network. The format for an IP address is 999.999.999.999. (16)

JAVA   An object-oriented programming language that allows dynamic applications to be run on the Internet. (11)

JOINT APPLICATION DESIGN (JAD)   IBM's proprietary approach to panel interviews conducted with analysts, users, and executives to accomplish requirements analysis jointly. (4)

KEY   One of the data items in a record that is used to identify a record. (13) *See also* primary key, secondary key.

LEVEL 0 DIAGRAM   The explosion (or decomposition) of a context-level data flow diagram that shows from three to nine major processes, important data flows, and data stores of the system under study. (7)

LOCAL AREA NETWORK (LAN)   The cabling, hardware, and software used to connect workstations, computers, and file servers located in a confined geographical area (typically within one building or campus). (16)

LOGICAL DATA FLOW DIAGRAM   A diagram that focuses on the business and how the business operates; describes the business events that take place and the data required and produced by each event. (7) *See also* data flow diagram, physical data flow diagram.

LOWER CASE TOOLS   CASE tools that analysts use to generate computer source code, eliminating the need for programming the system. (1) *See also* CASE tools.

MAINTENANCE   A phase of the SDLC in which maintaining the information system to improve it or to fix problems begins. Maintenance continues through the life of a system. Some maintenance can be done automatically through connecting to the vendor's website. (1)

MANAGEMENT INFORMATION SYSTEM (MIS)   A computer-based system composed of people, software, hardware, and procedures that share a common database to help users interpret and apply data to the business. (1)

MASHUP   A new application created by combining two or more Web-based APIs together. (14)

METHOD   In UML, an action that can be requested from any object of the class; the processes that a class knows how to carry out. (10)

MNEMONIC   A device (often using a combination of letters and symbols) that helps a data entry person remember how to correctly enter data or helps the user remember how to use the information. (15)

NATURAL-LANGUAGE INTERFACE   An interface that permits a user to interact with a computer by speaking or writing in human language. (14)

NORMALIZATION   The transformation of complex user views and data stores to a set of smaller, stable data structures. Normalized data structures are more easily maintained than complex structures. (13)

OBJECT   In the object-oriented approach, a computer representation of some real-world thing or event; can have both attributes and behaviors. (10)

OBJECT CLASS   A category of similar objects. Objects are grouped into classes. A class defines the set of shared attributes and behaviors found in each object in the class. (10)

OBJECT DIAGRAM   A diagram that is similar to a class diagram but that portrays the state of class instances and their relationships at a point in time; shows objects and their relationships. Also shows optionality (customer can have zero or more rental contracts) and cardinality (rental contract can have only one customer). (10)

OBJECT THINK   Elementary statements an analyst writes on CRC cards to begin thinking in an object-oriented way. (10)

OPEN-ENDED QUESTION   A type of question used in interviews or on surveys that opens up the possible response set available to respondents. (4) *See also* bipolar question, closed question.

OPEN SOURCE SOFTWARE (OSS)   A development model and philosophy of liberating software from certain licensing restrictions and publishing its source code, which can then be studied, shared, and modified by users and programmers, sometimes in proprietary ways for profit. The Linux operating system is an example. (1)

OPEN SYSTEM   Part of general systems theory; a system that freely receives information, energy, people, or raw materials as input. Systems are never totally closed or totally open, but exist on a continuum from more closed to more open. (2) *See also* closed system.

OUTPUT   Information delivered to users through an information system by way of intranets, extranets, or the Web, on printed reports, on displays, or via audio. (11)

PACKAGE   In UML, a grouping of things that can be considered physical subsystems. Systems are implemented and deployed in packages. (10)

PAIR PROGRAMMING   A core practice of the agile approach in which two programmers who choose to work together both do programming, run tests, and talk to one another about ways to efficiently and effectively get the job done. (6)

PERT DIAGRAM   A tool used to determine critical activities for a project. It can be used to improve a project schedule and evaluate progress. PERT stands for Program Evaluation Review Technique. (3)

PHYSICAL DATA FLOW DIAGRAM   A DFD that shows how a system will be implemented, including the hardware, software, people, and files involved. (7) *See also* logical data flow diagram.

PLANNING GAME   A game used in agile development that spells out rules that can help formulate an agile development team's relationship with its business customers. (1).

PLUG-IN   Additional software (often developed by a third party) that can be used with another program; for example, RealNetworks's Real Player and Adobe Flash are used as plug-ins in Web browsers to play streaming audio or video and view vector-based animation. (11)

PODCASTING   The technique of putting downloadable audio files on the Web. (11)

POLYMORPHISM   Alternative behaviors among derived classes in object-oriented approaches. When several classes inherit both attributes and behaviors, the behavior of a derived class might be different from its base class or its sibling-derived classes. (10)

PRESENT VALUE The total amount that a series of future payments is worth now; a way to assess the economic outlays and revenues of the information system over its economic life and compare costs today with future benefits. (3)

PRIMARY KEY A key that uniquely identifies a record. (13) *See also* key, secondary key.

PROBES Follow-up questions primarily used during interviews between analysts and users. (4) *See also* closed question, open-ended question.

PROBLEM DEFINITION A formal statement of a problem, including (1) the issues of the present situation, (2) the objectives for each issue, (3) the requirements that must be included in all proposed systems, and (4) the constraints that limit system development. (3)

PROCESS The activities that transform or change data in an information system. They can be either manual or automated. Signified by a rounded rectangle in a data flow diagram. (2)

PRODUCTIONIZING PHASE The phase in agile development when the software is released and feedback to improve the software product is received. Product releases can happen as often as every week. (1)

PROJECT CHARTER A written document that describes the expected results of a systems project (deliverables) and the time frame for delivery; it essentially becomes a contract between the chief analyst (or project manager) and his or her analysis team with the organizational users requesting the new system. (3)

PROJECT MANAGEMENT The art and science of planning a project, estimating costs and schedules, managing risk, and organizing and overseeing a team. Many software packages exist to support project management tasks. (3)

PROJECT MANAGER A person responsible for overseeing the planning, costing, scheduling, and team organization of a project. Frequently, it is a role played by a systems analyst. (3)

PROTOTYPING A rapid, interactive process between users and analysts to create and refine portions of a new system; it can be used as part of the SDLC for requirements determination or as an alternative to the SDLC. (6) *See also* rapid application development.

PSEUDOCODE A technique to create computer instructions that are the intermediate step between English and program code; used to represent the logic of each module on a structure chart. (16) *See also* structure chart.

PULL-DOWN MENU A GUI design element that provides an onscreen menu of command options that appear after the user selects the command name on a menu bar. (14). *See also* drop-down list.

QUERIES Questions users pose to a database concerning data within it. Each query involves an entity, an attribute, and a value. (14)

QUICK RESPONSE CODE (QR CODE) A popular type of about 30 two-dimensional code (2D) bar codes in existence. Microsoft Tags are another example of a 2D bar code. They are optical and machine-readable codes that appear in print and on the Web and can direct users to encoded websites. (15)

RADIO BUTTON A GUI design element that provides a round option button in a dialog box. Buttons are mutually exclusive because a user can choose only one radio button option within the group of options displayed. (8)

RECORD A collection of data items that have something in common with the entity described. (13)

RELATIONAL DATABASE MODEL A model that represents data in a database as two-dimensional tables called relations. As long as both tables share a common data element, the database can relate any one file or table to data in another file or table. (13)

RELATIONSHIP An association between entities (sometimes referred to as data association); can take the form of one-to-one, one-to-many, many-to-one, or many-to-many. (13)

REPEATING GROUP The existence of many of the same elements in a data structure. (8) *See also* data structure.

RUBY ON RAILS A combination programming language and code generator for creating Web applications. (1)

SAMPLING Systematically selecting representative elements of a population. Analysts sample hard data, archival data, and people during information requirements determination. (5)

SECOND NORMAL FORM (2NF) A stage of data normalization in which an analyst ensures that all nonkey attributes are fully dependent on the primary key. All partial dependencies are removed and placed in another relation. (13) *See also* first normal form, third normal form.

SECONDARY KEY A key that cannot uniquely identify a record; can be used to select a group of records that belong to a set. (13)

SEQUENCE DIAGRAM In UML, a diagram that illustrates a succession of interactions between object instances over time. Often used to illustrate the processing described in use case scenarios. (10)

SERVICE-ORIENTED ARCHITECTURE (SOA) An architecture in which individual software services that are unassociated or only loosely coupled to one another are available as applications or parts of applications to users, often using the Web as a platform. (16)

SIX SIGMA A culture built on quality; the goal is to eliminate all defects. (16)

STATECHART DIAGRAM In UML, a diagram that enables further refinement of requirements. (10)

STICKINESS A property of Web pages, particularly discussed in ecommerce, that indicates how well the pages draw and hold customers. Features that increase the stickiness of a Web page are those that entice customers to stay on the page longer, help them to navigate back to the page if they click on a link, and increase the likelihood they will complete a purchase. (14).

STRUCTURE CHART   A tool for designing a modular, top-down system consisting of rectangular boxes and connecting arrows. (16). *See also* control flag, data couple.

STRUCTURED ENGLISH   A technique for analyzing structured decisions based on structure logic and simple English statements, such as add, multiply, and move. (9)

STRUCTURED OBSERVATION OF THE ENVIRONMENT (STROBE)   A systematic observational method for classifying and interpreting organizational elements that influence decision making. Based on mise-en-scène film criticism. (5)

STRUCTURED WALKTHROUGH   A systematic peer review of a system's programming and overall development that points out problems and allows a programmer or an analyst to make suitable changes. (16)

SUPPLY CHAIN MANAGEMENT   An organization's effort to integrate its suppliers, distributors, and customer management requirements into one unified process. Ecommerce applications can improve supply chain management. (16)

SWIMLANES   Zones used in activity diagrams to indicate partitioning; can show which activities are done on which platform and by which user group; can also depict system logic. (10)

SYSTEM   A collection of subsystems that are interrelated and interdependent, working together to accomplish predetermined goals and objectives. All systems have input, processes, output, and feedback. Examples are a computer information system and an organization. (2) *See also* closed system, open system.

SYSTEMS ANALYST   A person who systematically assesses how businesses function by examining the inputting and processing of data and the outputting of information with the intent of improving organizational processes and the quality of work life for users. (1)

SYSTEMS DEVELOPMENT LIFE CYCLE (SDLC)   A seven-phase approach to systems analysis and design that holds that systems are best developed through the use of a specific cycle of analyst and user activities. (1)

SYSTEMS DEVELOPMENT METHODOLOGY   Any accepted approach for analyzing, designing, implementing, testing, maintaining, and evaluating an information system. SDLC, agile approaches, and object-oriented systems analysis and design are examples of methodologies. (1) *See also* systems development life cycle.

SYSTEMS PROPOSAL   A written proposal that summarizes a systems analyst's work in a business up to the current point and includes recommendations and alternatives to solve the identified systems problems. (3)

SYSTEMS TESTING   The sixth phase in the SDLC (along with maintenance). Uses both test data and eventually live data to measure errors, timeliness, ease of use, proper ordering of transactions, acceptable down time, understanding procedure manuals, and other aspects of the new system. (16)

TANGIBLE BENEFITS   Advantages measurable in dollars that accrue to an organization through the use of information systems. (3) *See also* intangible benefits, intangible costs, tangible costs.

TANGIBLE COSTS   The costs in dollars that can be accurately projected by a systems analyst, including the cost of computers, resources, analysts' and programmer's time, and other employees' salaries, to develop a new system. (3) *See also* intangible benefits, intangible costs, tangible benefits.

THINGS   In UML, the objects of object-oriented analysis and design. The two most often used groupings of things are structural things and behavioral things. (10)

THIRD NORMAL FORM (3NF)   A form in which any transitive dependencies are removed. A transitive dependency is one in which nonkey attributes are dependent on other nonkey attributes. (13) *See also* first normal form, second normal form.

TRANSACTION PROCESSING SYSTEM (TPS)   A computerized information system developed to process large amounts of data for routine business transactions, such as payroll and inventory. (1)

UNIFIED MODELING LANGUAGE (UML)   A language that provides a standardized set of tools to document the object-oriented analysis and design of a software system. (10)

USABILITY   A way for designers to evaluate the systems and interfaces they create, with an eye toward addressing as many HCI concerns as they can as thoroughly as possible. (14)

USE CASE   In UML, a sequence of transactions in a system; the purpose is to produce something of value to an actor in the system; focuses on what the system does rather than on how it does it. The use case model is based on the interactions and relationships of individual use cases. In a use case, an actor using the system initiates an event that begins a related series of interactions in the system. (10)

VALIDATION SOFTWARE   Software that checks whether data input to the information system is valid. Although validating input is largely done through software that is the programmer's responsibility, it is the analyst's responsibility to know what common problems might invalidate a transaction. (15)

VOICE OVER INTERNET PROTOCOL (VoIP)   The routing of voice data over the Internet. (3)

WEB 2.0 TECHNOLOGIES   Collaborative technologies such as blogs, email, and chat rooms that can be added to websites to improve interaction and collaboration. (11)

WEBMASTER   The person responsible for updating and maintaining a website; often initially the systems analyst during development of ecommerce applications. (11)

WORK BREAKDOWN STRUCTURE (WBS)   A system of breaking down a project into simpler tasks, usually by decomposition. A WBS can be either product or process oriented. Information systems projects tend to be process oriented. (3)

XML SCHEMAS   A tool for precisely defining the content of an XML document; may include the exact number of times an element can occur, the type of data within elements, limits on the data, and the number of places to the left and right of a decimal number. (8)

XP   *See* Extreme Programming. (6)

Acronyms

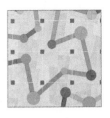

1NF	first normal form
2NF	second normal form
3NF	third normal form
AJAX	Asynchronous JavaScript and XML
API	application programming interface
ASP	application service provider
BI	business intelligence
BYOD	bring your own device
BYOT	bring your own technology
B2B	business-to-business
B2C	business-to-consumer
CASE	computer-aided software engineering
COTS	commercial off-the-shelf
CRUD	create, read, update, and delete
CSS	Cascading Style Sheets
DBMS	database management system
DDE	dynamic data exchange
DFD	data flow diagram
DHTML	Dynamic HTML
DLL	dynamic link library
DSL	digital subscriber line
DTD	document type definition
EDI	electronic data interchange
E-R	entity-relationship
ERD	entity-relationship diagram
ERP	enterprise resource planning
FAQ	frequently asked questions
FTP	File Transfer Protocol
GIF	Graphic Interchange Format
GUI	graphical user interface
HCI	human–computer interaction
HTML	Hypertext Markup Language
HTTP	Hypertext Transfer Protocol
IaaS	Infrastructure as a Service
ICTS	information and communication technology
ISDN	Integrated Services Digital Network
ISP	Internet service provider
JAD	joint application design
JPEG	Joint Photographic Experts Group
KDD	knowledge data discovery
KWS	knowledge work systems
LAN	local area network
MIS	management information system
MRP	materials requirements planning
NFC	near field communications
OCR	optical character recognition
OID	object identifier
OLAP	online analytical processing
OLE	Object Linking and Embedding
OOA	object-oriented approach
OSS	open source software
PERT	Program Evaluation and Review Technique
PHP	Hypertext Preprocessor; an open source programming language
PKI	Public Key Infrastructure
PaaS	Platform as a Service
QBE	Query by Example
QR CODE	Quick Response code
SaaS	Software as a Service
SAN	storage area network
SDLC	systems development life cycle
SET	Secure Electronic Transaction
SLA	software license agreement
SOA	service-oriented architecture
SQL	Structured Query Language
SSL	Secure Sockets Layer
STROBE	Structured Observation of the Environment
TQM	total quality management
UML	Unified Modeling Language
URL	uniform resource locator
VoIP	Voice over Internet Protocol
VPN	virtual private network
WAN	wide area network
WAP	Wireless Application Protocol
WBS	work breakdown structure
WiMAX	Worldwide Interoperability for Microwave Access
WLAN	wireless local area network
WMP	Windows Media Photo
WWW	World Wide Web
XP	Extreme Programming
XSLT	Extensible Stylesheet Language Transformations

INDEX

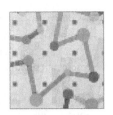